AF598013

METHODS IN MOLECULAR BIOLOGY™

For further volumes:
http://www.springer.com/series/7651

Claudins

Methods and Protocols

Edited by

Kursad Turksen

Regenerative Medicine Program, Sprott Centre for Stem Cell Research, The Ottawa Hospital Research Institute, Ottawa, ON, Canada

Editor
Kursad Turksen, Ph.D.
Regenerative Medicine Program
Sprott Centre for Stem Cell Research
The Ottawa Hospital Research Institute
Ottawa, ON
Canada
kturksen@ohri.ca

ISSN 1064-3745 e-ISSN 1940-6029
ISBN 978-1-61779-184-0 e-ISBN 978-1-61779-185-7
DOI 10.1007/978-1-61779-185-7
Springer New York Dordrecht Heidelberg London

Library of Congress Control Number: 2011931283

Printed on acid-free paper

Humana Press is part of Springer Science+Business Media (www.springer.com)

Preface

Since the initial discovery of the Claudins in 1998, the field of tight junctions and cell polarity has become energized. This is mainly due to the fact that, four decades after the morphological identification of tight junctions by Farquhar and Palade, the Claudins have given an opportunity to understand the molecular basis of the tight junction function and their tissue specific roles. The current volume is the very first protocol volume on Claudins. It is very timely, and the contributors include those who led the way in Claudins. I take this opportunity to thank all of the contributors for very graciously providing their protocols, making this new volume possible.

I thank Dr. John Walker, the Editor-in-Chief of the Methods in Molecular Biology series for his continued support.

Patrick Marton, the Editor of the Methods in Molecular Biology series at Springer, also deserves thanks for always being available to answer my questions, for patiently listening to my suggestions, and for supporting this volume during its maturation stages. A very special thank you goes to David Casey for his invaluable help during the production stages of this volume.

Ottawa, Canada *Kursad Turksen*

Contents

Contributors

EDUARDO LEME ALVES DA MOTTA • *Departamento de Ginecologia, Universidade Federal de São Paulo, São Paulo, SP, Brazil*
JAMES M. ANDERSON • *National Heart, Lung and Blood Institute, NIH, South Drive, Bethesda, MD, USA*
ANUSKA V. ANDJELKOVIC • *Department of Pathology, Neurosurgery, and Molecular and Integrative Physiology, University of Michigan, Ann Arbor, MI, USA*
IBOLYA E. ANDRÁS • *Molecular Neuroscience and Vascular Biology Laboratory, Department of Neurosurgery, University of Kentucky Medical School, Lexington, KY, USA*
NANCY C. APPLEBY • *Division of Biomedical Sciences, University of California, Riverside, CA, USA*
YOSHIAKI ARIMURA • *First Department of Internal Medicine, Sapporo Medical University, Sapporo, Japan*
EDMUND CHADA BARACAT • *Disciplina de Ginecologia, Universidade de São Paulo, Sao Paulo, Brazil*
PHUONG BUI • *Department of Biology, York University, Toronto, ON, Canada*
SARA K. CAMPBELL • *Division of Infection, Inflammation and Immunity, University of Southampton Medical School, Southampton, UK*
YAN-HUA CHEN • *Department of Anatomy and Cell Biology, Brody School of Medicine, East Carolina University, Greenville, NC, USA*
KATAYUN COHEN-KASHI-MALINA • *Department of Neurobiology, The Weizmann Institute of Science, Rehovot, Israel*
JANE E. COLLINS • *Division of Infection, Inflammation, and Immunity, University of Southampton Medical School, Southampton, UK*
MICHELLE M. COLLINS • *Departments of Pediatrics and Human Genetics, McGill University, Montréal, QC, Canada*
ITZIK COOPER • *Department of Neurobiology, The Weizmann Institute of Science, Rehovot, Israel*
DANIEL G. CYR • *INRS-Institut Armand-Frappier, Laval, QC, Canada*
LEI DING • *Department of Anatomy and Cell Biology, Brody School of Medicine, East Carolina University, Greenville, NC, USA*
ÉVEMIE DUBÉ • *INRS-Institut Armand-Frappier, Laval, QC, Canada*
JULIE DUFRESNE • *INRS-Institut Armand-Frappier, Laval, QC, Canada*
HOLLY A. ECKELHOEFER • *Division of Biomedical Sciences, University of California, Riverside, CA, USA*
JUNKO FUJITA-YOSHIGAKI • *Department of Physiology, Nihon University School of Dentistry at Matsudo, Chiba, Japan*
MIKIO FURUSE • *Division of Cell Biology, Department of Physiology and Cell Biology, Kobe University, Graduate School of Medicine, Kobe, Japan*

MAKOTO FURUTANI-SEIKI • *Centre for Regenerative Medicine, Department of Biology and Biochemistry, University of Bath, Bath, UK*
ERIKA GARAY • *Department of Physiology, Biophysics, and Neuroscience, Center for Research and Advanced Studies (CINVESTAV), México DF., México*
LORENZA GONZÁLEZ-MARISCAL • *Department of Physiology, Biophysics, and Neuroscience, Center for Research and Advanced Studies (CINVESTAV), México DF., México*
MARY GREGORY • *INRS-Institut Armand-Frappier, Laval, QC, Canada*
JULIAN A. GUTTMAN • *Department of Biological Sciences, Simon Fraser University, Burnaby, BC, Canada*
MARY HAMER • *Division of Biomedical Sciences, University of California, Riverside, CA, USA*
JEFF HARDIN • *Department of Zoology, Biology Core Curriculum, University of Wisconsin, Madison, WI, USA*
WALTER HUNZIKER • *Institute of Molecular and Cell Biology (IMCB), Republic of Singapore, Singapore*
MASAYO HOSOKAWA • *First Department of Internal Medicine, Sapporo Medical University, Sapporo, Japan*
GUOCHANG HU • *Departments of Pharmacology and Anesthesiology, and Center for Lung and Vascular Biology, University of Illinois at Chicago, Chicago, IL, USA*
TETSUICHIRO INAI • *Department of Morphological Biology, Fukuoka Dental College, Fukuoka, Japan*
WEN G. JIANG • *Metastasis & Angiogenesis Research Group, University Department of Surgery, Cardiff School of Medicine, Cardiff University, Cardiff, UK*
P. JAYA KAUSALYA • *Epithelial Cell Biology Laboratory, Institute of Molecular and Cell Biology, A*STAR, Singapore*
RICHARD F. KEEP • *Department of Pathology, Neurosurgery, and Molecular and Integrative Physiology, University of Michigan, Ann Arbor, MI, USA*
SCOTT P. KELLY • *Department of Biology, York University, Toronto, ON, Canada*
ADAM KIRK • *Division of Infection, Inflammation, and Immunity, University of Southampton Medical School, Southampton, UK*
TAKASHI KOJIMA • *Department of Pathology, Sapporo Medical University School of Medicine, Sapporo, Japan*
MICHAEL KOVAL • *Division of Pulmonary, Allergy, and Critical Care Medicine, Emory University School of Medicine, Atlanta, GA, USA*
TONG SENG LIM • *Singapore Immunological Network, A*STAR, Singapore*
CHWEE TECK LIM • *Division of Bioengineering and Department of Mechanical Engineering and Research Centre of Excellence in Mechanobiology, National University of Singapore, Singapore*
JUN LING • *Division of Biomedical Sciences, University of California, Riverside, CA, USA*
JANET LIVERSIDGE • *Immunology & Infection, Division of Applied Medicine, University of Aberdeen, Aberdeen, UK*
DAVID D. LO • *Division of Biomedical Sciences, University of California, Riverside, CA, USA*

TRACEY A. MARTIN • *Metastasis & Angiogenesis Research Group, University Department of Surgery, Cardiff School of Medicine, Cardiff University, Cardiff, UK*
JUAN MASON • *Wessex Renal and Transplant Service, Queen Alexandra Hospital, Cosham, Portsmouth, UK*
BRUCE A. MCCLANE • *Department of Microbiology and Molecular Genetics, University of Pittsburgh, Pittsburgh, PA, USA*
RICHARD D. MINSHALL • *Departments of Pharmacology and Anesthesiology, and Center for Lung and Vascular Biology, University of Illinois at Chicago, Chicago, IL, USA*
TATSUO MIYAMOTO • *Department of Genetics and Cell Biology, Research Institute for Radiation Biology and Medicine, Hiroshima University, Hiroshima, Japan*
KANNA NAGAISHI • *Second Department of Anatomy, Sapporo Medical University, Sapporo, Japan*
MAKOTO OSANAI • *Department of Pathology, Kochi University School of Medicine, Nankoku, Japan*
MIGUEL QUIRÓS • *Department of Physiology, Biophysics, and Neuroscience, Center for Research and Advanced Studies (CINVESTAV), México DF., México*
THEJANI E. RAJAPAKSA • *Division of Biomedical Sciences, University of California, Riverside, CA, USA*
SUSAN L. ROBERTSON • *Department of Microbiology and Molecular Genetics, University of Pittsburgh, Pittsburgh, PA, USA*
ANDRÉ MONTEIRO DA ROCHA • *Cell and Developmental Biology, University of Michigan, Ann Arbor, MI, USA*
GARY A. ROSENBERG • *Department of Neurology, University of New Mexico, Albuquerque, NM, USA*
AIMEE K. RYAN • *Department of Pediatrics, McGill University, Montreal, QC, Canada*
NORIMASA SAWADA • *Department of Pathology, Sapporo Medical University School of Medicine, Sapporo, Japan*
PAULO SERAFINI • *Huntington Center for Reproductive Medicine of Brazil, Sao Paulo, Brazil*
JEFFREY S. SIMSKE • *Rammelkamp Center for Education and Research, Cleveland, OH, USA*
GARY DANIEL SMITH • *Ob/Gyn, Urology and Physiology Departments and Reproductive Sciences Program, University of Michigan, Ann Arbor, MI, USA*
SVETLANA M. STAMATOVIC • *Department of Pathology, Neurosurgery, and Molecular and Integrative Physiology, University of Michigan, Ann Arbor, MI, USA*
YU SUN • *Departments of Pharmacology and Anesthesiology, and Center for Lung and Vascular Biology, University of Illinois at Chicago, Chicago, IL, USA*
RODNEY TATUM • *Department of Anatomy and Cell Biology, Brody School of Medicine, East Carolina University, Greenville, NC, USA*
VIVIAN I. TEICHBERG • *Department of Neurobiology, The Weizmann Institute of Science, Rehovot, Israel*

MICHAL TOBOREK • *Molecular Neuroscience and Vascular Biology Laboratory, Department of Neurosurgery, University of Kentucky Medical School, Lexington, KY, USA*
KURSAD TURKSEN • *Regenerative Medicine Program, Sprott Centre for Stem Cell Research, The Ottawa Hospital Research Institute, Ottawa, ON, Canada*
CHRISTINA M. VAN ITALLIE • *National Heart, Lung and Blood Institute, NIH, South Drive, Bethesda, MD, USA*
SRI RAM KRISHNA VEDULA • *Research Centre of Excellence in Mechanobiology, National University of Singapore, Singapore*
JING WANG • *Division of Biomedical Sciences, University of California, Riverside, CA, USA*
CHRISTINA WARD • *Division of Pulmonary, Allergy, and Critical Care Medicine, Emory University School of Medicine, Atlanta, GA, USA*
SUSAN J. WILSON • *Histochemical Research Unit, University of Southampton Medical School, Southampton, UK*
HEPING XU • *Centre for Vision and Vascular Science, Queen's University Belfast, Belfast, UK*
YI YANG • *Department of Neurology, University of New Mexico, Albuquerque, NM, USA*
KEN-TYE YONG • *Institute for Lasers, Photonics, and Biophotonics, University at Buffalo, The State University of New York, Buffalo, NY, USA*
ALAN S.L. YU • *Harry Statland and Solon Summerfield Professor of Medicine Director, Division of Nephrology and Hypertension and the Kidney Institute, University of Kansas Medical Center, Kansas City, KS, USA*
YUGUO ZHANG • *Department of Anatomy and Cell Biology, Brody School of Medicine, East Carolina University, Greenville, NC, USA*

Chapter 1

Measuring Size-Dependent Permeability of the Tight Junction Using PEG Profiling

Christina M. Van Itallie and James M. Anderson

Abstract

Tight junctions restrict the paracellular movement of ions, solutes, drugs, and larger material across epithelia and endothelia. For practical purposes, the barrier can be modeled as having two components. The first is a system of small 4 Å radius pores lined or created by claudins. The pores show variable ionic charge selectivity and electrical resistance based on the pattern of claudin proteins expressed in a particular junction. Transport of compounds that are larger than 4 Å are not subject to discrimination based on size or charge; they are likely passing through transient breaks in the tight junction barrier. The magnitude of the first and second pathways varies among epithelia and is altered in response to physiological and pathological stimuli. Unfortunately, most studies of permeability use few tracer sizes and thus provide limited information on size-dependent changes in permeability. Here we describe a method for simultaneously measuring the size-dependence of apparent permeability using a continuous series of polyethylene polymers which allows quantification of both the pore and leak pathways.

Key words: Tight junction, Claudin, Paracellular permeability, Apparent permeability (P_{app}), Polyethylene glycol (PEG) profiling, Epithelial transport

1. Introduction

Tight junctions create paracellular barriers which, depending on local transport requirements, differ in electrical conductance, ionic charge preference, and the level of permeability for uncharged solutes; these properties are collectively referred to as permselectivity (1). The barrier between cells is formed where continuous rows of transmembrane proteins, notably claudin proteins, adhere to seal the intracellular space (2). Permeability for solutes smaller than about 4 Å occurs through charge-selective pores, while larger solutes appear to go through a nonselective leak pathway. There is interest in defining how permeability varies as a function of

Kursad Turksen (ed.), *Claudins: Methods and Protocols*, Methods in Molecular Biology, vol. 762,
DOI 10.1007/978-1-61779-185-7_1, © Springer Science+Business Media, LLC 2011

solute size during both physiological and pathological situations since this has implications for normal transport and for entry of toxic compounds and antigens (3).

Tight junction permeability is usually measured by flux of one or several hydrophilic tracers such as urea, mannitol, inulin, and/or fluorescent dextrans of various sizes (4, 5). Use of a single tracer does not provide any information about how permeability changes as a function of size; even the comparison of flux of two different tracers provides at best a qualitative comparison, since the molecular radii for mannitol and 10 kDa dextran for example, differ by nearly sixfold (4.2 and 23 Å, respectively (6, 7)). Ideal tracers used to characterize paracellular permeability should show no transcellular transport and be noncharged so that they are not subject to charge discrimination by the pores. Further, they should have similar hydrodynamic characteristics for consistent behavior and should be available in a large size range. Use of continuous series of polyethylene glycol oligomers (PEGs, dimer to 30 mer) would appear to fulfill these criteria and reveal permeability at sub-Ångstrom increments. Although PEGs have been used extensively for permeability measurements, the most careful recent analysis was published by Watson et al. in 2001 (8) in which PEGs were separated by liquid chromatography and quantified by mass spectrometry. LC-MS is expensive and may not be available to some investigators. Consequently, we developed a method for quantifying PEGs after derivatization with a fluorescent regent. In this method, we derivatize the hydroxyl groups on the PEGs with 1-napthyl-isocyanate, followed by separation by HPLC and quantification of each species by fluorescence emission (9). We apply a two phase permeability model and application of Rankin sieving function that can be used to estimate the apparent permeability (P_{app}) for solutes which are either smaller or larger than the claudin-based pores.

2. Materials

2.1. Cell Culture and Permeable Supports

1. Culture media appropriate for each cell line:
   MDCK, T84, LLC-PK_1: Dulbecco's Modified Eagle's Medium (DMEM) containing 4.5 g/l glucose, glutamine, and sodium pyruvate (Invitrogen, Carlsbad, CA, or Cellgro, Mediatech, Inc., Hernon, VA) supplemented with pen/strep (Invitrogen) and 10% fetal bovine serum (SAFC Biosciences, Lenexa, KS or other suitable source).

   Caco-2: as above, additionally supplemented with nonessential amino acids.
2. 0.05%Trypsin/1 mM EDTA (Invitrogen).
3. Ca^{2+}, Mg^{2+}-free phosphate-buffered saline (PBS, Invitrogen or Cellgro).

4. Permeable supports
 24 mm, 0.4 μm pore size Transwell filters (4312, polycarbonate, Corning Life Sciences, Wilkes-Barre, PA)
 12 mm, 0.4 μm pore size filters (3401, polycarbonate) (see Note 1).
5. Hank's balanced salt solution (Invitrogen or Cellgro).
6. Epithelial volt-ohmmeter EVOM-G (World Precision Instruments, Sarasota, FL).
7. Plate electrodes for 12-well (Endohm-12) and 6-well filters (Endohm-24Snap) (World Precision Instruments, Sarasota, FL, see Note 2).

2.2. PEG Permeability Assays and Derivatization Reagents and Equipment

1. Polyethylene glycols 200, 400, and 900 (Fluka, Ultra grade, Sigma Aldrich Chemical Co., St. Louis, MO, USA; see Note 3), made up as 100 mg/ml stock solutions in HPLC grade H_2O, filter sterilized and stored at 4°C.
2. Purified PEG28 internal standard (Polypure AS, Oslo, Norway), made up as a 10 mg/ml solution in HPLC grade H_2O, filter sterilized and stored at 4°C.
3. 1-Napthyl isocyanate (1-NIC, Acros Organic, Sigma Aldrich Chemical Co.) (see Note 4).
4. Acetone [HPLC grade (see Note 5)].
5. Diethyl ether (Chromasolv, HPLC grade).
6. Methanol (HPLC grade).
7. H_2O (HPLC grade).
8. N-EVAP nitrogen evaporator and N_2 tank (see Note 6).
9. HPLC vials.

2.3. HPLC Separation of PEGs

1. Bare silica HPLC column (Waters Spherisorb 5 μm Silica 4.6 × 150 Analytical column, Waters Co., Milford, MA).
2. HPLC grade H_2O.
3. HPLC grade acetonitrile.
4. HPLC with automatic injector, programmable gradient capability and post-column fluorescent detector with software to quantify concentrations.

3. Methods

3.1. Cell Culture

1. Cells are cultured under standard conditions, washed with Ca^{2+}, Mg^{2+}-free phosphate-buffered saline (PBS), passaged with trypsin/EDTA and plated onto permeable membranes at just below confluent density (e.g., ~2×10^5 cells/4.7 cm^2 filter for MDCK cells); filters are plated at least in triplicate.

2. Monolayers are grown to confluence and cultured until permeability characteristics are stable. Generally, MDCK and LLC-PK_1 cells are cultured for 4–7 days on filters, while Caco-2 and T84 cells are cultured for 21 days on filters before permeability assays.
3. Media should be changed at least every 3 days and great care should be taken not to damage the monolayer or underlying filter.
4. TER measurements can be made to verify the presence of an intact barrier, but these measurements should not be taken at the same time as the flux measurements to avoid further potential disruption to the monolayer. In addition, we have used TER measurements of duplicate monolayers to determine the effects, if any, of incubation with varying concentrations of PEGs or any other additives for the period of time required to make flux measurements.

3.2. Measurement of Transepithelial Electrical Resistance

1. Rinse (sterilize) plate electrodes with 100% ethanol, then twice with warm PBS.
2. Replace PBS with warmed complete culture media.
3. Measure transepithelial electrical resistance (TER) on empty filter to determine background.
4. Measure TER on cell monolayers on filters cultured as above.
5. Calculate TER (ohms × cm^2) by subtracting background reading and multiplying by the surface area of the filter.

3.3. Permeability Assay

1. Cell monolayers are rinsed twice in 37°C Hank's balanced salt solution (Cellgro) and preincubated in Hanks' for 30 min; this preincubation period is critical to obtaining a linear flux measurement.
2. After 30 min, both apical and basal solutions are removed and replaced either with fresh warmed Hanks' or with warmed Hanks' containing a mixture of PEGs. We normally use 5 mg/ml (total combined PEG weights) mixture of PEG200, PEG400, and PEG900 at a ratio of 2:0.5:1 (by weight) added to the basolateral compartment.

   For each new cell line or experimental condition, flux should be tested in both the apical to basolateral and basolateral to apical directions, since lack of directional polarity is a defining characteristic of paracellular as opposed to transcellular transport. In practice, once lack of polarity has been determined, we usually measure flux from the basolateral (in this case the "donor compartment" to apical direction "acceptor compartment"), since there is less dilution of PEGs into the smaller volume of the apical chamber.

3. After the initiation of the flux assay, samples are removed from the acceptor compartments at 60, 120, and 180 min, adding additional warm Hanks to replace the volume removed for assay; 0.5–1 ml is placed into a polypropylene tube with a tight fitting cap (either screw top or a boil-proof snap-top tube). It is good practice to centrifuge any samples taken from the apical chamber to avoid possible contamination from floating cells. Remove the supernatant to a clean tube and discard any pelleted cells.

   After we have verified that flux is linear over the time period of the experiment, we normally just take a single 1 ml sample at either 90 or 180 min from the acceptor compartment; each sample is one of triplicate wells.
4. An internal standard (20 μg purified PEG28, Polypure AS, Oslo, Norway) is added to all samples to assess recovery during the subsequent derivatization and purification steps.
5. Blank samples (Hanks' of equal volume to the sample volume plus internal standard) are prepared in duplicate.
6. An estimate of the concentration of the donor compartment is prepared by making a dilution (1:40) of the starting PEG solution in Hanks plus internal standard; these samples are prepared in triplicate.

   The concentration of the donor solution will decrease over the time of the assay; it is thus useful to remove an aliquot from the donor compartment at the last time point of the assay to determine the change in the concentration of the donor compartment over the time period of the assay. In practice, this decrease in concentration is normally minor (a few percent) and can be ignored. After samples are removed for PEG assay, filters can be rinsed with Hanks' or PBS and cut out and used for immunoblot or immunofluorescent analysis if required; filters can be placed directly into SDS-sample buffer for immunoblot analysis.
7. After the addition of the internal recovery standard, samples can be stored at –20°C until further processed.
8. Fresh or thawed samples are dried in a 55°C water bath under a stream of N_2; this takes 2–4 h; dried samples can be stored indefinitely until they are derivatized. In practice, we normally dry samples the afternoon of an experiment, derivatize them, and start injecting them on the HPLC the next day.

3.4. Derivatization

PEGs are conjugated with 1-naphthyl isocyanate (1-NIC) (9) after the flux assay is performed. Each PEG is modified on the two terminal hydroxyl groups and the resulting fluorescent signal after HPLC separation of each PEG oligomer is directly proportional to the number of molecules, regardless of size (see Note 7).

1. Dried PEG-containing samples (including blanks and donor compartment samples) are derivatized by the addition of 10 μl 1-NIC (Acros Organics) plus 100 μl of acetone.
2. If using Snap-cap tubes, wrap the closure of a snap-top tube in parafilm, close screw-cap tubes tightly to prevent loss of sample; vortex for 4 h. We use a standard bench top vortex with a multi-tube adaptor.
3. Centrifuge samples briefly (1 min, 5,000 × *g*) to concentrate liquid in the bottom of the tubes.
4. Add 50 μl of methanol and 450 μl H_2O to samples to hydrolyze/quench excess reagent.
5. Vortex vigorously.
6. Centrifuge at 10,000 × *g* for 5 min to sediment particulate material.
7. Carefully remove 500 μl of supernatant and place into a clean 1.5 ml polypropylene screw-top tube.
8. Add 300 μl of diethyl ether.
9. Cap tubes and vortex vigorously.
10. Separate phases by brief (5 min) centrifugation at 5,000 × *g*.
11. Carefully remove 450 μl of the lower, aqueous phase containing the PEGs, to a clean screw cap tube.
12. Re-extract samples with 300 μl of diethyl ether.
13. Cap tubes and vortex vigorously.
14. Separate phases by brief (5 min) centrifugation at 5,000 × *g*.
15. Carefully remove 400 μl of the lower, aqueous phase containing the PEGs, to a clean screw cap tube.
16. Transfer 200 μl of this aqueous phase into an HPLC vial.
17. Reserve remainder of sample at 4°C in case samples need to be reanalyzed.

3.5. HPLC Analysis

1. 100 μl of aqueous phase (from Subheading 3.4, step 16 above) is analyzed by HPLC (Agilent Technologies) using a bare silica column (Waters Spherisorb 5.0 μm Silica column, 4.6 mm × 150 mm, Note 8).
2. Individual PEG oligomers are separated using a linear H_2O:acetonitrile gradient from 5 to 60% acetonitrile for 40 min, followed by 15 min wash 95% H_2O and 5% acetonitrile; flow rate 1.5 ml/min, at room temperature.
3. Peaks are quantified by fluorescence emission (Agilent HPCHEM station, Ex = 232 nm, Em = 358 nm).

Peaks should appear in the chromatogram as evenly spaced and well separated (Fig. 1). It is usually possible to assign the

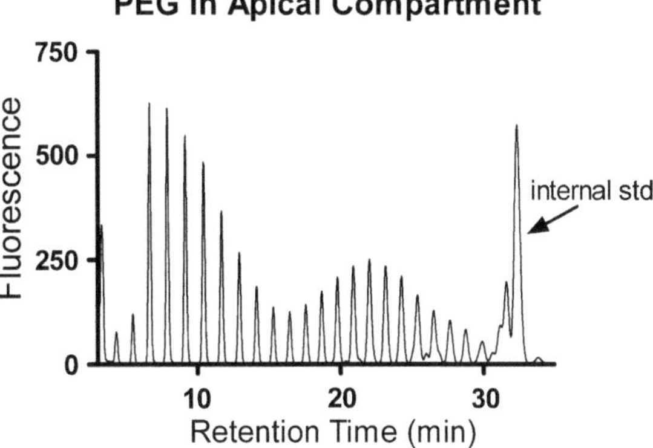

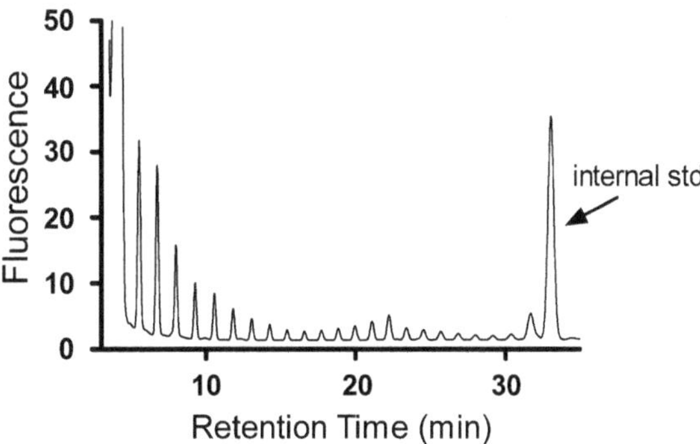

Fig. 1. Chromatogram of "donor" compartment (*top*) and "acceptor" compartment (*bottom*); each peak represents a single-derivatized PEG oligomer. The donor compartment (*top*) contains comparable concentrations of a continuous series of PEG sizes (PEG_2–PEG_{25}); with elution time proportional to size. After 90 min, the profile on the acceptor side (*bottom*) shows strong size discrimination above an HPLC retention time corresponding to about 4 Å radius. Purified PEG_{28} is added before derivatization as an internal recovery standard (*far right peak in figure*); a component of the peaks at 21–22 min corresponds to derivatized glucose in the Hanks' buffer. The smallest PEG reliably detected is 2.8 Å (PEG3, HPLC position verified with triethylene glycol) and the largest is ≈7 Å (PEG25).

identity of each peak by comparing the blank sample (containing only the PEG28 internal recovery standard) and the "donor compartment" samples and to count backwards from PEG28 (the internal standard) to PEG3 or 4, depending on the quality of the chromatogram (Note 9). PEG2 is lost in an earliest eluting peak of fluorescent material (Fig. 1).

Investigators may wish to determine the effect on P_{app} of exposing cells to various pharmacologic agents, cytokines or after manipulating the levels of different tight junction proteins through transfection or knock-down technologies. Those applications of the profiling method are beyond the scope of this review but are easily adapted to the basic permeability assay described here. Examples include our own manipulation of claudins which affect pores (<4 Å) (10) and knockdown of ZO-1 which only affects the leak pathway (>4 Å) (11).

3.6. Quantitative Interpretation of Flux Data

1. The area of each PEG fluorescence peak in each sample is divided by the area for a known amount of the internal standard (PEG28) in that sample and multiplied by the volume of the sample. This is a measure of the quantity of each PEG that crossed the filter (dQ).
2. dQ is divided by the time in seconds over which flux occurred (dt) giving the flux (dQ/dt).
3. The value of each peak in the donor sample is divided by the internal standard in that sample and multiplied by the dilution factor (C). This is the concentration gradient driving flux. C in the donor compartment only drops by only a few percent over the time of the flux assay and to a first approximation the concentration gradient does not change throughout the assay and dQ/dt is linear over the 60–90 min assay times.
4. The value of (C) is multiplied by the area of the filter (A).
5. The values of $(C)\times(A)$ for each peak of the three duplicate donor samples are averaged.
6. The values of dQ/dt for each peak in each sample are divided by the averaged value for $(C)\times(A)$; this value is the P_{app} for each PEG size: $(dQ/dt)/A\cdot C_0$, where A is the filter area and C_0 is the initial concentration in the donor compartment.
7. The hydrodynamic radii of PEG oligomers are related to molecular mass (M) by the relationship $r\,(\text{Å}) = 0.29\,M^{0.454}$ P_{app} can then be graphed as a function of PEG radius (Fig. 1) clearly revealing the different behavior of solute less and greater than 4 Å in radius.
8. Solutes small enough to pass through the pores can also pass through the larger breaks. This contribution must be subtracted to define the real permeability for small solutes. To correct P_{app} of the first phase, the second phase is extended after linear regression and subtracted from points in the first phase. Points are used on the portion of the graph above 4.2 Å to define the linear portion (Fig. 2).
9. As solute size approaches pore size, it is restricted in a nonlinear way leading to underestimation of the real pore size. The real pore radius can be estimated from the ratio of the corrected paracellular permeabilities of pairs of two small PEG species (2.8, 3.2, 3.5, or 3.7) applying a sieving function as described in Eqs. (7), (12), and (8) in Knipp et al. (13). Pore size [Renkin (sieving) function]: $A/A_0 = (1 - a/r)2[1-2.104(a/r)+2.09(a/r)3-0.95(a/r)5]$, where A = effective opening,

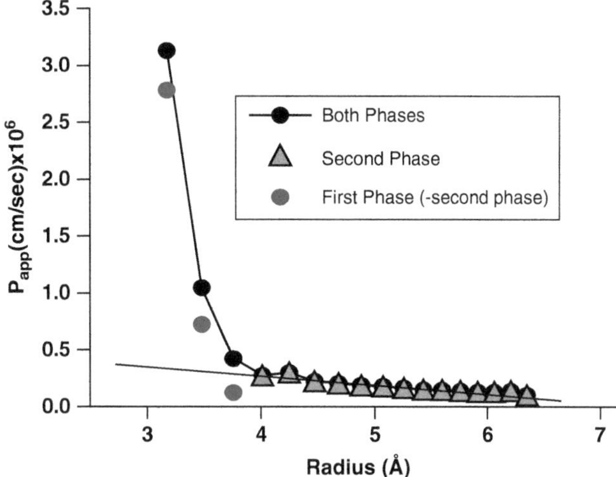

Fig. 2. Typical data for P_{app} as a function PEG size across a monolayer of Caco-2 cells and quantitative interpretation of the two transport phases. (*Filled circle*) Raw data for P_{app} as function of PEG size. Points above 4 Å are fit by linear regression and subtracted from the uncorrected values below 4 Å to give the P_{app} values for the PEGs passing through only the small pores.

A_0 = total cross-sectional area of the pore, r = pore radius, and a = PEG radius. In practice, this method for correcting the measured to actual pore radius is not required when investigators are simply interested in measuring the magnitudes of the pore and leak pathways. We refer readers to previous publications if they see the need to apply this level of rigor to their work (8, 10).

4. Notes

1. Polyester or polystyrene filters should work as well as polycarbonate filters; however, the use of a new filter should be tested without cells to verify that filter alone shows high permeability without size-selectivity. Occasional filter failures (possibly a manufacturing problem) or incomplete monolayers can be recognized by low TER or a lack of size selectivity in the PEG assay.
2. Although it is possible to use chopstick electrodes to measure TER, we find that the plate electrodes are considerably easier to use and give more reproducible readings.
3. Polyethylene glycol sizes represent a mixture of oligomers with an average size that approximates the stated molecular weight. We have been unable to find a reliable mixture of the smallest PEGs (fluorescent derivatization of PEG100, for example, did

not give us a usable chromatogram centered around 100 Da nor did the addition of tri-tetra and penta-ethylene glycols improve the sensitivity of the chromatogram in the lower molecular weight ranges). We found the mixture of 200, 400, and 900 mw PEGs gave us a broad range of peak sizes, but would recommend some experimentation if mixture does not cover the size range of interest.

Each PEG has two reactive hydroxyl groups. Single- and double-labeled forms of each PEG will resolve at different positions on the chromatogram and be quantitatively uninterpretable. If incomplete derivatization is suspected, use longer labeling times or increased NITC concentrations until the pattern reaches a final simple series of equally spaced peaks.

4. Buy 1-NIC in the smallest aliquots available. It oxidizes and hydrolyzes after the bottle is opened, so check for the presence of precipitate. If there is precipitate, discard the bottle. This reagent is relatively inexpensive, so it is better to start with fresh material if there is any question. We have tried gassing it each time after opening with N_2 but that does not seem to significantly prolong the usability.
5. It is best to purchase the acetone in small volumes since once it is opened it will start to absorb water; as this happens, the amount of insoluble material in the derivatization reaction increases.
6. Complete drying of PEG solutions before derivatization is critical, otherwise the 1-NIC will react with water rather than the PEG. We tried using a Speed Vac but found that the viscosity of the PEG samples inhibited complete drying. Some flocculent material is always present in this assay, but residual moisture from incompletely dried samples leads to a large amount of flocculent precipitate which will decrease recovery.
7. In this part of the assay, reproducible pipetting is absolutely critical. It is necessary to pre-wet pipette tips with ether when pipetting ether or the aqueous phase below the ether layer.
8. Column half-lives vary, but we normally replace the column after approximately 500 sample injections.
9. Glucose in the Hank's balanced salt solution is the major contaminant seen normally in the PEG chromatograms. It appears as multiple small peaks in the size range of PEG18-20. It can be distinguished from the PEG peaks by its appearance in the blank sample, containing only Hanks' and the PEG28 internal standard.

References

1. Powell, D.W. (1981) Barrier function of epithelia. *Am. J. Physiol.* 241:G275–G288.
2. Van Itallie, C.M. and Anderson, J.M. (2006) Claudins and epithelial pracellular transport. *Annu. Rev. Physiol.* 68:403–429.
3. Laukoetter, M.G., Nava, P., and Nusrat, A. (2008) Role of the intestinal barrier in inflammatory bowel disease. *World J. Gastroenterol.* 14:401–407.
4. Ghandehari, H., Smith, P.L., Ellens, H., Yeh, P.Y., and Kopecek, J. (1997) Size-dependent permeability of hydrophilic probes across rabbit colonic epithelium. *J. Pharmacol. Exp. Ther.* 280:747–753.
5. Sanders, S.E., Madara, J.L., McGuirk, D.K., Gelman, D.S., and Colgan, S.P. (1995) Assessment of inflammatory events in epithelial permeability: a rapid screening method using fluorescein dextrans. *Epithelial Cell Biol.* 4:25–34.
6. Schultz, S.G. and Soloman, A.K. (1996) Determination of the effective hydrodynamic radii of small molecules by viscometry. *J. Gen. Physiol.* 44:1189–1199.
7. Lang, I., Scholz, M., and Peters, R. (1986). Molecular mobility and nucleocytoplasmic flux in hepatoma cells. *J. Cell Biol.* 102: 1183–1190.
8. Watson, C.J., Rowland, M., and Warhurst, G. (2001) Functional modeling of tight junctions in intestinal cell monolayers using polyethylene glycol oligomers. *Am. J. Physiol Cell Physiol.* 281:C388–C397.
9. Rissler, K., Wyttenback, N. and Bornsen, K. (1998) High-performance liquid chromatography of polyethylene glycols as their 1-napthylurethane derivatives and signal monitoring by fluorescence detection. *J. Chromatog. A* 822:189–206.
10. Van Itallie, C.M., Holmes, J., Bridges, A., Gookin, J.L., Coccaro, M.R., Proctor, W., Colegio, O.R., and Anderson, J.M. (2008) The density of small tight junction pores varies among cell types and is increased by expression of claudin-2. *J. Cell Sci.* 121: 298–305.
11. Van Itallie, C.M., Fanning, A.S., Bridges, A., and Anderson, J.M. (2009) ZO-1 stabilizes the tight junction solute barrier through coupling to the perijunctional cytoskeleton. *Mol. Biol. Cell* 20:3930–3940.
12. Renken, E. (1954) Filtration, diffusion, and molecular sieving through porous cellulose membranes. *J. Gen. Physiol.* 20:225–243.
13. Knipp, G.T., Ho, N.F., Barsuhn, C.L., and Borchardt, R.T. (1997) Paracellular diffusion in Caco-2 cell monolayers: effect of perturbation on the transport of hydrophilic compounds that vary in charge and size. *J. Pharm. Sci.* 86:1105–1110.

Chapter 2

Biochemical Analysis of Claudin-Binding Compatibility

Christina Ward and Michael Koval

Abstract

Tissue barrier function is directly mediated by tight junction transmembrane proteins known as claudins. Cells that form tight junctions typically express multiple claudin isoforms, which suggests that heterotypic (head-to-head) binding between different claudin isoforms may play a role in regulating paracellular permeability. To test whether claudins are heterotypically compatible, we developed an assay system using HeLa cells, a claudin-null cell line which expresses other tight junction proteins, including occludin, junction adhesion molecule A, and zonula occludens-1, -2, and -3. HeLa cells stably transfected to express different claudins are cocultured, then subsequently analyzed for the ability to coimmunopurify. Using this approach, we have found that claudin-1, claudin-3, and claudin-5 are heterotypically compatible. In contrast, two closely related claudins, claudin-3 and claudin-4, are incompatible. Differential claudin-binding specificity is likely to have downstream effects on the regulation of tight junction composition and permeability.

Key words: Tight junction, Epithelia, Immunopurification, Intercellular interactions, Claudins

1. Introduction

Polarized cells produce selective barriers by regulating the movement of water, ions, and proteins across the monolayer through tight junctions (1). While tight junctions require the coordinated activity of several different proteins, the specificity of tight junction permeability is regulated by a family of transmembrane proteins known as claudins (2–5). There are almost two dozen different claudins and different cells simultaneously express several claudin genes. Claudins form what are functionally equivalent to charge selective pores which either promote or restrict the permeability to specific ions. The selectivity is in large part due to the structure of the extracellular loop domains (6). However, little is known about how claudin heterogeneity influences paracellular permeability.

Kursad Turksen (ed.), *Claudins: Methods and Protocols*, Methods in Molecular Biology, vol. 762,
DOI 10.1007/978-1-61779-185-7_2, © Springer Science+Business Media, LLC 2011

In part, this is due to a lack of knowledge of the molecular basis for the control of interactions between different claudins.

Claudins can potentially interact in two different ways: laterally in the plane of the membrane (heteromeric binding) and between adjacent cells by head-to-head (heterotypic) binding. Several methods have been developed to examine heteromeric claudin–claudin interactions, including coimmunopurification (7–9), fluorescence resonance energy transfer (FRET) (10, 11), and yeast two hybrid assays (12). Here, we describe a method using claudin-transfected HeLa cells with a claudin-null background to study claudin heterotypic compatibility (9). HeLa cells offer an advantage over other claudin-null cells, such as fibroblasts, since they express other tight junction proteins, such as occludin, which could have a regulatory role in controlling claudin–claudin interactions (13).

2. Materials

2.1. Cell Culture/Transfection

1. Dulbecco's Modified Eagle's Medium (Sigma Aldrich, St. Louis, MO), supplemented with 10% fetal bovine serum (FBS; Atlanta Biological, Lawrenceville, GA), 100 IU/ml penicillin, and 100 μg/ml streptomycin (DME/10).
2. Medium for selection of transfected HeLa cells: DME containing 2 mg/ml active Geneticin (G-418; Invitrogen, Carlsbad, CA) (see Note 1).
3. Medium for maintenance of transfected HeLa cells: DME containing 0.2 mg/ml active G-481.
4. Calcium- and magnesium-free phosphate-buffered saline (PBS; Sigma Aldrich).
5. Solution of 0.25% trypsin–1 mM EDTA (Sigma Aldrich).
6. 35-mm, 60-mm, 100-mm and 48-well tissue culture dishes (BD Biosciences, San Jose, CA).
7. Glass coverslips, #1 thickness, 25 mm diameter (Fisher Scientific, Pittsburgh, PA).
8. Claudin cDNAs in pcDNA 3.1. For each 60-mm dish to be transfected, use 8 μg of plasmid DNA (see Note 2).
9. OptiMEM I medium (Invitrogen) for HeLa cell transfection.
10. Lipofectamine 2000 (Invitrogen).

2.2. Immunofluorescence

1. For immunofluorescence, use PBS containing 2 mM calcium chloride and 1 mM magnesium chloride.
2. Goat serum (Sigma Aldrich) stored at −20°C. Before use, centrifuge in 1.5-ml Eppendorf tubes at 16,000 × *g* for 5 min to remove aggregates.

3. Fresh methanol/acetone (1:1).
4. PBS/TX: PBS + 0.5% Triton X-100.
5. PBS/TX/GS: 20 ml PBS + 0.5% Triton X-100 + 2% Goat serum.
6. PBS/GS: 20 ml PBS, 2% Goat serum.
7. 25 mm × 75 mm × 1 mm glass slides (Fisher).
8. Antibodies: Rabbit and mouse anti-claudin antibodies are from Invitrogen. Cy2-conjugated goat anti-rabbit and Cy3-conjugated goat anti-mouse are from JacksonImmuno (Malvern, PA) (see Note 3).
9. Mowiol mounting solution.
   (a) Place 6 mg glycerol in a 50-ml plastic disposable centrifuge tube containing a small magnetic stir bar.
   (b) Add 2.4 g Mowiol 4-88 (Calbiochem/EMD, Darmstadt, Germany), and then mix using a magnetic stirring plate.
   (c) While stirring add 6 ml ultrapure water and leave at room temperature for 2 h.
   (d) Add 12 ml 0.2 M Tris–HCl, pH 8.5 (Sigma Aldrich).
   (e) Incubate in a glass beaker with hot water at 50°C for 30–60 min with frequent stirring to dissolve the Mowiol.
   (f) Centrifuge at 5,000 × *g* for 15 min to clarify. Store as 1 ml aliquots in 1.5-ml Eppendorf tubes at –20°C. These are stable for about 2 years.
   (g) Warm to room temperature prior to use. Store remaining Mowiol solution at 4°C. Use within 1 month.

2.3. Coimmuno-purification

1. Lysis Buffer: Dissolve 1 Complete Protease Inhibitor Mini tablet (Roche, Nutley, NJ) in 50 ml PBS containing 0.5 ml 1 M sodium fluoride (NaF) and 0.5 ml 0.1 M sodium orthovanadate ($NaVO_4$).
2. Detergent stock: 20% Triton X-100 – diluted in MilliQ-purified H_2O (Millipore, Billerica, MA). Store at room temperature (see Note 4).
3. Lysis Buffer + 0.1% Triton X-100: 10 ml Lysis Buffer + 50 μl 20% Triton X-100.
4. Narrow tip XL-2000 probe Misonix sonicator (Qsonica, Newtown, CT).
5. Blocking Buffer: 0.25% bovine serum albumin (SeraCare Life Sciences, Milford, MA), 0.20% gelatin (Sigma Aldrich) diluted in PBS w/o Ca/Mg. Store at 4°C for up to 6 months (see Note 5).
6. Primary antibodies: Rabbit and mouse anti-claudin antibodies are from Invitrogen BioMag: Goat anti-rabbit IgG magnetic beads (QIAGEN, Valencia, CA). Store at 4°C (see Note 6).

7. Co-IP antibody/bead suspension: For each co-IP reaction, mix 80 μl Blocking Buffer solution, 40 μl goat anti-rabbit IgG Biomag beads, 7 μl rabbit anti-claudin IgG. Thoroughly mix by flicking with your finger, do not vortex. Incubate at 4°C for 30 min with intermittent mixing. Prepare separate batches using a different rabbit antibody for each claudin to be analyzed.
8. DynaMag™-2 magnetic isolation rack (Invitrogen).
9. 2× Sample Buffer (2× SB): 2 ml glycerol, 4 ml 10% sodium dodecyl sulfate (SDS), 1 ml 1 M Tris–HCl pH 6.7, final volume 10 ml with H_2O, 0.5 mg bromophenol blue. This is stable 4°C or room temperature (w/o DTT) for a few months. Add 100 mg dithiothreitol (DTT) to 1 ml 2× SB to make working stocks, unused portions can be stored at −20°C up to 1 month.

2.4. SDS–PAGE/ Immunoblot

1. Ammonium persulfate (APS), 10% solution in H_2O, store at 4°C up to 1 month.
2. Mini-PROTEAN electrophoresis apparatus, Trans-Blot SD Semi-Dry Transfer Cell and BioRad Pack 100 power supply (BioRad, Hercules, CA).
3. Resolving gel: For each gel, mix 2.8 ml 30% acrylamide/bis (29:1 mixture; BioRad), 3.0 ml 1 M Tris Base pH 8.9, 8 μl TEMED (BioRad), 2.2 ml H_2O, 80 μl 10% SDS, 80 μl 10% APS. Prepare just prior to pouring the gel (see Note 7).
4. Stacking gel: For each gel, mix 0.25 ml 30% acrylamide, 0.25 ml 1 M Tris Acid pH 6.7, 3 μl, TEMED, 1.46 ml H_2O, 21 μl 10% SDS, 21 μl 10% APS. Prepare just prior to pouring the gel.
5. Running buffer: Dissolve 45 g Tris Base, 216 g glycine, and 15 g SDS in 3 L H_2O. Bring to 15 L final volume with H_2O, store at room temperature.
6. Low range prestained molecular weight markers (BioRad).
7. Transfer buffer: Dissolve 5.8 g 50 mM Tris Base, 29 g 380 mM glycine, and 0.25 g 0.025% SDS in 500 ml H_2O. Bring to 800 ml with H_2O, then add 200 ml methanol. Store at room temperature.
8. Blotto with 5% milk: Dissolve in 400 ml H_2O: 2.42 g 40 mM Tris Base, 0.5 ml Tween 20. Bring the pH to 7.5 with HCl, bring volume to 500 ml, and then add 25 g Carnation brand powdered milk. Store at 4°C for up to 2 days.
9. Horseradish peroxidase goat anti-mouse IgG (JacksonImmuno).
10. Immobilon-P transfer membrane (Millipore, Billerica, MA).
11. Extra thick blot paper (BioRad).
12. ECL-enhanced chemiluminescence reagent (Perkin Elmer, Waltham, MA).

3. Methods

The key to this method is analysis of cocultures of cells expressing distinct claudins. Since the HeLa cells provide a claudin-null platform, this enables the heterotypic interactions to be analyzed for any pair of claudins, provided the cDNAs are available for transfection. Homotypic interactions can also be assessed by taking advantage of tagged claudin constructs. For instance, we used this approach to demonstrate a homotypic interaction between ECFP-claudin-5 and EYFP-claudin-5 using the coculture approach (9).

Examining heterotypic compatibility of claudin pairs using transfected claudin-null cells offers the advantage of analyzing the interaction in the absence of other claudins which could confound the analysis through heteromeric interactions. For example, claudin-3 and claudin-4 are heteromerically compatible when expressed in the same cell, however, they do not heterotypically interact despite having extracellular loop domains that are highly conserved at the amino acid level (9). Thus, heterotypic and heteromeric interactions cannot be biochemically distinguished in epithelial cells where both claudins are present in the same cell membrane.

Since heterotypic interactions between claudins are dictated by the extracellular loop domains (9, 13), analysis of transfected HeLa cells expressing point mutated claudins enabled the identification of a key amino acid, Asn-44, which is required to maintain the specificity of heterotypic binding between claudin-3 and claudin-4. A comparable analysis will enable other motifs to be identified at the molecular level.

3.1. Cell Culture and Transfection

1. HeLa cell stocks on 100-mm dishes are passed twice a week, every 3 or 4 days. Cells are released from the plates by washing twice in PBS, rinsed once with 5 ml trypsin/EDTA, which is then removed to leave a thin layer of fluid. The dishes are incubated for 5 min at 37°C, then the cells are resuspended in 10 ml DME/10, transferred to a 50-ml polyethylene tissue culture tube and counted using a hemocytometer. For maintenance, cells are plated at 1×10^5 and 7×10^5 cells/100-mm dish for stock cultures. Untransfected HeLa cells are cultured in DME/10, stably transfected HeLa cells are cultured in maintenance medium containing G-418 (see Note 8).
2. For transfection, HeLa cells need to be plated so they are 70–80% confluent on the day of transfection. For one 60-mm tissue culture dish, plate 4×10^5 HeLa cells/dish to achieve ~80% confluence on day 3.
3. The day before transfection change the media to 4 ml/dish OptiMEM I without antibiotics containing 4% FBS.

4. At least 1 h prior to transfection, change the media to 1 ml/dish serum-free OptiMEM I.
5. DNA/Lipofectamine complexes:
   (a) For each 60-mm dish to be transfected, add 8 μg DNA to 500 μl OptiMEM I in a sterile 5-ml round-bottomed polyethylene tissue culture tube.
   (b) In a separate sterile 5-ml polyethylene tube, add 16 μl Lipofectamine 2000 to 500 μl Optimem I.
   (c) Mix by flicking the side of the tubes with your finger. Do not vortex or triturate.
   (d) Incubate for 5–10 min at room temperature.
   (e) Add the DNA mixture to the Lipofectamine mixture and flick the tube to mix.
   (f) Incubate for 20 min at room temperature (see Note 9).
6. Add 1 ml of DNA/Lipofectamine 2000 complex to each 60-mm dish dropwise to several different areas of the plate. Swirl the plate to mix.
7. After 5 h, add 1 ml OptiMEM containing 4% FBS to each dish.
8. The next day, change to DME/10.
9. The cells should incubate for approximately 48 h after transfection before changing to selective medium. After 2 weeks in selective medium, isolate stably transfected clones. Cells are trypsinized and sparsely cultured by plating at 500–1,000 cells/100-mm dish in selective medium. After about 2–4 weeks, individual colonies will be visible by eye as white spots. Remove the medium from the plate and mechanically remove some of the larger, individual clones with a sterile 200 μl micropipette tip and transfer each to a single well in a 48-well tissue culture dish containing 1 ml selective medium. Allow the clones to expand and then screen for claudin expression by immunofluorescence microscopy. Once stable clones are established, they can be cultured in maintenance medium containing reduced G-418 (see Note 10).
10. For coculture experiments, HeLa transfectants expressing different claudins are trypsinized, counted, and equal amounts are mixed in a 50-ml polyethylene tissue culture tube by trituration ($1\text{–}5 \times 10^6$ cells of each type). Plate 7.5×10^5 cells total/100-mm dish or 3×10^5 total cells/35-mm dish for coimmunopurification or immunofluorescence, respectively. Cells will be ready for analysis 3 days after plating (see Note 11).

3.2. Immunofluorescence Microscopy

1. Cell cocultures on 25-mm glass coverslips in 35-mm dishes should be nearly confluent prior to processing for immunofluorescence. Include monocultures as controls.

2. Use PBS + Ca + Mg when processing cells for immunofluorescence.
3. Wash cell cocultures 3 × 1 ml PBS at room temperature.
4. Remove the PBS and replace with 1 ml 1:1 methanol/acetone (made fresh). Incubate for precisely 2 min at room temperature.
5. Wash the cells with 3 × 1 ml PBS three times. Make sure that the methanol/acetone solution is thoroughly rinsed off, particularly from under the coverslip, prior to adding PBS/TX.
6. Wash with 1 ml PBS/TX for 5 min at room temperature.
7. Wash with 2 × 1 ml PBS/TX/GS for 5 min at room temperature.
8. Dilute two primary antibodies into PBS/GS (1 ml per dish to be labeled). The antibody dilution will vary, depending on the antibodies used, but 1:1,000 dilution is a good starting point. Microfuge the solution for 5 min at high speed to clear aggregates.
9. Add primary antibody solution to cultured dish and incubate on a shaker for 1 h at room temperature.
10. Wash with 3 × 1 ml PBS/GS for 5 min at room temperature.
11. Dilute Cy2-goat anti-rabbit IgG and Cy3 goat anti-mouse IgG 1:2,000 into PBS/GS (1 ml per dish to be labeled). Microfuge the solution for 5 min at high speed to clear aggregates.
12. Add 1 ml/dish of secondary antibody solution and incubate on a shaker for 1 h at room temperature.
13. Wash with 3 × 1 ml PBS/GS for 5 min at room temperature.
14. Wash with 3 × 1 ml PBS at room temperature.
15. Place 30 μl Mowiol solution onto a slide using a 200 μl pipettor using a tip with 0.2 mm of the end cut off. Use a forceps to remove the coverslip from the culture dish, blot with a wipe to remove excess PBS, being careful not to disrupt the cell layer and invert onto the drop of Mowiol on the slide. Let the slides dry for at least an hour before microscopy, overnight drying is optimal.
16. Imaging is done by fluorescence microscopy, using filter packs which allow visualization of Cy2 (green) and Cy3 (red) fluorescence. Cocultures of cells containing compatible claudins show double-labeled cell–cell interfaces, these are absent in cultures containing cells expressing incompatible claudins. Examples of compatible and incompatible claudins are shown in Fig. 1 (see Note 12).

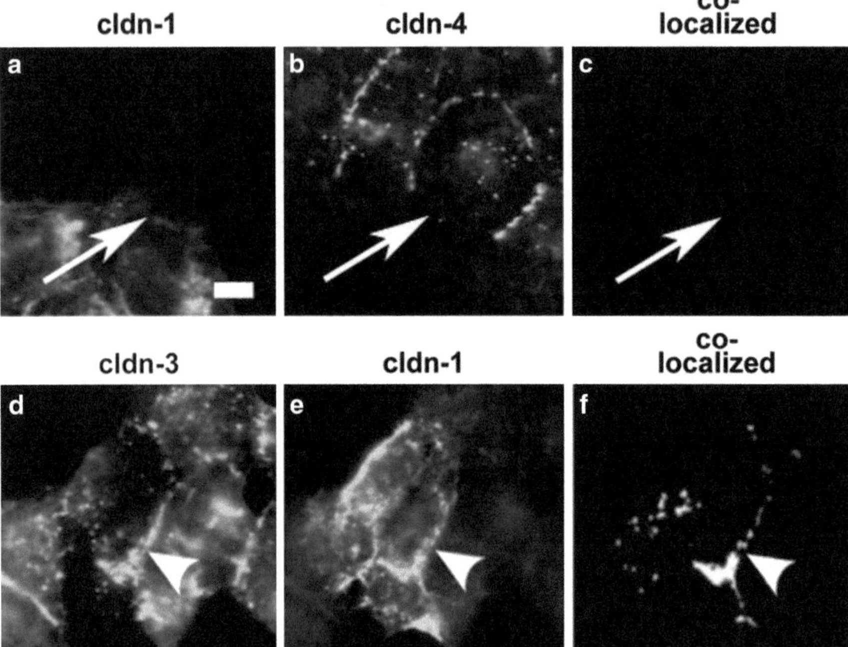

Fig. 1. Heterotypic claudin compatibility revealed by colocalization at cell–cell interfaces. HeLa cells transfected with different human claudins were cocultured and analyzed by double label immunofluorescence (**a**, **b**, **d**, and **e**). Colocalization (**c**, **f**) was determined by calculating the logical "and" function of paired immunofluorescence images, where high-intensity pixels correspond to areas where both claudins colocalize. Claudins did not colocalize in cocultures of HeLa/cldn-1 cells and HeLa/cldn-4 cells (**a**–**c**), indicating that these two claudins were incompatible (*arrow*). Bar, 10 μm. However, HeLa/cldn-1 cells cocultured with HeLa/cldn-3 cells (**d**–**f**) showed claudin immunofluorescence colocalization at areas of cell–cell contact and in intracellular vesicles containing both claudins (*arrowhead*) providing evidence of heterotypic (*head-to-head*) binding at sites of cell–cell contact. Adapted from ref. (9).

3.3. Coimmunopurification

1. Cell cocultures in 100-mm tissue culture dishes should be nearly confluent for co-IP analysis. Include monoculture samples as controls.
2. The entire experiment is best performed in a 4°C cold room, but can also be done on ice. Wash plates twice with cold PBS.
3. Add 5 ml of cold Lysis Buffer and incubate the plates for 5 min at 4°C.
4. Scrape the cells off the plates into the Lysis Buffer and transfer to a 15-ml conical polyethylene tissue culture tube.
5. Centrifuge at $500 \times g$ for 5 min at 4°C.
6. Aspirate the supernatant and resuspend the pellet in 1 ml Lysis Buffer containing 0.1% Triton X-100. Transfer to a 1.5-ml Eppendorf tube (see Note 13).
7. Incubate on ice for 10 min.
8. Sonicate while on ice with a narrow tip XL-2000 probe Misonix sonicator. Use 3×3 s pulses at 80% to shear the nuclear material.

9. Microfuge at 16,000 × *g* for 10 min.
10. Transfer 50 μl of the lysate supernatant to a separate 1.5-ml Eppendorf tube containing 50 μl of 2× sample buffer. This is the total lysate sample. Incubate the total lysate samples at 65°C for 20 min, then quick freeze by immersion in liquid nitrogen. Store at −80°C until ready to analyze by immunoblot.
11. Transfer 950 μl of the lysate supernatant to 100 μl co-IP antibody/bead suspension in 1.5-ml Eppendorf tube. Do not vortex, mix by flicking with your finger.
12. Incubate at 4°C for 1 h on a rotator platform.
13. Place tubes in the magnetic separator at 4°C for at least 2 min. The magnetic beads will form a brown line on the side of the tube aligned with the magnet. Open the tube while still mounted in the separator and remove the supernatant liquid. It should be clear.
14. Remove from the magnet, add 1 ml PBS and mix by flicking with your finger. Isolate the beads with the magnetic separator and remove the wash solution. Repeat two more times.
15. Resuspend the magnetic beads in 50 μl PBS and add 50 μl 2× Sample Buffer. Incubate the co-IP samples at 65°C for 20 min, then quick freeze by immersion in liquid nitrogen. Store at −80°C until ready to analyze by immunoblot.

3.4. SDS–PAGE/Immunoblot

1. Assemble plates using a 1.5 mm spacer and insert the gel comb; use a Sharpie to mark where the top of the resolving gel should be (~ 0.25 cm below the bottom of the comb). Remove the comb, prepare the resolving gel, and add until the level reaches the mark (~9 ml/gel) and then fill the remaining space with H_2O. Allow 20 min for the gel to solidify.
2. Once the gel hardens remove the distilled water by aspiration. Prepare the stacking gel mixture, add to the chamber, and then insert the sample comb and allow 20 min to solidify.
3. Remove the combs, place the gels in the running chamber and add running buffer to the chamber.
4. Using gel loading tips, add size standards and samples to the gel. Load 2–5 μl/well of total lysate sample to each well, leave a blank well containing sample buffer alone, and then load 20 μl/well of co-IP sample to each well.
5. Run at 100 V constant voltage for 90 min.
6. When the gels have completed running, remove from the apparatus, separate the plates, and cut a small part of one corner off to mark the gel orientation. Remove from the glass plate and place in transfer buffer in a small dish. Incubate on a shaker for ~30 min.

7. Activate a piece of PVDF transfer membrane cut to match the gel in 100% methanol for 2 min, rinse with H_2O, then incubate in transfer buffer for ~30 min. Pre-soak two pieces of pre-cut blotting paper for each gel in transfer buffer.
8. Assemble the one piece of blotting paper, PVDF membrane, gel and then another piece of blotting paper in the semi-dry transfer apparatus. Run for 1 h at 24 V constant voltage.
9. After the transfer is complete, place the PVDF membrane in 20 ml of Blotto and incubate at either room temperature for 1 h or overnight at 4°C. If the transfer was successful, the prestained markers should be clearly visible on the PVDF membrane.
10. Dilute mouse anti-claudin antibodies 1:1,000 in Blotto and incubate the PVDF membrane for 1 h at room temperature.
11. Wash PVDF membranes for 3 × 5 min in Blotto.
12. Dilute secondary HRP-conjugated goat anti-mouse IgG 1:2,000 in blotto and incubate PVDF membranes for 1 h at room temperature.
13. Wash PVDF membranes for 3 × 5 min in Blotto, then 2 × 5 min in PBS.
14. Incubate PVDF membranes in ECL solution for 2 min at room temperature, wrap in Saran wrap then detected using either X-ray film (Kodak) or Molecular Imager Gel Doc XR+System (BioRad). Examples of compatible and incompatible claudins are shown in Fig. 2 (see Note 13).

4. Notes

1. G-418 is provided as a mixture of active and inactive compound, typically with 85% activity. This needs to be taken into account when preparing selective and maintenance medium.
2. There are several options for cDNAs in expression vectors. Commercially available cDNAs for several claudins are available in a ready use formats (e.g., Ultimate ORF™ clones from Invitrogen). It is important to verify the cDNA by sequencing prior to use, regardless of source, to avoid pitfalls related to unanticipated mutations or an incorrect cDNA insert.
3. Primary antibodies are available from several vendors in addition to Invitrogen. For experiments examining claudin–claudin interactions, controls are particularly critical to avoid pitfalls related to antibody cross-reactivity (14). Every coculture experiment should include positive and negative controls consisting of monocultures to insure that detection by immunofluorescence, immunopurification, and immunoblot is specific.

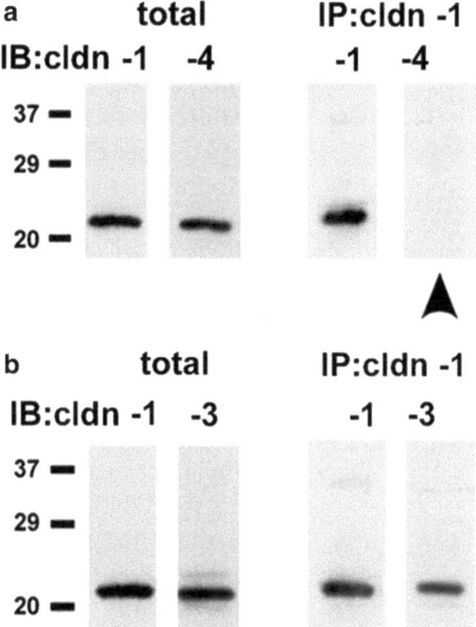

Fig. 2. Heterotypic claudin-binding specificity. Cocultures of cells expressing different claudins were solubilized in Triton X-100 and examined for the ability to coimmunopurify (IP) as determined by immunoblot (IB). Shown are representative immunoblots from cocultures containing incompatible claudins: HeLa/cldn-1 + HeLa/cldn-4 (**a**) or compatible claudins: HeLa/cldn-1 + HeLa/cldn-3 (**b**). The *arrowhead* denotes a lack of interaction between claudin-1 and claudin-4. Adapted from ref. (9).

4. All water used for solutions is Milli-Q purified (Millipore) or equivalent.
5. If the Co-IP blocking buffer becomes cloudy, discard, and make a fresh stock.
6. Magnetic coimmunopurification offers an advantage over coimmunoprecipitation by avoiding co-sedimentation of aggregated proteins with immune complexes during centrifugation. We have found that using rabbit polyclonal antibodies for immunoisolation routinely provides more reproducible recovery than mouse monoclonal antibodies. Also, using mouse monoclonal antibodies for immunoblots from samples where rabbit IgG was used for coimmunopurification reduces detection of IgG band. Since the IgG light chain migrates close to the position where claudins migrate (~22 kDa), this cross-species approach improves the interpretability of coimmunopurification immunoblots.
7. We have found that hand cast SDS–PAGE gels provide more consistent analysis for claudins than pre-cast gels.
8. HeLa cell lines show significant clonal variation; several lines have a more fibroblastic morphology and are less suitable for

the analysis of claudins than HeLa cells with more epitheloid morphology. Before beginning analysis, confirm that the HeLa cells have a cuboidal morphology and the cells should be analyzed for expression of non-claudin tight junction proteins (9). Note that despite their epitheloid morphology, we have found that HeLa cells do not form high resistance barriers since they tend to grow as discrete colonies. Thus, they are not a good platform for barrier function measurements, even if transfected to express one or more claudins. Stable HeLa cell transfectants are also critical for proper analysis; cell lines need to be selected that transport the claudin of interest to the plasma membrane and are not overexpressed, which makes them prone to misfolding and intracellular retention.

9. Mixing is a critical step, adding the Lipofectamine 2000 solution to the DNA solution does not yield reproducible complexes that the transfection efficiency will be poor. The fully formed DNA/Lipofectamine 2000 complex solution should become slightly cloudy. If the solution remains clear, the Lipofectamine 2000 stock should be replaced.

10. HeLa cell lines stably transfected with claudins can spontaneously lose their claudin expression over time (approx. 60–70 passages) and yet still remain resistant to G-418. This effect may be related to a role for claudins in the regulation of cell growth (15), although further work is needed to determine whether this is the case. Periodically check stably transfected HeLa cell clones by immunofluorescence to insure that the cells have retained claudin expression.

11. If the cocultures do not show an adequate number of heterotypic interfaces, they can be propagated and re-trypsinized prior to use for experiments. Also, claudin expression can have a differential effect of the growth of cell clones. For instance, we have found that HeLa/claudin-3 cells tend to be slower growing than other HeLa/claudin transfectants. If so, then compensate for this by altering the ratio of cells initially plated in the cocultures. For instance, for HeLa/claudin-3 cells, a 3:1 ratio gave better cocultures than a 1:1 ratio.

12. Double label immunofluorescence provides a quick, visual method to identify heterotypically compatible claudin pairs. Claudin co-localization can also be quantified by scoring the fraction of cell–cell interfaces between cells expressing different claudins that showed regions meeting thresholds for size and intensity. For instance, we have previously used a minimum size of 100 contiguous pixels with fluorescence intensity values greater than 100 for both channels (9). Confocal or deconvolution immunofluorescence microscopy are ideal to avoid interpreting the overlap between two signals at different *z*-axis depth as true colocalization. However, a high-quality

objective with an iris diaphragm enables achieving numerical apertures which can provide near confocal *z*-axis resolution.

13. To quantify coimmunopurification, densitometric scans or directly obtained luminescence signals of the amount of immunopurified claudin is first normalized to the total amount of claudin present in the samples. Then the ratio of the amount of coimmunopurified claudin divided by the amount of directly precipitated claudin is determined.

Acknowledgments

Figures 1 and 2 were adapted from data originally published in the Journal of Biological Chemistry, Daugherty, B.L., Ward C., Smith T., Ritzenthaler J.D., and Koval M. Regulation of heterotypic claudin compatibility, 2007; 282:30005–30013. © The American Society for Biochemistry and Molecular Biology. This work was supported by National Institutes of Health grants HL-083120 and AA-013757.

References

1. Schneeberger, E.E., and Lynch, R.D. (2004) The tight junction: a multifunctional complex. *Am J Physiol Cell Physiol* **286**, C1213–1228.
2. Angelow, S., Ahlstrom, R., and Yu, A.S. (2008) Biology of claudins. *Am J Physiol Renal Physiol* **295**, F867–876.
3. Heiskala, M., Peterson, P.A., and Yang, Y. (2001) The roles of claudin superfamily proteins in paracellular transport. *Traffic* **2**, 93–98.
4. Turksen, K., and Troy, T.C. (2004) Barriers built on claudins. *J Cell Sci* **117**, 2435–2447.
5. Koval, M. (2006) Claudins: Key pieces in the tight junction puzzle. *Cell Commun Adhes* **13**, 127–138.
6. Anderson, J.M., and Van Itallie, C.M. (2009) Physiology and function of the tight junction. *Cold Spring Harbor Perspect Biol* **1**, a002584.
7. Coyne, C.B., Gambling, T.M., Boucher, R.C., Carson, J.L., and Johnson, L.G. (2003) Role of claudin interactions in airway tight junctional permeability. *Am J Physiol Lung Cell Mol Physiol* **285**, L1166–1178.
8. Furuse, M., Sasaki, H., and Tsukita, S. (1999) Manner of interaction of heterogeneous claudin species within and between tight junction strands. *J Cell Biol* **147**, 891–903.
9. Daugherty, B.L., Ward, C., Smith, T., Ritzenthaler, J.D., and Koval, M. (2007) Regulation of heterotypic claudin compatibility. *J Biol Chem* **282**, 30005–30013.
10. Blasig, I.E., Winkler, L., Lassowski, B., Mueller, S.L., Zuleger, N., Krause, E., Krause, G., Gast, K., Kolbe, M., and Piontek, J. (2006) On the self-association potential of transmembrane tight junction proteins. *Cell Mol Life Sci* **63**, 505–514.
11. Piontek, J., Winkler, L., Wolburg, H., Muller, S.L., Zuleger, N., Piehl, C., Wiesner, B., Krause, G., and Blasig, I.E. (2008) Formation of tight junction: determinants of homophilic interaction between classic claudins. *FASEB J* **22**, 146–158.
12. Hou, J., Renigunta, A., Gomes, A.S., Hou, M., Paul, D.L., Waldegger, S., and Goodenough, D.A. (2009) Claudin-16 and claudin-19 interaction is required for their assembly into tight junctions and for renal reabsorption of magnesium. *Proc Natl Acad Sci U S A* **106**, 15350–15355.
13. Mrsny, R.J., Brown, G.T., Gerner-Smidt, K., Buret, A.G., Meddings, J.B., Quan, C., Koval, M., and Nusrat, A. (2008) A key claudin extracellular loop domain is critical for epithelial barrier integrity. *Am J Pathol* **172**, 905–915.

14. Nitta, T., Hata, M., Gotoh, S., Seo, Y., Sasaki, H., Hashimoto, N., Furuse, M., and Tsukita, S. (2003) Size-selective loosening of the blood-brain barrier in claudin-5-deficient mice. *J Cell Biol* **161**, 653–660.

15. Osanai, M., Murata, M., Chiba, H., Kojima, T., and Sawada, N. (2007) Epigenetic silencing of claudin-6 promotes anchorage-independent growth of breast carcinoma cells. *Cancer Sci* **98**, 1557–1562.

Chapter 3

Electrophysiological Characterization of Claudin Ion Permeability Using Stably Transfected Epithelial Cell Lines

Alan S.L. Yu

Abstract

Claudins are tight junction membrane proteins that act as paracellular pores and barriers and regulate epithelial permeability to small ions. A key step in understanding the function of any claudin isoform is the in vitro measurement of its ion permeability and selectivity. Herein, we describe methods to generate clonal lines with stable inducible overexpression of claudins in Madin–Darby canine kidney epithelial cells, measure conductance and diffusion potentials in Ussing chambers, correct for liquid junction potentials, and derive quantitatively accurate values for individual ion permeabilities.

Key words: Claudin, Tight junction, Paracellular, Diffusion potential, Conductance, Ion permeability, Junction potential

1. Introduction

Claudins are a family of tight junction membrane proteins that are believed to form the paracellular pore and barrier in epithelial tissues (1–3). In this review, we describe our approach to investigate the permeability properties of individual claudin isoforms in vitro, by overexpressing in Madin–Darby canine kidney (MDCK) cells, which are a well-characterized renal epithelial cell model (4, 5). The resultant phenotype is dependent on the background permeability (6, 7). Claudins that primarily increase ion permeability are best observed on a background of high transepithelial resistance (TER), so we use the MDCK I strain of cells. Claudins that primarily act as barriers to ion permeation are best observed in cells with low TER, so we usually use the MDCK II strain of cells. We find that we get the most reproducible results when we generate stable clonal cell lines (so that there is 100%

Kursad Turksen (ed.), *Claudins: Methods and Protocols*, Methods in Molecular Biology, vol. 762,
DOI 10.1007/978-1-61779-185-7_3, © Springer Science+Business Media, LLC 2011

expression, and a reproducible phenotype with repeated experiments) and use an inducible expression system (so that for each clone, the uninduced cells act as isogenic negative controls for cells induced to express the transfected claudin). The Tet-Off system is a widely used and quite robust inducible expression system (8). Gene expression is suppressed in the presence of doxycycline and induced in the absence of doxycycline. Tet-Off cells that already have the regulatory plasmid and express the Tet repressible transcriptional activator, tTA, are widely available. MDCK II Tet-Off cells are available commercially, and we have generated our own MDCK I Tet-Off cell line (7).

Because claudins determine paracellular ion permeability and selectivity, electrophysiological techniques are particularly powerful for characterizing their permeability properties. We find diffusion potential measurements to be particularly useful. Since they are measured under equilibrium conditions, they can be used to derive quantitative estimates of ion permeability without making any assumptions as to the mechanism by which the ions traverse the pore. A number of simplified electrode systems have been developed to be used on cells grown on filter inserts directly in their tissue culture plate. While these have the advantage of being convenient and inexpensive, measurements of conductance, voltage, and current with such systems can be grossly inaccurate because of many factors: lack of mixing leading to large pockets of unstirred fluid, geometry considerations leading to large series resistance due to the fluid itself, geometry considerations leading to nonuniform electrical field/current across the surface of the monolayer, use of naked electrodes without salt bridges leading to large uncorrected junction potentials, and poor control of temperature. For these reasons, such measurements are useful only for drawing qualitative conclusions. For quantitative work, electrophysiological measurements must be performed in a well-designed Ussing chamber and using agar salt bridges with each electrode. Even with such a system, asymmetries in liquid junction potentials can potentially introduce large and unforeseen errors into the calculation of permeabilities (9). We describe our simplified approach to correct for junction potentials and derive accurate estimates of claudin ion permeability (10).

2. Materials

2.1. Generating a Stable PT67 Packaging Cell Line

1. Retroviral Tet-response vector, pRevTRE (Clontech, Mountain View CA) or pRevTREP (Fig. 1).
2. RetroPack PT67 packaging cells (Clontech, Mountain View CA).
3. Culture medium: Dulbecco's Modified Eagle's Medium (D-MEM) with high glucose and L-glutamine supplemented

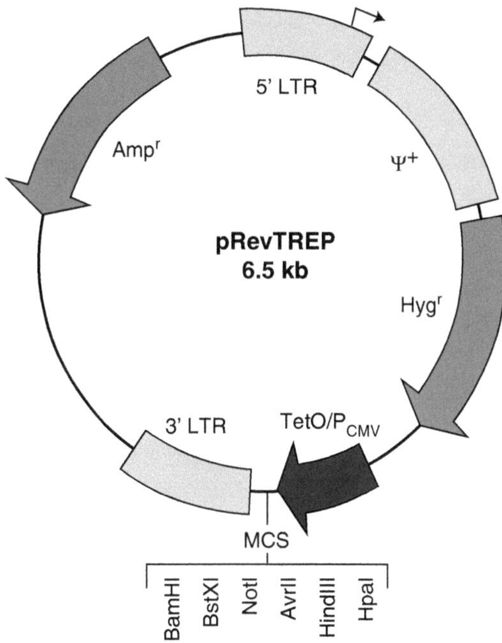

Fig. 1. Vector map of pRevTREP. *TetO/P$_{CMV}$* Tet operator sequence upstream of minimal CMV promoter, *LTR* viral long terminal repeat, ψ^{+} extended viral packaging signal, *Hyg^{r}* hygromycin resistance gene, *Amp^{r}* ampicillin resistance gene.

with 10% fetal bovine serum (FBS). Optional: Add 1/50th volume of penicillin–streptomycin solution (5,000 U/mL penicillin and 5,000 μg/mL streptomycin) (see Note 1).

4. Opti-MEM (Invitrogen, Carlsbad CA).
5. Lipofectamine 2000 (Invitrogen, Carlsbad CA).
6. 0.05% Trypsin and 0.5 mM EDTA.
7. Phosphate-buffered saline (PBS).
8. Hygromycin Selection Medium for PT67: D-MEM with 10% FBS and 0.3 mg/mL hygromycin (Invitrogen, Carlsbad CA).
9. Millex sterile PVDF syringe filters, 0.45 μm pore (Millipore, Billerica MA) (see Note 2).
10. Polybrene (hexadimethrine bromide), 100× stock solution. Make a 0.4 mg/mL solution in water, filter-sterilize, and store at 4°C.

2.2. Infecting Target MDCK Cells and Clonal Selection

1. MDCK I Tet-Off (7) and MDCK II Tet-Off cells (Clontech, Mountain View CA).
2. Culture medium (Dox–): D-MEM with high glucose and L-glutamine supplemented with 5% FBS. Optional: Add 1/50th volume of penicillin–streptomycin solution (5,000 U/mL penicillin and 5,000 μg/mL streptomycin).

3. Dox+ medium: Culture medium plus 20 ng/mL doxycycline (see Note 3).
4. Sterile cloning cylinders, 8 mm × 8 mm (Millipore Specialty Media, Billerica MA).
5. Trypsin–EDTA (0.05% Trypsin and 0.5 mM EDTA).
6. Phosphate-buffered saline (PBS).
7. Hygromycin Selection Medium for MDCK: D-MEM with 5% FBS and 0.3 mg/mL hygromycin (Invitrogen, Carlsbad CA).

2.3. General Setup of Ussing Chambers

1. Six-channel voltage–current clamp device (VCC MC6), EasyMount Ussing chamber system with sliders designed to fit Snapwell filters, plastic electrode tips, Acquire & Analyze 2.0 data acquisition software (all from Physiological Instruments, San Diego CA).
2. Snapwell cell culture filters (Corning, Inc., Corning NY, cat. #3801).
3. 100% O_2 gas cylinder.
4. Ringer solution: 150 mM NaCl, 2 mM $CaCl_2$, 1 mM $MgCl_2$, 10 mM glucose, 10 mM Tris–HEPES, pH 7.4 (see Note 4). This should be warmed to 37°C and oxygenated by bubbling with 100% O_2.
5. Agar bridge solution: 3% w/v agar in either 3 M KCl or 150 mM NaCl. Boil until the agar is dissolved.

2.4. Measurement and Interpretation of Diffusion Potentials

1. 75 mM NaCl solution: 75 mM NaCl, 150 mM mannitol, 2 mM $CaCl_2$, 1 mM $MgCl_2$, 10 mM glucose, 10 mM Tris–HEPES, pH 7.4 (see Note 5).
2. 150 mM XCl solution: 150 mM XCl, 2 mM $CaCl_2$, 1 mM $MgCl_2$, 10 mM glucose, 10 mM Tris–HEPES, pH 7.4, where X^+ represents an alkali metal cation (Li, K, Rb, and Cs).
3. 75 mM XCl/75 mM NaCl solution: 75 mM NaCl, 75 mM XCl, 2 mM $CaCl_2$, 1 mM $MgCl_2$, 10 mM glucose, 10 mM Tris–HEPES, pH 7.4, where X^+ represents a monovalent organic cation (e.g. methylamine, ethylamine, tetramethylammonium, tetraethylammonium, and *N*-methyl-D-glucamine).

3. Methods

3.1. Generating a Stable PT67 Packaging Cell Line

1. Clone the coding sequence of the gene of interest into pRevTRE or similar retroviral vector (see Note 6).
2. Plate PT67 cells into a 6-well plate (one well for each gene and one negative control) and grow in culture medium

(D-MEM with 10% FBS) until they reach 90–95% confluence (see Note 7).

3. For each well of cells to be transfected, dilute 4 μg of linearized plasmid DNA into 250 μl of Opti-MEM at room temperature. Dilute 10 μl of Lipofectamine 2000 into 250 μl of OptiMEM and incubate for 5 min. Include a negative control in which the DNA is omitted.
4. Add DNA to liposomes, mix gently, and incubate 20 min at room temperature.
5. Rinse cells in each well with 2 mL of antibiotic-free D-MEM with 10% FBS, then add the DNA–liposome mixture.
6. After 48 h, passage the cells from each well by washing with PBS and incubating with 0.25 mL trypsin–EDTA (37°C for 2–5 min), then seeding them onto a 15-cm plate in Hygromycin Selection Medium. Change media every 3–4 days.
7. Generally after 7–10 days, the negative control plate should be devoid of cells. The remaining cells in the transfected plates are stably transfected and should be grown to confluency as a polyclonal culture. Generally there is no need to isolate individual clones. Once confluent, the culture should be expanded and frozen stocks prepared.
8. To collect viral supernatant, plate the stably transfected cells on a 10-cm tissue culture dish. Once the cells reach confluency, add 5 mL antibiotic-free medium per 10-cm plate (see Note 8). After 24 h, harvest the medium, which contains viral particles. If needed, this can be repeated at 24 h intervals until the cells are no longer viable. Filter the viral supernatant through a 0.45-μm syringe filter and add polybrene to a final concentration of 4 μg/mL. The viral supernatant can now be used immediately to transduce MDCK cells, or stored frozen as a single-use aliquot at −80°C (see Note 9).

3.2. Infecting Target MDCK Cells and Clonal Selection

1. Seed the target cells (i.e., MDCK I or II TetOff) on a 24-well plate the day before infection, so that they are about 40–50% confluent at the time of infection.
2. Remove medium from the wells containing the MDCK cells and add 2 mL of viral supernatant per well of target cells. Include a negative control well that has no virus added.
3. After 24–48 h, passage cells into 15-cm plates at split ratios 1:600, 1:2,000, 1:6,000, and 1:20,000, in hygromycin selection media (see Note 10).
4. In the negative control plates, the cells should all die off within 4–5 days (except in the plates with too high a cell density). Colonies should appear in the transduced plates and be ready to pick after 7–10 days (see Note 11).

5. To pick the colonies, select colonies that are well spaced apart from each other and circle them on the bottom of the plate with a marker pen. Wash the plate with PBS, then aspirate it dry. Using sterile forceps, place a cloning cylinder over each colony. Add one drop of trypsin–EDTA into each cylinder and incubate at 37°C for 15 min or until the cells appear to round up under the microscope. Then resuspend the cells in the cylinder in 200 μl of media and transfer to a 96-well plate. Picking 16–24 colonies is usually sufficient.
6. Once the cells in the 96-well plate are confluent, split each clone into three wells: one well in a regular 96-well master plate and two wells (containing Dox+ and Dox– culture medium) in a 96-well glass coverslip-bottomed plate that will be used for immunofluorescence screening.
7. When the cells in the glass-bottomed plate are confluent, fix them, and immunostain them with anti-claudin antibody. Image the stained wells with an inverted epifluorescence microscope and identify clones that have robust expression of the transfected claudin at the tight junction in the induced well (Dox–) and absence of expression in the uninduced well (Dox+). The best five to ten clones can then be trypsinized from the master plate and expanded to generate frozen stocks (see Note 12).

3.3. General Set-up of Ussing Chambers

1. Culture cells on Snapwell filters until they would be expected to reach stable transepithelial resistance (TER). Using MDCK I and MDCK II cells plated at confluent density (~1–2 × 10^5 cells/cm^2), we generally find that it takes 7–8 days and 4–5 days, respectively, to reach a stable TER (see Fig. 2 and Note 13). The cells should be grown for this entire time either in 20 ng/mL doxycycline (Dox+, uninduced) or in the complete absence of doxycycline (Dox–, induced).

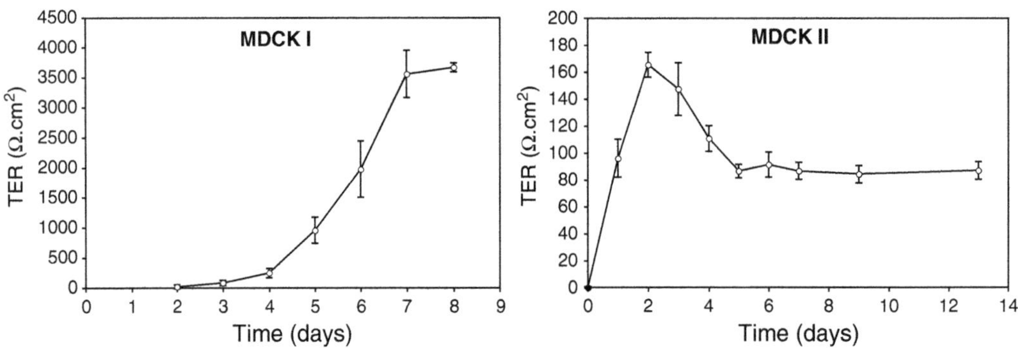

Fig. 2. Time-course of transepithelial resistance (TER) in MDCK I (*left*) and II (*right*) cells. Time is shown in days after plating the cells at confluent density. Note that the TER of MDCK I cells typically climbs monotonically to a peak at about 7 days, whereas the TER of MDCK II cells peaks early and then declines to a steady-state value at 4–5 days.

2. Fill plastic electrode tips with molten agar bridge solution containing either 3 M KCl or 150 mM NaCl to generate agar bridge pipettes. Allow the agar to cool and solidify, then top up with either the 3 M KCl or 150 mM NaCl solution as appropriate, and insert either a Ag/AgCl pellet voltage-sensing electrode or a silver wire current-passing electrode, ensuring that there are no trapped bubbles (see Note 14).
3. To set up the clamp, mount an empty Snapwell ring (created by cutting out the filter with a razor blade) in an Ussing chamber, insert the electrode/bridges, and fill each hemichamber with 4 mL of prewarmed Ringer solution. Measure the voltage in open circuit mode and adjust the offset between the two voltage-sensing electrodes until it reads zero. This adjusts for any asymmetry between the two electrodes.
4. Press the "Fluid Resistance Compensation" test button to inject a current of 60–68 μA. Verify that a current of appropriate magnitude is passing (if not, the current-passing electrodes need to be changed), and then switch the meter to "Voltage" and adjust the Fluid Resistance Compensation dial until the meter reads zero (see Note 15).
5. Take clamp offline and mount a Snapwell filter with a cell monolayer. Replace 4 mL Ringer saline in each hemichamber, taking special care on the apical side so as not to disrupt the cells, and then connect 100% O_2 to the gas lifts. Allow cells to stabilize for 15–30 min.
6. Read the spontaneous transepithelial voltage in open-circuit mode (see Note 16).
7. Using Acquire and Analyze software, switch to current clamp mode. Set the baseline current to zero (equivalent to an open circuit) and the command current to 10 μA, and start acquiring data at 1 s intervals. Conductance (G_T, calculated from the voltage deflection in response to a bipolar 10 μA current injection) and voltage data will be acquired at 1 s intervals.
8. Correct total G_T for the background values obtained with a blank filter (G_F) to obtain the true epithelial monolayer conductance (G_M) using:

$$G_M = \frac{1}{1/G_T - 1/G_F}. \quad (1)$$

3.4. Measurement and Interpretation of Diffusion Potentials

1. In current clamp mode, readjust the voltage offset to zero, then monitor the transepithelial voltage by acquiring data online at 1 s intervals (see Note 17). In this article, *all transepithelial voltages represent the apical potential referenced to the basolateral side.*
2. To measure the NaCl dilution potential by basolateral dilution, remove the solution from the basolateral hemichamber

by aspirating with a Pasteur pipette connected to a vacuum pump, and replace rapidly with 4 mL of 75 mM NaCl Ringer solution. The apical hemichamber cannot be aspirated dry in this way because this disrupts the cells on the filter and drastically reduces TER. To measure the NaCl dilution potential by apical dilution, the apical solution must be exchanged by infusing 5 volumes of 75 mM NaCl Ringer solution into the bottom of the apical hemichamber, while continuously aspirating from the surface.

3. The transepithelial voltage should change rapidly and, within 30 s to 1 min, reach a new equilibrium, which represents the apparent monolayer diffusion potential, V_T (Fig. 3). The solution should then be changed back to Ringer saline immediately (see Note 18).

4. To measure the biionic potential for Na^+ and a monovalent cation, X^+, steps 1–3 are repeated, but using either 150 mM XCl solution or 75 mM XCl/75 mM NaCl solution (see Note 19).

5. To determine the true diffusion potential across the epithelial monolayer, V_M, it is necessary to correct V_T for any asymmetry in the liquid junction potentials (agar bridge-Ussing chamber solution) between the basolateral (V_L^b) and apical (V_L^a) side, using:

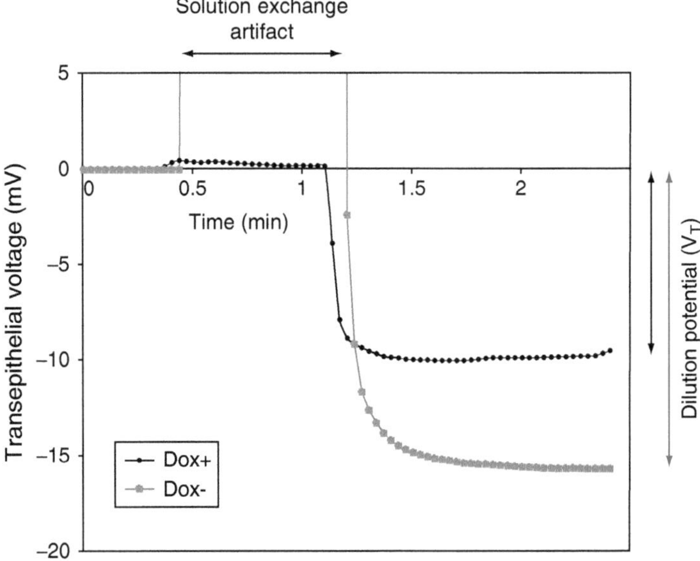

Fig. 3. Typical dilution potential curves. Replacement of the basolateral Ringer saline (150 mM NaCl) with a solution containing 75 mM NaCl induces a negative voltage (apical minus basolateral) in control, uninduced MDCK I TetOff claudin-2 cells (Dox+), indicating that transepithelial permeability is cation-selective. Induction of claudin-2 expression (Dox–) induces a more negative voltage, reflecting a further increase in cation-selectivity.

Table 1
Differential junction potential ($V_L^b - V_L^a$) across 3 M KCl or 150 mM NaCl bridges

Diffusion potential experiment		$V_L^b - V_L^a$	
Apical solution	**Basolateral solution**	**3 M KCl bridges**	**150 mM NaCl bridges**
150 mM NaCl	75 mM NaCl	-3.12	3.52
150 mM NaCl	150 mM LiCl	2.67	-2.54
150 mM NaCl	150 mM KCl	-4.71	4.53
150 mM NaCl	150 mM RbCl	-5.44	5.29
150 mM NaCl	150 mM CsCl	-5.18	5.18
150 mM NaCl	75 mM NaCl, 75 mM methylamine	-2.68	2.86
150 mM NaCl	75 mM NaCl, 75 mM ethylamine	-1.21	0.95
150 mM NaCl	75 mM NaCl, 75 mM tetramethylammonium	-0.36	0.05
150 mM NaCl	75 mM NaCl, 75 mM tetraethylammonium	0.83	-1.36
150 mM NaCl	75 mM NaCl, 75 mM arginine	0.80	-1.71
150 mM NaCl	75 mM NaCl, 75 mM *N*-methyl-D-glucamine	1.29	-2.36

Liquid junction potential is defined as the potential of the solution with respect to the pipette. Values for $V_L^b - V_L^a$ (in mV) are taken from Table S3 in ref. (10) and were determined empirically for 3 M KCl bridges and by theoretical calculation for 150 mM NaCl bridges

$$V_M = V_T - (V_L^b - V_L^a). \tag{2}$$

Table 1 lists values of $V_L^b - V_L^a$ for the most commonly used diffusion potential protocols.

6. The permeability to Cl^- relative to Na^+, P_{Cl}/P_{Na} (β), can be derived from the dilution potential using the Goldman–Hodgkin–Katz constant field voltage equation (11). For dilution of the *basolateral* solution to 75 mM, the equation simplifies to:

$$V = -\frac{RT}{F}\ln\left[\frac{(\alpha\beta+\beta)}{1+\alpha\beta}\right]; \tag{3}$$

where α is the activity ratio of NaCl in apical compared with basolateral compartments. For dilution of the *apical* solution, simply reverse the sign of V (form negative to positive, or vice versa). Equation 3 can further be rearranged to:

$$\beta = \frac{\alpha - x}{\alpha x - 1}, \tag{4}$$

where $x = e^{-(VF/RT)}$. The activity coefficient for NaCl at 150 mM is 0.752 and at 75 mM is 0.797, so α is 1.89. *RT/F* is 26.71 mV at 37°C.

7. The absolute permeability to Na^+ (P_{Na}) at 150 mM NaCl can then be derived from the Kimizuka–Koketsu equation (12):

$$P_{Na} = \frac{RT}{F^2} \times \frac{G_M}{a(1+\beta)}, \tag{5}$$

where a, the Na^+ activity, is 112.8 mM and RT/F^2 is 2.768×10^{-7} V mol/C. Using values of G_M in mS and a in mM, P_{Na} will be calculated in units of 10^{-1} m/s and can be multiplied by 10^9 to yield units of 10^{-6} cm/s. The absolute permeability to Cl^- (P_{Cl}) can simply be determined from:

$$P_{Cl} = \beta \times P_{Na} \tag{6}$$

8. Similarly, the permeability to X^+ relative to Na^+, P_X/P_{Na} (γ), can be derived from the biionic potential using the Goldman–Hodgkin–Katz equation. Assume that the activity coefficient of X^+ is identical to that of Na^+. For *basolateral* exchange with 150 mM XCl:

$$V = \frac{RT}{F} \ln\left(\frac{\gamma+\beta}{1+\beta}\right) \tag{7}$$

$$\gamma = (1+\beta) \cdot e^{V/26.71} - \beta \tag{8}$$

For basolateral exchange with 75 mM XCl/75 mM NaCl, the equivalent equation is:

$$\gamma = \alpha(1+\beta)e^{V/26.71} - \alpha\beta - 1, \tag{9}$$

where α is, again, 1.89. The absolute permeability to X^+ (P_X) can then be determined from:

$$P_X = \gamma P_{Na}. \tag{10}$$

Table 2 shows data from a typical experiment and sample calculations.

4. Notes

1. For antibiotic-free medium to be used during lipofection and viral harvest, omit penicillin and streptomycin.
2. It is important that the pore size is large enough for viral particles to pass through the filter (standard 0.2-μm sterilization units

Table 2
Sample raw conductance and diffusion potential data to demonstrate permeability calculations

	Units	Source or derivation	Dox+	Dox−
Raw data				
Total conductance (G_T)	mS		0.88	9.33
Blank filter conductance (G_F)	mS		474	474
NaCl dilution potential (dilution V_T)	mV		-9.47	-14.78
CsCl biionic potential (biionic V_T)	mV		-5.62	-9.50
Liquid junction potential correction				
Dilution potential ($V_L^b - V_L^a$)	mV	Table 1	-3.12	-3.12
Biionic potential ($V_L^b - V_L^a$)	mV	Table 1	-5.18	-5.18
Corrected data				
Monolayer conductance (G_M)	mS	Eq. 1	0.88	9.52
Corrected dilution potential (V_M)	mV	Eq. 2	-6.35	-11.66
Corrected biionic potential (V_M)	mV	Eq. 2	-0.44	-4.32
Permeability calculations				
P_{Cl}/P_{Na}		Eq. 4	0.44	0.18
P_{Na}	$\times 10^{-6}$ cm/s	Eq. 5	1.50	19.84
P_{Cl}	$\times 10^{-6}$ cm/s	Eq. 6	0.67	3.51
P_{Cs}/P_{Na}		Eq. 8	0.98	0.82
P_{Cs}	$\times 10^{-6}$ cm/s	Eq. 10	1.46	16.36

Raw data are from MDCK I TetOff claudin-2 cells that were uninduced (Dox+) or induced (Dox−) to overexpress claudin-2. Dilution potential was measured after basolateral solution change to 75 mM NaCl, and ionic potential after basolateral change to 150 mM CsCl. Measurements were performed with 3 M KCl agar bridges

are unsuitable) and that the filter material has low protein binding capacity (so nitrocellulose would not be suitable).

3. Doxycycline in solution is light-sensitive and unstable. We usually make a stock solution of 50 μg/mL in water and freeze at −20°C in single-use aliquots. Once thawed and added to culture media, we store it at 4°C in the dark and use within 2 weeks. The culture medium bathing the cells needs to be changed every 3–4 days. For cells maintained in doxycycline that are subsequently plated in the absence of doxycycline to induce claudin expression, it is important to rinse them carefully to remove any traces of doxycycline, as concentrations as low as 2 pg/mL are sufficient to inhibit maximal gene expression.
4. The accuracy of the data is directly related to the precision with which the solutions are made, so use of volumetric flasks,

an analytical balance, and analytical (ACS) grade chemicals is recommended. We generally make stock solutions of all the chemicals and store them either at room temperature on the bench (1 M NaCl, 0.1 M $CaCl_2$, 1 M $MgCl_2$, and 1 M mannitol), at room temperature shielded from light (1 M Tris–HEPES) or at 4°C (1 M glucose). Once the Ringer saline is made, it is fairly stable and unused solution can be stored at 4°C (to prevent bacterial growth) for weeks to months.

5. MDCK cell conductance and ion permeability are relatively insensitive to small changes in extracellular osmolality. Thus, to simplify solution-making, the concentration of mannitol used to balance the osmolality is calculated on the assumption that all solutes exhibit ideal behavior (i.e., salts are fully dissociated in solution, the osmotic coefficient of mannitol is 1, and osmolality and osmolarity are equivalent).

6. The commercial vector contains Moloney murine leukemia virus-based retroviral transcription and processing elements, together with a Tet-responsive promoter, and hygromycin resistance gene for selection in mammalian cells. We use a modified version in which we inserted a polylinker to add more convenient restriction sites (Fig. 1). The insert should contain the entire coding region of the gene of interest, and either the native translation initiation sequence or an optimized Kozak sequence (e.g. GCCACC<u>ATG</u>G where the start codon is underlined). Inclusion of 5′ and 3′ untranslated regions is optional, but do not include a polyadenylation sequence as this will cause premature cleavage of the viral genomic RNA. You do not need to include a polyadenylation sequence because the 3′ long terminal repeat in the retroviral vector already acts as a polyadenylation signal.

7. PT67 cells, once transfected, shed retrovirus that is capable of infecting nearly all dividing mammalian cells, including human cells, and could potentially be hazardous to humans depending on the inserted gene. Thus the transfected PT67 cells, subsequent transduced MDCK cells, and any media, plasticware and equipment that comes into contact with these cells, must all be handled in a Biosafety Level 2 facility with appropriate safety equipment and practices.

8. To maximally concentrate the viral particles, use the smallest volume that will adequately cover the bottom of the dish.

9. We find that the retroviral titer of the frozen stock is almost equivalent to fresh supernatant, so long as it has undergone no more than one freeze–thaw cycle. It is generally unnecessary (and overly time-consuming) to quantitate the viral titer prior to using it for transduction.

10. It is important to have several plates at different densities so that at least one will be at the optimum density. If the density is too high, the plate becomes confluent too quickly and the cells will no longer be susceptible to selection by hygromycin. If the plate density is too low, it will contain too few or no hygromycin-resistant clones. To calculate the plating density, one must take into account that the growth area of a well in a 24-well plate is 2 cm^2, and for a 15-cm plate is 135 cm^2. For example, to achieve a split ratio of 1:600, plate 1/90th of the total volume of cell suspension from one well onto a 15-cm plate.
11. An ideal plate should have 10–30 colonies: fewer would be insufficient for screening, and too many would make them crowded together and difficult to pick cleanly. At this density, small colonies are very difficult to find simply by scanning the plate under the microscope. However, if one waits long enough until the colonies have grown fairly large, one can see them with the naked eye just by inspecting the bottom of the plate (they appear as white spots). A colony is large enough to pick, when it fills the entire field of view under the microscope with a low power (10×) objective.
12. Further characterization of these cells, which will not be discussed in detail here, include confirming proper expression and induction/suppression of the transfected claudin protein by western blotting, confirming targeting to the tight junction by immunofluorescence staining with confocal microscopy, checking for secondary effects on the levels of expression of endogenous MDCK cell claudins, and assessing the effect on the morphology and particularly the strand number of the tight junction by freeze fracture electron microscopy.
13. The time needed in culture can be determined by monitoring TER with a Millicell-ERS volt ohm meter with "chopstick-style" electrodes (Millipore, Billerica MA). While TER measurements with such a device are notoriously inaccurate, they are very convenient and perfectly adequate for qualitatively monitoring the functional maturation of the tight junction.
14. 3 M KCl bridges are widely used because they exhibit very small liquid junction potentials. However, their liquid junction potentials are not easily predictable on theoretical grounds, and also change during the course of an experiment due to diffusion of salts from the chamber solution (9). 150 mM NaCl electrodes have large liquid junction potentials, but these are very consistent and so can be predicted by theoretical calculation, and there is little drift since their composition is similar to that of the Ussing chamber solution in which they are immersed for most of the experiment. See ref. (10) for a careful analysis of these issues.

15. The fluid resistance in Ringer saline is typically 55–65 Ω. Note that this procedure compensates only for the resistance of the Ringer at 37°C. The fluid resistance of other solutions, or of Ringer saline at different temperatures, would have to be compensated for either by readjusting the "Fluid Resistance Compensation" knob while in these solutions or by calculating the theoretical change in resistance.
16. In MDCK cells bathed in solution without K^+, the Na-K-ATPase is inactive and the spontaneous voltage is usually negligible.
17. It is critically important to ascertain the polarity of the voltage that is being read. This depends on the orientation of the filter in the chamber, the orientation of the electrode connections, and whether the hardware or software has been set up to invert the signal.
18. The rate of change of voltage varies somewhat and depends on the rate of stirring of the solution in the Ussing chamber. After the voltage reaches a plateau, prolonged exposure to the 75 mM NaCl solution often causes the voltage to decay slowly back to baseline. This may reflect unwanted effects induced by the high transepithelial voltage itself and by the low salt concentration (e.g., Na^+ efflux from the cells). To minimize these unwanted effects (as well as drift in the junction potentials of the agar bridges), it is important to keep the cells exposed to 75 mM NaCl for the shortest time possible and return them to Ringer saline as soon as the voltage plateau has been recorded. Similar considerations apply for the solutions used for biionic potential measurement.
19. The biionic potential protocol using 150 mM XCl solutions yields large diffusion potentials. This can be advantageous to obtain the most accurate measurements for the alkali metal cations. However, for organic cations, which are large and tend to be much less permeable, the voltage excursions can be too large (20 mV or more) leading to unwanted effects on the cell and unstable readings. We, therefore, recommend using 75 mM XCl/75 mM NaCl solutions, which cause smaller diffusion potentials, for the organic cations.

References

1. Tsukita S, Furuse M (2002) Claudin-based barrier in simple and stratified cellular sheets. Curr Opin Cell Biol 14:531–536
2. Van Itallie CM, Anderson JM (2006) Claudins and epithelial paracellular transport. Annu Rev Physiol 68:403–429
3. Angelow S, Ahlstrom R, Yu AS (2008) Biology of claudins. Am J Physiol Renal Physiol 295: F867–876
4. Saier MH, Jr. (1981) Growth and differentiated properties of a kidney epithelial cell line (MDCK). Am J Physiol 240:C106–109

5. Cereijido M, Gonzalez-Mariscal L, Borboa L (1983) Occluding junctions and paracellular pathways studied in monolayers of MDCK cells. J Exp Biol 106:205–215
6. Van Itallie C, Fanning AS, Anderson JM (2003) Reversal of charge selectivity in cation or anion selective epithelial lines by expression of different claudins. Am J Physiol Cell Physiol 286:F1078–1084
7. Angelow S, Schneeberger EE, Yu AS (2007) Claudin-8 expression in renal epithelial cells augments the paracellular barrier by replacing endogenous claudin-2. J Membr Biol 215: 147–159
8. Gossen M, Bujard H (1992) Tight control of gene expression in mammalian cells by tetracycline-responsive promoters. Proc Natl Acad Sci U S A 89:5547–5551
9. Barry PH, Diamond JM (1970) Junction potentials, electrode standard potentials, and other problems in interpreting electrical properties of membranes. J Membr Biol 3:93–122
10. Yu AS, Cheng MH, Angelow S, et al. (2009) Molecular basis for cation selectivity in claudin-2-based paracellular pores: identification of an electrostatic interaction site. J Gen Physiol 133:111–127
11. Hille B (2001) Ionic channels of excitable membranes. Sunderland: Sinauer.
12. Kimizuka H, Koketsu K (1964) Ion transport through cell membrane. J Theoret Biol 6:290–305

Chapter 4

The Tight Junction, Intercellular Seal as a Cell Signaling Player: Protocols for Examination of Its Status

Makoto Osanai

Abstract

Tight junctions (TJs) are intercellular structures in epithelial and endothelial cells, primarily playing critical roles in cell–cell adhesion. Among their molecular components, claudins are the main constituents as integral membrane proteins, encoded by at least 24 members of a single gene family. Accumulated evidence has demonstrated that TJ proteins such as claudins are directly involved in the regulation of cellular functions such as proliferation, differentiation, and apoptosis, due to their ability to recruit various signaling molecules that have proliferative and differentiative capacities, including transcription factors, lipid phosphatases, and cell cycle regulators. It is thus clear that TJs are not simple static constituents to establish cell adhesion structures, rather also functioning in cell signaling component that has functions in receiving environmental cues and transmitting signals inside the cells.

Key words: Tight junction, Claudin, Signal transduction, Apoptosis, Cancer, Tumorigenesis

1. Introduction

Tight junctions (TJs) are essential for the tight sealing of cellular sheets, thus functioning as major determinants of paracellular permeability and therefore maintaining tissue homeostasis (Fig. 1). TJs also play a crucial role in the determination of cell polarity by forming a fence preventing lateral diffusion of membrane proteins and lipids, thereby creating a boundary between the apical and the basolateral plasma membrane domains (1). Recent evidence has demonstrated that TJs are directly involved in the regulation of cellular functions such as proliferation, differentiation, and apoptosis, due to the ability of TJ proteins to recruit various types of signaling molecules that have proliferative and differentiative capacities, including transcription factors, lipid phosphatases, and cell cycle regulators (Fig. 2) (2, 3). Occludin was the first requisite integral protein identified in the TJ, but occludin-deficient

Kursad Turksen (ed.), *Claudins: Methods and Protocols*, Methods in Molecular Biology, vol. 762,
DOI 10.1007/978-1-61779-185-7_4, © Springer Science+Business Media, LLC 2011

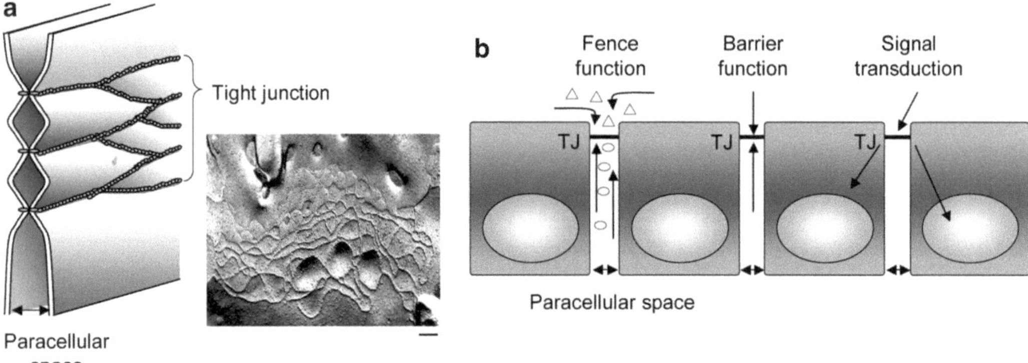

Fig. 1. Schematic representation of TJ. (**a**) *Left panel*, in this structural model of TJ, there are a number of intercrossing TJ strands (depicted as *small dots*) and three so-called "kissing points". *Right panel*, freeze-fracture replica of a TJ, consisting of an anastomosing network of strands forming irregular interstrand compartments and composed of a large number of protein components, including membrane proteins such as occludin and claudins, as well as cytoplasmic scaffolding proteins such as ZO-1. *Scale bar*, 50 nm. (**b**) In polarized cells, TJs are positioned at the boundaries of the apical and basolateral plasma membrane domains to maintain cell polarity by forming a fence. TJs also seal cells together to generate the primary barrier and prevent diffusion of solutes through paracellular pathways. In addition, certain TJ proteins such as occludin are signaling molecules that function in receiving environmental cues and transmitting signals inside cells.

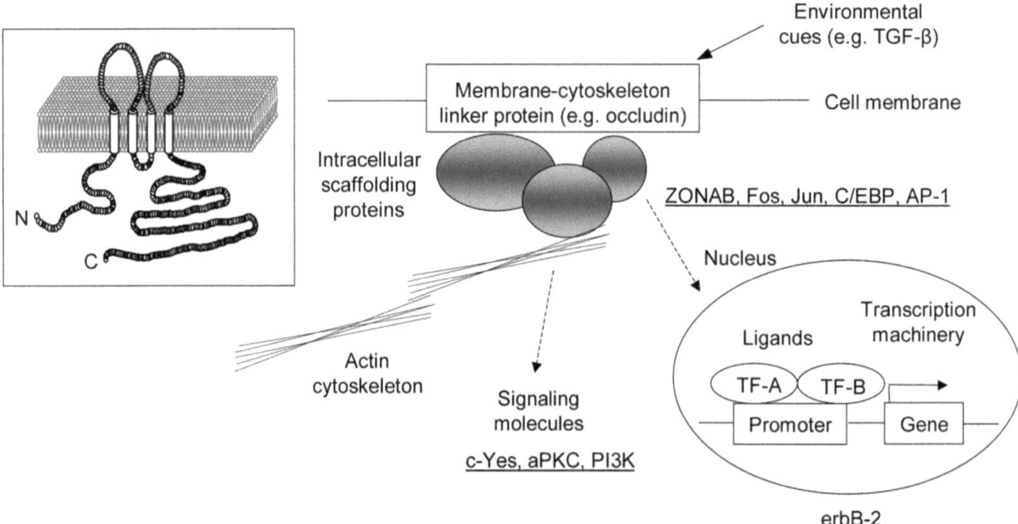

Fig. 2. Membrane-anchored TJ proteins such as occludin are signal transducers and transmitters for cells. Inset, schematic presentation of the structure of occludin, with four transmembrane domains and a long cytoplasmic tail. The COOH-terminal region of occludin has many phosphorylated sites, potentially associated with a number of signal transduction pathways. Many of the cytoplasmic TJ components are signaling proteins that have functions in transmitting cellular signals such as c-Yes, aPKC, Rho proteins, PKC-ξ, and PI3K. Intracellular scaffolding proteins such as ZO-1 and ZO-2 also associate with ZONAB, Fos, Jun, C/EBP, and AP-1, and functionally interact in the regulation of transcription factors (TF-A and TF-B) such as erbB-2 promoter activity in cells, pointing to direct participation in the control of gene expression. In addition, occludin associates with cytoplasmic scaffolding proteins, thereby providing direct linkage to the actin cytoskeleton.

cells and animals have fully developed well-organized TJs structurally (4, 5). Thus, it is now well accepted that the claudin family, encoded by at least 24 members of a single gene family, provides the main constituent membrane proteins of TJs, rather than occludin (1–3).

We have studied possible relationships between functional integrity of TJ proteins and cancer cell biology (6–10). There is a growing body of evidence that cells often exhibit loss of functional TJs in association with neoplasia (10, 11). Disruption of TJ structures has been shown to be associated with cancer development, which may be causally involved in malignant phenotype such as local tumor growth, invasion, and metastasis to distant sites (10–13). Interestingly, we have also demonstrated that epigenetic silencing of TJ proteins, including claudins, contributes to enhanced tumorigenic, invasive, and metastatic properties of cancer cells (7). In parallel, decreased and/or impaired TJ formation and deregulated expression of claudins has been reported for various types of cancer, and genes having an oncogenic character are known to disrupt TJs (14, 15). A number of immunohistochemical studies have also shown deregulated expression of claudins in a variety of cancer tissues (16, 17). From these observations, it is clear that TJ proteins are not simple static constituents of cell adhesion structures, rather also functioning as a cell signaling molecules that has functions in receiving environmental cues and transmitting signals inside the cells. Here, we describe protocols that we have applied for assessment of various functional aspects.

2. Materials

2.1. Cell Culture and Transfection

1. Dubecco's Modified Eagle's Medium (DMEM) (Life Technologies, Grand Island, NY) supplemented with 10% fetal bovine serum (FBS) (Life Technologies), 10 mM HEPES, 100 U/mL penicillin, and 100 μg/mL streptomycin (Sigma, St. Louis, MO).
2. Solution of trypsin (Sigma) diluted with DMEM lacking any supplement (such as antibiotics) to give working solutions ranging from 0.125 to 0.25%.
3. Phosphate-buffered saline (PBS) for cell culture, usually referred to as "PBS (−)", is used as a Ca^{2+}- and Mg^{2+}-free isotonic solution, autoclaved or filtered (pore size 0.22 μm) for sterilization. 1× PBS for 1 L: 8.0 g NaCl, 2.88 g Na_2HPO_4–$12H_2O$, 0.20 g KH_2PO_4, 0.20 g KCl in water (pH 7.5). Alternatively, "PBS (−)" is commercially available to simply dissolve as a premixed powder or tablets in water (from Sigma, for example).

4. Apoptosis can be stimulated with various death-inducing agents, including H_2O_2 (Sigma), etoposide (VP16; Nippon Kayaku, Tokyo, Japan), cisplatin (Nippon Kayaku), heat shock, gamma-irradiation, and death ligands such as TRAIL (R&D Systems, Minneapolis, MN), agonistic Fas antibody (clone CH-11, Panvera, Madison, WI), or TNF-α (R&D Systems).
5. All-*trans* retinoic acid (atRA; Sigma) dissolved at 1 mM in dimethyl sulfoxide (DMSO; Sigma) and stored in single use aliquots at −20°C is used to prepare working solutions by dilution in DMSO ranging from 1 nM to 1 μM.
6. The demethylating agent, 5′-aza-2′deoxycytidine (5′Aza-dC, at 1 μM as a default concentration in the dose range from 0 to 5 μM, Sigma), is stored in single use aliquots at −20°C.
7. The histone deacetylase inhibitor (HDAI), trichostatin A (TSA, at 100 nM as a default concentration in the dose range from 0 to 300 nM, Sigma), is stored in single use aliquots at −20°C.
8. FuGENE 6 (Roche, Basel, Switzerland) is employed as a transfection reagent.
9. Small-interfering RNAs (siRNA; Santa Cruz Biotechnologies, Santa Cruz, CA) should be purchased if commercially available. Control siRNA is routinely used in siRNA studies as a negative control.

2.2. Cell Death Analyses

1. The lysis buffer for MTT [3-(4,5-Dimethylthiazol-2-yl)-2,5-diphenyltetrazolium bromide, a tetrazole] assay: 10 mM Tris–HCl (pH 8.0), 10 mM EDTA, and 0.5% Triton X-100.
2. MTT Dye Kit: many manufacturers provide this type of kit, ask your providers. Storage temperature, 2–8°C. Alternatively, all solutions can be easily made in a standard laboratory as follows.
   (a) The MTT solution is 5 mg/mL MTT in PBS and must be filter sterilized after adding MTT.
   (b) The MTT solvent is 4 mM HCl, 0.1% Nonidet P-40 (NP-40) in isopropanol.
   (c) The lysis buffer for fragmentation assay is 10 mM Tris–HCl (pH 8.0), 10 mM EDTA, and 0.5% Triton X-100.
   (d) RNaseA (0.1 mg/mL, Roche) stored at −20°C.
   (e) Proteinase K (0.2 mg/mL, Boehringer Mannheim) stored at −20°C.
   (f) A phenol:chloroform 1:1 solution (Sigma) stored at −20°C.

2.3. Terminal Deoxynucleotidyl Transferase-Mediated Nick End Labeling (TUNEL) Assay

1. Collagen-coated glass coverslips: After autoclaving (sizes dependent on the well size of tissue culture plate), glass coverslips are coated with solutions of collagen or gelatin and allowed to dry overnight.
2. These instructions assume the use of an In Situ Cell Death Detection Kit (Roche) to histochemically detect dead cells undergoing apoptosis.
3. Fixative: 10% neutral-buffered formalin (Sigma, 4% formaldehyde) is employed.
4. All immunoreactions and the color reaction should be performed in a moist (humidified) chamber, since drying results in salt precipitation that produces artifacts.
5. Staining racks are used to keep moisture on the slides.
6. A light microscope is used to count positive cells.
7. Hematoxylin is applied for counterstaining.
8. A permanent mounting medium is used.

2.4. Reverse Transcription-Polymerase Chain Reaction Analysis

1. Double-distilled water (DDW): Autoclave-distilled water after running it through a Millipore Q filter system (see Note 1).
2. Extract total RNA with TRIzol (Invitrogen, Carlsbad, CA) for storage at 4°C.
3. Reverse transcription is performed using M-MuLV reverse transcriptase (Applied Biosystems, Foster City, CA), with storage of products at −20°C until just before use for the assay.
4. Alternatively, reverse transcription (RT) kits (for example, Super Script II RT, Invitrogen) for reactions can be applied for convenience.
5. For polymerase chain reaction (PCR) reactions, employ Taq DNA polymerase (TaKaRa, Shiga, Japan), stored at −20°C until just before use: 10× PCR buffer and a dNTP (deoxynucleotide triphosphate) mixture of satisfactory quality are provided with the enzyme.
6. A local distributor should be asked to find a specialized company to synthesize oligonucleotides for PCR primers.
7. Employ an appropriate Thermal Cycler (PCR Themocycler) for PCR reactions.
8. Agarose: Electrophoresis grade.
9. 5× TBE (Tris–Bolate–EDTA) buffer (stock solution): 54 g Tris base, 27.5 g boric acid, 20 mL of 0.5 M EDTA (pH 8.0), with the volume brought to 1 L.
10. 0.5× TBE buffer (working solution): 100 mL 5× TBE buffer stock solution, 900 mL, distilled water.

11. Ethidium bromide: 10 mg/mL as a stock solution.
12. Loading buffer: 0.25% bromophenol blue, 0.25% xylene cyanol in 40% (w/v) sucrose in water.
13. Ultraviolet (UV) illuminator.
14. Apply Scion Image 1.62 (Scion Corporation, Frederick, MD) for densitometric analysis of the results. This is also useful for western blot analysis.

2.5. Western Blot Analysis

1. Lysis buffer of cells and tissues: 50 mM Tris–HCl (pH 8.0), 150 mM NaCl, 0.1% sodium dodecyl sulfate (SDS), 1% Triton X-100, stored at 4°C. Just before use, 100 μg/mL phenylmethylsulfonyl fluoride (PMSF) in isopropanol.
2. Conduct Bradford assays (BCA assay; Bio-Rad) to quantitate the amounts of protein in samples.
3. Purchase polyacrylamide gel: precast gel (Ready Gels; Bio-Rad, Hercules, CA).
4. 2× Sample buffer: 130 mM Tris–HCl, pH 8.0, 20% (v/v) Glycerol, 4.6% (w/v) SDS, 0.02% Bromophenol blue, and 2% DTT.
5. Protein size marker: there are wide varieties of markers, so be sure to ensure use of an appropriate marker for the size of the target protein.
6. 10× Running buffer: 0.25 M Trizma base, 1.92 M Glycine, 1% SDS, without adjustment of the pH.
7. 10× Blotting buffer: 0.25 M Trizma base, 1.92 M Glycine at pH 8.3, without adjustment.
8. To make 1 L of 1× Blotting buffer: mix 200 mL methanol, 100 mL 10× Blotting buffer, and 700 mL water.
9. Polyvinylidene fluoride (PVDF) membrane: Millipore Immobion-P (Millipore, Billerica, MA).
10. Blocking buffer: add 0.05% Tween 20 to 3% bovine serum albumin in PBS, and store at 4°C to prevent possible contamination.
11. Employ Transblot for membrane transfer (Bio-Rad).
12. Antibody: appropriate primary and corresponding secondary antibodies should be employed for the detection of target protein, including β-actin (Santa Cruz Biotechnologies) as a primary antibody.
13. These instructions assume the use of the enhanced chemiluminescence (ECL) system to visualize immunoreactions (GE Healthcare, Buckinghamshire, UK).
14. Stripping buffer: 0.2 M Glycine (pH 2.5) and 0.05% Tween 20. Store at 4°C.

2.6. Methylation-Specific PCR

1. 10 mM Hydroquinone (Sigma).
2. 40.5% Sodium bisulfite (Sigma): 9.1 g sodium bisulfite dissolved in 17 mL cold water, adjusted to pH 5.0 with 10 N NaOH, and brought to a volume of 20 mL with water. Prepare immediately before use, and keep solutions cold and in the dark as much as possible.

2.7. cDNA Microarray Analysis

1. GEArray DNA Microarray (SuperArray Biosciences, Bethesda, MD): This pathway-specific cDNA array designed to determine the expression profile of genes in a wide variety of biological pathways such as apoptosis and cancer.
2. Since manufacturers basically supply the required materials in this kit, there is no need to purchase additional materials and agents.
3. TrueLabeling-AMP 2.0 kit: used to make probes with the protocol recommended by the manufacturer.
4. 20× SSC: dissolve 174.3 g NaCl and 88.2 g sodium citrate dihydrate in 900 mL DDW, adjust the pH to 7.0 with 1 M HCl. Dilute to 1 L with DDW, and store at room temperature.
5. 20× SDS: dissolve 200 g SDS in 1 L DDW with heating to 65°C if necessary, and store at room temperature.
6. A PCR Thermal Cycler (PCR Themocycler) was employed.
7. A routine hybridization oven and bottles was used.
8. X-ray film was used.
9. GEArray Analyzer software was provided by the manufacturer.

3. Methods

To examine the impact of TJ proteins such as claudins on cell behavior, it is important to use a wide variety of assays. This is based on the notion that TJ proteins are not simple static constituents of cell adhesion structures, but rather that they also contribute as cell signaling molecules that have functions in receiving environmental cues and transmitting signals inside the cells. It is now clear that the altered expression of TJ proteins in cancer cells affects a vast array of signaling pathways that may be important for cancer development.

3.1. Cell Death Analysis

1. Plate cells at 1×10^5 cells per six-well plates and incubate for 24 h.
2. Treat cells with 0.1 μM atRA as an apoptotic sensitizer (see Note 2).

3. Wash twice with PBS (−) followed by replacement with fresh DMEM supplemented with 10% charcoal–dextran-treated FBS (see Note 3).
4. Subsequently, stimulate apoptosis with various agents, including H_2O_2 in the dose range from 0 to 100 μM, 50 μM etoposide, 10 μg/mL cisplatin, heat shock at 42°C for 30 min, γ-irradiation (50 or 100 Gy), death ligands such as TRAIL at a dose range of 0–20 ng/mL for 0.5–24 h, agonistic Fas antibody (0.025–10 μg/mL), and TNF-α (1–100 ng/mL).
5. Evaluate the type of cell death and the rate of cell death.

3.2. Cell Death Analysis: DNA Fragmentation Assay

1. Fragmented DNA from floating dead cells is isolated according to a method that preferentially extracts low molecular weight cellular DNA.
2. Pellet down floating dead cells by centrifugation and wash with PBS.
3. Resuspend the pellet in lysis buffer and centrifuge at 12,000 × *g* for 10 min at 4°C.
4. Treat the supernatant with 0.1 mg/mL RNaseA for 1 h at 37°C. After the addition of 0.2 mg/mL proteinase K, further incubate the supernatant for 1 h at 37°C.
5. Extract each supernatant, preferentially containing low molecular weight cellular DNA, once with phenol/chloroform (1:1; optional) and ethanol precipitate the fragmented DNA.
6. Perform gel electrophoresis of the DNA samples on 2.5% agarose gels and visualize ladder formation due to fragmentation with ethidium bromide under ultraviolet illumination (Fig. 3).
7. Alternatively, proceed to MTT assay.
8. Harvest total cells in tissue culture wells, wash with PBS, and lyse with lysis buffer.
9. Centrifuge the cells at 12,000 × *g* for 20 min at 4°C to separate intact chromatin in the pellet from DNA fragments in the supernatant.
10. Determine the DNA amounts in each pellet and supernatant using diphenylamine reagent and express the fragmentation rate (%) as (fragmented DNA/total DNA) × 100.

3.3. Cell Death Analysis: MTT Assay

1. The MTT assay is a standard colorimetric assay for measuring the activity of an enzyme reducing MTT to give a purple formazan in the mitochondria of living cells, reflecting cell viability, and proliferation.
2. For best results, cell numbers should be determined during the log growth phase. Each test should include a blank

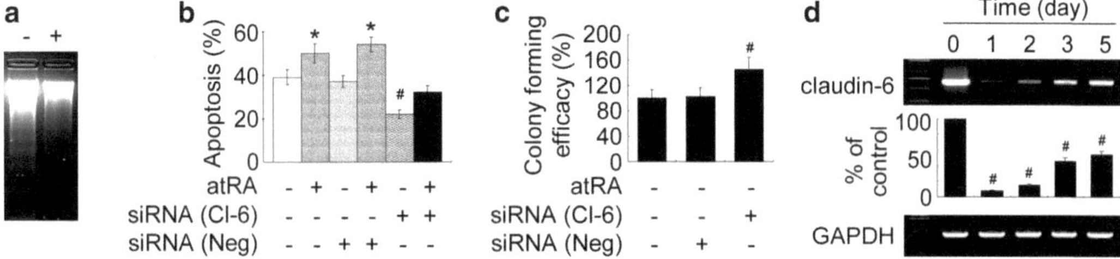

Fig. 3. Decreased expression of claudin-6 enhances colony-forming efficacy in MCF-7 human breast carcinoma cells. (**a**) Evasion of and resistance to apoptosis is a characteristic of cancers, and breakdown of anoikis may be a predominant contributor to oncogenic progression. We, therefore, examined whether siRNA-mediated knockdown of claudin-6 expression inhibited anoikis in MCF-7 cells. Anoikis was induced after adding cells to agarose-coated dishes to avoid cell attachment. DNA fragmentation analysis demonstrated that decreased expression of claudin-6 clearly showed the limited sensitivity to anoikis, and confirmed apoptotic cell death. (**b**) Quantification of anoikis in the presence or absence of atRA with or without claudin-6-specific (Cl-6) or negative control (Neg) siRNA transfection. (**c**) Suppression of endogenous claudin-6 expression in MCF-7 cells provided significant resistance to anoikis, causing significant promotion of colony formation in three-dimensional cultures. The number of colonies formed from control cells was defined as 100%. (**d**) Quantitative RT–PCR analysis confirmed the silencing effect of claudin-6-specific siRNA during the course of the experiment. Densitometric analyses for independent triplicate experiments are shown below the representative image. The signal of claudin-6 was defined as 100% in MCF-7 cells without the transfection. Since claudin-6 siRNA was transfected transiently in this experiment, it was clear that endogenous expression of claudin-6 was increased with cell division. We should note here that MCF-7 cells required 24 h for maximum reduction of claudin-6 after being transfected with claudin-6-specific siRNA, and significant suppression of claudin-6 was observed for 5 days. *$p<0.05$ vs. cells without atRA treatment; $^{\#}p<0.05$ vs. cells without claudin-6 siRNA transfection.

containing complete culture medium without cells, and each condition should be done in triplicate or more.

3. Day 1: Dilute the cells to 5×10^3 cells/mL with complete medium followed by the trypsinization.
4. Add 100 μL of cells (i.e., 5×10^2 cells per wells) to each well of 96-well plates and incubate overnight.
5. Day 2: Treat cells with apoptosis-inducing agents.
6. Day 3: Add 20 μL of 5 mg/mL MTT each well, including one set of wells with MTT but no cells as a blank (see Note 4).
7. Incubate for 3 h at 37°C in a culture hood.
8. Carefully remove media, but do not disturb cells and do not rinse cells with PBS (−).
9. Add 150 μL of MTT solvent (see Note 5).
10. Cover with foil to protect from room light and agitate cells on a shaker for 15 min at room temperature. In some cases, pipetting up and down may be required to completely dissolve the MTT formazan crystals, especially with dense cultures.
11. Read absorbance at 590 nm with a reference filter of 620 nm (Figs. 3 and 4).

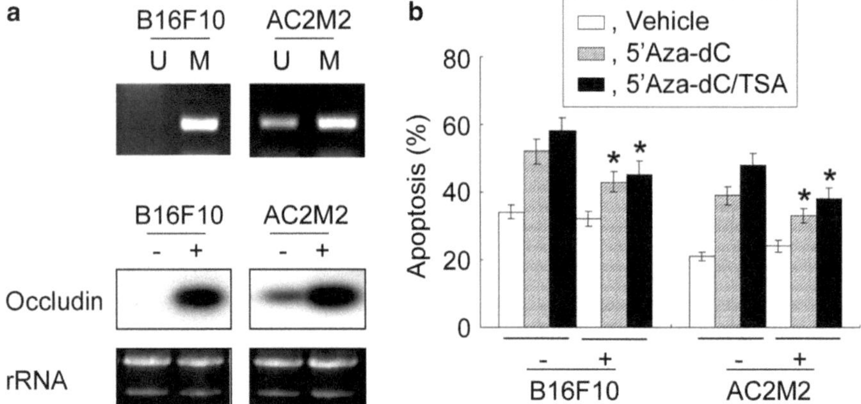

Fig. 4. Expression of TJ proteins such as occludin is epigenetically silenced by CpG island promoter methylation. Aberrant promoter region CpG island hypermethylation is associated with transcriptional silencing of many genes in a wide variety of cancers. Indeed, CpG islands exist in the promoter regions both of human and of murine occludin. (**a**) MSP analysis to detect unmethylated (U) and methylated (M) alleles on CpG islands of the occludin promoter showed CpG islands in the occludin promoter region to be densely methylated in B16F10 murine melanoma cells (*upper left*). A metastatic variant of murine breast carcinoma cell line, AC2M2, was partially methylated in this region and endogenous occludin expression was detected (*upper right*). Synergy with (+) or without (–) 1 μM demethylating agent (5′Aza-dC) and 100 nM histone deacetylase inhibitor (HDAI; trichostatin A, or TSA) for the induction of endogenous occludin (*lower panel*). (**b**) The synergistic effect with 5′Aza-dC and TSA proved sufficient for apoptotic sensitization in the presence (+) or absence (–) of 100 nM occludin-specific siRNA. *$p < 0.05$ vs. cells without siRNA transfection.

3.4. Cell Death Analysis: TUNEL Assay

1. Cells cultured on collagen-coated glass coverslips can be subjected to the TUNEL assay.
2. Apoptotic cells are visualized using an In Situ Cell Death Detection Kit (Roche). Although many manufacturers provide similar types of kit, this kit consistently gives stable results. The protocol is available on the web site of the manufacturer.
3. Staining should be carried out other than for terminal deoxynucleotidyl transferase (TdT) as a negative control.
4. Positive cells stained by the specified procedures are scored under a light microscope, counting numbers of cells under low magnification (×100) in ten separate arbitrarily selected fields in each section.

3.5. RT–PCR Analysis

1. Extract total RNA using TRIzol (see Note 6).
2. Ribosomal RNA (rRNA) should be routinely electrophoresed to confirm the same amounts of RNA for RT reactions, and to check the quality of total RNA. Well-separated 28S and 18S bands should be visible at a ratio of 2:1.
3. Reverse-transcribe total RNA (1 μg) using reverse transcriptase with a programmed thermal cycler for the PCR reaction. The program usually depends on the activity of the reverse transcriptase, but it is generally sufficient to incubate RT aliquots at 50°C for more than 30 min.

4. For analysis of gene expression, the gene of interest is amplified from diluted cDNA using specific sense and antisense primers for up to 40 cycles, with various cycling parameters for each PCR experiment examined to define optimal conditions for linearity to allow semiquantitative analysis of signal intensity.
5. Set up a 20 μL reaction mixture in a 0.2-mL microcentrifuge tube: 1 μL, template cDNA; 2 μL, 5 μM each primer; 2 μL, 10× PCR buffer; 1.6 μL, dNTP mixture (50 mM each dNTP: dCTP, dATP, dGTP, and dTTP); 11.3 μL, DDW.
6. The reaction mixture in steps 4 and 5 must be on ice to reduce the incidence of primer–dimer artifacts.
7. Add 0.1 μL Taq DNA polymerase.
8. Spin down briefly in a benchtop centrifuge.
9. Run the PCR reactions in a thermal cycler using the following profile: initial denaturation, 94°C for 3 min; denaturation, 94°C for 30 s; annealing, 55°C for 30 s (the annealing temperature is dependent on your primer design, and the optimal annealing temperature for your primers is the melting temperature so-called Tm − 5°C.); extension, 72°C for 1 min; repeat step of denaturation–annealing–extension for up to 40 cycles. Cycling should conclude with a final extension at 72°C for 10 min, and store products at 4°C.
10. Electrophorese on 0.8–2.0% agarose gels containing 100 μg/mL ethidium bromide at 100 V for approximately 30 min in 0.5× TBE electrophoresis buffer.
11. View results under UV illumination (Fig. 3).
12. Triplicate-independent PCR reactions should be carried out to ensure reproducibility of quantification.
13. With densitometry, signals can be quantitated using image-processing software such as Scion Image 1.62.

3.6. Western Blot Analysis

1. Preparation of samples: collect cancer cells from 10-cm tissue dishes by trypsinization and spin.
2. Lyse pellets with 1 mL of lysis buffer for western blot analysis on ice for 10 min.
3. Spin at 12,000 × g in a centrifuge for 10 min at 4°C.
4. Transfer supernatant to new tubes and discard the pellets.
5. Determine the protein concentration by BCA assay.
6. Take an appropriate amount of protein (up to 20 μg/sample) and mix with 2× sample buffer.
7. Boil for 5 min and cool on ice for 5 min.
8. Flash spin to bring down condensation prior to gel loading.

9. Polyacrylamide gel electrophoresis: a wide variety of precast gels with sufficient quality are now commercially available (see Note 7).
10. Setup the electrophoresis apparatus: assemble gels in gel rings.
11. After flash spinning the samples, load into wells.
12. Be sure to use protein size markers.
13. Run with a constant current (20–30 mA with voltage set at more than 200 V).
14. Usual running time is about 2–3 h for a minigel.
15. During the electrophoresis, cut a piece of PVDF membrane.
16. Wet for about 30 min in methanol at room temperature.
17. Remove methanol and add 1× blotting buffer until ready to use (see Note 8).
18. Membrane transfer: assemble "sandwich" for Bio-Rad's Transblot.
19. Prewet the sponge and filter papers in 1× blotting buffer as follows: sponge, filter paper, gel, membrane, filter paper, and sponge.
20. Transfer for 1 h at up to 20 mA on a stir plate. Larger proteins might take longer to transfer.
21. When the transfer is completed, immerse membranes in blocking buffer for at least 1 h.
22. Incubate with primary antibodies diluted in blocking buffer for 1 h at room temperature as a default condition. You may need incubation overnight at 4°C to decrease possible non-specific reaction to proteins.
23. Wash three times for 5 min with 0.01% Tween 20 in PBS (PBS-T).
24. Incubate with secondary antibody diluted in blocking buffer for 1 h at room temperature.
25. Wash membrane three times for 5 min with PBS-T.
26. Detect binding with an ECL kit.
27. You should also react antibodies targeting β-actin for the verification of equal loading of samples.
28. (Optional) You can remove the reacted antibody and again use the same membrane. First, rinse blot off with PBST and incubate in stripping buffer for 20 min at 80°C.
29. Rinse blot off with PBS-T.
30. Block for about 1 h with 5% bovine serum albumin (BSA)/Tween 20.
31. The preparation is ready for incubation with another primary antibody.

3.7. Colony Forming Assay in Three-Dimensional Cultures

1. To determine the colony forming capacity of transformed cells in vitro, this assay should be considered to evaluate tumorigenic characteristics. This procedure is also called the soft agar assay.
2. Base agar: autoclave 1% agar (DNA grade) for sterilization, and cool to 55°C in a water bath.
3. Warm 2× complete DMEM to 40°C in a water bath, and allow at least 30 min for the temperature to equilibrate.
4. Mix equal volumes of the two solutions to give 0.5% agar and 1× DMEM.
5. Add 3 mL to a 6-cm tissue culture dish and allow the mixture to solidify. Plates can be stored at 4°C for up to 1 week.
6. Top agar: autoclave 0.7% of agar (DNA grade) for sterilization, and cool to 40°C in a water bath (see Note 9).
7. Warm 2× complete DMEM at 40°C for at least 30 min.
8. Trypsinize cells and count. You require 1×10^4 cells/dish; therefore adjust cell counts to 1×10^5 cells/mL.
9. Add 0.1 mL of cell suspension to a 15-mL centrifuge tube, and gently mix 3 mL 0.7% agar and 3 mL 2× complete DMEM.
10. For plating, add this mixture to a plate, and wait until the top agar is solidified.
11. Incubate the plates at 37°C in a humidified incubator for at least 2 weeks or until visible colonies develop (Fig. 3).
12. (Optional) Stain plates with 1 mL of 0.005% crystal violet for more than 1 h.
13. Colonies developing on cell culture dishes in soft agar suspension are scored using phase contrast microscopy by counting colony numbers under low magnification (×100) in ten separate arbitrary fields in each plate. Cell clusters of more than approximately 100 μm in diameter are defined as positive results in the soft agar assay.
14. It is a good idea to prepare triplicate plates under same conditions to confirm the reproducibility of the results.

3.8. Tumor Growth In Vivo

1. To determine the growth capacity of transformed cells in vivo, this assay should be considered to evaluate tumorigenic characteristics.
2. Preparation of tumor cells: grow cells in complete medium in log growth phase.
3. Trypsinize cells and make counts using a hemocytometer.
4. Cells should be suspended in a volume so that up to a 200 μL aliquot contains the required number of cells per injection.

5. Preparation of animals: mice should be 4–7 weeks old (see Note 10). Allow at least 7 days acclimation period after mice have arrived in the animal care facility.
6. Preparation for injections: clean and sterilize the inoculation areas of the mice with 75% ethanol.
7. Use 1-mL syringes and less than 26-guage needles.
8. Mix cells and draw cells into the syringes without needles to avoid strong negative pressure that can cause cell damage and lysis.
9. Inject cells (2×10^5 cells/mouse) into an appropriate area such as subcutaneous tissues in the lower flank or the axillary region of the back. In a case of orthotopic transplantation, cells should be injected into the site from which the transplanted cells were originally derived.
10. Therapy can be started after 1 week when the tumors have reached an average volume of 50 mm^3.
11. Tumor diameter should be measured with digital calipers, and the volume (V) of the primary tumors calculated with the following equation: $V=(\pi/6)\times[(L+W)/2]$, where L is the length and W is the width (Fig. 5).
12. The maintenance and handling of animals should be carried out using protocols approved by the Animal Care Facility in the University and Institution.

3.9. MSP Analysis

1. DNA methylation is a commonly occurring modification of human DNA. It involves the addition of a methyl group to cytosine residues at CpG dinucleotides, a reaction that is

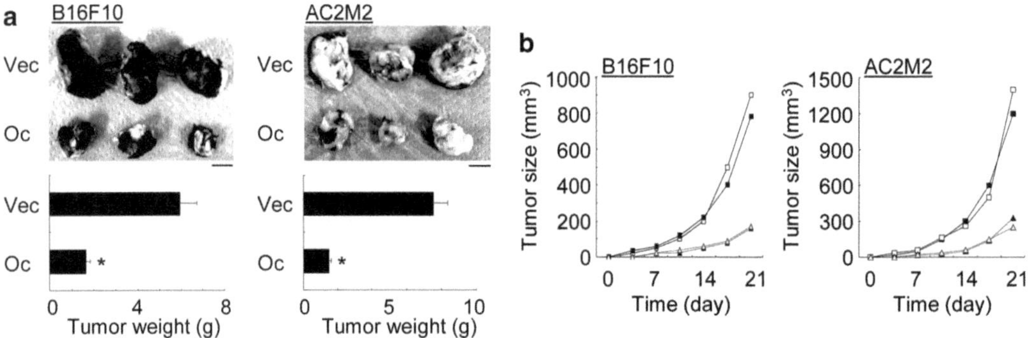

Fig. 5. Expression of TJ proteins such as occludin is sufficient to abrogate tumorigenicity, showing a markedly decelerated rate of tumor formation in vivo. Based on the in vitro observations (Fig. 4), we investigated the tumor-forming capacity of occludin-transfected cells in a syngenic mouse transplantation model. B16F10 and AC2M2 cells transfected with occludin (Oc) grew more slowly than empty vector transfected cells (Vec) in allograft transplantation models. (**a**) Representative tumors and their weights are shown. *Bar*, 1 cm. (**b**) Tumor volume was calculated in two dimensions in a time-dependent manner after inoculation of the vector (*square symbol and solid line*) or occludin-transfected (*triangle symbol and dotted line*) cells in the first (*filled*) and second (*open*) lines of experiments (B16F10 [2×10^4 cells/animal]: Vec, $n=8$; Oc, $n=10$; AC2M2 [2.5×10^4 cells/animal]: Vec, $n=16$; Oc, $n=15$; AC2M2 [7.5×10^3 cells/animal]: Vec, $n=8$; Oc, $n=12$. *$p<0.05$ vs. control group.

catalyzed by DNA methyltransferase enzymes. CpG dinucleotides are gathered in clusters called CpG islands, which are equally distributed across the human genome. There are approximately 30,000 CpG islands in the genome and 50–60% of these are found within the promoter region of genes. CpG islands are primarily unmethylated in normal tissues, and aberrant methylation of CpG islands is clearly related to disease, such as cancers.

2. The methylation status of the promoter CpG islands of genes of interest can be analyzed by methylation-specific PCR (MSP) on sodium bisulfite-converted DNA.
3. For MSP experiments, two pairs of primers are needed – one pair specific for methylated DNA (M) and the other for unmethylated DNA (U). To achieve discrimination of methylated and unmethylated DNA, one or more CpG sites are included in each primer (or at least one of the primer) sequence.
4. CpG islands of DNA can be analyzed by the CpG Islands Searcher at http://cpgislands.usc.edu, for example. In addition, primers for MSP for the amplification of both methylated alleles and unmethylated alleles can be designed using web-based free software, for example at http://www.changbioscience.com.
5. Digest 2 μg of high molecular weight genomic DNA overnight with a restriction enzyme that does not digest the DNA within the region of interest.
6. Extract the DNA with phenol:chloroform (1:1) and precipitate with 1/10 volume of 5 M sodium acetate and 2 volumes ethanol at −80°C for 15 min.
7. Centrifuge at room temperature or at 4°C for 10 min (more than 12,000 × *g*).
8. Remove the supernatant and wash the pellet twice with 70% ethanol.
9. Dry the pellet and resuspend in 100 μL TE buffer (10 mM Tris–HCl, pH 7.5, 1 mM EDTA).
10. Denature the DNA by adding freshly prepared NaOH (3 M) to a final concentration of 0.3 M.
11. Incubate at 42°C for 30 min.
12. To a microcentrifuge tube, add: 1,020 μL 40.5% sodium bisulfite, 60 μL 10 mM hydroquinone, 110 μL DNA (+NaOH), and 10 μL water.
13. Gently mix and cover the tube with aluminum foil to shield from room light.
14. Incubate at 55°C for 16–18 h.
15. After purification of DNA, resuspend DNA and add TE buffer to a final volume of 100 μL.

16. Denature the sample with freshly prepared NaOH (3 M) and incubate at 37°C for 15 min.
17. Neutralize by adding ammonium acetate (pH 7.0) to 3 M.
18. Precipitate the DNA with the same volume of isopropanol, centrifuge for 10 min at room temperature (more than 12,000 ×*g*), wash twice with 70% ethanol and dry under a vacuum.
19. Resuspend in 20 μL TE buffer and store at −20°C wrapped in foil.
20. Treated DNA should be used within 1 month as degradation may occur in cleaned and frozen samples (see Note 11).
21. Use aliquots (2 μL) as templates for PCR (Fig. 4).
22. Additional methods for PCR reactions and electrophoresis should be standard as described above.

3.10. cDNA Array Analysis

1. Various kinds of microarray analyses are now commercially available; however, this procedure is easy to perform in the standard laboratory without the need of specific apparatus and agents. Manufacturers supply all required materials in sets as kits, so that your laboratory basically does not need to purchase additional materials.
2. Pathway-specific cDNA arrays are designed to determine the expression profile of genes in a wide variety of biological pathways such as those involved in apoptosis and cancer development. If you would like to know detailed information for procedures and materials, visit http://www.sabiosciences.com.
3. To get started with this assay, good quality of total RNA extracted by TRIzol is required (see Note 12).
4. Generation of biotin-labeled cRNA: aliquots of RNA (up to 5 μg) are processed to generate cRNA probes: visit the official website and see the protocol in detail, because this protocol and required (but supplied) agents are highly specific for kits.
5. Hybridization and detection: see also official protocols at http://www.sabiosciences.com, because they are highly specific for kits.
6. Image-processing software allows analysis of digitally captured results (film).
7. The relative amounts of a given gene are normalized to signals derived from GAPDH on the same membrane.
8. Signal intensity should be expressed in arbitrary units (therefore the expression of GAPDH is 1 AU), to consider possible variation in RNA quantities between samples.
9. Average signal intensities of each gene are calculated using data from at least two independent experiments to verify the reproducibility of the results.

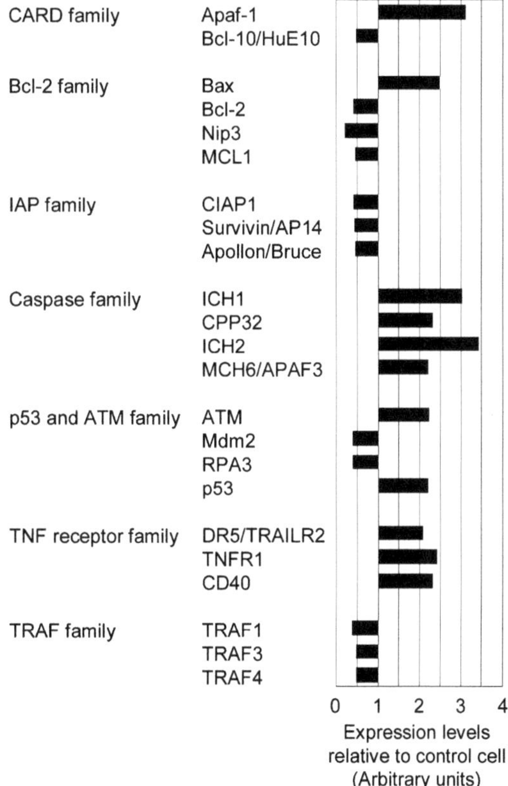

Fig. 6. Expression of TJ proteins such as occludin affects multiple signaling pathways. To explore the impact of occludin expression, we screened gene-expression profiles with the GEArray system (an apoptosis pathway-specific gene array). The profile in occludin-overexpressing HeLa cells clearly showed that occludin affects a wide variety of genes associated with the components of the apoptotic machinery. A number of genes demonstrating increase or decrease by an average of ≥100% or ≤−100% ($n=2$) are shown in the graph in functional gene groups.

10. Finally, gene-expression levels in cells that are treated with a certain drug or that are transfected with cDNA are divided by the values from that observed in control cells to represent the change of gene expression in the event of treatment of cells (Fig. 6).

4. Notes

1. Unless otherwise specified, we use ultra pure water that has a resistivity of 18.2 MΩ cm and total organic content of less than five parts per billion. This standard is referred to as “water” in this chapter.
2. Apoptotic sensitivity is easily modulated by pretreatment with atRA, because this agent has been shown to enhance apoptotic

sensitivity of cells (7, 8). Working solutions of atRA are prepared by dilution in DMSO ranging from 1 nM to 1 μM; 0.1 μM is a default concentration without any cell cytotoxicity. In some cases, it is important to optimize the duration of atRA treatment, because apoptotic cell death can be induced in certain types of cells possibly by a number of metabolites of atRA. To avoid the significant accumulation of RA metabolites in the media, atRA treatment should be very short (up to 5 min), which is known to be sufficient to switch on the RA-driven transcriptional machinery (7).

3. FBS treated with charcoal–dextran does not contain endogenous atRA. Its use, therefore, excludes possible effects on the cells by physiological concentrations of atRA.
4. The MTT reagent must be kept at 4°C in the dark. MTT solution is stable when stored frozen. Storage at 4°C may result in decomposition and yield erroneous results. Development of a dark color or formation of crystals usually indicates product deterioration.
5. The detergent reagent can be stored at either 4°C or ambient temperature. If the detergent reagent is kept at 4°C, prewarm the bottle for 5 min at 37°C and gently mix by inverting before use.
6. Since a good quality of total RNA is necessary to give satisfactory result, keep in mind measures to protect against possible contaminants that have RNase activity. For this purpose, it is important to deactivate RNases of plastic labware such as pipette tips and use gloves throughout experimentation.
7. Once you have prepared protein samples, you can immediately start the experiments. It is also important to avoid contact with toxic substances for neurons (i.e., polyacrylamide solution) before polymerization.
8. This procedure is very critical for this assay. If hydration of membranes is not appropriate, protein transfer is insufficient and gives unsatisfactory results.
9. It is important not to exceed 40°C, otherwise cells will be killed.
10. The species of mouse should be selected on the basis of which cancer cells are to be transplanted. If the cancer cells are derived from murine tissue, it is easiest to use the same species of mice. However, if the cancer cells are derived from another source, as with human material, immune deficient mice should be used, such as nude or severe combined immunodeficiency (SCID) mice.
11. It is important that DNA is completely denatured prior to and in the presence of the bisulfite solution. To ensure

complete denaturation, no more than 5 μg of DNA should be used. DNA digestion with restriction enzymes can be omitted.

12. The most important prerequisite for any gene expression analysis experiment is consistent, high-quality RNA from every experimental sample before starting the labeling process. Therefore, the sample handling and RNA isolation procedures are critical to the success of experiments. For the best results with kits, all RNA samples should be suspended in RNase-free water (not DEPC-treated water), and it should be confirmed that there is a sharp distinction with both the 18S and 28S ribosomal RNA bands and peaks. Any smearing and shoulders to the rRNA bands or peaks indicate that degradation has occurred in the RNA sample.

Acknowledgments

This study was supported by a Grant-in-Aid for Scientific Research from the Japan Society for the Promotion of Science.

References

1. Tsukita, S., Furuse, M., and Itoh, M. (2001) Multifunctional strands in tight junctions. *Nat Rev Mol Cell Biol* **2**, 285–293.
2. Matter, K., and Balda, M.S. (2003) Signalling to and from tight junctions. *Nat Rev Mol Cell Biol* **4**, 225–236.
3. Matter, K., Aijaz, S., Tsapara, A., and Balda, M.S. (2005) Mammalian tight junctions in the regulation of epithelial differentiation and proliferation. *Curr Opin Cell Biol* **17**, 453–458.
4. Furuse, M., Hirase, T. Itoh, M., Nagafuchi, A., Yonemura, S., and Tsukita, S. (1993) Occludin: a novel integral membrane protein localizing at tight junctions. *J Cell Biol* **123**, 1777–1788.
5. Saitou, M., Furuse, M., Sasaki, H., Schulzke, J.D., Fromm, M., Takano, H., et al. (2000) Complex phenotype of mice lacking occludin, a component of tight junction strands. *Mol Biol Cell* **11**, 4131–4142.
6. Sawada, N., Murata, M., Kikuchi, K., Osanai, M., Tobioka, H., Kojima, T., et al. (2003) Tight junctions and human diseases. *Med Electron Microsc* **36**, 147–156.
7. Osanai, M., Murata, M., Nishikiori, N., Chiba, H., Kojima, T., and Sawada, N. (2006) Epigenetic silencing of occludin promotes tumorigenic and metastatic properties of cancer cells via modulations of unique sets of apoptosis-associated genes. *Cancer Res* **66**, 9125–9133.
8. Osanai, M., Murata, M., Nishikiori, N., Chiba, H., Kojima, T., and Sawada, N. (2007) Occludin-mediated premature senescence is a fail-safe mechanism against tumorigenesis in breast carcinoma cells. *Cancer Sci* **98**, 1027–1034.
9. Osanai, M., Murata, M., Chiba, H., Kojima, T., and Sawada, N. (2007) Epigenetic silencing of claudin-6 promotes anchorage-independent growth of breast carcinoma cells. *Cancer Sci* **98**, 1557–1562.
10. Osanai, M. (2007) Tight junctions and cancer development, Encyclopedia of Cancer. M. Schwab. Springer.
11. Morin, P.J. (2005) Claudin proteins in human cancer: promising new targets for diagnosis and therapy. *Cancer Res* **65**, 9603–9606.
12. Michl, P., Barth, C., Buchholz, M., Lerch, M.M., Rolke, M., Holzmann, K.H. et al. (2003) Claudin-4 expression decreases invasiveness and metastatic potential of pancreatic cancer. *Cancer Res* **63**, 6265–6271.
13. Hoevel, T., Macek, R., Swisshelm, K., and Kubbies, M. (2004) Reexpression of the TJ protein CLDN1 induces apoptosis in breast

tumor spheroids. *Int J Cancer* **108**, 374–383.

14. Tobioka, H., Isomura, H., Kokai, Y., Tokunaga, Y., Yamaguchi, J., and Sawada, N. (2004) Occludin expression decreases with the progression of human endometrial carcinoma. *Hum Pathol* **35**, 159–164.
15. Li, D., and Mrsny, R.J. (2000) Oncogenic Raf-1 disrupts epithelial tight junctions via downregulation of occludin. *J Cell Biol* **148**, 791–800.
16. Tokes, A.M., Kulka, J., Paku, S., Szik, A., Paska, C., Novak, P.K., et al. (2005) Claudin-1, -3 and -4 proteins and mRNA expression in benign and malignant breast lesions: a research study. *Breast Cancer Res* 7, R296–305.
17. Resnick, M.B., Konkin, T., Routhier, J., Sabo, E., and Pricolo, V.E. (2005) Claudin-1 is a strong prognostic indicator in stage II colonic cancer: a tissue microarray study. *Mod Pathol* **18**, 511–518.

Chapter 5

Interactions Between *Clostridium perfringens* Enterotoxin and Claudins

Susan L. Robertson and Bruce A. McClane

Abstract

Clostridium perfringens enterotoxin (CPE), a single polypeptide of approximately 35 kDa in size, is associated with type A food poisoning and such non-foodborne gastrointestinal diseases as antibiotic-associated diarrhea and sporadic diarrhea. CPE action begins with binding of the toxin to a claudin receptor, forming a ~90 kDa small complex that then rapidly oligomerizes into a hexamer of ~450 kDa termed CH-1 (CPE hexamer-1). CH-1 is essentially a pore through which calcium gains entry to the cytoplasm, altering cell permeability and resulting in cell death by oncosis or apoptosis. Additionally, tight junctions are disrupted, allowing CPE access to the basolateral membrane so it can produce additional CH-1 complexes and also the CH-2 complex (~600 kDa) that contains occludin. We have recently demonstrated the presence of claudins-3 and -4 in both the CH-1 and CH-2 CPE complexes formed after CPE treatment naturally sensitive Caco-2 cells. Interestingly, claudin-1, which binds CPE poorly (if at all), was also present in these complexes.

Key words: *Clostridium perfringens* enterotoxin, Claudins, CPE hexamer-1, CPE hexamer-2, Coimmunoprecipitation, Electroelution

1. Introduction

Clostridium perfringens enterotoxin (CPE) has a complex mode of action that starts with binding to a claudin receptor, forming an SDS-sensitive complex of ~90 kDa. This small complex, which is noncytotoxic, rapidly oligomerizes at physiological temperatures into a ~450 kDa SDS resistant hexamer named CPE hexamer-1 (CH-1). CH-1 is unique among pore-forming toxins (1) by forming a pre-pore without apparently utilizing membrane rafts (2). The abundance of claudin CPE receptors on the cell surface may reduce the need for membrane rafts to cluster the receptors (2).

Kursad Turksen (ed.), *Claudins: Methods and Protocols*, Methods in Molecular Biology, vol. 762,
DOI 10.1007/978-1-61779-185-7_5, © Springer Science+Business Media, LLC 2011

The CH-1 pre-pore then inserts into the membrane to form an active pore that allows an influx of calcium ions. This calcium influx triggers apoptosis or oncosis, depending on the CPE concentration used (3, 4). Additionally, cell morphology is altered and tight junctions are disrupted (5), which exposes the basolateral surface of the cells. This allows CPE to bind to additional claudin receptors, producing more CH-1. It also provides this newly formed CH-1 access to the tight junction protein occludin, resulting in the formation of a second CPE complex of ~600 kDa that is named CPE hexamer-2 (CH-2) and contains both claudin and occludin (6, 7).

There is increasing interest in using CPE for cancer therapeutics since several cancers overexpress CPE receptor claudins, possibly allowing the toxin itself to selectively target and destroy these tumor cells (8). Other studies have looked at a noncytotoxic, but binding-proficient, mutant of CPE as a drug delivery agent (9–12). Furthermore, recent studies have shown that receptor claudins, or their fragments, can bind to CPE in vitro and prevent cell binding and cytotoxicity (13), suggesting possible use of claudins receptor decoys to treat antibiotic-associated diarrhea (AAD) or sporadic diarrhea (SD), which are common in hospital environments. Therefore, it is important to understand interactions between CPE, its receptor, and other tight junction proteins. Techniques such as electroelution and coimmunoprecipitation are important tools to study these interactions (14).

2. Materials

2.1. Cell Culture

1. Minimum essential Medium Eagle (EMEM) from Sigma, supplemented with 3.8 ml of 10% bicarbonate, 10% fetal bovine serum (Mediatech), 100 U/ml penicillin (Gibco), 100 μg/ml streptomycin (Gibco), and 2 mM glutamine (Gibco) and 1% nonessential amino acids (Sigma).
2. Solution containing trypsin (0.25%) and ethylenediamine tetraacetic acid (EDTA) (0.53 mM) (Gibco).
3. Dulbecco's phosphate-buffered saline (DPBS) without Ca^{2+}/Mg^{2+} (iDPBS) or with Ca^{2+}/Mg^{2+} (DPBS) (Mediatech).
4. Teflon cell scrapers (BD Falcon).
5. Erythrosine B, 0.4% in iDPBS.
6. Hemocytometer.

2.2. Treatment of Cells with CPE

1. CPE: prepared as described (15), stored as a lyophilized power –20°C. Resuspend powder in dH_2O, quantify by Lowry (16), and store for 1 month at –80°C.

2. Hanks Balanced Salt Solution (HBSS) (Mediatech) with Ca^{2+} and Mg^{2+} (cHBSS).
3. 25× Complete Protease Inhibitor cocktail (Roche). One tablet diluted into 2 ml dH_2O, stored at −20°C.
4. SDS–Polyacrylamide Gel Electrophoresis (SDS–PAGE) cell lysis buffer: 126 mM Tris–HCl pH 6.8, 20% glycerol, 4% sodium dodecyl sulfate (SDS), and 0.02% bromophenol blue; if desired to obtain reducing conditions, add 5% β-mercaptoethanol (Bio-Rad) before use.
5. Dilute Benzonase (Novagen, EMD Biosciences, Inc.), 1:10 and store at −20°C (see Note 1).

2.3. SDS–PAGE

1. Separating buffer (4.25% SDS–PAGE) 14.9 ml water, 6 ml 1.5 M Tris–HCl pH 9.3, 3.5 ml ProtoGel (National Diagnostics) 30% (w/v) acrylamide: 0.8% (w/v) Bis-Acrylamide stock solution, 300 μl 10% (w/v) SDS, and 250 μl 10% (w/v) ammonium persulfate (see Note 2), 30 μl Temed (Bio-Rad) prepared immediately before pouring gel.
2. Running buffer (10×): 250 mM Tris–HCl, 1.92 M glycine, and 1% (w/v) SDS.
3. Prestained SDS–PAGE standards (Bio-Rad), Broad Range 7,100–209,000 MWM mix of nine proteins (see Note 3).

2.4. Western Blotting for CPE Complexes

1. Trans-Blot Transfer medium; Pure Nitrocellulose Membrane (NCM) (0.45 μm) (Bio-Rad).
2. Filter paper (VWR).
3. Electrotransfer buffer (1×): 25 mM Tris–HCl, 192 mM glycine, and 20% MeOH pH to 8.3 (see Note 4).
4. Tris-buffered saline (TBS) (10×): 20 mM Tris–Hcl, 500 mM NaCl pH to 7.5. Dilute to 1× prior to use and add 0.01% Tween 20 (polyoxyethylene sorbitan monolaurate) (Bio-Rad) if desired (TBS-T) (see Note 5).
5. Blocking buffer: 5% (w/v) nonfat dry milk to TBS (MTBS) or TBS-T (MTBS-T).
6. Rabbit polyclonal anti-CPE serum, 100 μl serum to 200 ml MTBS (see Note 6).
7. Horseradish peroxidase-conjugated goat anti-rabbit IgG secondary antibody (Sigma), 15 μl aliquots stored at −20°C (see Note 7).
8. SuperSignal West Pico substrate or SuperSignal West Femto Maximum Sensitivity Substrate (Pierce) (see Note 8).
9. Kodak X-Omat cassette and Kodak X-Omat scientific Imaging Film.

2.5. Stripping and Reprobing

1. Stripping buffer: 62.5 mM Tris–HCl, pH 6.8, 100 mM β-mercaptoethanol, and 2% (w/v) SDS. Warm to 37°C for working solution.
2. Washing buffer TBS with 0.05% Tween 20.
3. Blocking buffer 5% (w/v) nonfat dry milk to TBS (MTBS).
4. Rabbit polyclonal antibodies against claudin-3, mouse monoclonal antibodies against claudin-1 or -4, and mouse monoclonal antibody against occludin (Zymed, Invitrogen). Peroxidase-conjugated goat anti-rabbit IgG and peroxidase-conjugated rabbit anti-mouse IgG (Sigma) (see Note 9).

2.6. Electroelution

1. Resuspension buffer: 100 μl of benzonase and 1.4 ml of cDPBS (DPBS with Ca^{2+}, Mg^{2+}).
2. Spectra-Pore 3 dialysis membrane (Spectrum) molecular mass cut-off, 3,500 Da.
3. Electroelution running buffer (10×): 25 mM Tris–Hcl, 0.192 M glycine, and 0.1% (w/v) SDS, pH 8.3.
4. Little Blue Tank, Isco, Inc. USA.
5. Acetone (HPLC Grade ≥ 99.9%), Sigma.
6. 6 M urea.

2.7. Coimmunoprecipitation

1. RIPA buffer: 50 mM Tris–Hcl (pH 8.0), 150 mM NaCl, 0.1% (w/v) SDS, 1% NP-40, and 0.5% sodium deoxycholate.
2. Protein A-coupled acrylic beads from *Staphylococcus aureus*, 250 μm (Sigma): supplied as lyophilized powder, resuspend in 5% MTBS.

3. Methods

3.1. CPE Treatment of Caco-2 Cells to Identify Complexes by Western Blotting

1. Wash CPE-sensitive Caco-2 cells (an immortalized line of human epithelial colorectal adenocarcinoma cells) with iDPBS, passage with trypsin–EDTA, and count on a hemocytometer after Erythrosine B viability staining. Seed at 2.5×10^5 cells/ml in a 100-mm^2 tissue culture dish 5–7 days before the assay. Once confluent, wash the monolayer twice with 10 ml of warm (see Note 10) cHBSS to remove cell culture medium.
2. Add CPE (diluted to a final concentration of 2.5 μg/ml in 10 ml cHBSS) to the Caco-2 cell monolayer and incubate at 37°C for 45 min.
3. Use a cell scraper to harvest the cells (most of which have already detached from the surface). Collect cells in a 50-ml conical tube (Corning) and centrifuge at 855 × *g*; resuspend pellet in 10 ml of cHBSS. Repeat this step twice to remove

the unbound enterotoxin. After the final spin, resuspend the pellet in 50 µl chilled cHBSS with 2 µl 25× Protease Inhibitor Cocktail (see Note 11). Add 5 µl of Benzonase, along with 50 µl 2× loading buffer (no βME); leave the mixture at RT until ready to separate by SDS–PAGE.

3.2. SDS–PAGE

1. Carry out Electrophoresis using the Bio-Rad Protean II xi Cell (see Note 12). Pour a separating gel (no stacking gel required) 0.5 mm thick, 4.25% gel by mixing 14.9 ml dH_2O, 6 ml 1.5 M Tris–HCl pH 9.3, 3.5 ml 30% Acrylamide/Bis, 300 µl 10% SDS, 250 µl 10% APS, and 30 µl Temed. Add a 10-well comb and leave the gel to polymerize for ~30 min.
2. Fill the Protean II tank one-third full with 1× electrode running buffer (dilute 100 ml of 10× electrode buffer in 900 ml dH_2O). Remove the comb from the polymerized gel and rinse wells with running buffer using a 100 µl Hamilton syringe. Fit the gel onto the central cooling core, place inside the buffer tank, and fill the upper chamber with 1× electrode running buffer.
3. Using a 100-µl syringe, load 20 µl of prestained molecular weight marker (MWM) in the first well, followed by 80 µl of untreated Caco-2 cells in cHBSS and then 80 µl CPE-treated Caco-2 cells. Place the lid on, connect the tank to a power supply and run the gel at ~25 mA (see Note 13).

3.3. Western Blotting for CH-1 and CH-2 Detection

1. Carry out western blotting using the Trans-Blot system with super cooling coil (Bio-Rad). Fill a large tray half-full with chilled transfer buffer and place the transfer cassette in with the black side nearest to the bottom of the tray. Next to this, a piece of foam is placed followed by one piece of 3-mm filter paper (cut to the size of the cassette). Place the NCM into the transfer buffer (using gloved hands) until wet.
2. Disconnect Protean II cell from the power supply and remove the gel from the tank. Remove the side clamps, spacers, and top glass plate to expose the gel. Using a razor blade, cut the top corner of the gel for proper orientation. Using excess transfer buffer lift the gel from the plate and place on top of the filter paper. Place the wet NCM on top of the gel, followed by another sheet of filter paper, then the foam, and finally close the white side of the cassette.
3. Place the transfer cassette into the transfer tank with the black side of the cassette facing the anode. Place the cooling rod into the tank and fill the entire unit with transfer buffer. Place on the assembly lid and connect to the power pack.
4. Switch on the cooling bar and activate the magnetic stir bar to keep cool water circulating around the cassette. Turn on the power supply; transfer takes 1 h at 600 mA.

5. Once transfer is complete, turn off the power supply and remove and disassemble the cassette. Remove the NCM, place into 50 ml of 5% MTBS-T, and incubate for 20 min at RT with gentle rocking to block the membrane.
6. Remove and discard blocking buffer. Wash the NCM blot briefly with 50 ml TBS. Add Rabbit polyclonal anti-CPE serum and incubate blot overnight at 4°C with gentle rocking.
7. Following overnight incubation, remove the primary antibody and retain (if stored at −20°C it can be used for up to five different blots). Wash NCM five times in 50 ml TBS-T for 10 min each, followed by five 10 min washes in 50 ml TBS.
8. Add Goat anti-rabbit IgG and incubate the blot at RT for 1 h with rocking.
9. Remove the secondary antibody and wash blot five times in 50 ml TBS-T and five times in 50 ml TBS, with 10 min for each wash.
10. Warm SuperSignal West Pico substrate to RT before mixing equal parts of stable Peroxide solution with Luminol enhancer solution (total of 10 ml). Add to the washed blot and agitate the substrate by hand for 1–2 min to ensure proper coverage of the membrane.
11. Remove the membrane from the substrate, blot to remove excess substrate, and sandwich the membrane between two 8 × 10 in. pieces of overhead transparencies. In the darkroom, place transparencies containing the blot into an X-ray film cassette and expose to the film for the appropriate time, typically between 1 and 5 min but can be as long as overnight (see Note 14). An example of the results produced is shown in Fig. 1.

3.4. Stripping and Reprobing Blots for Claudins and Occludin

1. Once the locations of CH-1 and CH-2 have been established, strip the membrane of antibodies and reprobe for the presence of claudins associated with CH-1 and CH-2 or occludin with CH-2.
2. Prepare stripping buffer on day of use. Pre-warm this buffer to 37°C, apply it to the blot, and incubate at 37°C with vigorous shaking for 30 min.
3. Remove stripping solution and wash the blot with 100 ml of TBS-T six times for 10 min each wash. Reprobe the membrane with the appropriate claudin or occludin primary antibody (see Subheading 2.4, step 4), diluted to 1 μg/ml in MTBS using the steps described above (see Subheading 3.3).

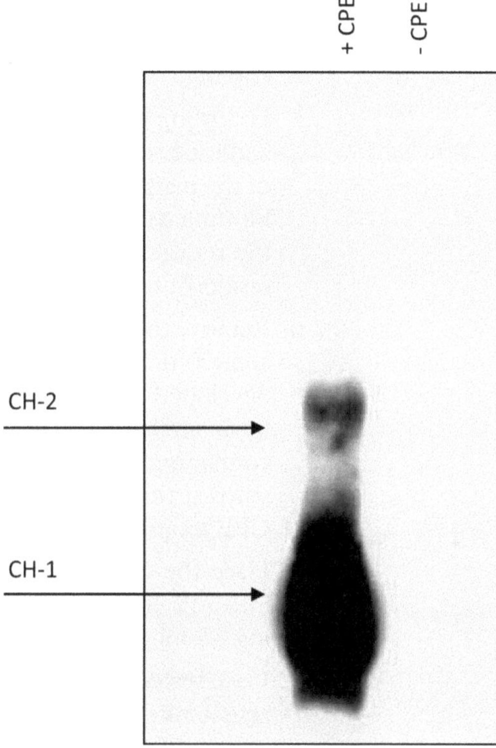

Fig. 1. Western immunoblot analysis of CPE complex formation in Caco-2-treated cells. Caco-2 cells were treated with (+) or without (–) 2.5 μg/ml CPE for 45 min at 37°C. Complexes were separated by electrophoresis, transferred to NCM, and detected using rabbit polyclonal anti-CPE serum. *Arrows* indicate the CH-1 (~450 kDa) and CH-2 (~600 kDa) complexes (see Note 15). Reproduced with permission from *Cellular Microbiology*.

3.5. Electroelution of Proteins for CPE Complexes

1. Prior to the experiment, prepare a 1.5 mm thick (see Note 16) 4.25% gel by mixing 44.7 ml dH_2O, 18 ml 1.5 M Tris/HCl pH 9.3, 10.5 ml 30% Acrylamide/Bis, 900 μl 10% SDS, 750 μl 10% APS, and 90 μl Temed. Insert a comb with one large center well and two small wells at either end and leave the gel left to polymerize for ~30 min.
2. Treat Caco-2 cells in monolayer with 2.5 μg/ml CPE, as described previously (see Subheading 3.1), except that ten 175-cm^2 tissue culture dishes are used.
3. Following treatment, harvest the Caco-2 cells by centrifugation (855 × *g*) and wash the pellet twice in DPBS before a final resuspension in 1.4 ml DPBS with 100 μl of 1:10 Benzonase. Following a 10 min incubation at RT, add 1.5 ml 2× loading buffer (without β-mercaptoethanol) is added (see Note 17).
4. Load this sample onto the SDS gel containing 4.25% polyacrylamide and also load 30 μl of broad-range MWM into

each of the end wells. Electrophorese the samples as described previously (see Subheading 3.2) at 8 mA overnight with cooling.

5. Following electrophoresis, carefully remove the top glass plate and use a scalpel to excise a small vertical strip that includes one of the marker wells. Process this gel strip for western immunoblotting as described under Subheading 3.3. Carefully wrap the remaining gel in saran wrap and store at −80°C until the results of the western blot are obtained (see Note 18).
6. Remove the gel from −80°C and, with the help of the MWM, align with the CH-1 and CH-2 complexes localized on the developed film. Using a scalpel, separately remove the two horizontal strips containing CH-1 and CH-2 from the gel containing samples from CPE-treated cells. Remove similar strips at the same location on a control gel (Caco-2 cells w/o CPE treatment).
7. Place the two gel strips from the treatment or control gels into ~12 cm Spectra Pore dialysis sacs (see Note 19) containing ~5 ml 1× running buffer.
8. Place sealed sacks (see Note 20) into a horizontal electrophoresis tank (Little Blue Tank) and cover with electrode running buffer. Add a magnetic stir bar in the tank to ensure circulation, and samples are electrophoresed at 50 mA for 4 h at 4°C.
9. Carefully remove buffer containing the electroeluted proteins from each of the tanks, concentrate the proteins by acetone precipitation (see Subheading 3.6), and resuspend the concentrated proteins in a final volume of 50 μl cDPBS.

3.6. Acetone Precipitation

1. Prior to use, prepare an acetone/water (4:1) solution and chill in a −20°C freezer.
2. Add to the sample containing eluted proteins (prepared in Subheading 3.5) a fourfold volume of chilled acetone solution and vortex. Leave sample at −20°C for 2 h.
3. Centrifuge at 16,000 × *g* for 10 min at 4°C and carefully pour off the supernatant. Wash pellet twice with chilled acetone solution.
4. After the final wash, SpeedVac the samples for 10–15 min and rehydrate in 50 μl DPBS (see Note 21).

3.7. Separation of Proteins from Electroeluted CH-1 and CH-2 Complexes

1. Carry out electrophoresis using a mini Protean 3 System (Bio-Rad). Pour a SDS gel containing 12% polyacrylamide by mixing 2.0 ml dH_2O, 1.25 ml 1.5 M Tris/HCl pH 9.3, 2 ml 30% Acrylamide/Bis, 50 μl 10% SDS, 25 μl 10% APS, 2.5 μl Temed. Overlay the gel solution with isopropanol and leave to polymerize (see Note 21).

2. Prepare the stacking gel by mixing 1.5 ml dH_2O, 630 μl 0.5 M Tris/HCl, pH 6.8, 330 μl 30% Acrylamide/Bis, 25 μl 10% SDS, 12.5 μl 10% APS, and 2.5 μl Temed.
3. Remove isopropanol layer and wash with dH_2O. Apply stacking gel, insert comb, and leave to polymerize.
4. To 50 μl of sample, directly add 50 μl 2× sample buffer with β-mercaptoethanol and then Urea (see Note 22) to a final concentration of 6 M Urea. Incubate at 90°C for 20 min and then boil for 5 min (see Note 23).
5. Electrophorese treatment and control samples, along with molecular weight markers, at 15 mA until dye front reaches the bottom of the gel ~45–60 min (see Note 24).

3.8. Western Blotting to Identify Proteins Electroeluted from CH-1 and CH-2 Complexes

1. Perform western as described earlier (see Subheading 3.3).
2. Following transfer, block the NCM for 20 min in 5% MTBS-T and briefly wash in TBS-T at RT with gentle rocking.
3. Add primary antibody, either rabbit polyclonal anti-CPE serum, mouse anti-claudin-4, mouse anti-claudin-1, rabbit anti-claudin-3, or mouse anti-occludin (see Note 25) and then incubate overnight at 4°C with gentle rocking.
4. Following overnight incubation, wash the blot (see Subheading 3.3, step 7) and then incubate with anti-rabbit IgG for claudin-3 or anti-mouse IgG for claudin-1, -4, or occludin (diluted 1:10,000 in a total of 100 ml 5% MTBS-T) at RT for 1 h with rocking.
5. Remove the secondary antibody, wash and develop the blot as described in Subheading 3.3. A typical result can be seen in Fig. 2.

3.9. Coimmunoprecipitation of CH-1 and CH-2 Complexes

1. Treat Caco-2 cells with (or without) CPE as described (see Subheading 3.1).
2. Resuspend the washed cell pellet, in 600 μl RIPA buffer and lyse cells on ice for 30 min.
3. After centrifugation at 16,000 × *g*, quantify protein in the cell lysate by Lowry (16) and adjust to a final concentration of 2 mg/ml in RIPA buffer.
4. To both the control and treatment lysate supernatants, add 20 μl of acrylic beads coupled with Protein A (see Note 26) and incubate for 1 h at 4°C with rotation to reduce nonspecific binding.
5. Pellet the beads by microcentrifugation (16,000 × *g*) and retain the supernatant. Split the 600 μl lysate supernatants (treatment and control) in equal volumes into three 1.5-ml tubes (200 μl per tube). To the first tube add no antibody,

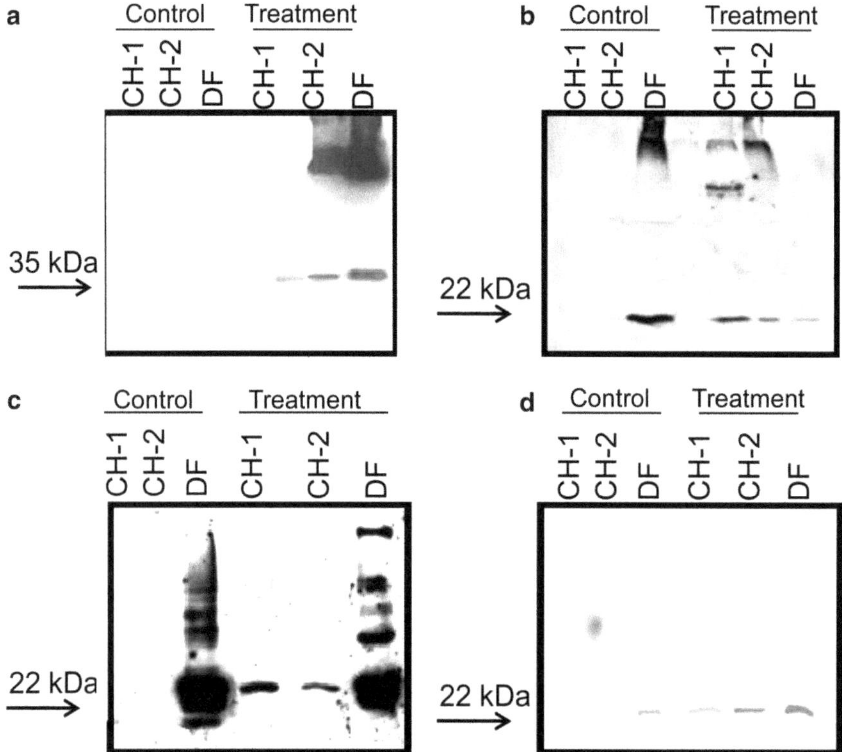

Fig. 2. Electroelution of Large CPE Complex Species from CPE-treated Caco-2 cells. Caco-2 cells were treated with 2.5 μg/ml CPE for 60 min at 37°C. Cells were lysed, separated by electrophoresis, transferred to NCM, and identified using rabbit anti-CPE serum. The bands corresponding to the CH-1 and CH-2 and dye front (DF) were cut out and electroeluted. Following electroelution, CH-1 and CH-2 were denatured using 6 M urea followed by the addition of SDS sample buffer containing βME. The proteins in each complex were then detected by separating on 12% SDS-containing PAGE, transferring to NCM, and detected using the appropriate claudin antibody. (**a**) Rabbit polyclonal anti-CPE serum, (**b**) mouse monoclonal anti-claudin-4, (**c**) rabbit polyclonal anti-claudin-3, and (**d**) mouse monoclonal anti-claudin-1. Reproduced with permission from *Cellular Microbiology*.

add to the second tube 2 μl rabbit serum IgG, and to third tube 2 μl affinity purified polyclonal rabbit IgG antibodies raised against purified CPE (see Note 27).

6. Incubate the tubes overnight at 4°C with gentle rotation. Add Protein A beads (60 μl) in 5% MTBS-T to each of the three tubes and incubate for a further 2 h at 4°C with rotation.
7. Microcentrifuge (16,000 × *g*) the beads containing the bound immunoprecipitate and then resuspend the pellet in 1,000 μl chilled RIPA buffer (with protease inhibitors). Repeat this wash step five times. On the final spin, resuspend the beads with bound immunoprecipitate in 50 μl DPBS to which 50 μl 2× sample buffer with βME is added for 10 min incubation at RT. Boil the bead/immunoprecipitate mixture for 5 min to elute the precipitated proteins from the beads.

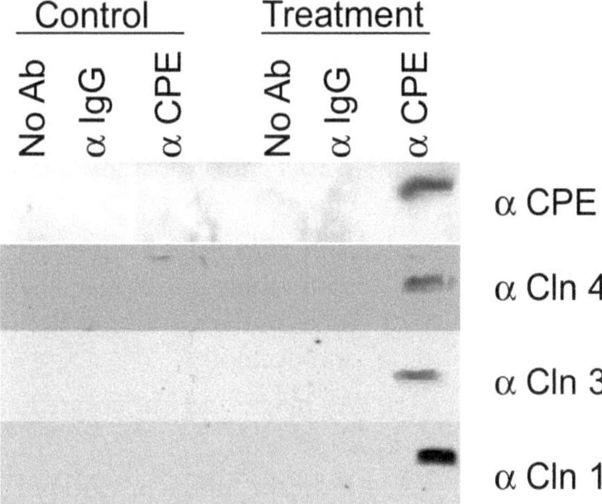

Fig. 3. Coimmunoprecipitation of Claudins from the CH-1 complex of CPE-treated Caco-2 cells. Caco-2 cells were treated with or without CPE and lysed. Cell lysates were incubated overnight at 4°C with either no antibody, rabbit anti-IgG, or rabbit anti-CPE. Finally, protein A acrylic beads were added and incubated for a further hour followed by the addition of SDS sample buffer (with βME). Boiling eluted the immunoprecipitated protein and these were separated by electorphoresis and detected by western blot analysis using rabbit anti-CPE serum, mouse anti-claudin-4, rabbit anti-claudin-3, and mouse anti-claudin-1. Reproduced with permission from *Cellular Microbiology.*

8. Electrophorese the samples on an SDS gel containing 12% acrylamide and western blotted for the appropriate claudins, as described under Subheadings 3.7 and 3.8. Figure 3 shows a representative gel result.

4. Notes

1. Benzonase is always diluted 1:10 for removing nucleic acid from the sample.
2. Ammonium persulfate (APS) is freshly prepared on day of use at 0.01 g/ml.
3. Prestained SDS–PAGE standards are stored at −20°C and thawed prior to use. They are referred to as MWM throughout chapter.
4. Electrotransfer Buffer is stored at 4°C and can be used up to five times.
5. Typically, all wash steps start with TBS-T followed by TBS unless otherwise stated.
6. Rabbit polyclonal anti-CPE serum is stored at −20°C and thawed prior to use; it can be reused up to five times.

7. A 15 μl aliquot of anti-rabbit IgG is thawed and diluted into 100 ml MTBS (1:10,000) immediately prior to use. Antibody is stored in 15 μl aliquots to avoid multiple freeze/thaw cycles.
8. West Femto substrate is used with reprobing blot unless otherwise stated.
9. Anti-Rabbit IgG secondary antibody is used for blots reacted with rabbit polyclonal claudin-3 and anti-mouse IgG is used as secondary antibody for blots reacted with claudin-3, -4, and occludin.
10. All media and buffers are heated to 37°C in a waterbath prior to use.
11. Protease Inhibitor Cocktail (Roche), one tablet is dissolved in 2 ml dH_2O to give a 25× stock solution stored at −20°C.
12. Bio-Rad Protean II xi cell is used for resolving the complexes.
13. Large complex gels are typically run until the dye front is ~1 cm from the gel bottom.
14. The blot is secured to the interior of the cassette so that proper alignment of the film and MWM can be made.
15. For all experiments, a control is run where the cells are treated exactly the same except for the omission of CPE.
16. The electroelution gel is 1.5 mm thick to allow a greater amount of sample to be loaded and the gel length is increased to 20 cm to maximize protein separation.
17. Gel is stored at −80°C as this minimizes distortion to the gel and allows for better alignment with the developed film.
18. Gel strips are sliced into smaller ~1 cm sections to optimize electroelution.
19. Sacks are sealed with a knot and then with SpectroPor closures (5 mm sealing width).
20. If SpeedVac is not an option, pellets can be left to dry at RT in a fume hood.
21. Isopropanol allows gel surface to remain flat.
22. Urea must be less than 30 days old.
23. Boiling for 5 min further denatures proteins. Do not exceed this boiling time or proteins may be carbamoylated.
24. Electrophoresis can be run between 10 and 20 mA depending on desired running time.
25. Primary antibody is diluted in 5% MTBS to 1 μg/ml final concentration and stored at −20°C. It is thawed on day of use and can be used up to five times. Occludin is only present in CH-2, therefore probing for it serves as a good positive and negative control.

26. Protein A beads are resuspended in 5% MTBS.
27. Three tubes are set up, with the first two being controls containing no antibody and nonimmune rabbit serum IgG and the third being the pull down with affinity purified polyclonal rabbit IgG antibody raised against purified CPE.

Acknowledgments

This work was generously supported by a grant from the National Institute of Allergy and Infectious Disease (R37-AI019844-28, B.A.M).

References

1. Lafont, F. and F.G. van der Goot (2005) *Bacterial invasion via lipid rafts.* Cell Microbiol. 7(5): p. 613–20.
2. Caserta, J.A., et al. (2008) *Evidence that membrane rafts are not required for the action of Clostridium perfringens enterotoxin.* Infect Immun. **76**(12): p. 5677–85.
3. Chakrabarti, G. and B.A. McClane (2005) *The importance of calcium influx, calpain and calmodulin for the activation of CaCo-2 cell death pathways by Clostridium perfringens enterotoxin.* Cell Microbiol. 7(1): p. 129–46.
4. Chakrabarti, G., X. Zhou, and B.A. McClane (2003) *Death pathways activated in CaCo-2 cells by Clostridium perfringens enterotoxin.* Infect Immun. **71**(8): p. 4260–70.
5. McClane, B.A. (1994) *Clostridium perfringens enterotoxin acts by producing small molecule permeability alterations in plasma membranes.* Toxicology. **87**(1–3): p. 43–67.
6. Singh, U., et al. (2001) *Comparative biochemical and immunocytochemical studies reveal differences in the effects of Clostridium perfringens enterotoxin on polarized CaCo-2 cells versus Vero cells.* J Biol Chem. **276**(36): p. 33402–12.
7. Singh, U., et al. (2000) *CaCo-2 cells treated with Clostridium perfringens enterotoxin form multiple large complex species, one of which contains the tight junction protein occludin.* J Biol Chem. **275**(24): p. 18407–17.
8. Michl, P. and T.M. Gress (2004) *Bacteria and bacterial toxins as therapeutic agents for solid tumors.* Curr Cancer Drug Targets. **4**(8): p. 689–702.
9. Deli, M.A. (2009) *Potential use of tight junction modulators to reversibly open membranous barriers and improve drug delivery.* Biochim Biophys Acta. **1788**(4): p. 892–910.
10. Harada, M., et al. (2007) *Role of tyrosine residues in modulation of claudin-4 by the C-terminal fragment of Clostridium perfringens enterotoxin.* Biochem Pharmacol. **73**(2): p. 206–14.
11. Kondoh, M., et al. (2006) *A novel strategy for a drug delivery system using a claudin modulator.* Biol Pharm Bull. **29**(9): p. 1783–9.
12. Takahashi, A., et al. (2008) *Domain mapping of a claudin-4 modulator, the C-terminal region of C-terminal fragment of Clostridium perfringens enterotoxin, by site-directed mutagenesis.* Biochem Pharmacol. **75**(8): p. 1639–48.
13. Robertson, S.L., et al. (2010) *Identification of a Claudin-4 residue important for mediating the host cell binding and action of Clostridium perfringens enterotoxin.* Infect Immun. **78**(1): p. 505–17.
14. Robertson, S.L., et al. (2007) *Compositional and stoichiometric analysis of Clostridium perfringens enterotoxin complexes in Caco-2 cells and claudin 4 fibroblast transfectants.* Cell Microbiol. **9**(11): p. 2734–55.
15. McDonel, J.L. and B.A. McClane (1988) *Production, purification, and assay of Clostridium perfringens enterotoxin.* Methods Enzymol. **165**: p. 94–103.
16. Lowry, O.H., et al. (1951) *Protein measurement with the Folin phenol reagent.* J Biol Chem. **193**(1): p. 265–75.

Chapter 6

Biophysical Methods to Probe Claudin-Mediated Adhesion at the Cellular and Molecular Level

Sri Ram Krishna Vedula, Tong Seng Lim, Walter Hunziker, and Chwee Teck Lim

Abstract

Claudins are a family of tetraspan membrane proteins that localize at tight junctions in an epithelial monolayer forming a selective barrier to diffusion of solutes via the intercellular spaces. It is widely accepted that the interaction between the extracellular loops of claudin molecules from adjacent cells is critical for this function. Though previous experiments utilizing traditional biological, biochemical, morphological, and electrophysiological approaches have provided significant insights into the role of claudins in regulating ion permeability, the interaction kinetics between these molecules has not been characterized. In this chapter, we describe two experimental procedures to study the adhesion forces imparted by claudins: (a) dual micropipette assay to quantify the adhesion forces at the cellular level and (b) single molecule force spectroscopy using atomic force microscopy to characterize the interaction kinetics at the molecular level. Though the experimental procedures are described for claudins, they can be easily modified for studying the interaction properties of a wide variety of other proteins.

Key words: Claudins, Dual micropipette assay, Intercellular adhesion, Single molecule force spectroscopy, Atomic force microscopy

1. Introduction

Characterizing adhesion between a pair of cells and interactions between two proteins have been a significant area of research in biology due to its importance in several biological processes such as selective permeability of epithelial barriers, tissue morphogenesis, and cancer metastasis just to name a few. Traditional approaches to understand these two phenomena (cell adhesion and protein interactions) have predominantly been qualitative. Quantitative understanding of these processes has been made possible now with the advent of new nano-technological tools like atomic force microscopy (AFM), optical traps, surface plasmon

Kursad Turksen (ed.), *Claudins: Methods and Protocols*, Methods in Molecular Biology, vol. 762,
DOI 10.1007/978-1-61779-185-7_6, © Springer Science+Business Media, LLC 2011

resonance, and micropipette assays. The main objective of this chapter is to describe in some detail, the application of dual micropipette assay and single molecule force spectroscopy (SMFS) to study cell–cell adhesion and protein–protein interactions.

1.1. Dual Micropipette Assay

The application of glass micropipettes in biology is not new. They have been used extensively for patch clamp studies to characterize ion channels. Other applications have been for studying the response of cells to externally applied forces (mechanotransduction) and characterizing the deformability of cells (micropipette aspiration). However, all these experiments involved the use of a single micropipette only. Dual micropipette assay for studying cell adhesion was first performed by Sung et al. in 1986 (1). They applied the dual micropipette assay to study the interaction between an activated cytotoxic T-cell and its target cell (Fig. 1a). A modification of this technique called the biomembrane force probe (BFP) was introduced by Evans et al. (2). The BFP uses a red blood cell or a liposome (both of which are highly deformable) with a bead attached to it as a force transducer (Fig. 1b). The bead is coated with the protein of interest and is made to interact with another cell of interest. The whole cell-bead system acts as a spring whose spring constant can be easily computed. Displacements of this "spring system" are tracked using reflection interference contrast to a nanometer resolution. This can then be used to compute the forces applied.

The use of dual micropipette assay was reestablished by Dufour et al. (3, 4) for studying the intercellular adhesion forces imparted by E-cadherins. Using the same technique, it has been

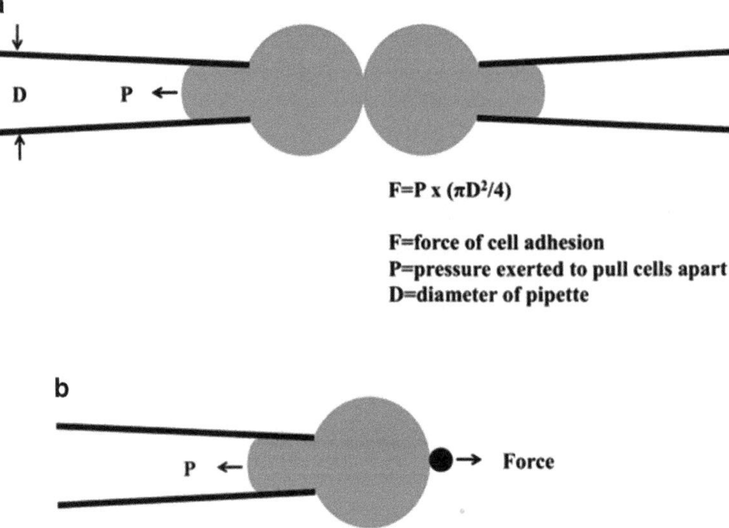

Fig. 1. Micropipette-based techniques for studying cell–cell adhesion: (**a**) Dual micropipette assay, (**b**) Biomembrane force probe (BFP) (15).

shown that adhesion mediated by claudins is much weaker than that mediated by E-cadherins (5). The basic experimental setup consists of two long and thin glass capillaries (micropipettes) with a diameter in the range of a few microns (typically 5–10 μm) connected to water columns. The pressure inside the micropipettes is regulated by changing the height of the connected water columns. The change in height of the water column can be accurately regulated by either using a micrometer, a precision infusion/withdrawl pump or a pressure transducer connected to the water column.

1.2. Single Molecule Force Spectroscopy

SMFS attempts to study the behavior of a molecular interaction under an externally acting load. SMFS can be performed using several tools one of which is AFM. AFM is a relatively new technique which was initially designed to image surfaces at high resolution. The basic principle is the use of a soft deflecting cantilever with a sharp tip at its end to probe a surface. The deflections of the cantilever are monitored by a laser reflecting off from the surface of a cantilever. These deflections are then used to map the topography of the surface (Fig. 2).

Recently, AFM has emerged as a powerful tool to implement SMFS experiments. Here, the protein or cell of interest is coupled to the cantilever tip using different strategies (6–8). The cantilever is then made to approach a surface containing the other interacting protein or cell. After allowing contact for a defined time

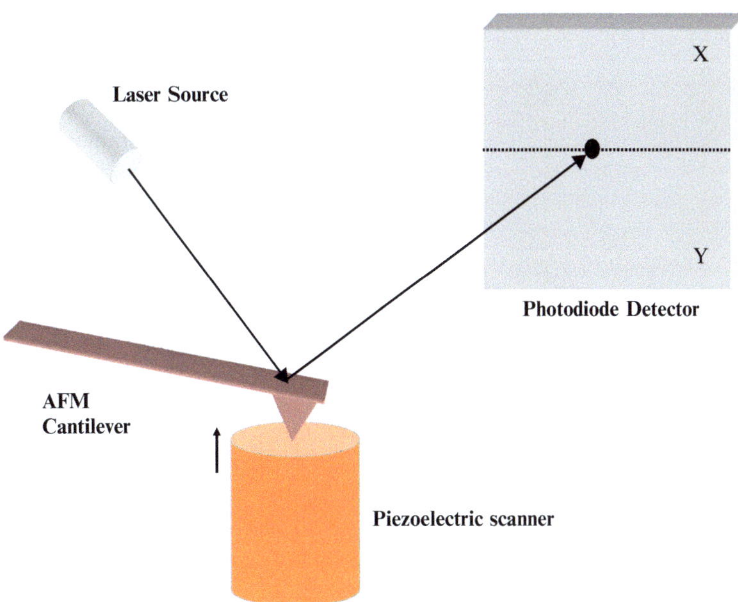

Fig. 2. Schematic depiction of the principle of AFM showing a laser reflecting off from the surface of flexible cantilever onto a photodetector.

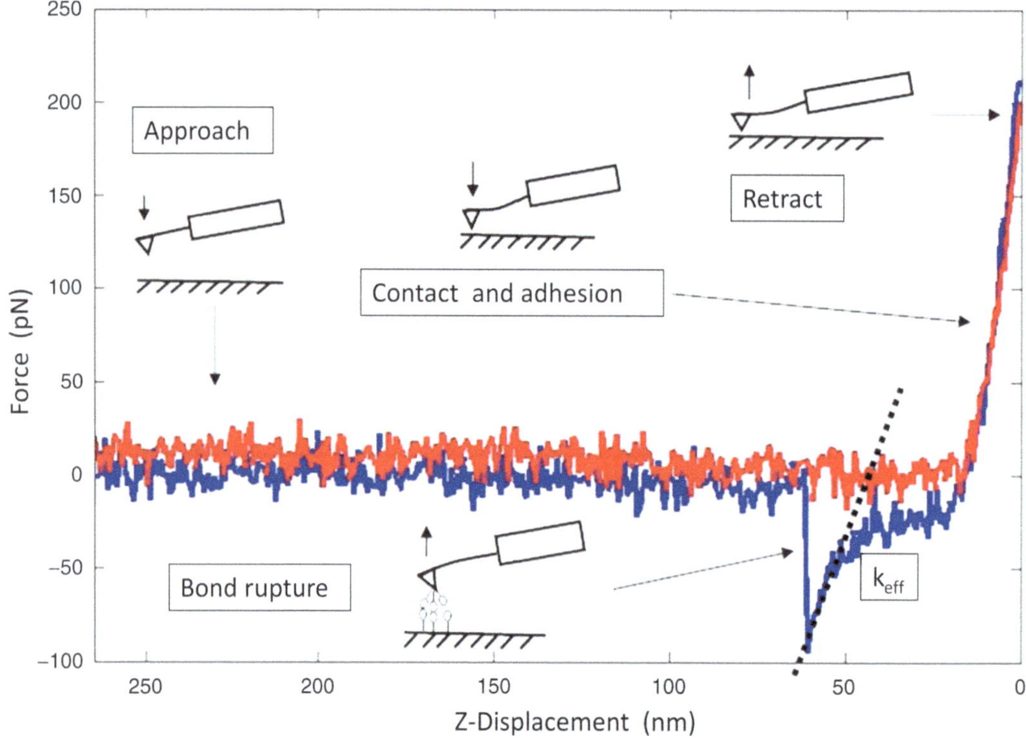

Fig. 3. A typical force–displacement curve showing the approach (*red*) and reproach curve (*blue*). Bond ruptures are observed as sharp jumps in the reproach curve. The effective spring constant of the system (k_{eff}) is computed as the slope of the reproach curve prior to the rupture of the bond (*dashed line*). The loading rate is obtained by multiplying the effective spring constant of the system and the reproach velocity of the cantilever (16).

period at a defined contact force, the cantilever is made to retract at a predefined velocity. The deflections of the cantilever during this whole approach-contact-reproach cycle constitute one force–distance curve. The force–distance curve is the basic data unit of SMFS (Fig. 3). During the retraction process, bonds formed between the proteins or cells rupture. The rupturing of bonds is observed as discrete jumps in the force–distance curve. The changes in the characteristics of bond rupture events over a spectrum of applied loading rates are analyzed to extract the kinetic nature of the interaction under investigation (9).

2. Materials

2.1. Dual Micropipette Assay

1. L-Fibroblasts stably transfected with Claudin-1 and Claudin-2 (see Note 1).
2. 1× Trypsin (Sigma).
3. Cell culture medium (DMEM containing 10% FBS and 1% penicillin and streptomycin).

4. Inverted microscope.
5. Micromanipulators (Eppendorf) mounted on the microscope.
6. Borosilicate glass capillaries (B100-75-10, Sutter instruments).
7. Micropipette filler (Microfil™, World Precision Instruments, Fl).
8. Micropipette holders.
9. 10-ml Glass pipettes.
10. Tubing (Cole-Parmer).
11. Three-way valves (Cole-Parmer).
12. Syringe pump for controlling suction (Cole-Parmer).
13. 5-ml Syringe (BD).
14. Hydrophobic glass cover slips (22 × 60 mm, Matsunami, Japan).
15. Micropipette puller (Model P-97, Sutter Instruments).
16. Micropipette forger (MF-900, Narishige, Japan).

2.2. Single Molecule Force Spectroscopy Using Recombinant Claudin Proteins

1. Atomic force microscope (MultiMode™ Picoforce™, Veeco, Santa Barbara, CA) with Nanoscope Controller IV coupled to an upright microscope.
2. AFM tips (MLCT AUNM tips, Veeco, Santa Barbara, CA).
3. Fluid cell (Veeco, Santa Barbara, CA).
4. 1-ml Syringe.
5. Recombinant C-terminal GST-tagged full-length claudin fusion protein (Abnova, Taiwan), monoclonal mouse anti-GST antibody (Invitrogen).
6. Bis-sulfosuccinimidyl suberate (BS^3, Pierce, Rockford, IL).
7. Acetone (Sigma).
8. Solution containing 70% H_2SO_4 and 30% H_2O_2.
9. 3-Aminopropyltriethoxysilane (APTES, Sigma).
10. 1 M Tris buffer (pH 7.5).
11. 1% Bovine serum albumin solution (Sigma).
12. 12-mm Glass cover slips (Deckglaser, Germany).
13. Metal disk 15 mm.

3. Methods

3.1. Dual Micropipette Assay

This section describes the use of dual micropipette assay for studying the global adhesion forces between two l-fibroblasts expressing Cldn-1 or Cldn-2. The method can be easily extended to study other cell adhesion proteins as well (see Note 2).

3.1.1. Prepare L-Fibroblast Suspension

1. Trypsinize stably transfected L-fibroblast (expressing Cldn1 or Cldn2) grown to confluence in a T-25 culture flask (see Note 3).
2. Centrifuge cells at 1,200 rpm (230 × *g*) for 5 min.
3. Resuspend cell pellet in 2 ml of culture medium.

3.1.2. Setup Water Columns and Micropipette Holders

1. Fill the 10-ml pipettes with DI H_2O and put them vertically on either side of the microscope stage with the help of a stand.
2. Connect a three-way valve to the bottom end of both these (right and left) water columns.
3. Connect one outlet of the three-way valve on the left side to a 5-ml syringe clamped on to an infusion/withdrawing pump.
4. Connect the other outlet of the three-way valve to the micropipette holder using a thin tube.
5. On the right side, the 5-ml syringe can be operated with hand and no infusion pump is needed. Make sure that there are no air bubbles in the whole system (syringes, tubing, pipettes, etc.). Figure 4 shows a schematic depiction of the set up.
6. Calibrate the decrease in the height of the water column (in cm) for every 1 ml of water sucked in by the pump.

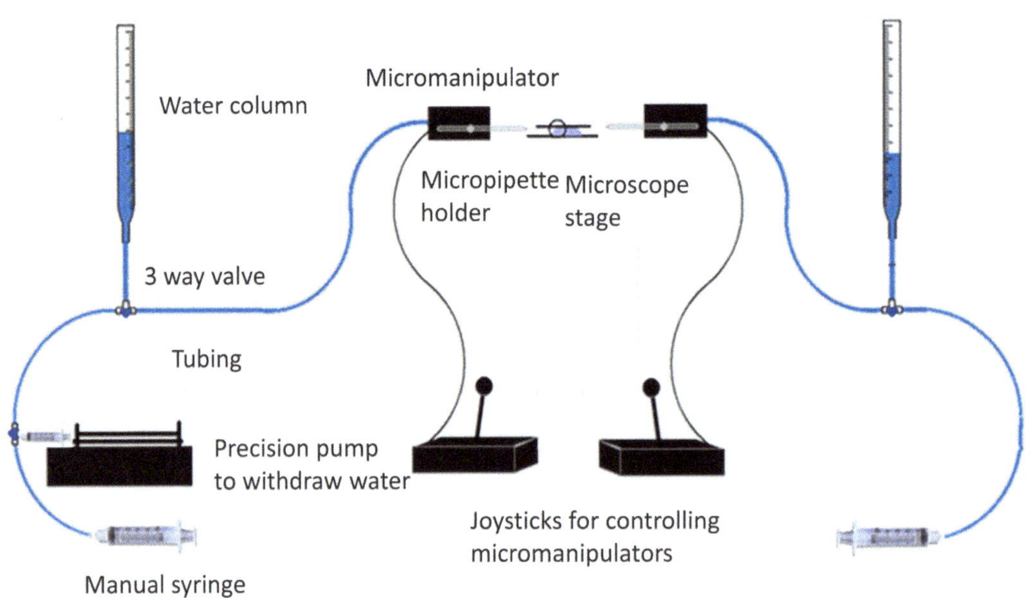

Fig. 4. Schematic depiction of the micropipette setup for conducting dual micropipette assay.

3.1.3. Adhesion Assay

1. Pull micropipettes and forge them to the required diameter (~5 μm).
2. Using micropipette filler, fill the micropipette with a solution of 3% BSA. Make sure that there are no air bubbles inside the micropipette.
3. Place the micropipette inside the micropipette holder carefully without breaking it and fasten the screw.
4. Place a cover slip on the microscope stage and put a large drop of culture medium (200 μl) on it. Make sure that the drop does not spread or spill over.
5. Using the micromanipulators, gently insert the micropipettes into the drop making sure that they do not break.
6. Add about 5 μl of cell suspension into the droplet and leave for about 1–2 min. Focus on the cells that have settled to the bottom.
7. Adjust the height of the micropipettes to bring them into the same plane as the cells. To confirm that the micropipettes are not blocked, generate a large positive pressure in the micropipette (by increasing the water column height using the manual syringe). Cells close to the micropipette will be pushed away if the lumen is not obstructed.
8. Obtain the height of the water column that represents zero pressure inside the micropipette. To do this, bring the micropipette close to a small floating particle (usually small cell debris) and vary the height of the water column using the manual syringe such that the particle neither gets sucked in (negative pressure) nor is pushed away (positive pressure). Record this height of the water column as the baseline.
9. Bring the right micropipette close to a single suspended cell. Decrease the height of the water column by sucking in ~2 ml of water using the manual syringe. The cell will get sucked into the micropipette partially and stay there firmly because of the large suction pressure.
10. Hold another single cell in the left micropipette using a small suction pressure and bring it close to the cell held in the right micropipette. Gently push them against one another and hold them. Make sure that the cells do not slip against one another or detach from one another at this stage. Increase the water column height in the left micropipette to make the pressure zero.
11. Leave the cell adhesion to develop for a defined period of time (~15 min). When the left micropipette is gently pulled away, the cells will remain adherent to one another. Now, decrease the pressure in the left micropipette by withdrawing ~0.1 ml of water using the syringe pump. Bring the left micropipette

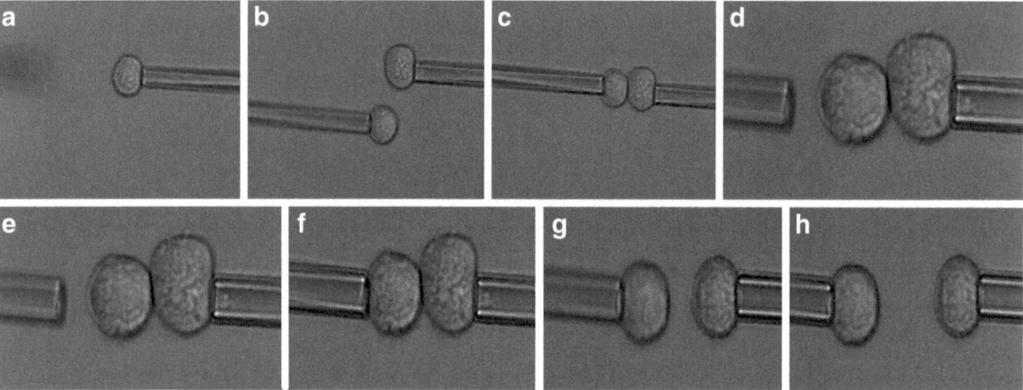

Fig. 5. Steps in the dual micropipette assay to measure cell–cell adhesion: (**a**) a cell is aspirated into the right pipette with a large pressure; (**b**, **c**) a second cell is moved close to the first till a contact is made; (**d**) the cell–cell adhesion is allowed to form for a specified amount of time; (**e**–**h**) pressure in the left pipette is increased step wise till the two cells detach from each other (17).

close to the left cell so that it gets sucked in. Try to pull the left cell away from the right cell. If the cells do not separate, decrease the pressure further by withdrawing another 0.1 ml. Repeat the procedure till the pressure in the left pipette is sufficiently strong enough to separate the two cells.

12. Record the amount of the water that has been withdrawn and calculate the drop in the height of the water column from the calibration done previously. The force required to separate the cells is now calculated as the separation pressure multiplied by the cross-sectional area of the micropipette (Fig. 5).
13. Repeat the experiment for different pairs of cells. Using a heating stage experiments can be carried out at a temperature of 37°C. However, at room temperature, it is advisable not to carry out the experiments for long durations (>1 h).

3.2. Single Molecule Force Spectroscopy Experiments

This section describes the use of SMFS for characterizing the interaction kinetics between recombinant Cldn-1 and Cldn-2 proteins at the single molecule level. There are a number of functionalization protocols available to couple the protein or cell of interest to the AFM cantilever and have been described in detail elsewhere (10) (see Note 4). Here we describe a modification of the protocol used by Hanley et al. (6, 7).

3.2.1. Functionalizing AFM Tips

1. Clean the tips in UV for 15 min and incubate them in a mixture of 70% H_2SO_4 and 30% H_2O_2 for 30 min. Wash them thoroughly with DI water and dry them (see Note 5).
2. Prepare a 4% solution of APTES in acetone. Incubate the tips in this solution for 3 min. Wash the tips with pure acetone and let them dry. Alternatively, tips can also be silanized using a dehumidifier. Place the tips in a petri dish in the dessicator

along with a small volume of APTES (10 μl) in a small eppendorf tube. Close the dehumidifier and purge out the air inside using a vacuum pump for 1 min. Close the valve of the dessicator and leave it for about 15–30 min for the APTES vapor to deposit on the tips. APTES is highly sensitive to moisture and should be placed in a dehumidifier.

3. Prepare a fresh solution of BS3 (2 mg/ml) in DI water. Place a large drop of the solution on a small strip of paraffin film. Incubate the tips in this drop for 30 min. Remove the tips and immerse them in a drop of anti-GST monoclonal antibody (10 μg/ml) for 2 h. Wash the tips by immersing them in PBS. Quench the reaction by immersing the tips in 1 M Tris buffer.
4. Wash the tips by immersing them in PBS and place them in a solution of freshly prepared recombinant GST-Cldn1 or GST-Cldn2 proteins (10 μg/ml) for 2 h. For control experiments, immerse the anti-GST antibody coupled tips in a solution of PBS (see Note 6). Remove the tips and place them in a drop of 1% BSA for 15 min. Finally, wash the tips by immersing the tips in PBS.
5. Functionalize 12-mm glass cover slips in a similar fashion as described for the AFM tips. Stick the functionalized cover slip onto a metal disk using a double-sided tape. Place the metal disk with the cover slip on the AFM scanner and a small amount of PBS to prevent it from drying.

3.2.2. Obtaining Force–Distance Curves

1. Place the functionalized AFM tip in the fluid cell so that it sits snugly in the groove and is clamped firmly. Align the laser onto the largest and softest cantilever (spring constant ~10 pN/nm) and place the fluid holder on top of the scanner. Inject a small amount of PBS into the fluid cell using a 1-ml syringe. Bring the cantilever close to glass cover slip but not in contact with it.
2. Obtain the deflection sensitivity and spring constant of the cantilever using the thermal tune module of the AFM.
3. Engage the AFM tip and set the following parameters to obtain force plots: contact force 200 pN, ramp size 2 μm, approach and retraction velocity 1 μm/s, and contact duration 1 ms. Optimize these parameters in such a way that the observed frequency of adhesion events is <30%. This optimization is based on trial and error. This would ensure that most of the recorded adhesion events represent the rupture of single bonds.
4. Obtain several force–distance curves ~10–30 randomly at different points on the cover slip. Acquire a total of at least 500 force curves for a given retraction velocity. Repeat the acquisition of force curves using a range of retraction velocities (100, 250, 500, and 750 nm/s).

3.2.3. Analyzing and Interpreting Force–Distance Curves

The deflections of the cantilever and the movement of the piezoelectric scanner during one cycle of approach, contact and retraction constitute a single force–displacement curve, which constitute a basic data unit in SMFS experiments (Fig. 3) (see Note 7).

1. Measure the bond rupture force (f, in pN) and loading rate (r_f, in pN/s) for each force–displacement curve showing a clear single rupture event. Bond rupture force is the magnitude of the sharp jump in the retract curve (blue, Fig. 3). The corresponding loading rate (r_f, in pN/s) is the product of retraction velocity (v, in nm/s) and the effective stiffness of the force transducer (k_{eff}, in pN/nm, slope of force–displacement curve immediately prior to bond rupture, Fig. 3).
2. Partition each measurement $\{f_i, r_{fi}\}$ into different bins based on the range of loading rate (7, 8, 11) (Fig. 6). An example of proper binning is to use windows of 50 pN/s for loading

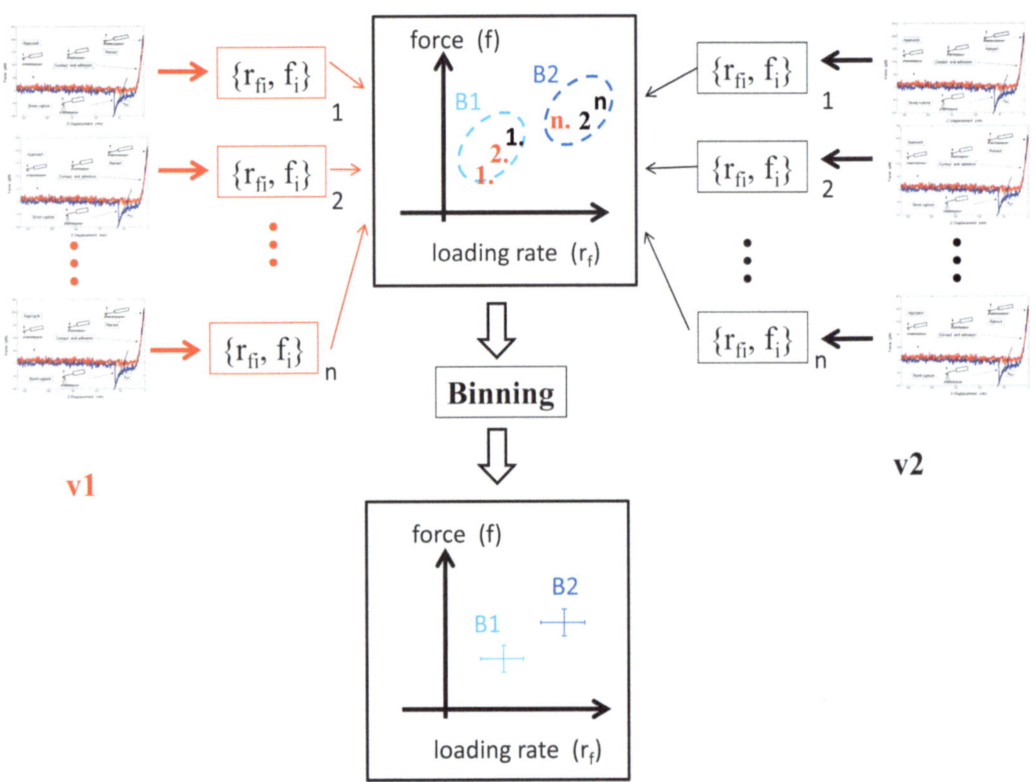

Fig. 6. Analysis of force spectroscopy data using binning method. Unbinding force (f_i) and the corresponding loading rate (r_{fi}) were determined directly from each force–displacement curves, regardless of which pulling speed was applied in the experiments. Binning method (7, 8) was applied to group the scattered data points $\{f_i, r_{fi}\}$. The corresponding peak (f^*) or the mean force ($\bar{f}$) and the associated loading rate in each bin was further mapped into the dynamic force spectroscopy and subsequently analyzed using Bell–Evans model (12, 13). $v1$ and $v2$ are two different retraction velocities in the illustration.

rates between 100 and 1,000 pN/s and by binning windows of 500 pN/s for loading rates between 1,000 and 10,000 pN/s (see Note 8).

3. Calculate mean rupture force ($\overline{f}$) or peak force (f^*) from each bin. Use Bell–Evans model to analyze the force spectrum (12, 13). The model links the relationship of rupture force and loading rate (r_f):

$$f^* = \frac{k_B T}{x_b} f_b \ln\left[\frac{x_b r_f}{k_{off}^0 k_B T}\right] \quad (1)$$

or

$$\overline{f} = \frac{k_B T}{x_b} \exp\left(\frac{k_{off}^0 k_B T}{x_b r_f}\right) E_i\left(\frac{k_{off}^0 k_B T}{x_b r_f}\right), \quad (2)$$

where k_{off}^0 is the unstressed off-rate, x_b is the reactive compliance of the molecules along force-driven dissociation path, k_B is the Boltzmann constant, T is the absolute temperature, and $E_i(z) = \int_z^{\infty} t^{-1} e^{-t} dt$ is the exponential integral.

4. Extract the kinetic parameters such as unstressed dissociation rate (k_{off}^0) and reactive compliance (x_b) for the molecular interactions by fitting the force spectroscopy data using Eq. (1) or (2). These parameters characterize the binding interactions at the single molecule level.

The above data analysis is suitable for molecular interactions that follows single-step energy activation barrier process (e.g., Claudin-1/Claudin-1 interactions (14)). In the case of molecular interactions that involve multiple dissociation barriers such as Claudin-2/Claudin-2 interactions (11), each barrier would have distinct kinetic parameters (i.e., k_{off}^0 and x_b) to be characterized. In this case, multiple curves with different gradient of force versus loading rate (in logarithm scale) are observed in different regimes of force spectrum in SMFS. For each regime (corresponding to an energy barrier), data analysis and extraction of kinetic parameters can be conducted using Bell–Evans model as described before.

4. Notes

1. L-Fibroblasts have been chosen because they do not normally express claudins or E-cadherins. Furthermore, the adhesion forces between wild-type L-fibroblasts is very low so that increase in cell adhesion due to any transfected proteins is easily detected.

2. It should be first confirmed (e.g., using immunostaining) that the transfected protein localizes well to the cell membrane when the cell is in a suspended state.
3. Preparation of cell suspension is dictated by the type of protein that is to be studied. If the protein is cleaved by trypsin (e.g. E-cadherin), cells can be obtained by gentle scraping followed by vigorous pipetting to break cell clumps. Claudins and occludin have been reported to be resistant to trypsin.
4. There are a few important points to take note while chemically coupling a protein to the AFM tip. Firstly, it should be confirmed that the recombinant protein is functional. Secondly, the coupling should be performed in such a manner that the interacting domain of the protein is free. For this reason, it is a good idea to use fusion proteins (e.g., Fc, GST, or His tag coupled to the protein of interest). Coupling the protein of interest utilizing antibodies to these tags would ensure that the interacting domain of the protein of interest is free. Finally, it should be ensured that the coupling is much stronger than the interactions being probed to avoid erroneous results.
5. The treatment of AFM tips in this way is bound to damage the gold coating on the cantilever surface leading to poor reflection of the laser over time. It is preferable not to use the tips more than two to three times. If the laser is not reflecting from the cantilever surface well (the photodetector sum will be very low), the tip should be changed.
6. As in any experiment, control experiments using tips without the protein of interest are essential for proper characterization of the molecular interactions. Since it is impossible to get rid of all nonspecific interactions, it is necessary to compare the frequency of rupture events in control experiments and test experiments. Contact force and contact duration have to be optimized such that few interactions are observed in the control experiments. Typically, while the test experiments show a frequency of rupture events close to ~20–30%, control experiments show a frequency of rupture events <5%.
7. Since there are thousands of force curves to be analyzed, automatic analysis using software routines is necessary. Most groups working on SMFS use custom written codes to extract the required data from the force curves. There are also several open source programs for this purpose (e.g., http://code.google.com/p/hooke/).
8. Alternative to the binning-based method is the velocity-based method to fit the obtained data points to Bell–Evans model. In the velocity-based method, the most probable rupture force and loading rate for a given retraction velocity is computed and analyzed (9).

References

1. Sung, K.L., L.A. Sung, M. Crimmins, S.J. Burakoff and S. Chien (1986) Determination of junction avidity of cytolytic T cell and target cell. Science. 234(4782): p. 1405–8.
2. Evans, E., K. Ritchie and R. Merkel (1995) Sensitive force technique to probe molecular adhesion and structural linkages at biological interfaces. Biophys J. 68(6): p. 2580–7.
3. Martinez-Rico, C., F. Pincet, E. Perez, J.P. Thiery, K. Shimizu, Y. Takai and S. Dufour (2005) Separation force measurements reveal different types of modulation of E-cadherin-based adhesion by nectin-1 and -3. J Biol Chem. 280(6): p. 4753–60.
4. Chu, Y.S., W.A. Thomas, O. Eder, F. Pincet, E. Perez, J.P. Thiery and S. Dufour (2004) Force measurements in E-cadherin-mediated cell doublets reveal rapid adhesion strengthened by actin cytoskeleton remodeling through Rac and Cdc42. J Cell Biol. 167(6): p. 1183–94.
5. Vedula, S.R.K., T.S. Lim, P.J.Kausalya, E.B. Lane, G. Rajagopal, W. Hunziker and C.T. Lim (2009) Quantifying forces mediated by integral tight junction proteins in cell–cell adhesion. Exp Mech. 49(1): p. 3–9.
6. Hanley, W., O. McCarty, S. Jadhav, Y. Tseng, D. Wirtz and K. Konstantopoulos (2003) Single molecule characterization of P-selectin/ligand binding. J Biol Chem. 278(12): p. 10556–61.
7. Hanley, W.D., D. Wirtz and K. Konstantopoulos (2004) Distinct kinetic and mechanical properties govern selectin-leukocyte interactions. J Cell Sci. 117(Pt 12): p. 2503–11.
8. Panorchan, P., M.S. Thompson, K.J. Davis, Y. Tseng, K. Konstantopoulos and D. Wirtz (2006) Single-molecule analysis of cadherin-mediated cell-cell adhesion. J Cell Sci. 119(Pt 1): p. 66–74.
9. Vedula, S.R.K., T.S. Lim, W. Hunziker and C.T. Lim (2008) Mechanistic insights into physiological functions of cell adhesion proteins using single molecule force spectroscopy. Mol Cell Biomech. 5(3): p. 169–82.
10. Hinterdorfer, P. and Y.F. Dufrene (2006) Detection and localization of single molecular recognition events using atomic force microscopy. Nat Methods. 3(5): p. 347–55.
11. Lim, T.S., S.R. Vedula, W. Hunziker and C.T. Lim (2008) Kinetics of adhesion mediated by extracellular loops of claudin-2 as revealed by single-molecule force spectroscopy. J Mol Biol. 381(3): p. 681–91.
12. Bell, G.I. (1978) Models for the specific adhesion of cells to cells. Science. 200(4342): p. 618–27.
13. Evans, E. and K. Ritchie (1997) Dynamic strength of molecular adhesion bonds. Biophys J. 72(4): p. 1541–55.
14. Lim, T.S., S.R. Vedula, P.J. Kausalya, W. Hunziker and C.T. Lim (2008) Single-molecular-level study of claudin-1-mediated adhesion. Langmuir. 24(2): p. 490–5.
15. Shao, J.Y., G. Xu and P. Guo (2004) Quantifying cell-adhesion strength with micropipette manipulation: principle and application. Front Biosci. 9: p. 2183–91.
16. Vedula, S.R., T.S. Lim, W. Hunziker and C.T. Lim (2009) Mechanistic insights into physiological functions of cell adhesion proteins using single molecule force spectroscopy. Mol. Cell Biomech. 111(1): p. 1–14.
17. Lim, C.T., E.H. Zhou, A. Li, S.R.K. Vedula and H.X. Fu (2006) Experimental techniques for single cell and single molecule biomechanics. Mater Sci Eng C. 26(8): p. 1278–1288.

Chapter 7

Detection of Tight Junction Barrier Function In Vivo by Biotin

Lei Ding, Yuguo Zhang, Rodney Tatum, and Yan-Hua Chen

Abstract

Tight junctions (TJs) are the most apical component of the junctional complexes in mammalian epithelial cells and form selective paracellular barriers restricting the passage of solutes and ions across the epithelial sheets. Claudins, a TJ integral membrane protein family, play a critical role in regulating paracellular barrier permeability. In the in vitro cell culture system, transepithelial electrical resistance (TER) measurement and the flux of radioisotope or fluorescent labeled molecules with different sizes have been widely used to determine the TJ barrier function. In the in vivo system, the tracer molecule Sulfo-NHS-Biotin was initially used in *Xenopus* embryo system and subsequently was successfully applied to a number of animal tissues in situ and in different organisms under the experimental conditions to examine the functional integrity of TJs by several laboratories. In this chapter, we will describe the detailed procedures of applying biotin as a paracellular tracer molecule to different in vivo systems to assay TJ barrier function.

Key words: Tight Junctions, Permeability barrier, Biotin tracer, Claudin-7, Epithelial cells

1. Introduction

Tight junction (TJ) is a gatekeeper of paracellular space between epithelial cells and prevents macromolecules and pathogens from entering the tissues while allowing the passive entry of ions and small molecules. Ample studies have reported that claudins, TJ integral membrane proteins, are essential for TJ barrier function (1–7). To study the TJ barrier function in vivo, Sulfo-NHS-Biotin has been commonly used as a tracer molecule. Biotins are small water-soluble molecules with their molecular weight ranging from 443 to 666 Da. Biotins are membrane impermeable reagents and allow efficient labeling of proteins and primary amine-containing macromolecules on the cell surface. Biotin reagents will not diffuse through the intercellular space if TJ is intact.

Kursad Turksen (ed.), *Claudins: Methods and Protocols*, Methods in Molecular Biology, vol. 762,
DOI 10.1007/978-1-61779-185-7_7, © Springer Science+Business Media, LLC 2011

However, if TJ structure/function is disrupted, the Biotin molecule will penetrate into the intercellular space. Sulfo-NHS-Biotin was first used in the early *Xenopus* embryo system by Chen et al. in 1997 to examine TJ integrity and barrier function after microinjecting various occludin deletion constructs into the 8-cell stage of blastomere (8). Subsequently, Furuse et al. (1) successfully applied it to the claudin-1-deficient mice to detect the epidermal barrier function. Nitta et al. (9) used this surface biotinylation technique to study blood brain barrier in claudin-5-deficient mice. TJ permeability assay using Sulfo-NHS-Biotin was also performed in pathogen-infected intestinal epithelia (10). More recently, Jeong et al. demonstrated that the blood brain barrier of adult zebrafish is functionally and molecularly similar to that of higher vertebrates by using Sulfo-NHS-Biotin and other molecular tracers and markers (11). Thus, it is clear that Biotin can be successfully applied to many different in vivo systems to examine TJ permeability barrier. In this chapter, we will describe the step-by-step protocol of applying Sulfo-NHS-Biotin to the *Xenopus* embryo system and to our recently developed claudin-7 knockout mouse to detect TJ barrier function in vivo.

2. Materials

2.1. Biotin Detection in Xenopus Embryo

1. Preparation for 25× MMR (Marc's Modified Ringer's) solution: For a total of 250 ml solution, add 36.525 g NaCl (2.5 M), 0.93 g KCl (50 mM), 0.75 g $MgSO_4$ (25 mM), 1.84 g $CaCl_2$ (50 mM), 1.25 ml of 0.5 M EDTA (2.5 mM), and 7.447 g HEPES (125 mM). The working solution is 0.1× MMR. To make the working solution, add 2 ml of 25× MMR to 498 ml dH_2O, and adjust pH to 7.8 using 1N NaOH.
2. 6-well and 60-mm petri dishes, forceps, plastic and glass transfer pipettes (see Note 1).
3. A refrigerated immersion cooler (Harvard Apparatus, Catalog # 724900).
4. To prepare 1 mg/ml EZ-Link Sulfo-NHS-Biotin (Pierce Chemical Co., Molecular weight: 443.43; Catalog # 21217), take Sulfo-NHS-Biotin out from –20°C freezer and let it sit at room temperature for 15 min before opening the cap. Dissolve Biotin in 0.1× MMR+10 mM HEPES to make 1 mg/ml solution and cool it at 10°C.
5. Preparation for 5% paraformaldehyde fixation solution: dissolve 1.5 g paraformaldehyde in 15 ml H_2O, bring up to 19.5 ml, and then add 10.5 ml of 0.2 M sodium cacodylate.

6. Preparation for blocking buffer: make 1% fish skin gelatin (Sigma, Catalog # G7765) and 1% BSA (Sigma, Catalog # A8022) in PBS.
7. RITC-avidin (Pierce Chemical Co.) was diluted 1:500 in blocking buffer containing 1% fish skin gelatin and 1% BSA in PBS.
8. Tissue-Tek O.C.T. compound (VWR International, Catalog # 25608-930).
9. 2-Methyl butane (Fisher Scientific, Catalog # O3551) and liquid nitrogen.
10. Superfrost Plus slides (Fisher Scientific, Catalog # 12-550-15).
11. Cryostat Microtome (Microm HM 505, Richard-Allan Scientific, Kalamazoo, MI).
12. Zeiss Axioskop fluorescence microscope (Carl Zeiss, Inc., Thornwood, NY).

2.2. Biotin Detection in Claudin-7 Knockout Mice

2.2.1. Biotin Injection

1. One-week old claudin-7 knockout (KO) mice recently generated in this laboratory were used for the biotin permeability assay.
2. Low pressure syringe pump (Harvard Apparatus, Catalog # 55-2219).
3. 1-ml syringe with 25G needle (see Note 2) connected to a polyethylene tubing (PE 50, VWR, Catalog # 427411).
4. Stainless steel injection needle (30G, Small Parts, Catalog # HTX-30R) connected to a polyethylene tubing (PE 10, VWR, Catalog # 427401). The other end of this tubing is inserted into the end wall of above PE 50 tubing connected with the 25G needle. The connecting site of two tubings is sealed by parafilm.
5. Ketamine (18 mg/ml) and Xylazine (2 mg/ml) mixture was made from the stock solution of 100 mg/ml of Ketamine and 100 mg/ml of Xylazine and diluted with physiological saline (0.9% NaCl).
6. EZ-Link Sulfo-NHS-LC-Biotin was obtained from Pierce Chemical Co (Molecular weight: 556.59; Catalog # 21335) and stored in –20°C freezer with desiccant. Dissolve 15 mg Sulfo-NHS-Biotin in 2.97 ml of PBS (pH 7.5) and 0.03 ml of 100 mM $CaCl_2$. This gives a final Biotin concentration of 5 mg/ml in PBS containing 1 mM $CaCl_2$ (see Note 3).

2.2.2. Biotin Detection

1. Tissue-Tek O.C.T. compound (VWR International, Catalog # 25608-930) for tissue embedding.
2. Disposable base mold (VWR International, Catalog # 60872-488) as tissue embedding container.

3. 2-Methyl butane (Fisher Scientific, Catalog # O3551) for freezing the tissue.
4. Superfrost Plus slides (Fisher Scientific, Catalog # 12-550-15) for mounting tissue sections.
5. A Cryostat Microtome (Microm HM 505, Richard-Allan Scientific, Kalamazoo, MI) for cutting frozen sections.
6. Ethanol and acetone (obtained from Fisher Scientific) for fixing the tissue sections on the slides.
7. Phosphate Buffered Saline (PBS): Prepare 10× stock with 1.38 M NaCl (80 g/L), 26.67 mM KCl (2 g/L), 80.6 mM $Na_2HPO_4{\cdot}7H_2O$ (21.6 g/L), 14.71 mM KH_2PO_4 (2 g/L), and adjust pH to 7.4 using 1N NaOH if necessary. Make up to 1 L with dH_2O and autoclave it before stored at room temperature. Prepare working solution by 1:10 dilution of 10× stock with dH_2O.
8. Blocking solution: 5% BSA in 1× PBS.
9. TEXAS RED-conjugated Streptavidin (CALBIOCHEM, Catalog # 189738).
10. The ProLong Antifade reagents (Molecular Probes, Inc., Catalog # P7481).
11. A Zeiss Axiovert S100 microscope (Carl Zeiss, Inc., Thornwood, NY) equipped with Metamorph Imaging Software (Molecular Devices, Downingtown, PA).

3. Methods

Both occludin and claudins are tetraspan membrane proteins localized at TJs of epithelial cells. Expression of truncated occludin mutants has been reported to cause malfunction of TJ functions in *Xenopus* embryos and cultured MDCK II cells (8, 12). Claudin-7 is a member of the claudin family and has been reported to be involved in modulating paracellular Cl^- permeability in cultures (6, 13–15). Our recently generated claudin-7 KO mice model showed that these mice displayed salt wasting, chronic dehydration, and distinct growth retardation phenotypes (16). Most of these pups died around 8–9 days after birth. Biotin, as a tracer molecule, has been used in both the above *Xenopus* embryo system and our claudin-7 KO mouse to detect TJ barrier function. We observed the biotin leakage through the paracellular space of *Xenopus* embryos injected with occludin mutant constructs as well as the renal tubular epithelial cells of 1-week old claudin-7 KO pups. The biotin leakage in kidney tubular epithelial cells may not be directly caused by claudin-7 deletion since it occurs in some of the KO pups later in their lives (such as 7/8 days), but not at a younger age.

3.1. Tight Junction Permeability Assay in Xenopus Embryos Using Surface Biotinylation Method

1. At the 8-cell stage, the embryos were injected with full length or C-terminally truncated occludin RNA constructs. After RNA injection, the embryos were incubated at room temperature for 6 h.
2. Turn on the refrigerated immersion cooler and keep the temperature of water in a container at 10°C.
3. Add 1 mg/ml Biotin solution to a 6-well dish and cool it at 10°C.
4. The embryos injected with WT and mutant occludin RNA constructs were placed into a 60-mm petri dish containing pre-cooled 0.1× MMR solution and kept at 10°C for 5 min (see Note 4).
5. The above embryos were carefully transferred into the 6-well petri dish containing pre-cooled 1 mg/ml NHS-LC-Biotin in 0.1× MMR solution using the L-shaped glass pipette and then were kept at 10°C for 12 min (see Note 5).
6. The embryos were washed twice with 0.1× MMR solution at 10°C and fixed in 5% paraformaldehyde overnight at 4°C. The fixed embryos were rinsed three times with PBS.
7. Cut the aluminum foil into about 3 cm^2 and fold it into a funnel shape. Add the TISSUE-TEK O.T.C. compound into this funnel-shaped small container half full and embed two embryos into it, and then add O.T.C. compound to 80% full of the funnel-shaped container. The samples were immediately frozen in 2-methyl butane/dry ice and then in liquid nitrogen.
8. The frozen blocks were trimmed and cut into 14-μm thick sections on a Cryostat Microtome. The frozen sections were put onto the Superfrost Plus slides.
9. The sections were dried at room temperature for 30 min and incubated with the blocking buffer (1% fish skin gelatin and 1% BSA in PBS) overnight or for at least 5 h at room temperature.
10. The sections and the control sections from the embryo without Biotin labeling were incubated with RITC-avidin diluted 1:500 in blocking buffer for 1 h at room temperature.
11. All slides were washed three times with blocking buffer and two times with PBS, and then mounted with mounting medium.

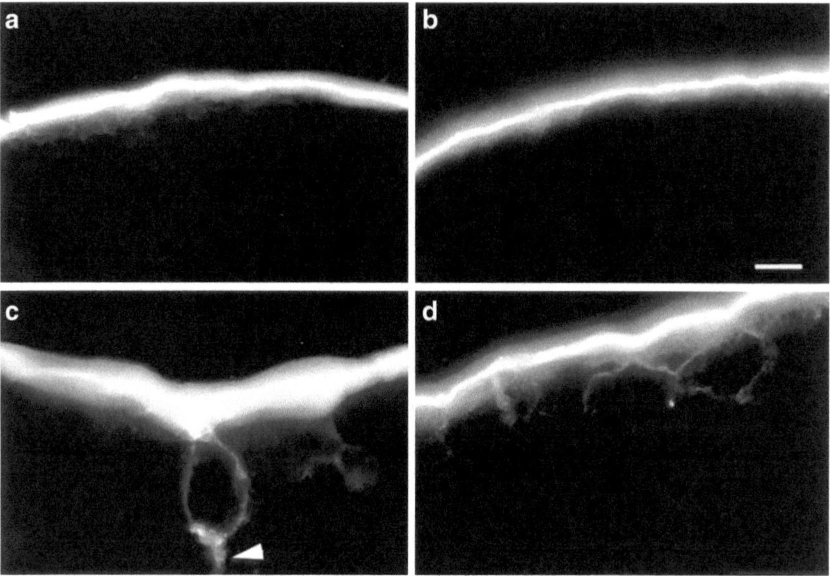

Fig. 1. Expression of mutant occludin disrupted the barrier function of TJ in *Xenopus* embryos. mRNAs transcribed from full-length or mutant occludins were microinjected into the antero-dorsal blastomere of eight-cell embryos. Six hours after injection (2,000 cell blastula), the embryos were labeled by incubation in 1 mg/ml NHS-LC-Biotin for 12 min at 10°C, then washed and fixed. Frozen sections were stained with RITC-avidin. The staining of NHS-LC-Biotin appeared as a thick continuous line on the surface of blastomeres. The TJs in the embryos injected with full-length (**a**, 504 amino acids), or the least COOH-terminally truncated (**b**, 486 amino acids) occludin mRNAs were impermeable to the biotin tracer. In contrast, in the embryos injected with two COOH-terminally truncated occludins (**c**, 386 amino acids, and **d**, 336 amino acids), the TJs were leaky and therefore, the biotin tracer penetrated into the intercellular spaces. The *arrowhead* (**c**) reveals that the membranes of internal cells beneath those at the embryonic surface have also been biotinylated, indicating that the tracer penetrated beyond the first tier of cells. *Bar*, 10 μm. (From Chen et al., J Cell Biol 1997; 138(4): 891–899 (8), by copyright permission of the Rockefeller University Press).

12. Sections were examined by epifluorescence using a Zeiss Axioskop microscope and photographed with TMAX-400 film (see an example in Fig. 1).

3.2. Biotin Injection Through the Cardiac Ventricle of Claudin-7 Wildtype and Knockout Mice

3.2.1. Injection Procedures

1. The vial containing Sulfo-NHS-Biotin powder was equilibrated to room temperature first before opening the cap. A small amount of Biotin was weighed at a time and directly dissolved in PBS (pH 7.5, Sulfo-NHS ester reacts with primary amines at pH 7.0–9.0) in the presence of 1 mM $CaCl_2$ to make 5 mg/ml Biotin solution.
2. One-week old claudin-7 wildtype (WT) and knockout (KO) pups were anesthetized by intraperitoneal injection with 0.05 ml/10 g of Ketamine (18 mg/ml) and Xylazine (2 mg/ml) mixture (see Note 6). After the pup was deeply anesthetized (see Note 7), thoracotomy was made to gain access to the thoracic cavity. All these procedures were approved by the East Carolina University (ECU) Animal Care and Use Committee and conducted in compliance with guidelines from the National Institute of Health and ECU on laboratory animal care and use.

3. The 1-ml syringe and its attached tubing were filled with 5 mg/ml of Sulfo-NHS-Biotin. The syringe was then connected to the low pressure pump and the injection speed was set at 100 μl/min. The 30G injection needle was inserted into the other end of the tubing.
4. The 30G injection needle with the blunt end was carefully inserted into the left ventricle of the heart and the biotin solution was perfused into the tissues through the blood stream (see Note 8). Five minutes after receiving 150 μl/g (body weight) Biotin injection, the pup was sacrificed by decapitation and the kidneys were removed from the body.
5. The kidneys were immediately rinsed with PBS and placed into a base mold containing O.C.T. compound. This base mold was half immersed in 2-methyl butane on dry ice in a container until the O.C.T. compound was completely frozen (change the color into white) and then transferred to the liquid nitrogen.
6. The O.C.T. blocks were trimmed and sectioned on a cryostat. The 5 μm frozen sections were mounted on superfrost plus slides and moved to the next step (see Note 9) or stored at −80°C until needed.

3.2.2. Biotin Detection Using Fluorescent Light Microscopy

1. The frozen sections of mouse kidneys mounted on slides were thawed out at room temperature for 5 min and then fixed in 95% ethanol at 4°C for 20 min followed by 100% acetone at room temperature for 5 min.
2. After fixation, the sections were rinsed three times in PBS and blocked with 5% BSA in PBS (blocking buffer) for 50 min.
3. After removing the blocking buffer, the sections were incubated with TEXAS RED-conjugated Streptavidin (1:200 dilution in the blocking buffer) for 30 min and then washed three times each with the blocking buffer for 5 min and briefly rinsed with PBS twice before adding the mounting medium.
4. Kidney sections from claudin-7 WT and KO pups without biotin injection were also incubated with TEXAS RED-conjugated Streptavidin to serve as a negative control as well as to test for endogenous biotin reactivity.
5. One hour before the slides were ready, powdered Prolong antifade reagent (Component A) and Prolong mounting medium (Component B) were taken from −20°C freezer (see Note 10). Approximately 1 ml of Component B was added to one of the brown vials containing powdered Component A. The two components were mixed by very gently pipetting the mounting medium up and down. After the antifade reagent was no longer adhered to the sides of the vial, the vial was left at room temperature for about 30 min to eliminate the small air bubbles.
6. After the last rinse with PBS, the residual liquid on the slide was removed by a Kimwipe prior to coverslipping the sample.

The antifade reagent/mounting medium mixture was applied to the sections with a 200-μl pipette. The sections were then covered by a coverslip. The slide was dried on a flat surface in the dark for at least 2–3 h before being viewed.

7. The sections were observed and photographed with an Axiovert fluorescent microscope. The images were taken using Metamorph Imaging Software as shown in Fig. 2.

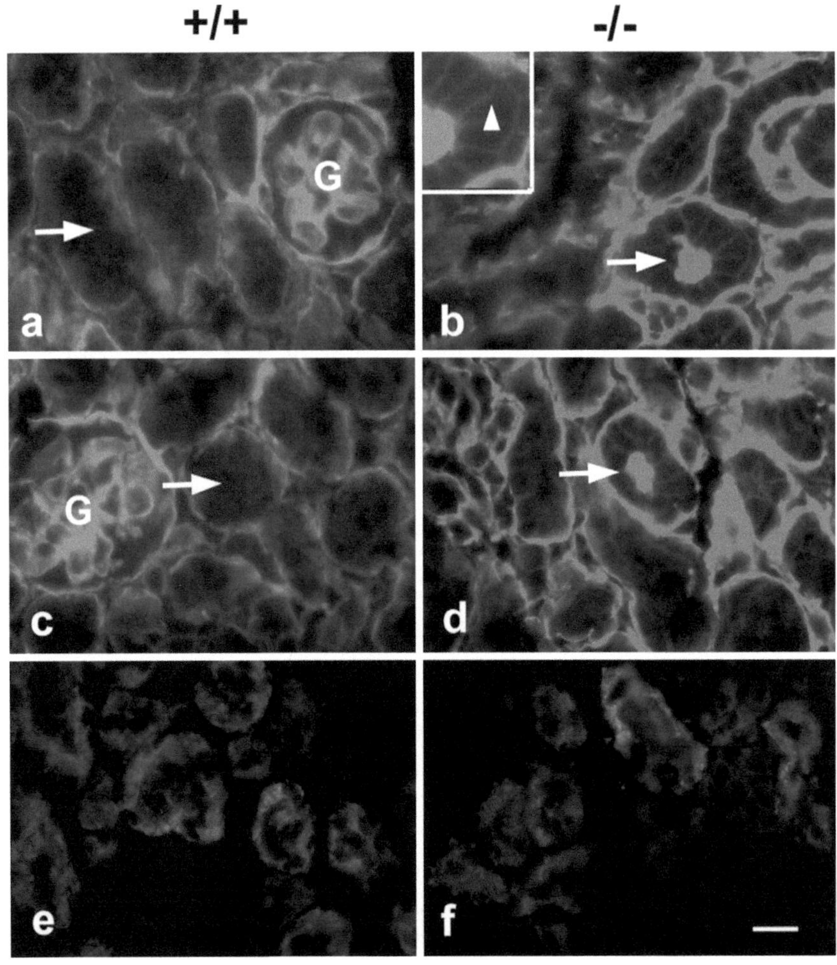

Fig. 2. Leakage of tight junctions in renal tubules of claudin-7 KO mice. Two 7/8-day old claudin-7 KO pups (**b** and **d**) and their two WT littermates (**a** and **c**) were anesthetized by intraperitoneal injection of Ketamine/Xylazine mixture. After deep anesthesia was achieved, Sulfo-NHS-LC-Biotin was perfused through the left ventricle of the heart. Five minutes after biotin perfusion, the pups were sacrificed and kidneys were removed from the body. Five-micrometer frozen sections were fixed in 95% ethanol at 4°C for 20 min and then fixed in 100% acetone at room temperature for 5 min. Sections were incubated with blocking buffer (5% BSA in PBS) for 50 min before incubating with TEXAS RED-conjugated Streptavidin for 30 min at room temperature. The distribution of injected biotin tracer was visualized by immunofluorescence microscopy. Images (**e**) and (**f**) were taken from WT and KO kidneys of the same litter without biotin injection to serve as a negative control for TEXAS RED-conjugated Streptavidin. All images were taken from the kidney cortex region and the *arrows* indicate the lumen of the renal tubule. The *arrowhead* in insert of image (**b**) points to the biotin labeling between tubular epithelial cells. **G** Glomerulus. *Bar*: 30 μm.

4. Notes

1. For an easy transfer of the embryos from one plate to another, use the glass pipette with its end bent into a 90° angle and polish its tip on the flame of a Bunsen Burner.
2. It is important to use the blunt end needle. The tip of the needle was cut by a Rotary Tool device (DREMEL, Racine, Wisconsin).
3. The Biotin solution needs to be freshly prepared for each use. Do not prepare stock solution for storage.
4. It is critical to keep the embryos at 10°C because at this temperature, the blastomeres will stop the cell division, which is required for the experiments, while allowing for the normal development of these embryos after biotin labeling procedures.
5. The 12-min labeling time was chosen since this was the maximum time at 10°C, which was 100% consonant with normal development of embryos to tadpole stages.
6. The correct dose of the drug is critical since overdose of the drug will kill the pup.
7. Whether the deep anesthesia is achieved or not can be tested by using a forceps to pinch the mouse tail. If there is no response after a sharp pinch, this means that a deep anesthesia is achieved.
8. To test whether the perfusion system is working or not, the syringe and its attached tubing can first be filled with 1% Evans Blue diluted in 0.9% NaCl. Five minutes after injecting the Evans Blue into the left ventricle of the heart, the skin of the whole body of the pup will turn blue. This indicates that the perfusion system is working.
9. To prevent the sections from coming off the slides, it is necessary to dry the slides at room temperature for about 30 min to make sure that the sections stick well to the slides.
10. If the Prolong mounting medium turns milky or is too viscous to manipulate, it should be placed in a 50°C water bath for 1 h before mixing it with Component A.

Acknowledgments

We would like to thank Joani T. Zary and Beverly G. Jeansonne for their technical assistance. Part of this work was supported by the North Carolina Biotechnology Center grant and the National Institutes of Health grant HL085752 to Y.H.C.

References

1. Furuse, M., Hata, M., Furuse, K., Yoshida, Y., Haratake, A., Sugitani, Y., et al. (2002) Claudin-based tight junctions are crucial for the mammalian epidermal barrier: a lesson from claudin-1-deficient mice. *J Cell Biol.* **156**, 1099–1111.
2. Colegio, O. R., Van Itallie, C. M., McCrea, H. J., Rahner, C., and Anderson, J. M. (2002) Claudins create charge-selective channels in the paracellular pathway between epithelial cells. *Am J Physiol Cell Physiol.* **283**, C142–147.
3. Turksen, K., and Troy, T. C. (2002) Permeability barrier dysfunction in transgenic mice overexpressing claudin 6. *Development.* **129**, 1775–1784.
4. Amasheh, S., Meiri, N., Gitter, A. H., Schoneberg, T., Mankertz, J., Schulzke, J. D., et al. (2002) Claudin-2 expression induces cation-selective channels in tight junctions of epithelial cells. *J Cell Sci.* **115**, 4969–4976.
5. Yu, A. S., Enck, A. H., Lencer, W. I., and Schneeberger, E. E. (2003) Claudin-8 expression in Madin-Darby canine kidney cells augments the paracellular barrier to cation permeation. *J Biol Chem.* **278**, 17350–17359.
6. Alexandre, M. D., Lu, Q., and Chen, Y. H. (2005) Overexpression of claudin-7 decreases the paracellular Cl^- conductance and increases the paracellular Na^+ conductance in LLC-PK1 cells. *J Cell Sci.* **118**, 2683–2693.
7. Hou, J., Paul, D. L., and Goodenough, D. A. (2005) Paracellin-1 and the modulation of ion selectivity of tight junctions. *J Cell Sci.* **118**, 5109–5118.
8. Chen, Y., Merzdorf, C., Paul, D. L., and Goodenough, D. A. (1997) COOH terminus of occludin is required for tight junction barrier function in early *Xenopus* embryos. *J Cell Biol.* **138**, 891–899.
9. Nitta, T., Hata, M., Gotoh, S., Seo, Y., Sasaki, H., Hashimoto, N., et al. (2003) Size-selective loosening of the blood-brain barrier in claudin-5-deficient mice. *J Cell Biol.* **161**, 653–660.
10. Guttman, J. A., Li, Y., Wickham, M. E., Deng, W., Vogl, A. W., and Finlay, B. B. (2006) Attaching and effacing pathogen-induced tight junction disruption in vivo. *Cell Microbiol.* **8**, 634–645.
11. Jeong, J. Y., Kwon, H. B., Ahn, J. C., Kang, D., Kwon, S. H., Park, J. A., et al. (2008) Functional and developmental analysis of the blood-brain barrier in zebrafish. *Brain Res Bull.* **75**, 619–628.
12. Balda, M. S., Whitney, J. A., Flores, C., Gonzalez, S., Cereijido, M., and Matter, K. (1996) Functional dissociation of paracellular permeability and transepithelial electrical resistance and disruption of the apical-basolateral intramembrane diffusion barrier by expression of a mutant tight junction membrane protein. *J Cell Biol.* **134**, 1031–1049.
13. Hou, J., Gomes, A. S., Paul, D. L., and Goodenough, D. A. (2006) Study of claudin function by RNA interference. *J Biol Chem.* **281**, 36117–36123.
14. Alexandre, M. D., Jeansonne, B. G., Renegar, R. H., Tatum, R., and Chen, Y. H. (2007) The first extracellular domain of claudin-7 affects paracellular Cl^- permeability. *Biochem Biophys Res Commun.* **357**, 87–91.
15. Tatum, R., Zhang, Y., Lu, Q., Kim, K., Jeansonne, B. G., and Chen, Y. H. (2007) WNK4 phosphorylates ser(206) of claudin-7 and promotes paracellular Cl^- permeability. *FEBS Lett.* **581**, 3887–3891.
16. Tatum, R., Zhang, Y., Salleng, K., Lu, Z., Lin, J. J., Lu, Q., et al. Renal salt wasting and chronic dehydration in claudin-7-deficient mice. *Am J Physiol Renal Physiol.* **298**, F24–34.

Chapter 8

The Coculture Method to Examine Interactions Between Claudin Isoforms in Tight Junction-Free HEK293 Cells and Tight Junction-Bearing MDCK II Cells

Tetsuichiro Inai

Abstract

The paracellular transport of water, ions, and small solutes is regulated by tight junctions (TJs) in epithelial, endothelial, and mesothelial cells. Both the prolonged increase and decrease of the paracellular permeability are involved in various diseases. Claudins, a family of at least 24 integral membrane proteins in TJs, are major components of TJs and usually more than two claudin isoforms are found in TJs. The combination and mixing ratios of claudin isoforms determine the paracellular permeability in TJs. To create the paracellular permeability barrier, claudins must interact laterally in one membrane (*cis*-interaction) and also must interact by head-to-head binding between adjacent cells (*trans*-interaction). Therefore, examination of claudin–claudin interactions provides insights into the mechanism of regulation of the paracellular permeability. This study introduced coculture systems using TJ-bearing MDCK II cells and TJ-free HEK293 cells to examine claudin–claudin interactions.

Key words: Tight junction, Claudin, *Trans*-interaction, *Cis*-interaction, Coculture, MDCK cells, HEK293 cells

1. Introduction

Tight junctions (TJs) regulate the paracellular transport of water, ions, and small solutes in epithelial, endothelial, and mesothelial cells (1). Claudins, a family of at least 24 integral membrane proteins in TJs in mice and humans, reconstitute TJ strands in fibroblasts (2). Claudins interact laterally in one membrane (homomeric and heteromeric *cis*-interactions) and also interact by head-to-head binding, probably via the two extracellular loops of claudins between adjacent cells (homotypic and heterotypic *trans*-interactions) (3–8). The exogenous expression of claudin-2 in

Kursad Turksen (ed.), *Claudins: Methods and Protocols*, Methods in Molecular Biology, vol. 762,
DOI 10.1007/978-1-61779-185-7_8, © Springer Science+Business Media, LLC 2011

MDCK I cells markedly decreased transepithelial electrical resistance (9). MDCK I cells, high-resistance clone, express claudin-1, -3, -4, and-7 but not claudin-2, while MDCK II cells, low-resistance clone, do express claudin-2 in addition to claudin-1, -3, -4, and-7. The coculture of fibroblasts expressing either claudin-1 or claudin-2 showed that claudin-2 does not interact with claudin-1 (5). These results suggest that the combination and mixing ratios of claudin isoforms may determine the paracellular permeability. The expression levels of claudins are altered in many types of tumor cells in comparison to normal cells and these alterations may be involved in tumor progression. The modulation of the constituent of claudins in tumor cells may alter the microenvironment around them by affecting the paracellular permeability (10). The expression of incompatible claudins, for example claudin-2 in MDCK I cells, increases in tumor cells, and growth factors may easily reach their receptors in basolateral membranes of tumor cells because of the increased paracellular permeability. Therefore, examination of the interactions between claudin isoforms provides insights into the mechanism of regulation of the paracellular permeability and tumor progression.

The coculture of claudin-null cells, which exogenously expressed a distinct claudin isoform, reveals the occurrence of heterotypic *trans*-interactions between claudin isoforms (4, 5, 11). The coculture of epithelial cells (usually expressing more than two claudin isoforms), which exogenously express a claudin isoform or not, reveals the occurrence of interactions between endogenous claudins plus the exogenously expressed claudin and endogenous claudins at the heterotypic cell–cell contacts (6, 8). We herein introduce two types of coculture system using TJ-bearing MDCK II cells and TJ-free HEK293 cells to examine the interactions between the claudin isoforms.

2. Materials

2.1. Cell Culture, Transfection, and Selection

1. 100× L-glutamine solution (200 mM; Gibco/BRL, Bethesda, MD).
2. Fetal bovine serum (FBS, Gibco/BRL).
3. Geneticin solution (Gibco/BRL): 50 mg/ml G418 sulfate in distilled water. Store in aliquots at −20°C.
4. Puromycin solution (InvivoGen, San Diego, CA): 10 mg/ml solution in HEPES buffer. Store in aliquots at −20°C.
5. Growth medium: Minimum Essential Medium Alpha (α-MEM; Invitrogen, Carlsbad, CA) supplemented with 2 mM L-glutamine, 10% FBS, and 100 μg/ml geneticin for HEK293 Tet-Off cells. Dulbecco's Modified Eagle's Medium

(DMEM with high glucose; Invitrogen) supplemented with 2 mM L-glutamine and 10% FBS, and 1 μg/ml puromycin for MDCK II Tet-Off cells.

6. Dulbecco's phosphate-buffered saline (DPBS; Invitrogen): Ca^{2+}- and Mg^{2+}-free PBS.
7. Trypsin–EDTA solution (Gibco/BRL): 0.25% trypsin, 1 mM EDTA·4Na in Hanks buffer.
8. Opti-MEM I reduced-serum medium (Gibco/BRL).
9. Lipofectamine 2000 (Invitrogen): Store at 4°C.
10. Doxycycline solution: Dissolve 20 mg doxycycline hydrochloride (Sigma, St. Louis, MO) in 10 ml distilled water (2 mg/ml solution) and sterile by Millex-LG (0.2-μm filter; Millipore, Bedford, MA). Store in aliquots at −20°C.
11. Hygromycin B solution (Gibco/BRL): 50 mg/ml in PBS. Store in aliquots at −20°C.
12. Cell culture dishes (Corning Incorporated, Corning, NY): Culture area of 35-mm, 60-mm, and 100-mm dishes and 15.6-mm well in 24-well plates are 8, 21, 55, and 1.9 cm^2, respectively. Add 0.2–0.3 ml medium per 1 cm^2 culture area.
13. Collagen solution: Dilute Cellmatrix Type I-C (Nitta Gelatin, Inc., Osaka, Japan) 10–30 fold with 1 mM HCl in distilled water (pH 3.0) sterilized by autoclaving and store at 4°C. For culture of HEK293 Tet-Off cells, prepare collagen-coated dishes and 10-well glass slides. Add collagen solution to cover the entire culture area of dishes and glass slides and incubate for 30–60 min at room temperature (RT). Remove the solution (the solution can be used several times) and completely dry the surface. Rinse twice with DPBS before use.
14. HEK293 Tet-Off Advanced cells (Clontech, Mountain View, CA): HEK cells, human embryonic kidney cells.
15. MDCK II Tet-Off cells (Clontech): MDCK II cells, Madin-Darby canine kidney cells (strain II).
16. pTRE and pTRE2hyg expression vectors (Clontech): The pTRE2hyg vector contains a hygromycin resistance gene, however, the pTRE vector does not contain any marker genes against selective antibiotics. Therefore, pTK-Hyg vector (Clontech) containing hygromycin resistance gene is cotransfected with pTRE vector. pTREM-EGFP-1CL, pTREM-EGFP-15CL, pTREM-15CL (8), and pTRE2hyg-RFP-2CL were used. The expression of EGFP-1CL, EGFP-15CL, 15CL, or RFP-2CL is induced by the removal of doxycycline (Dox) from growth medium when one of these expression vectors is stably transfected in MDCK II Tet-Off cells or HEK293 Tet-Off Advanced cells (see Note 1). pTREM,

modified pTRE vector whose multiple cloning site is altered; EGFP-1CL, mouse claudin-1 whose NH_2 terminus is tagged with enhanced green fluorescent protein (EGFP); EGFP-15CL, human claudin-15 whose NH_2 terminus is tagged with EGFP; 15CL, human claudin-15; RFP-2CL, mouse claudin-2 whose NH_2 terminus is tagged with red fluorescent protein (RFP).

2.2. Cell Cloning

1. Cloning rings (Minato Medical Corporation, Tokyo, Japan): Cloning rings, stainless cylinders (6 mm in inner diameter, 8 mm in outer diameter, and 10 mm high), are put on glass petri dishes coated with a thin layer of silicon grease and sterilized by autoclaving. Silicon grease is spread by a finger with latex gloves. Cloning rings can be reused by washing in a beaker containing xylene once and acetone twice in an ultrasonic bath.
2. Sterilized forceps.
3. Highly hydrophobic pattern print glass slide (10-well glass slide; Matsunami Glass Ind., Osaka, Japan): The glass slide has 10 wells (6 mm in diameter). The culture area of the well is 0.28 cm^2 and add 60 μl of growth medium per one well. There are three types of glass slides: non-coated, APS-coated, and MAS-coated types. Non-coated glass slides must be coated with collagen before use when HEK293 Tet-Off cells but not MDCK II Tet-Off cells are cultured.
4. Sterile plastic container (9.7 cm length × 13.7 cm width × 1.3 cm height; Eiken Chemical Co., Tokyo, Japan): Up to five 10-well glass slides can be incubated in this container for culture.

2.3. Cell Freezing

1. Dimethyl sulfoxide (DMSO): DMSO is light sensitive and should be kept in the dark. DMSO is used as a cryopreservative.
2. Freezing medium: Growth medium containing 10% (v/v) DMSO but not a selective antibiotic. Prepare just before use and keep it on ice.

2.4. Coculture

1. 10-well glass slides (see Subheading 2.2, item 3).
2. Sterile plastic container (see Subheading 2.2, item 4).

2.5. Immunofluorescent Staining

1. 10× Phosphate-buffered saline (PBS): Prepare 10× stock with 1.37 M NaCl, 27 mM KCl, 100 mM Na_2HPO_4, 20 mM KH_2PO_4 (adjust to pH 7.4 with HCl) in distilled water and autoclave before storage at RT. Prepare 1× solution by dilution of one part with nine parts distilled water.
2. 20% (w/v) paraformaldehyde (PFA) solution: Dissolve 20 g PFA in less than 70 ml distilled water at 60°C using a stirring

hot-plate in a fume hood. Keep temperature at 60°C with stirring and add 1N NaOH solution drop-wise (add one to two drops at once and wait for several minutes) with a Pasteur pipette. Usually ~10 drops of NaOH are necessary until the clouded solution becomes clear. Cooling down to RT. Dilute the solution to 100 ml with distilled water and filter out remained particles. The stock solution can be stored at 4°C up to 2 months. Prepare 1% PFA in PBS freshly by mixing 20% PFA (1 volume), 10× PBS (2 volume) and distilled water (17 volume) before use.

3. 1% (w/v) bovine serum albumin (BSA, Sigma) in PBS: This solution is used for blocking and dilution of primary and secondary antibodies.
4. 0.2% (v/v) Triton X-100 in PBS: Permeabilization solution
5. Primary antibodies: Rabbit anti-claudin-15 antibody and mouse anti-ZO-1 antibody (Zymed Lab., South San Francisco, CA) are diluted for use at a ratio of 1:200 and 1:1,000, respectively.
6. Secondary antibodies: Alexa Fluor 488 goat anti-rabbit Ig and Alexa Fluor 594 goat anti-mouse Ig (Molecular Probes, Eugene, OR) are diluted for use at a ratio of 1:400.
7. Vectashield mounting medium (Vector Laboratories, Burlingame, CA): Store in the dark at 4°C.
8. Coverslips (24×55×0.12–0.17 mm, Matsunami Glass Ind., Osaka Japan).
9. LSM 510 META confocal microscope (Carl Zeiss, Oberkochen, Germany).

3. Methods

Coculture of TJ-free HEK293 cells expressing claudin-X or -Y (X and Y are distinct claudin isoforms) reveals the occurrence of heterotypic *trans*-interactions between claudin-X and -Y, if both claudins are colocalized at the heterotypic cell–cell contacts. Although disturbance of *cis*-interaction may prevent claudin-X and -Y from concentrating at the heterotypic cell–cell contacts, it is very unlikely that the homomeric *cis*-interactions are disturbed. The occurrence of *cis*-interactions can be analyzed by fluorescence resonance energy transfer (FRET) analysis (7). MDCK II cells exogenously expressing claudin-X were cocultured with the parent cells which express at least claudin-1, -2, -3, -4, and -7 in coculture system using TJ-bearing MDCK II cells. These endogenous claudins affect the localization of claudin-X at both homotypic and heterotypic cell–cell contacts. Incompatibility of heteromeric *cis*-interactions between endogenous claudins and exogenous claudin-X can exclude claudin-X from TJs at both homotypic and heterotypic cell–cell contacts. The incompatibility of heterotypic *trans*-interactions between endogenous claudins and exogenous claudin-X might exclude claudin-X from TJs at the heterotypic cell–cell contacts, even though exogenous claudin-X could *cis*-interact with endogenous claudins. The explanation of the results obtained from coculture analysis using epithelial cells including MDCK II cells is complicated. However, it is worth examining the interactions between endogenous claudins and an exogenous claudin in TJ-bearing epithelial cells expressing various combinations of endogenous claudins.

Exogenously expressed claudin-15 could not be localized between the heterotypic cell–cell contacts in coculture using MDCK II Tet-Off cells (Fig. 1). *Trans*-interaction between claudin-2 and claudin-15 (Fig. 2) as well as between claudin-1 and claudin-15 (Inai et al., unpublished data) did not occur in a coculture using HEK293 Tet-Off cells. These incompatible *trans*-interactions between claudin-15 and endogenous claudins in MDCK II Tet-Off cells may result in the exclusion of exogenously expressed claudin-15 from the heterotypic cell–cell contacts in a coculture using MDCK II Tet-Off cells (Fig. 1).

3.1. Transfection and Selection

Warm growth medium, DPBS, and trypsin–EDTA at 37°C in a water bath before use. All tissue culture procedures must be manipulated in a clean bench to maintain sterility. The cells are passaged twice a week by splitting 1:10–20.

1. One-day before transfection, aspirate growth medium from a 100-mm culture dish containing cells at 80–90% confluence. Add 10 ml of DPBS (see Note 2) to rinse the cells and aspirate DPBS.
2. Add 1.0 ml of trypsin–EDTA and incubate the dish in a 37°C humidified incubator with 5% CO_2, observing the cultures periodically with a phase contrast microscope until the cells are rounded up (see Note 3).
3. Add 2 ml of new growth medium containing 10% serum to neutralize the trypsin, gently mix the rounded up cells with a 5-ml pipette and transfer the cell suspension into a 15-ml centrifuge tube. Count the cells with a hemocytometer and calculate the total number of cells.
4. Centrifuge the harvested cells at 1,000 rpm (180 × g) for 5 min to pellet the cells.
5. After aspirating the medium, add new growth medium to make 5×10^5 cells/ml or 3×10^5 cells/ml of cell suspension for HEK293 Tet-Off cells or MDCK II Tet-Off cells, respectively.

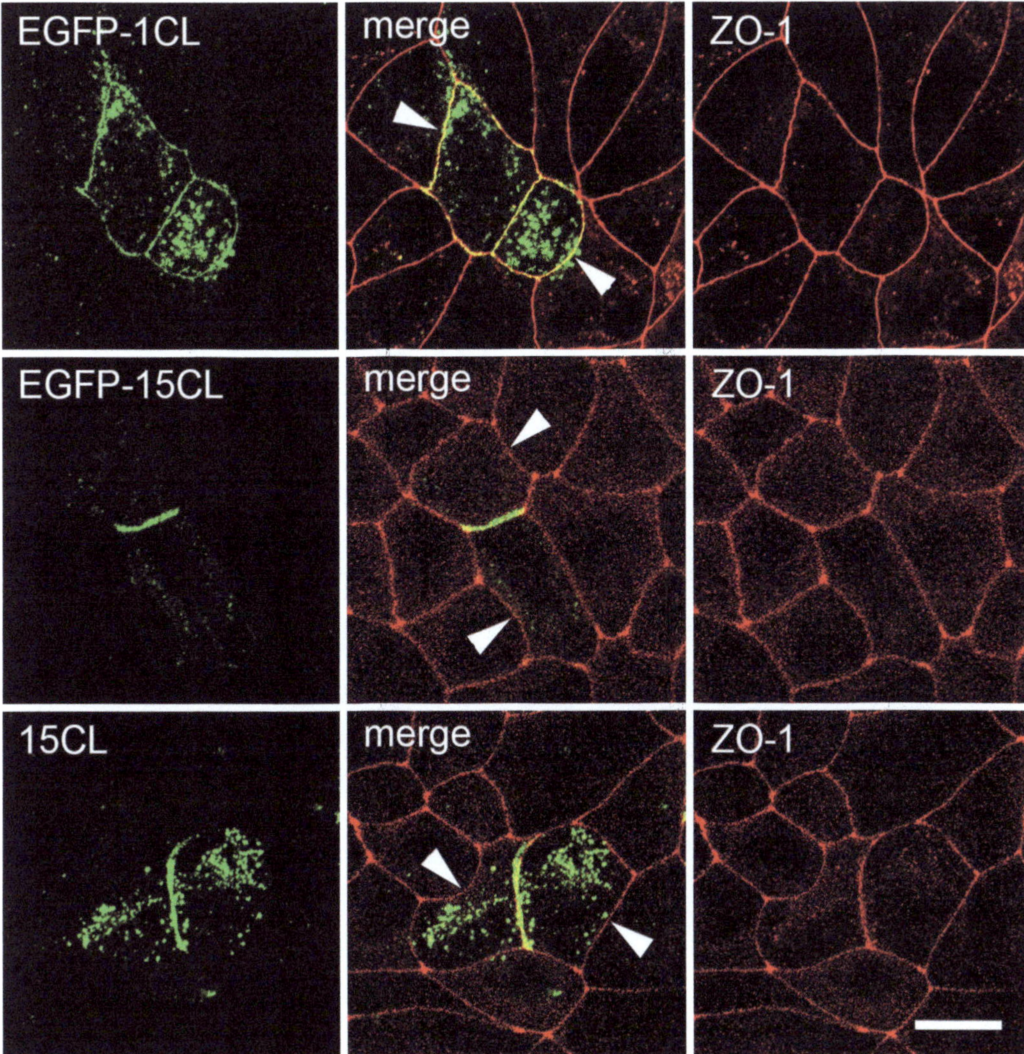

Fig. 1. Confocal laser scanning microscopy of MDCK II Tet-Off cells expressing EGFP-1CL, EGFP-15CL, or 15CL cocultured with the parent MDCK II Tet-Off cells. Cells expressing EGFP-1CL, EGFP-15CL, or 15CL were cocultured for 1 day with the parent cells at a ratio of 1:10. Cells were stained with anti-ZO-1 monoclonal antibody or with anti-claudin-15 polyclonal antibody and anti-ZO-1 monoclonal antibody. *Green signals* for EGFP-1CL, EGFP-15CL, or claudin-15, *red signals* for ZO-1, and the merged images are shown. EGFP-1CL is detected in the heterotypic cell–cell junctions (*arrowheads*) formed by EGFP-1CL-expressing cells and the parent cells in addition to the cell–cell junctions between EGFP-1CL-expressing cells. In contrast, EGFP-15CL is not detected in the heterotypic cell–cell junctions (*arrowheads*), but it is detected in the homotypic cell–cell junctions between EGFP-15CL-expressing cells. Similarly, 15CL is not detected in the heterotypic cell–cell junctions (*arrowheads*), but it is detected in the homotypic cell–cell junctions between the 15CL-expressing cells. *Bar*, 10 μm. (Reproduced partly from ref. 8 with permission from Springer).

6. Dispense 1, 1.5, or 2 ml of the cell suspension into 35-mm culture dishes containing 1, 0.5, or 0 ml of growth medium, respectively. Gently rock the dishes to evenly distribute the cells and culture for 1 day.

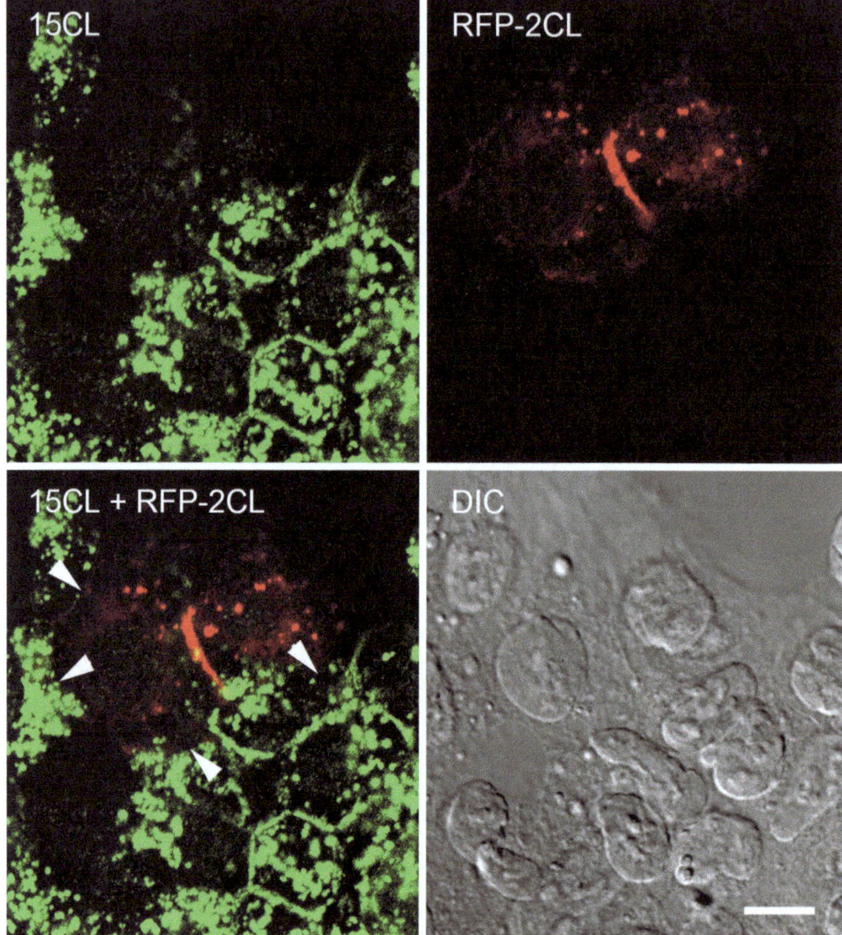

Fig. 2. Confocal laser scanning microscopy of HEK293 Tet-Off cells expressing 15CL cocultured with RFP-2CL-expressing cells. Cells expressing RFP-2CL were cocultured for 1 day with cells expressing 15CL at a ratio of 1:20. Cells were stained with anti-claudin-15 antibody and Alexa Fluor 488-labeled anti-rabbit Ig. *Green signals* for claudin-15, *red signals* for RFP-2CL, the merged image of 15CL and RFP-2CL, and differential interference contrast (DIC) image are shown. Both 15CL and RFP-2CL are detected in the homotypic cell–cell junctions. However, neither 15CL nor RFP-2CL is localized in the heterotypic cell–cell junctions (*arrowheads*). *Bar*, 10 μm.

7. On the day of transfection, select the dish containing 90–95% confluent cells.
8. Dilute 4.5 μg of plasmid DNA (see Note 4) into 250 μl of Opti-MEM I Reduced-Serum Medium without serum and mix gently.
9. Dilute 12 μl (HEK293 cells) or 18 μl (MDCK II cells) of Lipofectamine 2000 (LF2000) reagent into 250 μl of Opti-MEM I Reduced-Serum Medium without serum (see Note 5), mix gently, and incubate for 5 min at RT (see Note 6).
10. Combine the diluted DNA (from step 8) with the diluted LF2000 reagent (from step 9). Mix gently and incubate for 20 min at RT (see Note 7).

11. Aspirate growth medium, rinse cells once with 2 ml of DPBS, and add 1.5 ml of Opti-MEM I Reduced-Serum Medium (see Note 8).
12. Add 500 μl of DNA/LF2000 complexes drop-wise into the 35-mm dish and mix gently by rocking the dish back and forth.
13. Incubate the dish for 4 h in a CO_2 incubator.
14. Aspirate the medium containing the DNA/LF2000 complexes (see Note 9) and add 2 ml of new growth medium containing 10% FBS but not G418 and puromycin.
15. After 2–4 h, add 2 μl of 2 mg/ml Dox to suppress the expression of the gene of interest and mix gently by rocking the dish back and forth.
16. One day after transfection, rinse the cells with 2 ml of DPBS, add 0.2 ml of trypsin/EDTA, and incubate in a CO_2 incubator until cells are rounded up.
17. Add 1 ml of growth medium, mix gently with a 2-ml pipette, and transfer the cell suspension into a 15-ml centrifuge tube.
18. Centrifuge at 1,000 rpm (180×g) for 5 min to pellet the cells.
19. Aspirate the supernatant, add 2 ml (HEK293) or 4 ml (MDCK II) of growth medium (see Note 10) and mix gently with a 2-ml pipette.
20. Add 1 ml of the cell suspension to a 100-mm dish containing 9 ml of growth medium with 10 μl of 2 mg/ml Dox and mix gently by rocking the dish back and forth.
21. On the next day, add 20 μl (HEK293 Tet-Off cells) or 80 μl (MDCK II Tet-Off cells) of 50 mg/ml hygromycin per 10 ml medium. Replace the growth medium containing hygromycin and doxycyclin every 3 days.

3.2. Cell Cloning

The colonies are ready to harvest when they grow to 2–3 mm in diameter. The colonies can be seen by the naked eye as a white dot on the underside of the culture dish.

1. Circle the underside of the 100-mm dishes where the well-isolated colony is to be picked from.
2. Prepare 24-well plates with 0.5 ml of growth medium containing 20% FBS per well (see Note 11) but not hygromycin and Dox and incubate in a CO_2 incubator.
3. Aspirate the culture medium from the 100-mm dishes, add 10 ml of DPBS, and then aspirate DPBS.
4. Immediately place the cloning ring over the marked clones using forceps. The grease-coated end of the ring should be down against the culture surface so that a small amount of grease allows the ring to stick to the plate and creates a seal.

5. Add one drop (~50 μl) of trypsin–EDTA with a 2-ml pipette into the cloning ring and incubate in a CO_2 incubator until cells are rounded up.
6. Take 150 μl of medium from the well in the 24-well plate (from step 2) with a 200-μl pipettor, add it into the ring, and pipette the solution gently up and down five times with the 200-μl pipettor to suspend the detached cells completely.
7. Transfer 150 μl of the cell suspension by the 200-μl pipettor to the well in the 24-well plate (from step 2) for master cells and transfer the rest of the cell suspension (~50 μl) to the well in 10-well glass slides for immunofluorescent screening.
8. After cells adhere (2–4 h), aspirate medium and add 0.5 ml of growth medium containing hygromycin and Dox to the wells in the 24-well plates and 60 μl of growth medium containing hygromycin but not Dox to the wells in 10-well glass slides.
9. Cells cultured on 10-well glass slides without Dox are ready for immunofluorescent screening by 2–3 days after plating.
10. Change growth medium every 2 days.

3.3. Cell Freezing

After immunofluorescent screening (see Note 12), passage positive cells sequentially to 35-mm, 60-mm, and finally 100-mm dishes in growth medium containing Dox and hygromycin when cells are 80–95% confluent.

1. When positive clones are 90–95% confluent in a 100-mm dish, rinse cells once with 10 ml of DPBS, and add 1 ml of trypsin–EDTA.
2. Incubate the dish in a CO_2 incubator until cells are rounded up.
3. Add 2 ml of growth medium, pipette the solution gently up and down, and transfer the cell suspension to a 15-ml centrifuge tube. Count the cells with a hemocytometer and calculate the total number of cells.
4. Centrifuge at 1,000 rpm (180 × g) for 5 min and aspirate the supernatant.
5. Add ice-cold freezing medium to make 2×10^6 cells/ml of cell suspension and resuspend the cell pellet gently. Transfer the cell suspension into cryotubes (1 ml per vial) and freeze at –80°C (see Note 13). Transfer to liquid nitrogen after 24 h.

3.4. Coculture

1. Thaw vials containing frozen cells (2×10^6 cells/vial) in 37°C water bath with agitation and transfer the cells to 15-ml centrifuge tubes. Prepare claudin-X- and claudin-Y-expressing cells for coculture using HEK293 Tet-Off cells. Prepare the parent cells and claudin-X-expressing cells for coculture using MDCK II Tet-Off cells.

2. Centrifuge at 1,000 rpm for 5 min, aspirate the supernatant, add 5 ml of growth medium in the absence of Dox to express a gene of interest (see Note 14).
3. Plate the cell suspensions to 60-mm dishes and culture until cells are 80–95% confluent.
4. Wash cells once with DPBS, add 0.2 ml of trypsin–EDTA, and incubate the dishes in a CO_2 incubator until the cells are rounded up.
5. Add 1 ml of growth medium, gently mix the rounded up cells, and transfer the cell suspensions into 15-ml centrifuge tubes.
6. Centrifuge the harvested cells at 1,000 rpm (180×g) for 5 min to pellet the cells.
7. After aspirating the medium, add 3.6 ml of growth medium without Dox, if the 60-mm dish containing 80% confluent cells is used (see Note 15).
8. Mix 300 μl of one cell suspension and either 30 μl (10:1) or 15 μl (20:1) of another cell suspension for one 10-well glass slide.
9. Add 4, 8, 15, 30, and 60 μl of the mixed cell suspension in duplicate to the wells in a 10-well glass slide containing 56, 52, 45, 30, and 0 μl of growth medium without Dox, respectively. Prepare glass slides in duplicate and culture for 1 or 2 days to be processed for immunofluorescent staining.

3.5. Immuno-fluorescent Staining

The cells reach at 30–100% confluence in the wells of 10-well glass slides 2–3 days after cell cloning, and ready to be screened. Cocultured cells for 1 or 2 days are also processed for immunofluorescent staining. All incubation procedures should be done in a moist chamber in the dark if fluorescent proteins are expressed in cells.

1. Wash wells containing cells in 10-well glass slides once with two drops (~50 μl) of PBS by a Pasteur pipette.
2. Add ~50 μl of 1% PFA in PBS in wells and incubate for 10 min after aspirating PBS.
3. Wash twice with PBS briefly.
4. Add ~25 μl of 0.2% Triton X-100 in PBS in wells and incubate for 10 min.
5. Wash once with PBS, add ~25 μl of 1% BSA in PBS (BSA–PBS) in wells, and incubate for 15 min.
6. Aspirating BSA–PBA and add 20 μl of primary antibody in wells with a 20-μl pipettor and incubate for 1 h.
7. Wash twice with PBS briefly and further with three changes of PBS for 5 min each.

8. Add 20 μl of secondary antibodies corresponding to the primary antibodies in wells by a 20-μl pipettor and incubate for 30 min.
9. Wash with PBS as done in step 7.
10. Aspirate PBS from five wells at one time leaving a thin layer of PBS in the wells. Take 16 μl of Vectashield-mounting medium using a 20-μl pipettor and sequentially put a tiny drop, ~1 mm in diameter (~3 μl), of mounting medium on the surface of the five wells. The mounting medium easily spread over the entire surface of the well. Put the coverslips slowly on 10-well glass slides using forceps to prevent forming air bubbles. Put two-fold Kimwipe paper over the coverslip, hold down the long edges of both the coverslip and glass slide under the paper by fingers with latex gloves to prevent smashing cells by sliding the coverslip. Gently push the coverslips by forceps and absorb excessive mounting medium by gently pressing the paper over the edges of the coverslip by fingers (see Note 16).
11. Seal the coverslips with nail varnish. The samples can be observed when the varnish is dry (see Note 17). Excitation at 488 nm induces fluorescence of EGFP and Alexa Fluor 488 (green emission), while excitation at 543 nm induces fluorescence of RFP and Alexa Fluor 594 (red emission). Examples of coculture using MDCK II Tet-Off cells and HEK293 Tet-Off cells are shown in Figs. 1 and 2, respectively.

4. Notes

1. Expression of a gene of interest can be detected ~4 h after removal of Dox and the maximum induction levels are usually observed within 12–24 h. Passage of cells after the removal of Dox ensures the expression.
2. Pour PBS gently to the edge of the culture because HEK293 Tet-Off cells detach easily. MDCK II Tet-Off cells strongly attach to the surface of plastic and even glass.
3. HEK293 Tet-Off cells and MDCK II Tet-Off cells round-up within 1–3 and 5–15 min, respectively. It takes a longer period of time to detach the cells if they are 100% confluent, especially MDCK II Tet-Off cells. Longer incubation in trypsin–EDTA may damage the cells.
4. The following plasmid DNA mixtures are used: 3 μg of pTREM-EGFP-1CL plus 1.5 μg of pTK-Hyg, 3 μg of pTREM-EGFP-15CL plus 1.5 μg of pTK-Hyg, 3 μg of pTREM-15CL plus 1.5 μg of pTK-Hyg, and 4.5 μg of pTRE2hyg-RPF-2CL. In general, a selective vector such as pTK-Hyg are cotransfected with an expression vector at a ratio of 1:10 to ensure that

the most of selected cells are transfected with the expression vector. However, cotransfection with 0.4 μg of pTK-Hyg and 4 μg of pTREM-EGFP-1CL (1:10) did not yield any colonies. Usually 3–10 stable colonies are obtained when cotransfected at a ratio of 1:2.

5. LFA2000 should be diluted in polypropylene tubes because it might be absorbed in polystyrene.
6. Proceed to step 10 within 25 min.
7. Complexes are stable for 6 h at RT.
8. Do not add antibiotics such as penicillin and streptomycin during transfection as this causes cell death. Growth medium containing 10% FBS can be used without antibiotics in place of Opti-MEM I Reduced-Serum Medium.
9. Transfection can be done for 24 h in growth medium containing 10% FBS without removing the DNA/LF2000 complexes.
10. There should be less than 20 colonies in a 100-mm dish to obtain well-isolated colonies after selection by hygromycin. Usually 20–40 (HEK293) or 40–80 (MDCK II) colonies are obtained when 90–95% confluent cells in a 35-mm dish are transfected.
11. Use 20% FBS to neutralize trypsin because trypsinized cells are cultured for 2–4 h in step 7 without removing trypsin by centrifugation.
12. Positive clones at first screening occasionally lose the homogenous expression of a gene of interest in cells after several passages. Therefore, the positive cells should be screened again before cell freezing. When positive cells (master cells) cultured in the presence of Dox in a 24-well plate are passaged to 35-mm dishes, a small amount of positive cells are cultured without Dox in a 24-well plate and passaged at least twice. The cells are plated in 10-well glass slides and screened again. The positive cells that homogenously express a gene of interest are frozen.
13. Cells can be stored at −80°C for approximately 1 year. Cells can be stored in liquid nitrogen for a long period without loss of viability.
14. Do not maintain cells routinely in the absence of Dox because prolonged expression of a gene of interest may affect some properties of the cells. Therefore, do not reuse the cells for another experiment once the cells have been cultured in the absence of Dox. Make a large number of frozen stocks of cells grown in the presence of Dox and use one vial for one experiment.
15. The number of 80% confluent cells in a 60-mm dish (21 cm^2) corresponds to the number of 100% confluent cells in 60 wells

of 10-well glass slide (0.28 cm^2). Therefore, when 80% confluent cells in a 60-mm dish are suspended in 3.6 ml growth medium, plating 60 μl of the cell suspension into one well in a 10-well glass slide means plating 100% confluent cells in the well.

16. An excessive amount of mounting medium interferes with microscopic observation at high magnification because the working distance of the 40× and 63× objective lens (Carl Zeiss) is ~0.1 mm.
17. Samples can be stored in the dark at 4°C for several weeks.

Acknowledgments

This work was supported in part by Grants-in-Aid for Scientific Research from the Ministry of Education, Culture, Sports, Science, and Technology, Japan (Nos. 11770008, 13670018, 16590146, 18590187, and 20590193).

References

1. Gumbiner, B. M. (1993) Breaking through the tight junction barrier, *J Cell Biol* **123**, 1631–1633.
2. Furuse, M., Sasaki, H., Fujimoto, K., and Tsukita, S. (1998) A single gene product, claudin-1 or -2, reconstitutes tight junction strands and recruits occludin in fibroblasts, *J Cell Biol* **143**, 391–401.
3. Blasig, I. E., Winkler, L., Lassowski, B., Mueller, S. L., Zuleger, N., Krause, E., Krause, G., Gast, K., Kolbe, M., and Piontek, J. (2006) On the self-association potential of transmembrane tight junction proteins, *Cell Mol Life Sci* **63**, 505–514.
4. Daugherty, B. L., Ward, C., Smith, T., Ritzenthaler, J. D., and Koval, M. (2007) Regulation of heterotypic claudin compatibility, *J Biol Chem* **282**, 30005–30013.
5. Furuse, M., Sasaki, H., and Tsukita, S. (1999) Manner of interaction of heterogeneous claudin species within and between tight junction strands, *J Cell Biol* **147**, 891–903.
6. Inai, T., Sengoku, A., Hirose, E., Iida, H., and Shibata, Y. (2009) Freeze-fracture electron microscopic study of tight junction strands in HEK293 cells and MDCK II cells expressing claudin-1 mutants in the second extracellular loop, *Histochem Cell Biol* **131**, 681–690.
7. Piontek, J., Winkler, L., Wolburg, H., Muller, S. L., Zuleger, N., Piehl, C., Wiesner, B., Krause, G., and Blasig, I. E. (2008) Formation of tight junction: determinants of homophilic interaction between classic claudins, *FASEB J* **22**, 146–158.
8. Sengoku, A., Inai, T., and Shibata, Y. (2008) Formation of aberrant TJ strands by overexpression of claudin-15 in MDCK II cells, *Histochem Cell Biol* **129**, 211–222.
9. Furuse, M., Furuse, K., Sasaki, H., and Tsukita, S. (2001) Conversion of zonulae occludentes from tight to leaky strand type by introducing claudin-2 into Madin-Darby canine kidney I cells, *J Cell Biol* **153**, 263–272.
10. Tsukita, S., Yamazaki, Y., Katsuno, T., and Tamura, A. (2008) Tight junction-based epithelial microenvironment and cell proliferation, *Oncogene* **27**, 6930–6938.
11. Coyne, C. B., Gambling, T. M., Boucher, R. C., Carson, J. L., and Johnson, L. G. (2003) Role of claudin interactions in airway tight junctional permeability, *Am J Physiol Lung Cell Mol Physiol* **285**, L1166–1178.

Chapter 9

Claudin-4: Functional Studies Beyond the Tight Junction

Holly A. Eckelhoefer, Thejani E. Rajapaksa, Jing Wang, Mary Hamer, Nancy C. Appleby, Jun Ling, and David D. Lo

Abstract

Claudin-4 is an unusual member of the claudin family; in addition to its role in epithelial tight junction barrier function, it is a receptor for the *Clostridium perfringens* enterotoxin. We have also found that claudin-4 is regulated in mucosal epithelium M cells, both in increased expression of the protein and in redistribution into endocytosis vesicles. Our ongoing studies are studying the potential for developing ligands specific to claudin-4 for targeted delivery of cargo such as proteins and poly(DL-lactide-co-glycolide) nanoparticles to mucosal M cells. Methods for the study of claudin-4 movement within epithelial cells, and delivery of nanoparticles through targeted binding of claudin-4 are described.

Key words: Tight junction, Mucosal immunity, M cell, PLGA, Nanoparticles, Drug delivery, Vaccine

1. Introduction

The tight junction protein claudin-4 is one of the unusual members of the claudin gene family, sitting at a distinct locus with the closely related claudin 3, away from other family members, in both the mouse and human genome. Claudins 3 and 4 have similar c-terminal cytoplasmic tails, again distinct from other family members, and their second external domains bind to the *Clostridium perfringens* enterotoxin (1, 2). Thus, despite the observations that claudin-4 participates in epithelial tight junction barrier function, and indeed plays a major role in establishing transepithelial electrical resistance (3), this protein has distinct features that point to additional functions. This has been borne out in our studies identifying claudin-4 as a protein associated with the unique particle capture functions of M cells in mucosal

Kursad Turksen (ed.), *Claudins: Methods and Protocols*, Methods in Molecular Biology, vol. 762,
DOI 10.1007/978-1-61779-185-7_9, © Springer Science+Business Media, LLC 2011

epithelium (4–6). M cells are specialized epithelial cells that capture microparticles and antigens and transport them across the epithelium ("transcytosis") to underlying immune cells (e.g., dendritic cells) for stimulation of mucosal immunity (7). While we do not yet know exactly how claudin-4 participates in M cell transcytosis, it appears to be a structural component of transcytosis vesicles. Thus, M cell delivery can be targeted by ligands with affinity for the exposed domain of claudin-4. The methods described here provide details on how we are able to work with claudin-4 both in vitro in studies on its cell biology and in vivo in studies on its role in M cell transcytosis.

1.1. Claudin-4: Cell Biology Studies Using GFP-Tagged Claudin-4

In order to visualize claudin-4 in live cells, we tagged claudin-4 with green fluorescence protein (GFP). However, since the C-terminal domain of claudin-4 has a PDZ domain which binds to PDZ proteins such as ZO-1, we fused GFP to the N-terminal to avoid interference with normal claudin-4 function. A common cell line used for studying epithelial cell tight junctions is Caco-2, or the subclone Caco-2 BBe (8); it is not as easily transfected as other cell lines, so we also generated stable integrants with specific expression levels of GFP-claudin-4, using serial cell sorting.

2. Materials

2.1. Cell Biology Studies on Claudin-4

1. Caco-2 BBe cell line was obtained from ATCC.
2. Advanced Dulbecco's modified Eagle's medium (DMEM) (Invitrogen) supplemented with 10% fetal bovine serum (Biowest), 1% 4-(2-hydroxyethyl)-1-piperazineethanesulfonic acid (HEPES, 1 M), and pen/strep/glutamine (Invitrogen) were used for cell culture. Geneticin (Invitrogen) is added to the cell culture medium to maintain the stable transfected cell line.
3. Opti-MeM I reduced serum medium (Invitrogen) and lipofectamine 2000 (Invitrogen) were used for cell transfection.
4. Trypsin-EDTA (Invitrogen) and EDTA (Amresco).
5. The primary antibody mouse anti-ZO-1 and the secondary antibody goat anti-mouse Alexa647 were purchased from Invitrogen. Casein (Thermo Scientific), Tween-20 (Fisher), and 4% paraformaldehyde (Electron microscopy sciences).
6. The FACS machine used for cell sort is a FACSAria (BectonDickinson).

2.2. Surface Plasmon Resonance

1. Biacore X100.
2. *Sensorchip CM5* (Biacore Cat #BR-1000-12): Stored at 4°C, equilibrate to RT before use.

3. *Amine coupling kit* (#BR-1000-50): EDC and NHS are aliquoted and stored at −20°C. Ethanolamine is stored at 4°C.
4. *Coupling buffers*: 10 mM acetate buffer, pH 4.0, 4.5 (Cat# BR-1003-49 to 50), stored at 4°C.
5. *Running buffer*: HBS-EP + buffer (10× stock: 0.1 M HEPES, 1.5 M NaCl, 30 mM EDTA, and 0.5% surfactant P20) (Biacore Cat #BR-1008-26). Stored at RT, dilute to 1× before use. Use the exact same aliquot of buffer for sample preparation as is run through the instrument.
6. Regeneration solution: 2 M NaCl.
7. Sample tube & cap specific for instrument rack (Cat# BR-1002-87, BR-1004-11).
8. *Microcon centrifuge tube*: Millipore (Cat# 42412 for 100kD MWCO), used for buffer exchange.
9. *Ligand*: GST-Cldn4 R4 recombinant protein expressed and purified from bacteria (BL21), exchanged to HBS-EP + buffer for immobilization. Only GST protein is used as the control (Genscript Cat#Z02039-1).
10. *Analyte*: (a) Recombinant HA proteins containing Cldn4 Ecl2-binding sequences; (b) C-CPE protein (His tagged aa 184-319) used as the positive control. All proteins need to be buffer exchanged to HBS-EP + buffer using the Microcon tubes.
    - All solutions need to be filtered using 0.22-μm filter before use. Small volumes (e.g., ligand and analyte) can be filtered through UltraFree-MC centrifugal units (Millipore, Cat#UF30GVNB).

2.3. PLGA Nanoparticles

1. The poly(DL-lactide-co-glycolide) (PLGA 85:15, MW 50,000–75,000) and poly(vinyl alcohol) (PVA, MW 30,000–70,000, 87–90% hydrolyzed) were obtained from Sigma–Aldrich.
2. HEPES (1 M), phosphate-buffered saline (PBS, 1×), and sodium dodecyl sulfate solution (SDS, 10%), F-12 Kaighn's medium, and geneticin were purchased from Invitrogen.
3. Methylene chloride *optima*®, PBS (10× ready concentrate pouches), and sodium hydroxide (certified ASC) were obtained from Fisher Scientific.
4. Rhodamine 6 G was obtained from Fluka® Analytical and 16% paraformaldehyde was obtained from Electron Microscopy Sciences.
5. Prolong Gold anti-fade reagent with DAPI and 0.2 μm 505/515 (yellow-green) Neutravidin FluoSpheres® were purchased from Molecular Probes.

3. Method

3.1. Design DNA Construct and Transfection

We added the GFP coding sequence to the N-terminal of claudin-4 to make our fusion GFP-CLDN4 construct. The primers used for GFP and CLDN4 cloning were

GFP

5′-CCCTACCCAAGCTTGATAATATGGCCACCACC-3′ (For),

5′-CGCGGATCCTGCTGACTTGTACAGCTCATCCAT-3′(Rev),

CLDN4

5′-ATATAAACGCGGATCCATGGCCTCCATGGGG-3′ (For)

5′-CCGGAATTCTAACACGTAGTTGCTGG-3′ (Rev).

The GFP-CLDN4 was then ligated into the pcDNA43.1 (+) plasmid, which has a CMV promoter in the sequence. After the construct was established, we transfected Caco2-BBe with the construct. Transfection of Caco-2BBe cells is often not very efficient, so we have used a variety of methods to develop transfected cell lines.

1. Caco2-BBe cells were passaged into a six-well plate 2 days before transfection. On the day of transfection, the cells were around 90% confluent.
2. Plasmid transfection was accomplished as per the manufacturer's data sheet. Briefly, perform the following:
   (a) Dilute 4 μg DNA in 250 μl Opti-MEM I reduced serum medium and mix gently.
   (b) Mix 12 μl of lipofectamine 2000 in 250 μl of Opti-MEM I reduced serum medium. Incubate for 5 min at room temperature.
   (c) After 5-min incubation, combine the diluted DNA with the diluted lipofectamine 2000, mix gently by pipetting, and incubate at room temperature for 20 min.
   (d) Add 500 μl of complexes to the well containing cells and 2 ml of Opti-MEM I reduced serum medium. Mix gently by rocking the plate back and forth.
   (e) Incubate the cells at 37°C in a CO_2 chamber for 24 h before subjecting to geneticin selection (600 μg/ml). Change the medium after 4 h and then every 2 days.

3.2. FACS Selection

1. The cells were cultured until 100% confluent in the full medium with 600 μg/ml geneticin in two 75-cm^2 flasks.
2. The medium was removed from the flasks and the cells were washed twice with PBS. To detach the cells, 1 ml of

trypsin-EDTA was added into each flask and incubated at 37°C for 5 min.

3. The trypsinization was quenched with 5 ml sorting buffer (PBS 500 ml, EDTA 1 mM, HEPEs 25 mM, and 1%FBS), the cells were then transferred into a separate, 15-ml cornical tube, and spun at 1,000 rpm for 5 min.
4. After aspirating the supernatant, the cells were resuspended by pipetting in 1 ml sorting buffer and spun again at 1,000 rpm for 5 min.
5. The supernatant was again aspirated and resuspended by pipetting in 1 ml sorting buffer. The cells were then passed through the cell strainer and combined in a new 15-ml tube.
6. The cells were counted and then diluted with the sorting buffer at a concentration of 10 million/ml.
7. The cells were sorted using the FACSAria. The non-transfected Caco2-BBe cells were used as a negative control. A higher GFP expression may cause cell death, so we sorted for the medium expression of GFP cells (Fig. 1).
8. The sorted cells were spun down at 1,000 rpm for 5 min. The supernatant was aspirated and the cells were washed with 1 ml full medium twice before the sorted cells were seeded back to the flask and incubated at 37°C.

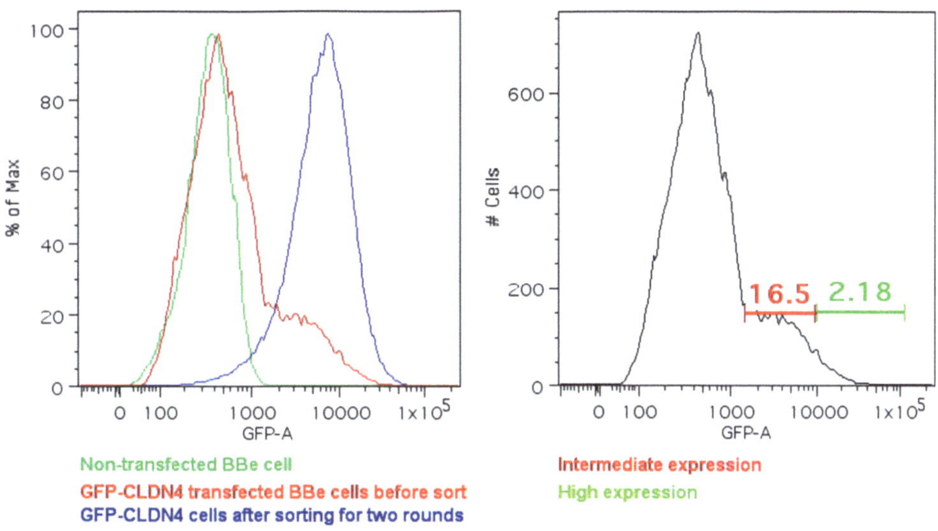

Fig. 1. Selection of Caco2-BBe cells expressing GFP-CLDN4 by FACS. Figure (**a**) demonstrates the expression of GFP in BBe control cells (*green*), GFP-CLDN4 transfected cells before sorting (*red*), and GFP-CLDN4 transfected cells after two rounds of sorting (*blue*). It is clear that after two rounds of sorting, the transfected cells significantly increased their GFP-CLDN4 expression. Figure (**b**) shows the gate of GFP expression. In order to avoid any toxic effects induced by overexpression of GFP, we sorted for medium GFP-expressing cells.

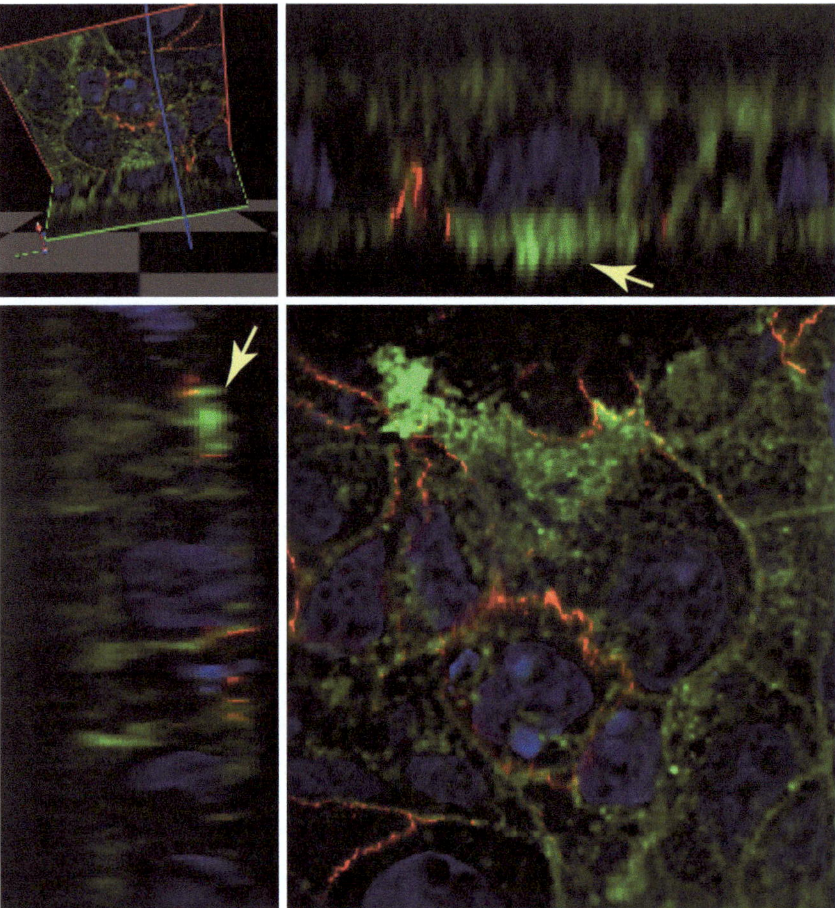

Fig. 2. Immunohistochemistry. Three-dimensional confocal image of stable GFP-CLDN4 transfected cells. Stable GFP-CLDN4 transfected Caco2-BBe after three rounds of sorting was used for immunostaining. *Green* indicates GFP, *red* indicates ZO-1, and *blue* indicates nuclei. In the GFP-CLDN4 transfected cells, GFP-CLDN4 can not only go to the tight junction where it co-localizes with ZO-1 (*red*) but can also be found at the apical membrane (*arrow*).

9. The cells were cultured with geneticin (600 μg/ml). When they were confluent, steps 1–9 were repeated twice.
10. After sorting the transfected cells, the GFP expression of the transfected cells was significantly shifted to the right (Fig. 1). We also performed histocytochemistry to further confirm the expression pattern of GFP-CLDN4 in the cells (Fig. 2). The sorted cells were then stored in liquid nitrogen (see Note 1).

3.3. Histocytochemistry

1. After the cells were sorted by FACS and cultured in the 75-cm^2 flask until confluent, 1×10^5 cells were passaged into the chamber slides and cultured with geneticin for 48 h before staining. During staining, the medium was removed and washed twice with PBS.

2. The cells were fixed with 1% paraformaldehyde in PBS at room temperature for 20 min and washed with 0.1% Tween in PBS (PBST) twice for 5 min. They were then permeabilized with 0.5% Tween in PBS for 10 min and then washed twice again with 0.1% Tween in PBS.
3. Before being incubated with the primary antibody (ZO-1) at room temperature for 90 min, the cells were blocked with casein plus Tween (CT) for 10 min.
4. After primary antibody incubation, the cells were washed with PBST twice for 5 min and then incubated with secondary antibody (goat anti-rabbit 568 1:1,000 in CT) for 45 min. They were then washed with PBST again and post-fixed with 4% paraformaldehyde for 15 min. The cells were then washed with PBS once, mounted with Prolong Gold anti-fade reagent with DAPI, and dried at room temperature for 24 h.
5. The cells were imaged with a BD CARV II spinning disc confocal microscope using IPLab software and analyzed with Volocity imaging processing software (Perkin–Elmer/Improvision).

3.4. Surface Plasmon Resonance Studies on Claudin-4: Binding to the Second External Domain

This assay is designed to measure recombinant HA protein containing claudin-4-binding sequences to the claudin-4 Ecl2 region (9). Purified GST-Cldn4 R4 is used as the ligand, and GST protein only is the control, both of which are immobilized onto a CM5 sensorchip using an amine-coupling reaction.

3.4.1. Protein Expression and Purification

Claudin-4: mouse claudin-4 (cldn4, NM_00903) was subcloned into pGEX4T-2 by PCR with high-fidelity DNA polymerase Pfu (Stratagene). The primers for the R4 deletion mutant (Ecl2.CT) were F: 5′-GGATCCTGGACCGCTCACAACG-3′ and reverse primer 1: 5′-CTCGAGTTACACATAGTTGCTGGCGGGG-3′. The constructs were confirmed by DNA sequencing. GST-Cldn4/pGEX4T-2 construct was transformed into *Escherichia coli* (BL21, pLysS) for protein expression. The soluble protein was purified by glutathione-agarose affinity chromatography (Pierce), and the co-purified GST protein was separated by gel filtration chromatography on FPLC with Superdex 200 column. For use as the analyte in Biacore assay, GST-Cldn4 was balanced to HBS-EP buffer by Microcon (Millipore) centrifugation.

3.4.2. Preparation of the Chip for Use

1. Turn on the Biacore ~1 h before use to allow temperature to stabilize to 25°C.
2. Take CM5 sensorchip out of 4°C, remove it from the sealed bag, and allow to warm to RT.
3. Disconnect the inlet tubes from the H_2O storage bottle and connect to the running buffer HBS-EP + bottle.
4. Dock new CM5 chip into the instrument.

5. Immobilize GST and GST-R4 to the chip by diluting each to 15 μg/ml in pH 4.0 acetate buffer, and 30 μg/ml in pH 4.5 acetate buffer, respectively. GST will be immobilized to the control channel, Fc1; GST-R4 will be immobilized to the assay channel, Fc2. The immobilization is accomplished by an amine-coupling mechanism that reacts the –NH2 on the protein with the –COOH of the dextran on the chip.
6. Prepare the required amounts of EDC, NHS, and ethanolamine (85 μl, 85 μl, and 237 μl, respectively), and load all of these reagents into Biacore tubes and into the rack.
7. Use the "amine coupling" tool in surface preparation Wizard to operate the immobilization, choose Fc2 for GST-R4 and Fc1 for GST, and select the "aiming for certain RU parameter" to 3,000RU to control the amount of ligand that will be bound. The procedure will continue automatically, and the resultant immobilized RU will be shown at completion.

3.4.3. Binding Assay

1. Prepare enough of each analyte (70 μl) in HBS-EP + buffer. Normally, recombinant HA proteins are prepared at a 1–5 μM concentration range. Prepare 5 μM of C-CPE as the positive control, which should give a clear positive binding result on the sensorgram, indicating the chip is active.
2. Load all of the analyte samples, positive control, and regeneration solution into sample rack.
3. Measure the binding by using the "Binding assay" tool from the "Wizard." Select the variable options: sample contact time 120 s, dissociation time 120 s, regeneration time 180 s, and 30 s stabilization. The procedure will run automatically using the wizard software (see Note 2).
4. Analyze the data by using the "Evaluation" software. Alignments and presentation modifications can be done very specifically, depending on the need.
5. Undock the active chip and store at 4°C in a 50-ml conical tube with ~2 ml of H_2O to keep it moist. The chip can be reused until noticeable activity is decreasing in regard to binding C-CPE.
6. Shut down the instrument.

3.5. Claudin-4 Targeting In Vivo: PLGA Nanoparticles for Targeting to M Cells

Our studies have suggested that claudin-4 is a component in mucosal M cell particle uptake as part of its role in immune surveillance at the mucosal epithelium. This raised the intriguing possibility that ligands targeting claudin-4 could be used to mediate delivery of cargo such as vaccines. We have begun to test this potential using PLGA nanoparticles incorporating a recombinant protein with a claudin-4-binding peptide, confirmed by Surface Plasmon Resonance to have high affinity binding (10).

3.5.1. Preparation of Claudin-4-Targeted Recombinant Protein-Loaded PLGA Nanoparticles

Given that the CPE30 targeting peptide can indeed mediate measurable uptake of fluorescent beads by M cells, we developed nanoparticles that could incorporate the recombinant HA-HT-CPE30 fusion protein (a fusion between the extracellular domain of influenza hemagglutinin and a C-terminal CPE30 peptide). It was critical that such particles would retain the proteins long enough for delivery to the target tissue, yet display enough of the protein on the surface to enable the CPE30 targeting peptide moiety to mediate uptake by M cells. PLGA polymer nanoparticles seemed to be an appropriate choice, since they could be produced with incorporated protein, yet they are biodegradable and can release the protein over time in vivo once delivered (11). PLGA nanoparticles containing targeting (HA-HT-CPE30) and nontargeting (HA-HT) peptides were prepared from 85:15 PLGA using solvent evaporation/double emulsion (also known as water-in-oil-in water, w/o/w) method.

1. Preparation of stock solutions
   (a) 4% PLGA polymer (85:15) solution was prepared by adding 0.18 g of PLGA into 4.5 ml of methylene chloride in a glass beaker and stirring until dissolved (see Note 3).
   (b) 2% PVA solution was prepared by dissolving 0.6 g of PVA in 30 ml 10 mM HEPES and adjusting the pH to 7.5 with NaOH (see Note 4).
   (c) Protein solutions: HA-HT-CPE30 or HA-HT protein in HEPES buffer at 3.0–4.5 mg/ml concentration (see Note 5).
   (d) For labeling experiments; a 40 mg/ml Rhodamine 6 G (R6G) solution was prepared by dissolving 1 mg of R6G in 25 μl of methylene chloride (see Note 6).
2. Preparation of first w/o emulsion:

   The reagents listed in the table were added to a 18 × 150 mm disposable glass tube in the order listed.

4% PLGA solution	4.25 ml
Protein	0.5 ml of HA-HT-CPE30 or HA-HT
2% PVA stabilizer	0.25 ml

   For labeled nanoparticles, 25 μl of 40 mg/ml R6G was added to the PLGA solution before adding the protein. The solution was emulsified by probe sonication (Branson Sonifier 450) for 20 s (duty cycle 20%, output control 3) to obtain w/o emulsion.
3. The resulting w/o emulsion was divided into two disposable glass tubes (see Note 7), and 12.5 ml of 2% PVA solution was added to each tube. The solution was emulsified by probe sonication for 30 s (duty cycle 20%, output control 3) to obtain the final w/o/w emulsion.

4. The final w/o/w was then combined in a 50-ml glass beaker and stirred uncovered for 20 h with a magnetic stirrer at 400 rpm at 4°C to allow solvent evaporation.
5. The solution was added to a 40-ml Oakridge tube and centrifuged at 3,800 rpm (RCF 1000 *g*) for 30 min. The supernatant was discarded and the pellet was resuspended gently in 20 ml of distilled water. The washing step was repeated with two 20-min and one 15-min centrifugation. The supernatant was discarded.
6. The resulting nanoparticle pellet was frozen in liquid nitrogen and lyophilized overnight at –88°C, 0.006 Torr.
7. The final product was stored at 4°C and kept dry with Dry-rite calcium sulfate pellets till ready to use.

3.5.2. Nanoparticle Characterization

Scanning Electron Microscopy

The morphology of the protein-loaded nanoparticles was visualized by scanning electron microscopy (SEM). A very small amount of nanoparticles was placed on a double-sided adhesive tape attached to an aluminum stub and sputter coated with gold/palladium beam for 2 min. The coated samples were imaged with Philips XL30-FEG SEM at 10 kV (Fig. 3).

Particle Size Measurements

The particle size of the nanoparticles was measured with ImageJ software using the obtained SEM images. The diameter of

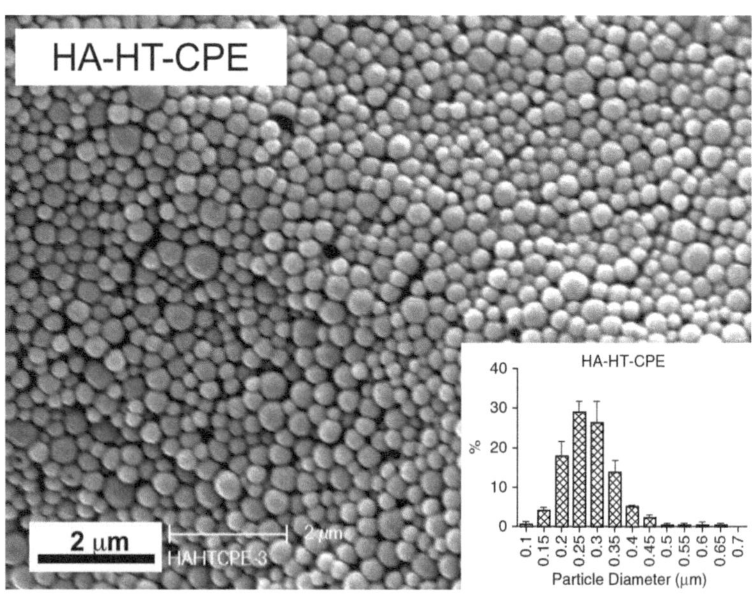

Fig. 3. SEM and size distribution of HA-HT-CPE nanoparticles showing a narrow range of particle size around 300 nm in diameter. Particle diameter was measured by ImageJ® software using SEM images obtained for three different preparations of nanoparticles. The diameter of approximately 150 particles was measured for each preparation. The frequency distribution of particle size was plotted with Prism software. *X*-axis represents the particle diameter in micrometer, while *y*-axis represents the percentage of particles at a given diameter.

approximately 150 nanoparticles was measured, and the size distribution was plotted using Prism software.

Determination of Protein Loading

Total protein loading was estimated using BCA assay. Approximately 5–8 mg of freeze-dried nanoparticles were accurately measured, added to 2 ml of 5% SDS in 0.1 M NaOH solution, and incubated with shaking for 24 h at room temperature until a clear solution was obtained. The protein content was measured in triplicates for each sample using BCA protein assay. The protein loading (%, w/w) was expressed as the amount of protein relative to the weight of the nanoparticles assayed (12).

3.6. In Vitro Uptake of Claudin-4-Targeted Protein-Loaded PLGA Nanoparticles

In vitro, we found that the nanoparticles with the HA-HT-CPE protein were readily taken up by GFP-claudin-4 CHO transfectants, showing both the function of the targeting peptide and the accessibility of the functional targeting peptide in the nanoparticles. The GFP-claudin-4 CHO cell transfectants were selected by serial FACS selection as described above for transfected Caco-2BBe cells.

1. In vitro uptake studies of R6G-labeled protein-loaded nanoparticles were performed in GFP-tagged claudin-4 transfected Chinese hamster ovary (CHO) cells (9).
2. The cells were maintained in F-12 Kaighn's medium supplemented with 10% fetal bovine serum and 0.8 mg/ml geneticin.
3. For the confocal studies, the cells were plated on cover slides placed in six-well plates and grown at 37°C in 5% CO_2 incubator for 48 h.
4. The cells were washed with PBS and the medium was replaced by 1 ml of nanoparticle solution in culture medium prewarmed to 37°C (10 μg of protein/well). The cells were incubated at 37°C in 5% CO_2 incubator for 1 h.
5. Upon incubation, the cells were washed three times with PBS to remove unbound nanoparticles.
6. The cells were then fixed with 4% paraformaldehyde in PBS for 20 min at room temperature and washed with PBS + 0.1% Tween20 for 3–5 min, two times.
7. The cover slides were mounted on glass slides with Prolong Gold anti-fade reagent with DAPI and incubated for 24 h at room temperature.
8. The cells were imaged using a BD CARV II spinning disc confocal microscope, using IPLab software (Fig. 4).
9. Histological analysis of particle uptake was performed by counting the number of particles taken up per cell in randomly selected fields of the slides for three different experiments.

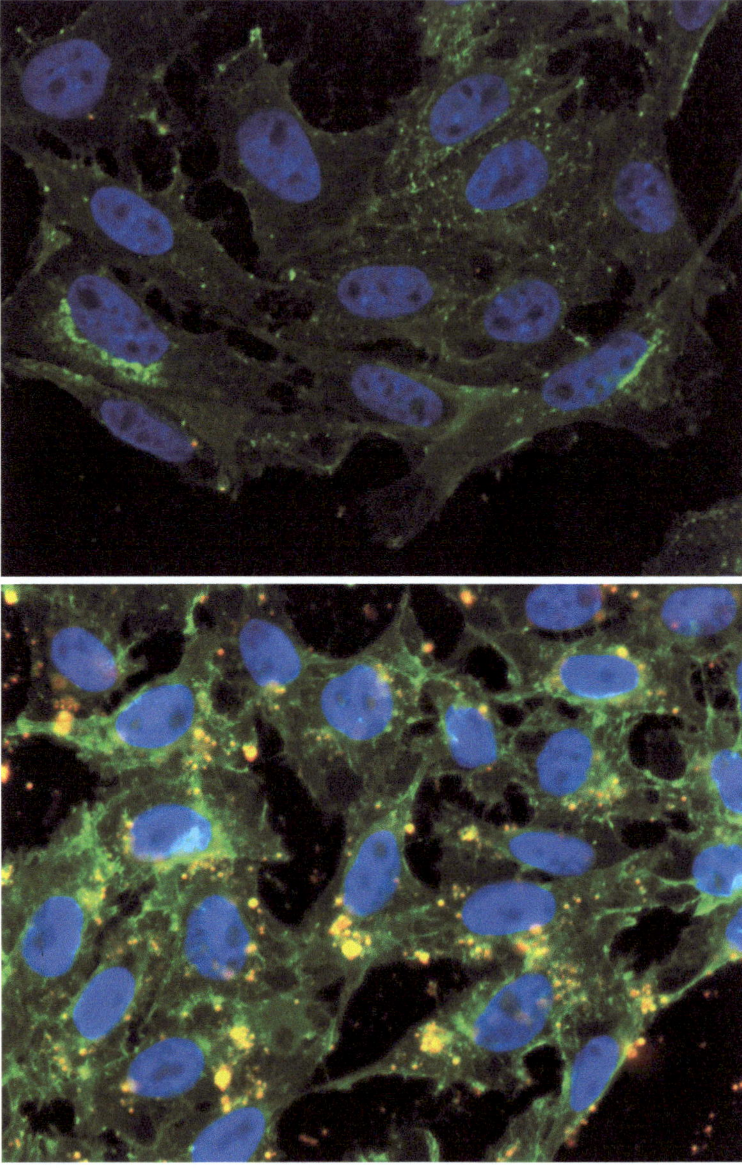

Fig. 4. PLGA nanoparticle uptake by GFP-claudin-4 transfected CHO cells. R6G-labeled nanoparticles were given to GFP-claudin-4 transfectants (*green*) for 1 h, and the cells were processed for confocal microscopy. Control HA-HT nanoparticles were not taken up (*top*), while HA-HT-CPE/R6G particles (*bottom*) were found to be readily bound and ingested, with particles shown to co-localize internally with GFP-claudin-4 (*yellow spots, arrows*). Cell nuclei, *blue.*

4. Notes

1. In general, the high level expression of the GFP is stable after three or more rounds of FACS selection.
2. To better account for free GST in the GST-R4 sample, set the target RU to 3,000 for that channel, but set the Fc1 GST RU

to 1,500. Roughly half of our GST-R4 sample is free GST, which can skew binding results when comparing Fc2 to Fc1 as the evaluation software will do. If immobilization is difficult, and the optimum RU cannot be reached, increase the problem sample concentration by 50% and repeat.

3. When dissolving PLGA polymer in methylene chloride, it is important to add the methylene chloride into the beaker first and then add PLGA. Otherwise, PLGA would get stuck on the walls of the beaker, making it harder to dissolve. In addition, glass pipettes should be used when measuring methylene chloride.
4. When making PVA stock solution, first add the buffer into the beaker and while stirring, add PVA gradually to avoid formation of clumps, which would take a longer time to dissolve.
5. When nontargeted HA-HT and claudin-4-targeted HA-HT-CPE recombinant fusion proteins are incorporated into nanoparticles, they should be in a low ionic buffer such as HEPES, or PBS:HEPES 1:10, compared to full strength PBS.
6. For labeled nanoparticles, R6G dye should be dissolved in an organic solvent such as methylene chloride, ethanol, or DMSO.
7. Glass pipettes should be used when dividing the w/o emulsion into two tubes as this solution contains methylene chloride.

Acknowledgments

This work was supported by a Grand Challenges in Global Health award from the Foundation for the National Institutes of Health (FNIH), and Grants R21 AI73689 and R01 AI63426 from the National Institutes of Health.

References

1. Sonoda, N., Furuse, M., Sasaki, H., Yonemura, S., Katahira, J., Horiguchi, Y., Tsukita, S. (1999) *Clostridium perfringens* enterotoxin fragment removes specific claudins from tight junction strands: evidence for direct involvement of claudins in tight junction barrier. *J. Cell Biol.* **147**, 195–204.
2. Katahira, J., Sugiyama, H., Inoue, N., Horiguchi, Y., Matsuda, M., Sugimoto, N. (1997) *Clostridium perfringens* enterotoxin utilizes two structurally related membrane proteins as functional receptors in vivo. *J. Biol. Chem.* **272**, 26652–8.
3. Colegio, O.R., Van Itallie, C., Rahner, C., Anderson, J.M. (2003) Claudin extracellular domains determine paracellular charge selectivity and resistance but not tight junction fibril architecture. *Am. J. Physiol. Cell Physiol.* **284**, C1346–54.
4. Lo, D., Tynan, W., Dickerson, J., Scharf, M., Cooper, J., Byrne, D., Brayden, D., Higgins, L., Evans, C., O'Mahony, D.J. (2004) Cell

culture modeling of specialized tissue: Identification of genes expressed specifically by Follicle Associated Epithelium of Peyer's Patch by expression profiling of Caco-2/Raji co-cultures. *Int. Immunol.* **16**, 91–99.

5. Clark, R.T., Hope, A., Lopez-Fraga, M., Schiller, N., Lo, D.D. (2009) Bacterial particle endocytosis by epithelial cells is selective and enhanced by tumor necrosis factor-receptor ligands. *Clin. Vacc. Immunol.* **16**, 397–407.
6. Wang, J., Lopez-Fraga, M., Rynko, A., Lo, D.D. (2009) TNFR and LTβR agonists induce Follicle-Associated Epithelium and M cell specific genes in rat and human intestinal epithelial cells. *Cytokine* **47**, 69–76.
7. Neutra, M.R., Pringault, E., Kraehenbuhl, J-P. (1996) Antigen sampling across epithelial barriers and induction of mucosal immune responses. *Annu. Rev. Immunol.* **14**, 275–300.
8. Peterson, M.D., Mooseker, M.S. (1993) An in vitro model for the analysis of intestinal brush border assembly. I. Ultrastructural analysis of cell contact-induced brush border assembly in Caco-2BBe cells. *J. Cell Sci.* **105**, 445–460.
9. Ling, J., Liao, H., Clark, R., Wong, M. S., Lo, D. D. (2008) Structural constraints for the binding of short peptides to claudin-4 revealed by surface plasmon resonance. *J. Biol. Chem.* **283**, 30585–30595.
10. Rajapaksa, T. E., Stover-Hamer, M., Fernandez, X., Eckelhoefer, H. A., Lo, D. D. (2009) Claudin 4-targeted protein incorporated into PLGA nanoparticles can mediate M cell targeted delivery. *J. Control Release.*
11. Zhao, A., Rodgers, V.G. (2006) Using TEM to couple transient protein distribution and release for PLGA microparticles for potential use as vaccine delivery vehicles. *J. Control. Release.* **113**, 15–22.
12. Coombes, A. G., Yeh, M. K., Lavelle, E. C., Davis, S. S. (1998) The control of protein release from poly(DL-lactide co-glycolide) microparticles by variation of the external aqueous phase surfactant in the water-in oil-in water method. *J. Control. Release.* **52**, 311–320.

Chapter 10

Methods to Analyze Subcellular Localization and Intracellular Trafficking of Claudin-16

P. Jaya Kausalya and Walter Hunziker

Abstract

The integral tight junction protein Claudin-16 (Cldn16) is predominantly expressed in renal epithelial cells of the thick ascending limb of Henle's loop where, together with claudin-19, it forms a cation-selective pore that allows influx of Na^{+} from the interstitial fluid into the lumen of the kidney tubule. This leads to an electrochemical gradient that drives the reabsorbtion of Mg^{2+} and Ca^{2+} ions from the renal filtrate. Mutations in the Cldn16 gene have been identified in patients suffering from familial hypomagnesemia with hypercalciuria and nephrocalcinosis, with excessive renal wastage of Mg^{2+} and Ca^{2+} being a hallmark of this condition. Studies into the mechanism by which mutations impair Cldn16 function have shown that although several mutations affect paracellular ion transport, many interfere with intracellular trafficking of Cldn16, ultimately compromising its localization to TJs. Here, we describe the experimental approaches that can be used to monitor intracellular localization and trafficking of Cldn16. These methods can easily be adapted to study other claudins, provided suitable antibodies are available.

Key words: Claudin, Familial hypomagnesemia with hypercalciuria and nephrocalcinosis, Intracellular transport, Kidney, MDCK cells, Paracellular ion transport, Subcellular localization, Tight junction

1. Introduction

Familial hypomagnesemia with hypercalciuria and nephrocalcinosis (FHHNC, OMIM248250) is an autosomal recessive disorder, with progressive renal Ca^{2+} and Mg^{2+} wasting leading to impaired renal function and kidney stone formation, which is one of its hallmarks (1). Positional cloning initially identified mutations in the gene for Cldn16, also referred to as paracellin-1, in patients suffering from this condition, and to date, more than 40 mutations in Cldn16 have been linked to FHNNC (1–6).

Kursad Turksen (ed.), *Claudins: Methods and Protocols*, Methods in Molecular Biology, vol. 762, DOI 10.1007/978-1-61779-185-7_10,

Cldn16 is a member of a family of transmembrane proteins that are part of the epithelial intercellular tight junction (TJ) barrier. Claudins span the membrane four times, with their N and C termini protruding into the cytosol. A C-terminal PDZ-binding motif interacts with PDZ domains of TJ scaffold or adaptor proteins, in particular ZO-1, ZO-2, and ZO-3 (7). The two luminal loops are thought to engage in homo- and/or hetero-typic interactions with other claudins on neighboring cells. Cldn16 is exclusively expressed in the kidney, where it is restricted to the thick ascending limb (TAL) segment and distal convoluted tubules (8, 9). Cldn16 associates with Cldn19 to form cation-selective paracellular pores (10, 11). Cldn16/Cldn19 containing TJs facilitate influx of Na^+ into the lumen of the TAL, with the resulting lumen positive charge helping to drive Ca^{2+} and Mg^{2+} resorption (10, 11). In addition, Cldn16 influences a Cl^- channel to further modulate the transepithelial potential that drives Mg^{2+} and Ca^{2+} transport (12, 13).

Mutations in Cldn16 have been postulated to interfere with ion transport properties mediated by Cldn16. However, recent studies have shown that many of the mutations identified in patients with FHHNC affect the subcellular localization and the intracellular transport of the mutant Cldn16 when expressed and analyzed in renal epithelial cell lines (2–4, 14, 15) (Fig. 1). Some of the Cldn16 mutants do not localize to the plasma membrane and TJs, but accumulate in intracellular compartments such as the endoplasmic reticulum (ER), the Golgi complex, or lysosomes. ER-retained Cldn16 mutants can be subject to ubiquitination and proteasomal degradation, suggesting that the mutant is misfolded and fails to pass ER quality control. Other Cldn16 mutants mislocalize to lysosomes. Taking advantage of antibodies to extracellular loops and endocytosis inhibitors, experiments on live cells showed that some of these mutants reach lyososomes without transiting through the plasma membrane, while others appear on the cell surface en route to lysosomes. The latter includes mutations that inactivate the C-terminal PDZ-binding motif in Cldn16, suggesting that tethering to ZO-1 and/or other scaffold proteins is critical for TJ localization of Cldn16 (3, 14).

Understanding how mutations in Cldn16 affect its intracellular trafficking may also open new therapeutic approaches for FHHNC. For example, chemical chaperones can rescue surface expression of several ER-retained Cldn16 mutants (14). In addition, the residence time in the plasma membrane can be increased for Cldn16 mutants that are not retained in the plasma membrane by inhibiting endocytosis (4, 14). Chemical chaperones for cases of cystic fibrosis where mutant CFTR is retained and degraded in the ER are already in clinical trials (16, 17) and if successful, may be applied to selected cases of FHHNC.

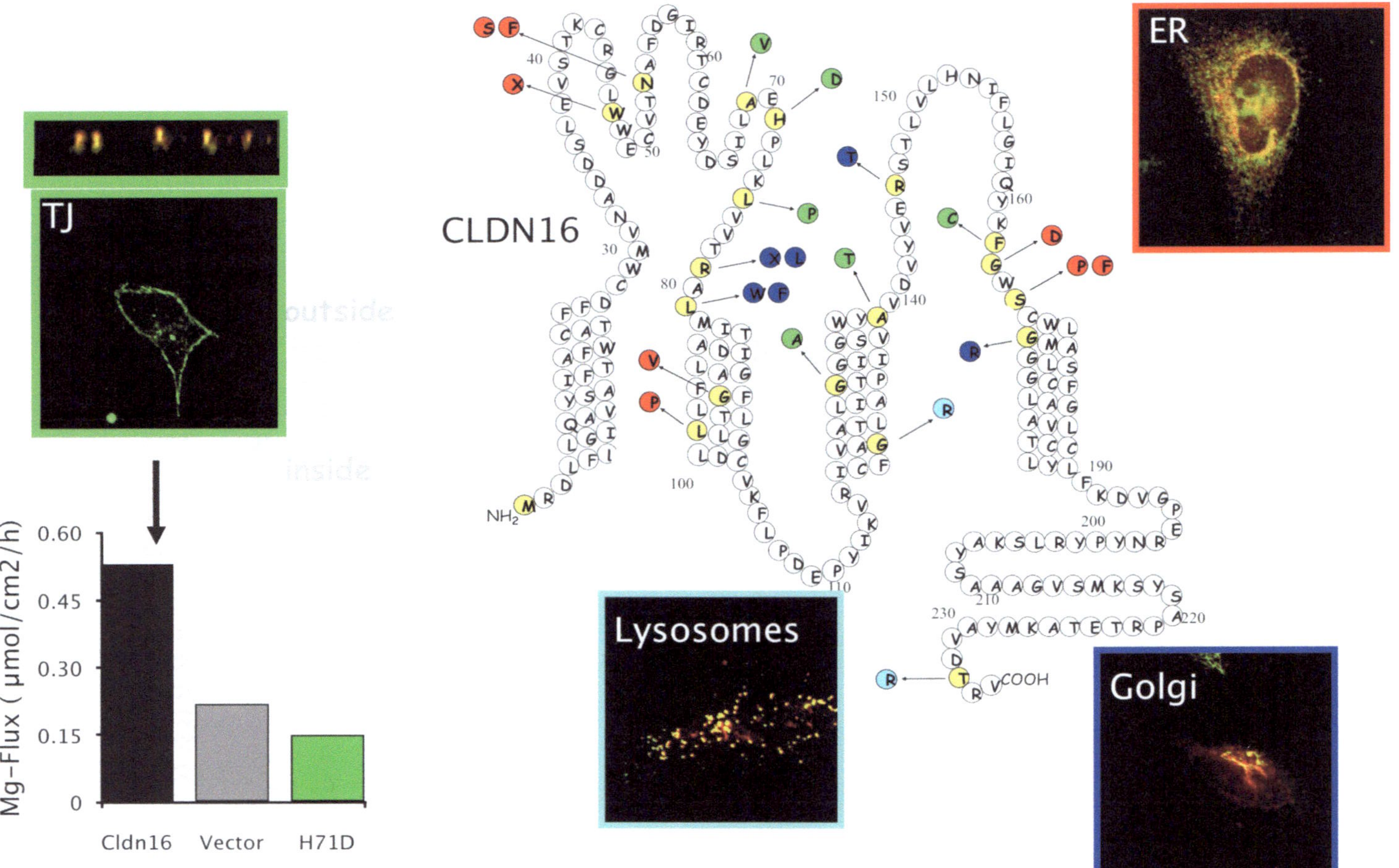

Fig. 1. Predicted topology of Cldn16 and the location of some of the reported mutations linked to FHHNC. Shown is the amino acid sequence in *yellow* and *arrows* indicating the change introduced by the mutation. The effect of the corresponding mutation on the pre-dominant steady-state distribution of Cldn16 is highlighted in different colors: *green*, cell surface; *red*, ER; *dark blue*, Golgi complex; *light blue*, lysosome; X, stop codon; fs, frameshift. The region (T52-S66) is used to generate the anti-loop antibody.

2. Materials

2.1. Plasmids and Recombinant DNA Reagents

1. pcDNA 3 mammalian expression vector (Invitrogen).
2. Advantage 2 PCR kit (Clontech).
3. T_4 DNA ligase (Promega).
4. DNA miniprep kit (Qiagen).
5. Midi prep kits (Nucleobond).
6. Gel extraction kit (Qiagen).
7. LB media and LB–ampicillin containing plates.
8. *Eco*RI and *Xba*I (Roche).

2.2. Cell Lines and Cell Culture

1. MDCK II and HEK293T cells are cultured in Dulbecco's modified Eagle's medium (DMEM) (Gibco/BRL) supplemented with 10% fetal bovine serum (Hyclone), 1% penicillin–streptomycin (Gibco/BRL), and 1% L-glutamine (Gibco/BRL).
2. HeLa cells are grown in RPMI medium (Gibco/BRL) supplemented with 10% fetal bovine serum (Hyclone) and 1% penicillin–streptomycin (Gibco/BRL) for culturing.
3. MDCK-C7 cells are cultured in MEM (Gibco/BRL, Bethesda, MD) supplemented with 10% fetal bovine serum (Hyclone), 1% penicillin–streptomycin (Gibco/BRL), and 1% L-glutamine (Gibco/BRL, Bethesda, MD) for culturing.
4. Tryspin/EDTA solution (2.5 g/L) (Gibco/BRL).
5. G418 sulfate (100 mg/ml stock, Calbiochem) is prepared in sterile water, filter-sterilized, and stored in single-use aliquots stored at –20°C. Final concentration is 0.4 mg/ml for selection of stably transfected MDCK II cells and 0.5 mg/ml for selection of stably transfected MDCK-C7 cells.
6. Lipofectamine and Lipofectamine Plus reagent (Invitrogen).
7. OPTI-MEM (Gibco/BRL).

2.3. Cell Lysis

1. Lysis buffer: 20 mM Tris–HCl, pH 7.4, 150 mM NaCl, 1 mM EDTA, 1% NP-40, and 0.25% deoxycholate (Note 1).
2. Protease inhibitor cocktail (Calbiochem; diluted 1:1,000).
3. Cell scraper (SPL Life Sciences).
4. PBS: 137 mM NaCl, 2.7 mM KCl, 4.3 mM Na_2HPO_4, and 1.47 mM KH_2PO_4, pH 7.4.
5. Bradford Protein assay solution (Bio-Rad).

2.4. SDS–PAGE

1. Separating buffer (4×): 1.5 M Tris–HCl, pH 8.7, 0.4% SDS.
2. Stacking buffer (4×): 0.5 M Tris–HCl, pH 6.8, 0.4% SDS.

3. 30% Acrylamide/bis-acrylamide solution (37.5:1 with 2.6% C) and *N*,*N*,*N*,*N*′-tetramethylethylenediamine (TEMED, Bio-Rad).
4. Ammonium persulfate: 10% solution in water, stored in 4°C.
5. 100% ethanol.
6. 4× SDS sample buffer: 0.25 M Tris–HCl, pH 6.8, 6% SDS, 40% glycerol, 0.04% bromophenol blue, and 20% mercaptoethanol.
7. Pre-stained protein molecular weight markers (Kaleidoscope markers, Bio-Rad).
8. 10× Running buffer: 0.25 M Tris (do not adjust pH), 1.92 M glycine, and 1.0% SDS.
9. Mini-Protean 3 Gel System (Bio-Rad).

2.5. Western Blot Analysis

1. Transfer buffer: 12.5 mM Tris (do not adjust pH), 95 mM glycine, and 20% (v/v) methanol.
2. PVDF membranes (Amersham) and 3MM Chr chromatography paper (Whatman).
3. PBS, 0.1% Tween-20.
4. Blocking buffer: 5% (w/v) nonfat dry milk in PBS and 0.1% Tween-20.
5. Primary antibodies: rabbit anti-canine ZO-1 (Zymed; 1:1,000), rat monoclonal anti-HA (Roche; 1:1,000), mouse monoclonal antihuman actin (Chemicon; 1:2,500), and mouse monoclonal anti-ubiquitin (Santa Cruz Biotechnology; 1:1,000). Antibodies are diluted in blocking buffer.
6. Secondary antibodies: HRP-conjugated goat anti-rat, HRP-conjugated goat anti-rabbit, and HRP-conjugated goat anti-mouse antibodies (Pierce) are diluted (1:5,000) in blocking buffer.
7. ECL chemiluminescence kit (Pierce).

2.6. Confocal Immunofluorescence Microscopy

1. Round microscope cover glass (12 mm), microscope slides (Marenfeld), and curved forceps.
2. 12-mm Costar Transwell polycarbonate filter units with 0.4-μm pore size (Corning).
3. PBS (see above).
4. Paraformaldehye (PFA; Sigma): prepare 2% (w/v) fresh PFA solution in PBS. The solution needs to be heated to 50°C to dissolve (use a hot plate in the fume hood) and is cooled to room temperature before use.
5. Quenching solution: 50 mM NH_4Cl in PBS.
6. Permeabilization solution: 0.2% TritonX-100 in PBS.

7. Methanol (kept at –20°C).
8. Blocking solution: 10% goat serum with 0.1% Triton X-100 in PBS.
9. Primary antibodies: rabbit anti-canine ZO-1 (Zymed), rat anti-HA (Roche), rabbit anti-calreticulin (ABR), mouse monoclonal anti-CD63 (Developmental Studies Hybridoma Bank), mouse anti-ubiquitin (Santa Cruz Biotechnology), mouse anti-EEA1 (BD Biosciences-Pharmingen), and mouse anti-GM130 (BD Biosciences-Pharmingen).
10. Secondary antibodies: Alexa Fluor 594-labeled goat anti-rabbit IgG, Alexa Fluor 594-labeled goat anti-rat IgG, Alexa Fluor 488-labeled goat anti-mouse IgG, and 594-labeled goat anti-mouse IgG (Invitrogen Corp).
11. Mounting medium: Aqua Poly Mount (Polysciences, Inc.).

2.7. Monitoring of Cell Surface Expression and Endocytosis

1. Ice-cold PBS containing 0.9 mM $CaCl_2$ and 0.5 mM $MgCl_2$.
2. Anti-Cldn16 loop antibodies. A peptide representing the first extracellular loop of Cldn16 (amino acids 52–56) was used to immunize rabbits and obtain affinity-purified antibodies (Biogenes).
3. Cldn16 loop peptide for competition was used at a final concentration of 100 μM.
4. Thapsiagarin (A.G. Scientific, Inc). A 10 mM stock is prepared in DMSO and stored at –20°C. Use at a final concentration of 1 μM in DMEM complete media.
5. Sodium 4-phenylbutyrate (Calbiochem). A stock of 100 mM is prepared in sterile water and stored at –20°C. Use at a final concentration of 1 mM in complete media.
6. Cell counter (Fisher).

2.8. Temperature Blocks and Inhibition of Protein Synthesis

1. Cyclohexamide (Sigma). A 2 mg/ml stock is prepared in sterile water, aliquoted, and stored at –20°C. Use at a final concentration of 20 μg/ml in complete media.
2. *N*-acetyl-leu-leu-norleucinal (ALLN; Calbiochem). A 20 mM stock is prepared in DMSO and stored at –20°C. Use at a final concentration of 0.2 mM in complete media.

2.9. Inhibition of Endocytosis

1. Sucrose: use at a final concentration of 0.45 M in RPMI complete media.
2. 20 mM 2-[*N*-moropholino]ethanesulfonic acid (Sigma) in RPMI complete media.
3. 20 mM succinic acid (Sigma) in RPMI complete media.
4. 2 U/ml cholesterol oxidase (Sigma) in RPMI complete media.

2.10. Biochemical Analysis of the Distribution of Internalized Anti-loop Antibodies

1. HRP assay solution: 50 mg of *O*-phenlyenediamine dihydrochloride dissolved in 50 ml PBS, pH 6.0, and 50 μl 30% H_2O_2.

2.11. Inhibition of Proteasomal Degradation

1. Mouse anti-ubiquitin antibody (dilute 1:100 in for immunofluorescence microscopy).
2. Rat anti-HA antibody (dilute 1:100 for immunofluorescence microscopy; 1:1,000 for Western blot analysis).

2.12. Rescue of Cldn16 Mutants Using Pharmacological Chaperones

1. Thapsiagarin (A.G. Scientific). A 10 mM stock is prepared in DMSO and stored at −20°C. Use at a final concentration of 1 μM in complete media.
2. Sodium 4-phenylbutyrate (Calbiochem). A 100 mM stock is prepared in sterile water and stored at −20°C. Use at a final concentration of 1 mM in complete media.

3. Methods

3.1. Subcloning of Wild-Type and Mutant Cldn16 cDNAs into Expression Vector

1. The full-length human CLDN16 cDNA is obtained by RT-PCR from a kidney cDNA library (Clontech) using suitable primers covering the 5′ and 3′ coding region of the cDNA using the Advantage 2 PCR kit (Clontech). The primers are designed to contain an *Eco*RI restriction site introduced at the 5′ site, an N-terminal HA epitope tag, and a 3′ Xba restriction site. The epitope tag contains a functional Kozak sequence and is fused in-frame to the first amino acid after the endogenous initiation methionine.
2. Mutations that have been identified in patients with FHHNC (Fig. 1) are introduced by PCR into the Cldn16 cDNA using suitable primers.
3. The PCR products are run on 1% agarose gel, the bands are cut out with a scalpel blade, and the PCR fragments are purified using Qiagen gel extraction kit (Qiagen).
4. The purified PCR products and pDNA3 vector are digested with the restriction enzymes *Eco*RI and *Xba*I (New England Biolabs) for 2 h and gel purified. DNA inserts are ligated to the vector with T4 DNA ligase (Promega) overnight at 16°C. Competent DH5α bacteria are transformed with the ligated plasmids and plated on LB–ampicillin plates. After incubation overnight at 37°C, bacterial colonies are picked, plasmids are isolated with a DNA miniprep kit (Qiagen), and digested with *Eco*RI and *Xba*I to screen for successful ligation. All constructs are verified by DNA sequencing.

5. Bacterial clones harboring verified cDNA constructs are expanded for plasmid DNA extraction with a Midi prep kit (Nucelobond).

3.2. Transfection and Selection of Mammalian Cells

1. MDCK II, MDCK-C7, HeLa, and HEK293T cells are maintained in 10-cm tissue culture dishes and passaged with trypsin/EDTA (2.5 g/L, Gibco/BRL) when reaching confluency.
2. The day before the transfection, cells are trypsinized with trypsin/EDTA, counted, and plated to reach 70–80% confluency on the day of transfection. For transfection in a 6-well plate, 2 μg of wild-type or mutant Cldn16 plasmid and 5 μl of Lipofectamine Plus reagent are added to 100 μl of OPTI-MEM in a sterile microfuge tube and incubated at room temperature for 15 min. In a second tube, 100 μl of OPTI-MEM is mixed with 9 μl of lipofectamine reagent. After 15 min, the lipofectamine reagent is added to the first tube and further incubated for 15 min at room temperature. In parallel, cells are rinsed with serum-free media and incubated for 15 min at 37°C in serum-free media. The serum-free media is then removed from the cells, 600 μl of OPTI-MEM is added to the transfection mixture, and the diluted tansfection mixture is added to the cells and incubated for 3 h at 37°C. After 3 h, complete media containing 20% FCS is added to the cells and the cells are incubated at 37°C in the CO_2 incubator overnight. The following day, fresh complete media is added. For transfection on 24-well plates, the amounts of DNA and reagents are scaled down accordingly.
3. For the selection of stable cell lines (Note 2), cells are trypsinized with 1 ml trypsin/ETDA 48 h post-transfection, split into two 15-cm dishes containing 25 ml of complete media supplemented with G418, and incubated at 37°C in the CO_2 incubator. Media containing G418 is replaced every 2–3 days. After 2–3 weeks, single G418 clones are visible. The clones are gently scrapped with a sterile 200-μl pipette tip and dislodged cells are aspirated and plated into 24-well plates for expansion and screening (Note 3). Screening for positive clones is carried out by immunofluorescence staining or Western blot analysis. Positive clones are expanded, frozen down in media supplemented with 20% FCS and 20% DMSO, and stored in liquid nitrogen.
4. For transient transfection, cells are harvested for Western blot analysis 48 h after transfection, or plated onto coverslips and processed for immunofluorescene staining 24 h post-transfection.

3.3. Cell Lysis

1. Transiently or stably transfected cells are washed twice with ice-cold PBS. Lysis is carried out on ice. Cold lysis buffer containing protease inhibitor is added to the cells and a cell

scraper is used to scrape the cells gently off the plate. Cells are transferred into a 1.5-ml microfuge tube and vortexed continuously for 20 min in the cold room. The lysate is then centrifuged in a microfuge at maximum speed at 4°C for 20 min. The resulting supernatant is collected into a clean 1.5-ml microfuge tube and kept on ice for Bradford assay for protein quantification, or stored at −20°C until use.

2. For Bradford protein quantification, 1–2 μl of the protein lysate is added to 200 μl of Bradford solution and 800 μl of H_2O, and the absorbance at 595 nm is measured. The protein concentration in the test sample is extrapolated from the absorbance of a series of BSA concentration standards.

3.4. SDS–PAGE and Western Blot

1. For SDS–PAGE, 10 μl of 4× sample buffer is added to 50–100 μg of protein in 40 μl PBS and heated for 10 min at 50°C in a thermal heat block (Note 4). Heated cell lysates are briefly centrifuged at maximum speed in a microfuge and subjected to SDS–PAGE.
2. Glass plates for a Mini-Protean 3 cell gel system are scrubbed clean, rinsed with deionized water followed by wiping with 70% ethanol, and assembled for a 1.5-mm thick gel.
3. Prepare 10% separating gel in a 50-ml falcon tube by mixing 2.5 ml lower gel buffer, 3.4 ml 30% polyacrylamide solution, 4.1 ml H_2O, 30 μl APS, and 15 μl TEMED. Pour the gel leaving space for the stacking gel. 95% ethanol is overlaid on the gel. The gel should polymerize in 15–20 min. Pour off the overlaid ethanol and rinse the top of the gel with water.
4. Prepare the 4% stacking gel by mixing 2.5 ml upper gel buffer, 1.5 ml polyacrylamide solution, 6 ml H_2O, 100 μl APS, and 10 μl TEMED. Pour the stacking gel solution up to the brim and quickly insert a 10-well comb. The stacking gel should polymerize in 10–15 min. Once polymerized, remove the comb and flush the wells with water to remove any unpolymerized gel.
5. Set up the gel tank and slide the gel cassette sandwich into the clamping frame. The inner chamber is filled with 1× running buffer, and 200 ml of 1× running buffer is added to the lower chamber. Samples are loaded into the wells with a pipette using gel-loading tips. A volume of 10 μl of the pre-stained Kaleidoscope marker is also loaded into one lane. Place the lid on the gel tank and run at 100 V for 2 h or until the dye reaches the bottom of the gel.
6. Remove the gel from the glass plates and cut one corner from the separating gel to mark the orientation. The gel is pre-soaked in cold transfer buffer for 10 min.
7. Whatman 3MM filter paper and PVDF membrane are cut slightly larger than the size of the gel. PVDF membranes are

labeled at the corner with pencil and are pre-soaked in methanol for 30 s before soaking in the transfer buffer together with the filter paper and sponge pads.

8. A tray is prepared with some transfer buffer to lay out the transfer cassette with one sponge pad laid on the white panel of the cassette, and in the following order: filter paper, PVDF membrane, gel, and filter paper. A glass pipette is rolled over the membrane to ensure that no bubbles are trapped. Another sponge pad is laid on the top (facing the black side of the cassette). The cassette is closed with the latch, locking the sandwich in place.
9. The cassette is placed in the mini-Transblot tank with the black panel facing the cathode. The ice-cooling pack (prepared in advance by adding deionized water and freezing at –20°C) is placed in the buffer chamber and the tank is filled with transfer buffer. The lid is put on the tank and the transfer carried out at 100 V for 3 h or 40 V overnight in the cold room.
10. Once the transfer is completed, the cassette is removed from the gel tank and the PVDF membrane retrieved. If the transfer is done correctly, the pre-stained protein markers should be visible on the membrane. The membrane is rinsed with PBS and rolled up into a 50-ml falcon tube with the side carrying the transferred protein facing the inside of the tube. A volume of 10 ml of blocking solution is added to the falcon tube and placed on a roller for 1 h at room temperature or at 4°C overnight.
11. The blocking solution is discarded and the membrane is incubated in a 1:1,000 dilution of rat-mononclonal anti-HA antibody in 5% nonfat milk and 0.1% Tween-20 in PBS on the roller for 1 h at room temperature or at 4°C overnight.
12. The primary antibody is removed and the membrane is washed four times for 15 min each with 30 ml of 0.1%Tween-20 in PBS on a roller.
13. The secondary antibody diluted in blocking buffer is added to the membrane and incubated for 1 h at room temperature on a roller.
14. The secondary antibody is removed and the membrane is washed four times for 15 min each with 30 ml of 0.1%Tween-20 in PBS on a roller.
15. From each of the different components of the ECL reagent, 1 ml is mixed, immediately added to the blot, and incubated for 5 min.
16. The membrane is blotted with Kimwipes to remove excess ECL solution, placed between a sheet protector and in a dark room exposed to an X-ray film for a suitable exposure time.

3.5. Steady-State Distribution of Cldn16 Determined by Confocal Immunofluorescence Microscopy

1. Cells stably or transiently transfected with cDNAs for wild-type or mutant Cldn16 are plated on coverslips placed in 24-well plates and grown to confluency before fixation. The coverslips are washed and autoclaved prior to use. Alternatively, 4×10^5 transfected MDCK cells are plated on 12-mm Costar Transwell polycarbonate filter units in 12-well plates (Fig. 2) and grown for 4–5 days until they are fully polarized. Media is changed every day for the cells grown on Transwell filters.
2. The cells are rinsed with PBS twice and fixed with 3.7% PFA in PBS for 30 min or with −20°C methanol for 2.5 min. The cells are washed twice with PBS. In the case of PFA fixation, residual PFA is quenched with 50 mM NH_4Cl in PBS for 10 min, cells are permeabilized with 0.2% Triton X-100 in PBS for 10 min, and rinsed twice with PBS at room temperature. The cells are then incubated in blocking solution for 30 min at room temperature.

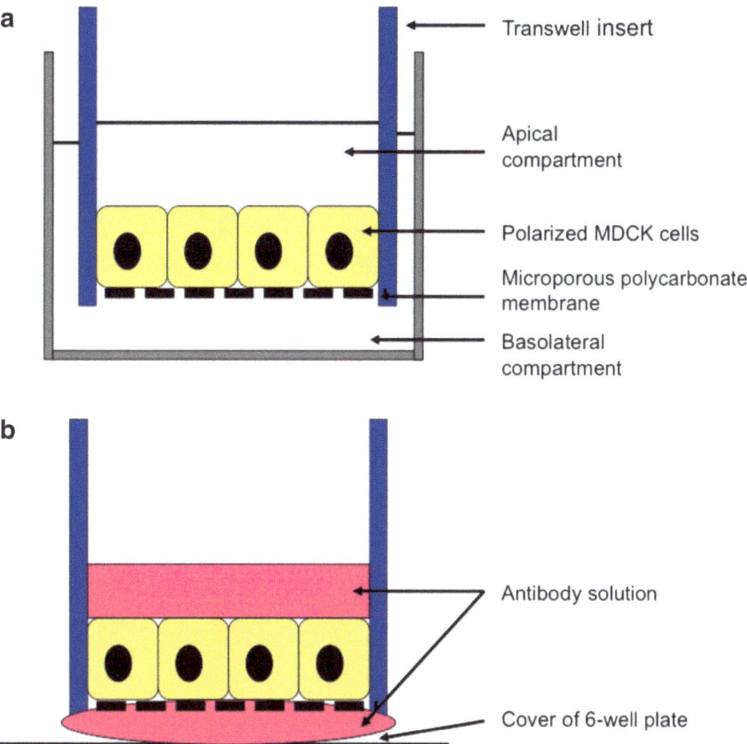

Fig. 2. Culturing and staining MDCK cells grown on Transwell polycarbonate filters. (**a**) Cell culture. Cells grown on Transwell filters polarize to form monolayers with functional TJs. This is characterized by an increase in the transepithelial electrical resistance during monolayer formation. The pores in the filter allow uptake of nutrients from the basolateral surface, mimicking the in vivo situation in epithelia. (**b**) Staining of cells grown on Transwell filters with antibodies for immunofluorescence microscopy. The depicted set up minimizes the amount of antibodies required. A volume of 20 μl of the antibody solution is placed on the cover plate of the 12-well plate and the filter is placed on top of the drop, allowing the antibody access to the basolateral surface of the cells across the pores of the Transwell filter. Another 100 μl of the antibody solution is added into the Transwell to the apical surface of the cells.

3. Primary antibodies are diluted in blocking buffer as follows: rabbit anti-ZO-1 (2.5 μg/ml), rat anti-HA (1 μg/ml), rabbit anti-calreticulin (2.5 μg/ml), mouse monoclonal anti-CD63 (a gift from Hong Wan Jin), mouse anti-ubiquitin (2.0 μg/ml), mouse anti-EEA1 (2.5 μg/ml), and mouse anti-GM130 (2.5 μg/ml).
4. The blocking solution is removed from the cells and the coverslips are centered in the 24-well plate using the curved forceps. A volume of 20 μl of the primary antibody solution is added to the coverslips and incubated for 1 h at room temperature. Placing the coverslips at the center of the well helps retain the drop with the primary antibody on the coverslip (Note 5). As for the Transwell filter units, 20 μl of the antibody is placed on the cover plate of the 12-well plate and the filter is placed on top of the drop, allowing the antibody access to the basolateral surface of the cells across the pores of the Transwell filter. Another 100 μl of the antibody solution is added into the Transwell to the apical surface of the cells (Fig. 2).
5. After 1-h incubation, the primary antibody solution is removed and the cells are washed three times each for 5 min with PBS.
6. The secondary antibody solution is prepared in blocking solution at a 1:1,000 dilution and added to the centered coverslips for 1 h at room temperature in the dark by wrapping the plate with aluminum foil.
7. The secondary antibody is removed by washing three times with PBS. A drop of mounting solution is added to a glass slide, the coverslip is carefully lifted from the 24-well with forceps, and the side with the cells is placed onto the mounting solution and left to air-dry overnight. As for the Transwell filter units, the polycarbonate membrane is removed from the Transwell insert by carefully cutting around the membrane edge with a scalpel blade. The membrane is placed on a drop of mounting media on a glass slide with the cells facing up. Another drop of mounting media is placed onto the cells and a coverslip is carefully lowered to avoid trapping of air bubbles. The slides can be stored in the dark at 4°C for several months.
8. The slides are viewed by confocal microscopy using appropriate excitation wavelengths (e.g., 488 nm and 594 nm for the secondary antibodies listed). For cells on Transwell filters, confocal sections along the *Z*-axis can be obtained, allowing the visualization of TJs at the apical side of the lateral plasma membrane. The steady-state subcellular localization of Cldn16 is established based on the co-localization of the labeling for Cldn16 with the staining for markers for different intracellular compartments (i.e., calreticulin for the ER, GM130 for the Golgi complex, EEA1 for endosomes, CD63 for lyososmes, and ZO-1 for TJs).

3.6. Monitoring Cell Surface Expression and Endocytosis of Cldn16

1. Generation of an antibody against the first extracellular loop of Cldn16 was outsourced (Biogenes). The ability of live cells to bind and internalize the antibody added to the medium is used as an assay to detect transient surface expression and internalization of integral membrane protein (18). The antibody is diluted to make sure no signal is observed in non-transfected control cells. Additional controls include the use of pre-immune serum and competition of the antibody with a 100-fold molar excess of the peptide used for the immunization.
2. MDCK or HeLa cells are seeded on 24-well containing coverslips to obtain subconfluent cultures the following day. Live cells are then incubated with anti-loop antibody in complete media for 1 h at 37°C. Serial dilutions showed that for the particular antibody batch, a 1:100 dilution gave a specific signal in transfected compared to non-transfected cells.
3. The cells are then chilled on ice and washed three times each for 10 min in ice-cold PBS containing 0.5 mg/ml $MgCl_2$ and 0.9 mg/ml $CaCl_2$. The cells are fixed in PFA, permeabilized, blocked with blocking buffer, stained with secondary Alexa Fluor 488-labeled goat anti-rabbit IgG, washed three times with PBS, and mounted on glass slides for viewing under the microscope.
4. This assay will selectively detect Cldn16 molecules that are present on the cell surface, or have transiently been exposed on the cell surface and were thus able to bind and internalize the anti-loop antibody present in the media.
5. Cells can also be co-stained with a rat anti-HA antibody for 1 h followed by Alexa Fluor 594-labeled goat anti-rat IgG to detect all exogenously expressed Cldn16 molecules. Alternatively, different intracellular compartments can be detected using appropriate antibodies to identify the organelles that contain the anti-loop antibody internalized by Cldn16.

3.7. Temperature Blocks and Inhibition of Protein Synthesis

1. At 20°C, proteins accumulate in the *trans*-Golgi complex (TGN) (19). To monitor the exit of Cldn16 from the TGN, cells can be incubated at 20°C to accumulate Cldn16 in the TGN and the temperature block can then be released to monitor by immunoflourescence microscopy the ability of wild-type or mutant Cldn16 to exit the TGN.
2. HeLa cells are plated on coverslips and transiently transfected with cDNAs for wild-type or mutant Cldn16.
3. After 24-h transfection, live cells are incubated for 3 h at 20°C to accumulate proteins in the TGN.
4. The cells are either fixed with PFA or transferred to 37°C in the presence of 20 μg/ml cyclohexamide for 1 h. Cyclohexamide,

an inhibitor of protein synthesis, will block synthesis of additional Cldn16. Incubation at 37°C causes the release of the accumulated proteins from the Golgi apparatus.

5. The cells are then fixed in PFA and processed for immunoflourescence staining with rat anti-HA and mouse anti-GM130 antibodies and suitable secondary antibodies to detect Cldn16 and the Golgi complex, respectively.

3.8. Inhibition of Endocytosis

1. If Cldn16 mutants are delivered to the plasma membrane and rapidly endocytosed, blocking endocytosis will result in the accumulation of sufficient Cldn16 molecules on the plasma membrane to allow their detection. Protocols are provided below to block clathrin-mediated endocytosis or caveolae-mediated uptake.
2. To block clathrin-mediated endocytosis, transiently transfected HeLa cells are pre-incubated with RPMI media containing 0.4 M sucrose (hypotonic conditions (20)) or 20 mM 2-[*N*-moropholino]ethanesulfonic acid and 20 mM succinic acid pH 2.2 (cytosol acidification (21)) for 30 min at 37°C.
3. The cells are then incubated with anti-loop antibody in RPMI media supplemented with 0.4 M sucrose or 20 mM 2-[*N*-moropholino]ethanesulfonic acid and 20 mM succinic acid (pH 2.2) (cytosolic acidification) for 1 h at 37°C.
4. To block caveolae-mediated uptake, transfected HeLa cells are pretreated with RPMI containing 2 U/ml cholesterol oxidase for 30 min.
5. The cells are then incubated with anti-loop antibody in RPMI media supplemented with 2 U/ml cholesterol oxidase for 1 h at 37°C.
6. The cells are subsequently chilled on ice and washed three times for 5 min each with ice-cold PBS containing 0.9 mM $CaCl_2$ and 0.5 mM $MgCl_2$, fixed in PFA, and processed for immunofluorescence microscopy, as detailed in Subheading 3.5. The cells are stained with rat anti-HA and mouse anti-EEA1 antibodies followed by the appropriate secondary antibodies.

3.9. Biochemical Analysis of the Internalization of Cldn16 Anti-loop Antibodies

1. Internalization of anti-loop antibodies can be quantified biochemically by comparing the amounts of anti-loop antibodies associated with non-permeabilized (anti-loop antibody bound to the surface) to that associated with permeabilized (anti-loop antibody on the surface and in intracellular endocytic compartments) cells, or by quantifying the amount of anti-loop antibodies associated with control cells compared to cells in which endocytosis has been inhibited.
2. Two sets of MDCK cells transiently transfected with cDNAs for wild-type or mutant Cldn16 in 6-well plates are prepared.

3. The live cells are incubated with anti-loop antibodies at 37°C for 1 h under normal conditions or in conditions where endocytosis is inhibited (see Subheading 3.6).
4. The cells are subsequently washed in ice-cold PBS containing 0.9 mM $CaCl_2$ and 0.5 mM $MgCl_2$.
5. One set of cells is fixed with 2% PFA for 30 min and not permeabilized, the second set is permeabilized with 0.2% Tx-100 in PBS for 10 min after fixation.
6. The cells are blocked with 1% BSA in PBS for 30 min at room temperature and then incubated with HRP-conjugated secondary anti-rabbit antibodies (1:400 dilution) for 1 h.
7. After washing the cells three times for 15 min each with PBS, HRP enzymatic activity associated with the cells is determined by adding 100 μl of HRP assay solution for 10 min at room temperature, or until the brown reaction product becomes visible. The reaction is then stopped by adding 5 μl of 6N HCl.
8. 50 μl aliquots of the supernatant containing the reaction product are transferred to a 96-well plate and the absorbance is measured at 490 nm using a Bio-Rad Model 680 microplate reader.
9. HRP activity associated with permeabilized cells is taken to be the total cell-associated anti-loop antibody (surface bound + internalized) and the value is set as 100%. The fraction associated with intact cells represents the cell surface-bound anti-loop antibody (surface bound). The fraction representing the internalized fraction is calculated by subtracting the surface bound from the total fraction.

3.10. Analysis of Proteasomal Degradation of Cldn16

1. Inhibition of proteasomal degradation will allow the detection of Cldn16 mutants that are retained in the ER and rapidly degraded.
2. After 24-h transfection, the cells are incubated for 10 h in the presence or absence of 100 μM ALLN in complete media at 37°C.
3. The cells are then processed for immunofluorescence analysis as described in Subheading 3.5 using rat-anti-HA and mouse anti-ubiquitin antibodies and the appropriate secondary antibodies.
4. To analyze Cldn16 protein turnover, HEK 293T cells in 6-well plates are transfected with wild-type or mutant Cldn16 cDNAs. For HEK 293T cells, growth on polylysine-coated surfaces may be required as they tend to detach from the plate when solutions are pipetted. Coverslips and culture plates are coated with 0.1% polylysine (sigma) for 10 min.

5. At 48 h post-transfection, the cells are incubated with proteosomal inhibitor ALLN (100 μM) for different periods of time up to 8 h. A carrier-treated control is included.
6. The cells are harvested at different time intervals and processed for SDS–PAGE and Western blot analysis using rat-anti-HA antibodies to detect exogenously expressed Cldn16 and mouse anti-actin antibodies to check for equal loading.

3.11. Rescue of Cldn16 Mutants Using Pharmacological Chaperones

1. Cells expressing Cldn16 mutants that are retained in the ER are treated with pharmacological chaperones such as 1 μM thapsigargin for 3 h or 1 mM 4-phenyl sodium butyrate (4-PBA) for 24–48 h at 37°C in complete medium.
2. The anti-loop antibodies are added to these cells during the last hour of the incubation for 1 h at 37°C.
3. The cells are then chilled on ice, washed three times for 15 min each with ice-cold PBS containing 0.9 mM $CaCl_2$ and 0.5 mM $MgCl_2$, fixed in PFA, and processed for immunofluorescence microscopy as detailed in Subheading 3.5. The cells are stained with rat anti-HA antibodies followed by Alexa Fluor 594-labeled goat anti-rat IgG antibodies to detect all Cldn16 molecules. The anti-loop antibody is detected with Alexa Fluor 488-labeled goat anti-rabbit IgG antibodies.
4. For quantification, 100 cells expressing Cldn16 are randomly selected and the fraction of cells showing cell surface expression (i.e., binding or uptake of anti-loop antibody) in the presence or absence of pharmacological chaperones is determined.

4. Notes

1. Unless otherwise stated, all solutions are prepared in sterile, deionized water.
2. MDCK II and MDCK-C7 cells are difficult to transfect, hence the generation of stable cell lines. Transient transfection will generate sufficient cells expressing the protein of interest for detection by immunofluorescence microscopy, but generally not produce sufficient protein for biochemical analysis.
3. Instead of pipette tips to pick up stably transfected clones, cloning rings can also be used. Glass or stainless steel cloning rings are autoclaved before use and dipped in autoclaved silicone before placing around the cell clone. A drop of trypsin–EDTA solution is added into the cloning ring and detachment of the cells from the dish monitored under a microscope. The detached cells are then carefully transferred with a wide-tip pipette into 24-well plates for expansion and screening.

4. Cldn16 protein tends to aggregate upon heating at 95°C for 10 min, thus samples for SDS–PAGE are heated to 50°C for 10 min.
5. Round glass 12-mm coverslips require only 20 μl of diluted antibody solution. Before adding the antibody solution to the coverslips, any excess liquid is completely aspirated and the coverslips are centered in the well with a forceps. This prevents the antibody solution from running off the coverslip during the incubation.

Acknowledgments

Our work is supported by the Agency for Science, Technology and Research, Singapore (A*STAR).

References

1. Simon, D. B., Lu, Y., Choate, K. A., Velazquez, H., Al Sabban, E., Praga, M., Casari, G., Bettinelli, A., Colussi, G., Rodriguez-Soriano, J., McCredie, D., Milford, D., Sanjad, S., and Lifton, R. P. (1999) Paracellin-1, a renal tight junction protein required for paracellular Mg^{2+} resorption, *Science* **285**, 103–106.
2. Muller, D., Kausalya, P. J., Bockenhauer, D., Thumfart, J., Meij, I. C., Dillon, M. J., van't Hoff, W., and Hunziker, W. (2006) Unusual clinical presentation and possible rescue of a novel claudin-16 mutation, *J. Clin. Endocrinol. Metab.* **91**, 3076–3079.
3. Muller, D., Kausalya, P. J., Claverie-Martin, F., Meij, I. C., Eggert, P., Garcia-Nieto, V., and Hunziker, W. (2003) A novel claudin 16 mutation associated with childhood hypercalciuria abolishes binding to ZO-1 and results in lysosomal mistargeting, *Am. J. Hum. Genet.* **73**, 1293–1301.
4. Muller, D., Kausalya, P. J., Meij, I. C., and Hunziker, W. (2006) Familial hypomagnesemia with hypercalciuria and nephrocalcinosis: blocking endocytosis restores surface expression of a novel Claudin-16 mutant that lacks the entire C-terminal cytosolic tail, *Hum. Mol. Genet.* **15**, 1049–1058.
5. Weber, S., Hoffmann, K., Jeck, N., Saar, K., Boeswald, M., Kuwertz-Broeking, E., Meij, II, Knoers, N. V., Cochat, P., Sulakova, T., Bonzel, K. E., Soergel, M., Manz, F., Schaerer, K., Seyberth, H. W., Reis, A., and Konrad, M. (2000) Familial hypomagnesaemia with hypercalciuria and nephrocalcinosis maps to chromosome 3q27 and is associated with mutations in the PCLN-1 gene, *Eur. J. Hum. Genet.* **8**, 414–422.
6. Weber, S., Schneider, L., Peters, M., Misselwitz, J., Rönnefarth, G., Böswald, M., Bonzel, K. E., Seeman, T., Suláková, T., Kuwertz-Bröking, E., Gregoric, A., Palcoux, J. B., Tasic, V., Manz, F., Schärer, K., Seyberth, H. W., and Konrad, M. (2001) Novel paracellin-1 mutations in 25 families with familial hypomagnesemia with hypercalciuria and nephrocalcinosis, *J. Am. Soc. Nephrol.* **12**, 1872–1881.
7. Itoh, M., Furuse, M., Morita, K., Kubota, K., Saitou, M., and Tsukita, S. (1999) Direct binding of three tight junction-associated MAGUKs, ZO-1, ZO-2, and ZO-3, with the COOH termini of claudins, *J. Cell Biol.* **147**, 1351–1363.
8. Lelievre-Pegorier, M., Merlet-Benichou, C., Roinel, N., and de Rouffignac, C. (1983) Developmental pattern of water and electrolyte transport in rat superficial nephrons, *Am. J. Physiol.* **245**, F15.
9. Simon, D. B., Lu, Y., Choate, K. A., Velazquez, H., Al-Sabban, E., Praga, M., Casari, G., Bettinelli, A., Colussi, G., Rodriguez-Soriano, J., McCredie, D., Milford, D., Sanjad, S., and Lifton, R. P. (1999) Paracellin-1, a renal tight junction protein required for paracellular Mg^{2+} resorption, *Science* **285**, 103–106.
10. Hou, J., Renigunta, A., Konrad, M., Gomes, A. S., Schneeberger, E. E., Paul, D. L., Waldegger, S., and Goodenough, D. A. (2008)

Claudin-16 and claudin-19 interact and form a cation-selective tight junction complex, *J. Clin. Invest.* **118**, 619–628.

11. Hou, J., Renigunta, A., Gomes, A. S., Hou, M., Paul, D. L., Waldegger, S., and Goodenough, D. A. (2009) Claudin-16 and claudin-19 interaction is required for their assembly into tight junctions and for renal reabsorption of magnesium, *Proc. Natl. Acad. Sci. U. S. A.* **106**, 15350–15355.
12. Gunzel, D., Amasheh, S., Pfaffenbach, S., Richter, J. F., Kausalya, P. J., Hunziker, W., and Fromm, M. (2009) Claudin-16 affects transcellular Cl^- secretion in MDCK cells, *J. Physiol.* **587**, 3777–3793.
13. Gunzel, D., Haisch, L., Pfaffenbach, S., Krug, S. M., Milatz, S., Amasheh, S., Hunziker, W., and Muller, D. (2009) Claudin function in the thick ascending limb of Henle's loop, *Ann. N. Y. Acad. Sci.* **1165**, 152–162.
14. Kausalya, P. J., Amasheh, S., Gunzel, D., Wurps, H., Muller, D., Fromm, M., and Hunziker, W. (2006) Disease-associated mutations affect intracellular traffic and paracellular Mg^{2+} transport function of Claudin-16, *J. Clin. Invest.* **116**, 878–891.
15. Hou, J., Paul, D. L., and Goodenough, D. A. (2005) Paracellin-1 and the modulation of ion selectivity of tight junctions, *J. Cell Sci.* **118**, 5109–5118.
16. Rubenstein, R. C., and Zeitlin, P. L. (1998) A pilot clinical trial of oral sodium 4-phenylbutyrate (Buphenyl) in deltaF508-homozygous cystic fibrosis patients: partial restoration of nasal epithelial CFTR function, *Am. J. Respir. Crit. Care Med.* **157**, 484–490.
17. Zeitlin, P. L., Diener-West, M., Rubenstein, R. C., Boyle, M. P., Lee, C. K., and Brass-Ernst, L. (2002) Evidence of CFTR function in cystic fibrosis after systemic administration of 4-phenylbutyrate, *Mol. Ther.* **6**, 119–126.
18. Höning, S., and Hunziker, W. (1995) Cytoplasmic determinants involved in direct lysosomal sorting, endocytosis, and basolateral targeting of rat lgp120 (lamp-I) in MDCK cells, *J. Cell Biol.* **128**, 321–332.
19. Matlin, K. S., and Simons, K. (1983) Reduced temperature prevents transfer of a membrane glycoprotein to the cell surface but does not prevent terminal glycosylation, *Cell* **34**, 233–243.
20. Ulloa-Aguirre, A., Janovick, J. A., Brothers, S. P., and Conn, P. M. (2004) Pharmacologic rescue of conformationally-defective proteins: implications for the treatment of human disease, *Traffic* **5**, 821–837.
21. Sandvig, K., Olsnes, S., Petersen, O. W., and van Deurs, B. (1987) Acidification of the cytosol inhibits endocytosis from coated pits, *J. Cell Biol.* **105**, 679–689.

Chapter 11

Claudin Family Proteins in *Caenorhabditis elegans*

Jeffrey S. Simske and Jeff Hardin

Abstract

In the last decade, the claudin family of integral membrane proteins has been identified as the major protein component of the tight junctions in all vertebrates. The claudin superfamily proteins also function to regulate channel activity, intercellular signaling, and cell morphology. Subsequently, claudin homologues have been identified in invertebrates, including *Drosophila* and *Caenorhabditis elegans*. Recent studies demonstrate that the *C. elegans* claudins, *clc-1* to *clc-5*, and similar proteins in the greater PMP22/EMP/claudin/calcium channel γ subunit family, including *nsy-1*-*nsy-4* and *vab-9*, while highly divergent at a sequence level from each other and from the vertebrate claudins, in some cases play roles similar to those traditionally assigned to their vertebrate homologues. These include regulating cell adhesion and passage of small molecules through the paracellular space. The claudin superfamily proteins also function to regulate channel activity, intercellular signaling, and cell morphology. Study of claudin superfamily proteins in *C. elegans* should continue to provide clues as to how core claudin protein function can be modified to serve various specific roles at regions of cell–cell contact in metazoans.

Key words: Claudin, Junctions, *Caenorhabditis elegans*, Epithelia, Morphogenesis, Actomyosin, Neuronal symmetry, VAB-9, CLC-1, NSY-4

1. Introduction: Claudins Regulate Barrier Functions in Vertebrate Epithelia

Epithelia function as regulated barriers in tissue and organ architecture, defining compartment boundaries within organisms and forming boundaries apposed to the external environment. The proper physiological state of epithelia depend on cellular contacts, or junctions, that mediate cell–cell adhesion, function as selective gates for the ingress or egress of specific ions and small molecules, and maintain unique membrane identities on apical and basal sides of the cellular layer, thus contributing to epithelial polarity. The tight junction is the most apical of cellular junctions in vertebrate epithelia. Freeze fracture TEM reveals one of the most striking features of tight junctions, a network of intertwined

Kursad Turksen (ed.), *Claudins: Methods and Protocols*, Methods in Molecular Biology, vol. 762,
DOI 10.1007/978-1-61779-185-7_11, © Springer Science+Business Media, LLC 2011

strands that occlude the intercellular or paracellular space at points of contact that often circumscribe cells. In this way, tight junction strands function as gaskets that separate and isolate the two sides of an epithelium (Fig. 1). The major protein component of the tight junction has been identified in the last decade as the claudin family of proteins. Claudins were originally identified in a search for the protein component of tight junction strands in vertebrate epithelia following a painstaking purification scheme (1, 2). Claudins 1 and 2 were shown to localize to tight junctions, to generate tight junction strands when expressed in fibroblasts, and to mediate homotypic and heterotypic cell–cell adhesion (1, 2). Consistent with these findings, loss of claudin activity results in paracellular barrier dysfunction (3–8). Overall tissue polarity does not appear to be heavily dependent on claudin activity,

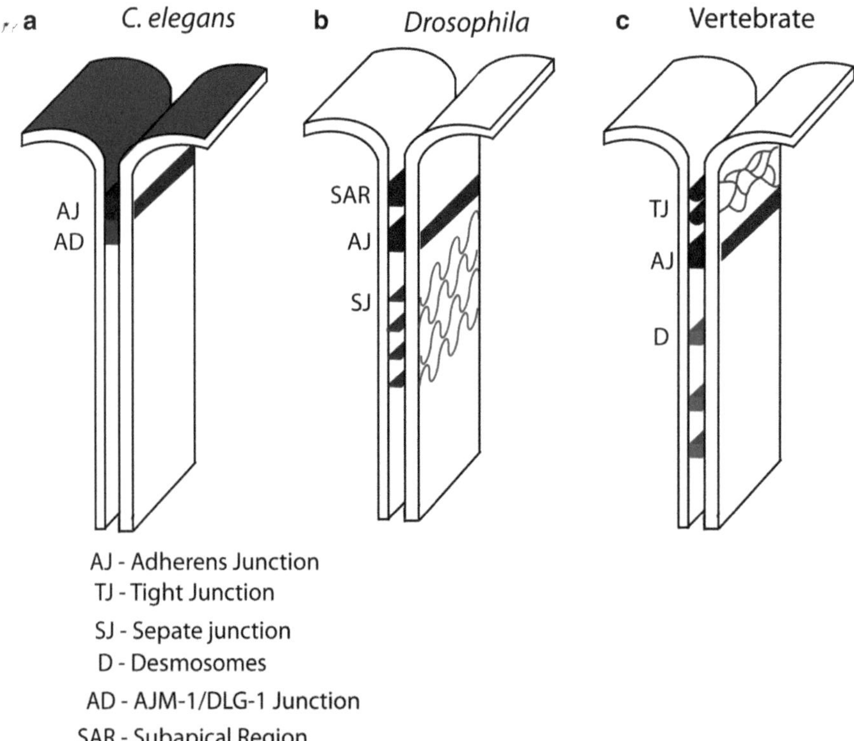

Fig. 1. Schematic of various cell junction arrangements from *C. elegans*, *Drosophila*, and vertebrates. Corresponding cell junction regions either in relative location, function, and/or molecular make-up are indicated with similar colors. The SAR-like region in *C. elegans,* the subapical region (SAR) in *Drosophila*, and tight junction (*TJ*), in vertebrates, respectively, are indicated with *black shading*. *AJ* indicates the cadherin-based adherens junction. *SJ* identifies the septate junction in *Drosophila*, and *AD* identifies the AJM-1/ DLG-1 region in *C. elegans.* Vertebrates have no similar structure in epithelia, but share a molecularly similar barrier at the paranodal junction (see text). Vertebrate desmosomes are shown *D* Lateral views (*cutaways*) show the nature of the junctional structures in the membranes. Adherens junctions appear as solid bands in the membrane, Pleated septate junctions are characterized by regular wave-like strands, and tight junctions appear as irregular but connected anastomosing strands. For simplicity, only the composition of cell junction components in the paracellular space is shown.

since gross cell polarity is maintained following the loss of claudin function (6). The vertebrate claudin family has expanded over the years to include about 24 members all of which are predicted four-pass integral membrane proteins with cytoplasmic N and C termini, two extracellular loops, and one short intracellular loop (9–11).

Claudins have been shown to form oligomers in vitro, with hexamers being favored (12–14). This organization suggests that claudins, like gap junctions, might generate pores through the lipid bilayer, although there is no clear evidence at present that this is the case. Rather, claudins appear to establish charged pores *within* the paracellular space that regulate the traffic of charged ions across the epithelial barrier. Charged residues in the extracellular domain are responsible for regulating charge selectivity. As a result, claudins with different charged amino acids can, by virtue of their expression in overlapping and nonoverlapping patterns, define the paracellular charge selectivity of a wide variety of tissues (15). For example, in the vertebrate kidney, different claudins are expressed along the length of the nephron. Their differential expression may help to generate differing charge selectivity within distinct regions of the nephron (16). Supporting this hypothesis, claudin 16 maintains a negative charge required for the reuptake of Mg^{2+} and Ca^{2+} ions in the thick ascending part of Henle's loop in the nephron and loss of claudin 16 activity results in magnesium and calcium wasting (17). Similarly, in the longitudinal axis of the intestine as well as the crypt to villus axis of the intestine, the claudins are expressed in complex patterns (18). Charge selectivity of claudins is most dramatically illustrated by experiments in which the charge of the extracellular loops is reversed by swapping amino acids in this loop; in this case, the charge selectivity is reversed, as was demonstrated for claudins 10 and 15 (19, 20). In general, amino acid charge in extracellular loop 1 dramatically influences paracellular charge selectivity.

Claudins are also targets of viruses, and binding of viruses to extracellular claudin loops can disrupt epithelial barriers, open the paracellular space, and allow viral entry (21–23). The claudin C terminus functions in trafficking of the claudin protein, binding to junctional plaque proteins through PDZ domain binding motifs, or regulation of the paracellular space through phosphorylation (24–29). While the PDZ-binding peptide is found in most claudins, this domain is absent from many of the C termini in related protein families of both vertebrates and invertebrates, suggesting that there may be as many functions as there are unique C-terminal domains.

While much is now known about the function of vertebrate claudins, less is known about similar molecules in invertebrates. Since tight junction structures in invertebrates are uncommon, claudin-like molecules in invertebrates will necessarily have functions distinct from those of their vertebrate relatives (30). This review will detail the known functions of this class of proteins in *Caenorhabditis elegans*.

2. Sequence Conservation Among Claudin Family Proteins

The sequence similarity of *C. elegans* and vertebrate claudin family proteins is so low that it was proposed that sequence similarity was insufficient to establish homology or that claudins are present only in chordates (9, 31). Despite these initial assessments, claudin family proteins were ultimately identified in *C. elegans* by sequence homology searches and also through the identification of claudin-related proteins participating in diverse processes. Here we will review what is known about *vab-9, nsy-4, clc-1–5*, and related claudin superfamily genes. The comparison of *C. elegans* sequences with each other and with claudin family proteins from vertebrates indicates that the sequences are indeed highly divergent (Fig. 2). A similar observation was made when *Drosophila* claudins, including *Sinuous* and *Megatrachea*, were placed into a claudin family phylogenetic tree (32). A previous analysis of *Stargazin* indicated that there were at least three distinct clades within this greater tetraspan family: PMP22/EMP, claudin, and the gamma subunits of voltage-gated calcium channels (33). Generally, PMP22/EMP/MP/Claudin-related proteins belong to pfam00822 (15). Analysis of these protein groupings, in conjunction with VAB-9 and VAB-9 orthologs from *Drosophila* and vertebrates, indicates that a separate VAB-9 clade exists, which is distinct from other subgroups (Fig. 2). Previously PERP, a tetraspan protein required for desmosomal structure and epithelial integrity, was recognized as the closest ortholog to *TM4SF10* (or *TMEM47*) in the mouse (34, 35) (JSS and L. Attardi, unpublished observations). In our phylogenetic analysis, PERP associated with the VAB-9 clade (Fig. 2). Despite being closer to VAB-9 in homology, expression of mouse *PERP* in *C. elegans* epithelia is incapable of rescuing *vab-9* phenotypes, while similar expression of *TM4SF10* rescues *vab-9* (36) (JSS and L. Attardi, unpublished observations). Interestingly, NSY-4 clustered with *C. elegans* claudins, whereas some vertebrate gamma subunits clustered separately. Bootstrap analysis shows that the frequency of particular association nodes for *C. elegans* claudins are typically low, highlighting the diversity of this protein family.

Previously, a consistent motif was found in the first extracellular loop of the claudin superfamily of proteins. Very roughly, this consensus corresponds to W(8 or >X)GLWXXC(8-10X)C (Fig. 3).

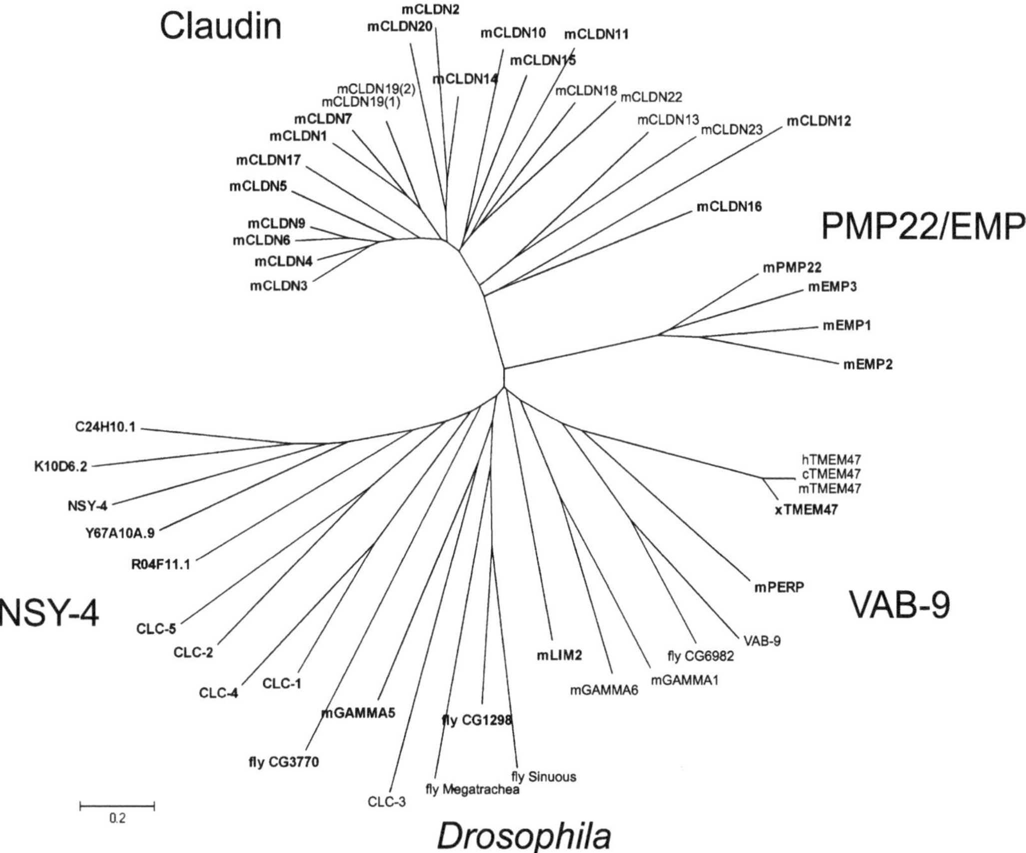

Fig. 2. Claudin family proteins loosely cluster in highly divergent clades. The phylogenetic tree resulting from ClustalW2 analysis. Several different subgroup clades are observed. Phylogeny was constructed using the MEGA (Molecular Evolutionary Genetic Analysis) software program available at http://www.megasoftware.net (115). Units are in number of amino acid substitutions per site.

Direct sequence comparison shows that VAB-9 is not entirely conserved in this region, since VAB-9 lacks the highly conserved tryptophan in the GLW tripeptide. Nevertheless, the overall topology of VAB-9, including many of the residues in the motif, suggests that it may share structural features with other members of the claudin superfamily. Other proteins from *C. elegans* tend to lack one or more key residues but retain a broad similarity. Whether this similarity translates to a conserved structure will require functional tests. Since the claudin family has undergone extensive divergence, the study of claudin-like proteins in *C. elegans* and other invertebrates is likely to reveal both the very basic essential functions of the claudin family of proteins as well as how different members of this divergent family are adapted to unique functions.

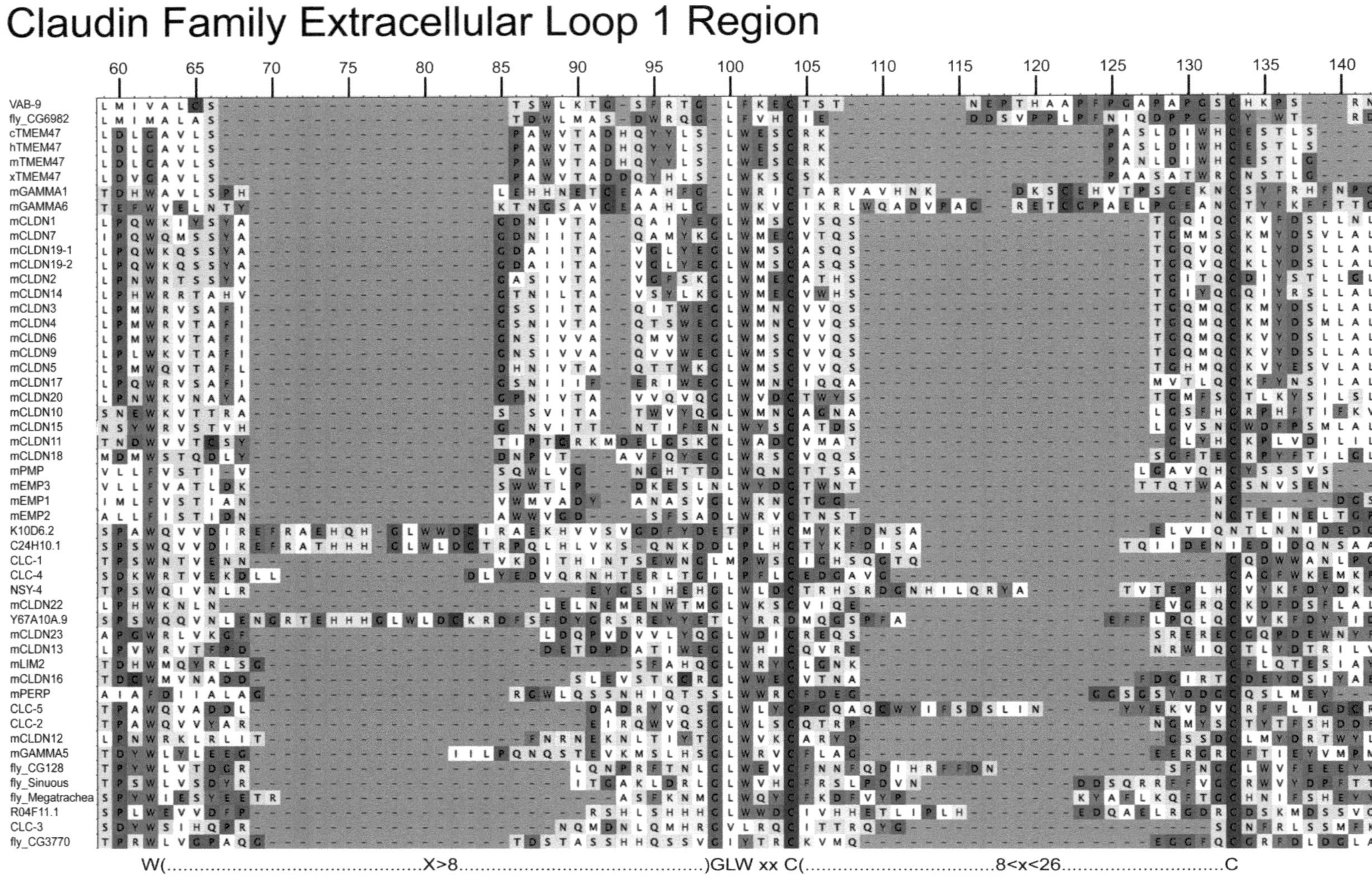

Claudin Family Extracellular Loop 1 Region
60
65
70
75
80
85
90
95
100
105
110
115
120
125
130
135
140
VAB-9
fly_CG6982
cTMEM47
hTMEM47
mTMEM47
xTMEM47
mGAMMA1
mGAMMA6
mCLDN1
mCLDN7
mCLDN19-1
mCLDN19-2
mCLDN2
mCLDN14
mCLDN3
mCLDN4
mCLDN6
mCLDN9
mCLDN5
mCLDN17
mCLDN20
mCLDN10
mCLDN15
mCLDN11
mCLDN18
mPMP
mEMP3
mEMP1
mEMP2
K10D6.2
C24H10.1
CLC-1
CLC-4
NSY-4
mCLDN22
Y67A10A.9
mCLDN23
mCLDN13
mLIM2
mCLDN16
mPERP
CLC-5
CLC-2
mCLDN12
mGAMMA5
fly_CG128
fly_Sinuous
fly_Megatrachea
R04F11.1
CLC-3
fly_CG3770
W(.........X>8.........)GLW xx C(.........8<x<26.........C

3. Septate Versus Tight Junctions: Cell Junctions in *C. elegans*, *Drosophila*, and Vertebrates

While cadherin-based junctions appear to be present in epithelia all throughout the animal kingdom, the position and morphology of occluding junctions vary among metazoans. The most apparent distinction is the location and structure of cell junctions (Fig. 1). In vertebrates, the tight junction is apical to the adherens junction, while invertebrate junctions typically have no significant occluding junction apical to the adherens junction. Although some lower chordates and invertebrates have tight junction structures, most invertebrate epithelia have instead a different junctional structure, the septate junction, localized basal to the adherens junction (37–39). Septate junctions consist of at least two types: pleated septate junctions, which are characterized by ladder-like bridges, and smooth septate junctions, in which bridges are not detected using electron microscopy (30, 40). Pleated septate junctions are often distributed all along the paracellular space, from just basal to the adherens junction to the basolateral surface (41). The separation across the paracellular space at the septate junction is about 10–20 nm and there are no "kissing points," as in vertebrate tight junctions, where there is effectively no paracellular separation and opposing membranes appear to directly touch (42). The actual paracellular space-spanning component of the pleated septate junction is currently unknown, although it is likely to be a protein or protein–carbohydrate complex, much like other cell junctions. Despite the continuous pleats in septate junctions, freeze fracture TEM of insect and *Ascaris* epithelia typically reveals isolated contacting puncta, rather than the continuous strands of vertebrates tight junctions (43, 44). Despite the freeze fracture appearance of a

Fig. 3. Claudin family extracellular loop 1 region. A short stretch of homology is shown between all family members in the region of the first extracellular loop. The first loop includes a consensus W(x>8)GLWxxC(8-26x)C motif. The two cysteines may form an intramolecular disulfide bridge important for claudin extracellular structure. Aligned sequence file is displayed using Cinema v1.4.5, Advanced Interfaces Group. Accession numbers for proteins included in the alignments are: NSY-4 (NP_500189.4), K10D6.2(NP_505843), R04F11.1 (NP_506087), C24H10.1(NP_508863), Y67A10A.9(NP_502746.1), mGamma5 (voltage-dependent calcium channel gamma subunit 5) (NP_542375), mGamma1 (NP_031608), mGamma6 (NP_573446.1), mGamma3 (NP_062303), CLC-1(NP_509847), CLC-2(NP_509257), CLC-3 isoform a(NP_001024993), CLC-4 (NP_509800), CLC-5(NP_509258), VAB-9 (NP_495836), CG6982 (dVAB-9) (NP_001097876), mTMEM47(NP_620090), cTMEM47 (NP_001003045.1), hTm47 (NP_113630.1), xtmem47 (NP_001085134.1), mclaudin1 (NP_057883), mclaudin2 (NP_057884), mclaudin3 (NP_034032), mclaudin4 (NP_034033), mclaudin5 (NP_038833), mclaudin6 (NP_0247), mclaudin7 (NP_058583), mclaudin9 (NP_064689), mclaudin10 (a) (NP_076367), mclaudin11 (NP_032796), mclaudin12 (NP_075028), mclaudin13 (NP_065250), mclaudin14 (NP_001159398), mclaudin15 (NP_068365), mclaudin16 (NP_444471), claudin17 (NP_852467), mclaudin18(NP_062789), mclaudin19(1) (NP_001033679), Claudin19(2)(NP_694745), mclaudin20 (NP_001095030), mclaudin22 (NP_083659), mclaudin23 (NP_082274), fly_Sinuous (a) (NP_647971), fly_Megatrachea (NP_726742), fly_CG3770 (NP_611985.), fly_CG1298 (NP_610179), mEMP2 (NP_031955), mPMP22 (NP_032911), mEMP3(NP_001139818), mEMP1(NP_034258), mPERP(NP_071315), LIM2 (NP_808361).

more porous and incomplete barrier, the septate junction structure still fulfills the function of a barrier, since the loss of proteins that localize to the septate junction in *Drosophila*, including claudin-like proteins *Megatrachea* and *Sinuous*, result in the disruption of junctional structure, epithelial cell adhesion, and the paracellular barrier gate function, based on Dextran-TRITC dye exclusion assays (32, 45–47).

More recently, molecular analysis of tight and septate junction components underscores the molecular differences between TJs and SJs, and suggests that the SJ may be more like the vertebrate lateral cell membrane than a vertebrate TJ. The vertebrate tight junction protein complexes containing PAR-3/PAR-6/aPKC/CDC-42 or CRB3/Pals1/PATJ have homologues in *Drosophila* that localize to the subapical region (also known as the apical marginal zone), a region *apical* to the adherens junction (48). In contrast, a distinct complex, including the MAGUKs Varicose and Dlg, the Erm protein Coracle, Neurexin IV, and other proteins localizes basal to the adherens junction at the SJ (46, 47, 49–52). Thus, although tight junctions and septate junctions share some common functions, they are not analogous structures at the structural and molecular levels.

Although there is a clear divergence in deployment of proteins to occluding junctions in vertebrates and invertebrates, recent evidence does suggest a conserved molecular complex in vertebrates that corresponds to the invertebrate SJ. Many of the protein components of the epithelial septate junction in flies, including the claudins *Sinuous* and *Megatrachea*, also localize to and are required for maintenance of the septate junction between glial and neuronal cells of the chordotonal sensory organ structure (46, 53–56). The chordotonal organ is analogous to the paranodal junction in vertebrates, and analogous cell junctions and similar proteins in both effectively establish the blood brain barrier (57, 58). For example, claudin-5 contributes to the blood brain barrier in the mouse brain (3, 4). The designation of analogous junctional structures from diverse organisms has been reexamined based on more recent and comprehensive structural, molecular, and functional studies of various cellular junctions. As more is learned about these complex structures, appreciation for their unique, specialized functions increases.

4. Function of Junctional Proteins in *C. elegans*: Variations on an Ecdysozoan Theme

Phylogenetically, nematodes are grouped with other ecdysozoans, including arthropods such as *Drosophila* (59). Unlike in *Drosophila*, however, in *C. elegans*, the epidermis contains a single discernable electron dense junctional region (60). Although the spatial distribution of specific molecular complexes within this single electron density has not been established, *C. elegans* epithelia contain many of the same proteins as in *Drosophila*, with similar spatial ordering. *C. elegans* possesses a single homologue of classical cadherins, HMR-1, and associated proteins HMP-2/β-catenin, HMP-1/α-catenin, and a divergent p120ctn, JAC-1 (61, 62). These proteins appear to localize to the nematode equivalent of adherens junctions. Similarly, in *C. elegans*, the discs large homologue DLG-1 and its binding partner AJM-1 localize basal to the adherens junction. Costaining demonstrates that these cell junction proteins localize to different regions of the lateral cell membrane, with the adherens junction proteins being more apical (63–66). A third localization domain, apical to the adherens junction, which may extend from the junctions across the apical surface, contains the familiar Par-3–Par-6–aPKC complex (Fig. 4) (63, 65–68). Other proteins, not part of the Par complex, such as

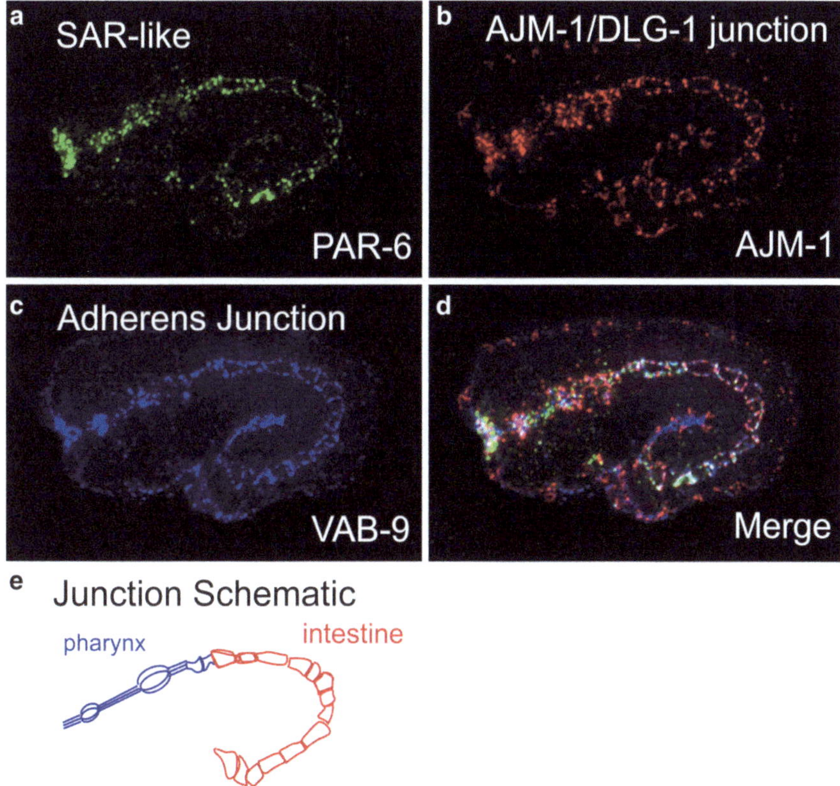

Fig. 4. Cell junctions in the embryonic alimentary system. Staining of the SAR-like region, the DLG-1/AJM-1 domain, and the adherens junction, with PAR-6 (**a**), AJM-1 (**b**), and VAB-9 (**c**), respectively, at pharyngeal and intestinal junctions, demarcating the apical–lateral regions of these cells, just below the apical surface that forms the lumen of the alimentary system. The staining of the markers only shows partial overlap (**d**, *white*), illustrating that the three junctional domains are all localized within a narrow lateral region, but remain largely distinct. At this stage, and in this preparation, most PAR-6 is closely associated with the cell junctions, and little is observed across the apical surface. (**e**) A schematic of cell junctions in the pharynx and intestine, determined by AJM-1 localization.

CHE-14 and EAT-20 (a crumbs homologue) are also localized across the apical surface (69, 70).

As in *Drosophila*, overall polarity of epithelial junctions is established and maintained by LET-413, the LAP (LRR and PDZ) domain containing protein homologue of *Drosophila* Scribble (71, 72). LET-413 localizes along the basolateral surface of epithelial cells, establishes global cell polarity and, therefore, is required for the localization of cellular junction components in epithelial cells and the structure of the intestinal terminal web, visualized by indirect immunofluorescence of cell junction components and by TEM of cell junctions (72, 73). Even though loss of *dlg-1* gene function can completely eliminate this cell junction as observed at the electron microscopy level, and cause the mislocalization of the AJM-1 protein, classical adherens junction proteins such as HMR-1 (cadherin), HMP-1 (α-catenin), and VAB-9 (TM4SF10) remain essentially localized, if not completely aggregated, in a narrow band. Similarly, loss of adherens junction proteins does not disrupt the localization of DLG-1 or AJM-1, suggesting that there are at least two independent junctional regions within or near the apical junction (63–67, 72). Furthermore, cell polarity and adhesion are only mildly disrupted in *ajm-1* and *dlg-1* RNAi animals, and these adhesion defects are enhanced by *vab-9* (67). Notably, the localization of apical proteins is not affected by the loss of adherens junction or DLG-1/AJM-1 domain proteins (66, 69). Loss of *dlg-1* can affect the localization of the *Drosophila* Crumbs homologue (CRB-1) protein, which is a component of a separate apical protein complex, but this protein is not required for polarity in *C. elegans* as is the case for *Drosophila Crumbs* (63). These findings indicate that there are at least three distinct apical cell junctional complexes in *C. elegans* epithelia that together contribute to cell adhesion and morphogenesis.

5. Spermathecal Junctions

Caenorhabditis elegans cellular junctions arising from the mesoderm, such as the cell junctions in the spermatheca of the somatic gonad, are distinct from the epidermal or intestinal epithelia. The spermatheca stores sperm in the hermaphrodite. During fertilization, the spermatheca expands dramatically to allow the passage of an oocyte from the syncytial gonad into the uterus (74–76). During this expansion, cell junctions must “unzip” on both sides of a stable cell junction so that the amount of cell membranes contributing to the expanded lumen is matched by an increase in the membranes comprising the basal membranes (77). Apical to the adherens junction is a junction with an appearance similar to a pleated septate junction, while basal is a junction with a smooth septate junction appearance. Immunostaining of TEM spermathecal preparations with the MH27 antibody, which recognizes AJM-1, shows that there are at least three unique junctions, one decorated by MH27 (Fig. 5) (77) (David Hall, personal communication). Pleated SJ are characterized by ladder-like crossbridges between cells and are found apical (luminal) to the adherens junction. Based on MH27 immunogold staining (A), and localization of AJM-1::Cherry (G), AJM-1 expression appears to be strongest in the region of the stable (adherens) junction, while MEL-11::GFP localizes to the pleated septate

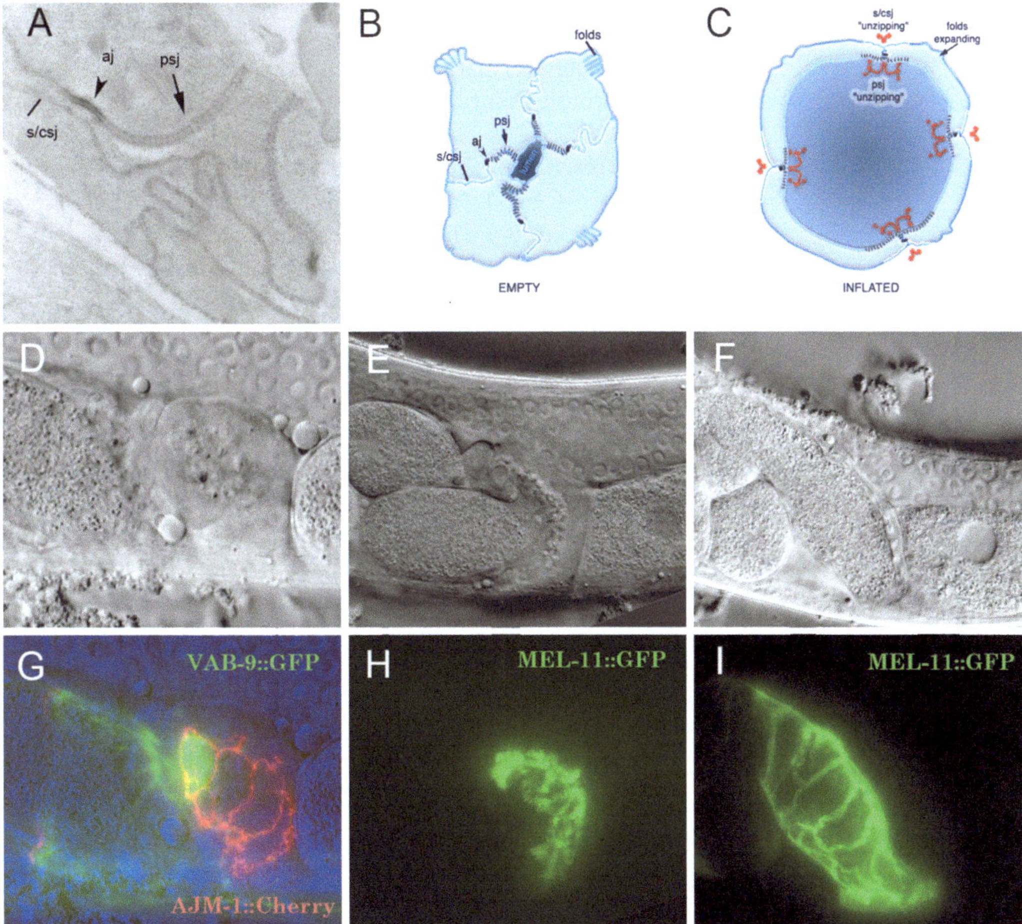

Fig. 5. Cell junctions in the spermatheca. (**a**) Immuno-TEM preparations of the spermatheca using the anti-AJM-1 antibody MH27 demarcates three distinct junctional regions: the pleated septate junction (pSJ), the adherens junction (AJ), and the smooth septate junction (sSJ). In this preparation, MH27 only decorates the adherens junction (**b**, **c**). Representative drawings of the spermatheca either collapased (**b**) or expanded (**c**) to allow an embryo to pass through and be fertilized (**b** and **c** are reprinted, with permission, from *C. elegans* Wormatlas) (77). (**e**, **f**) Expression of MEL-11::GFP at the pSJ is shown, matching the stages in the drawings above; (**f**) and (**i**) are the matching DIC images for (**e**) and (**f**), respectively. (**d**) Expression of AJM-1::Cherry (*red*) and VAB-9::GFP (*green*) are shown for a stage matching (b). Note that strong AJM-1 expression at the AJ is more basal than MEL-11 at the equivalent stage. In these adults, VAB-9::GFP expression in the spermatheca is reduced and is largely restricted to the spermatheca-uterine valve.

junction (Fig. 5h, i). MEL-11 regulates actomyosin-mediated contraction during cytokinesis and embryonic elongation, and also is required in the spermatheca for fertility (78–80). Since actomyosin at cellular junctions regulates both apical constriction and the organization of cellular junctions, it is exciting to speculate that *mel-11* regulates actomyosin contraction required for unzipping and zipping of the pleated spermathecal junctions during fertilization. How similar are the spermathecal junctions to those of *Drosophila* pSJs based on the expression of claudins and other SJ proteins? Of the claudin-like proteins in *C. elegans*, both VAB-9 and CLC-1 are expressed in the spermatheca, with VAB-9 being expressed during the larval stages of development when junctions are forming. As in epidermal tissues, VAB-9 expression in the spermatheca does not overlap with AJM-1 (67) and (JSS, unpublished observation). However, in adults, during fertilization, VAB-9 is absent from spermathecal junctions and instead is present in the spermatheca-uterine valve (Fig. 5). Although both CLC-1 (and DLG-1) are expressed in the spermatheca, it is unknown whether they colocalize with AJM-1 in or near the adherens junction, as they do in the epidermis, or whether they associate with MEL-11 in the septate junction (63, 64, 66, 81). Based on localization data, at least one claudin-like protein, VAB-9, is likely to be involved in the "unzippering" of the septate junction. The data described below suggests that among the claudin family proteins, such a role may be unique to VAB-9, however further analysis of claudin family protein function in the spermatheca will be required. In any event, understanding the role of VAB-9 in the rapid, dynamic reorganization of cellular junctions should prove fascinating, particularly if regulation occurs through VAB-9 influence on nonmuscle myosin, as we suggest below.

6. VAB-9: A Claudin-Like Protein at Adherens Junctions

VAB-9 is an approximately 22 kDa protein with similarity to the PMP22/EMP/claudin/gas3 family of four pass membrane spanning proteins. VAB-9 is expressed in all *C. elegans* epithelia and colocalizes with the adherens junction proteins HMR-1 (cadherin) and HMP-1 and 2 (α- and β-catenin, respectively). HMR-1 is required for VAB-9 membrane localization, and HMP-1 is required to maintain uniform circumferential VAB-9 distribution about the adherens junction. *vab-9* mutants have defects in body morphology, likely due to defective filamentous actin organization in epidermal cells. Mutations in *vab-9* enhance the morphological defects of weak *hmp-1* loss of function and enhance cell adhesion defects in *ajm-1* and *dlg-1* animals. Thus, *vab-9* participates in the organization of F-actin at the adherens junction and, alongside the AJM-1/DLG-1 complex, maintains proper epithelial adhesion.

Clues to the general functions of VAB-9 may come from the vertebrate ortholog, TM4SF10, previously known as BCMP1. *TM4SF10* is expressed strongly in the canine brain, and based on SAGE analysis, is expressed in human brain astrocytoma, ependymoma, and normal spinal cord. This finding, along with the genetic position of *TM4SF10* on the X-chromosome at p21.1, suggests an association with hereditary X-linked mental retardation loci; however, no disease-associated changes were detected in the *TM4SF10* locus in XLMR patients from 14 unrelated families (82, 83). Thus, mutations in *TM4SF10* are either rare among affected individuals or *TM4SF10* is involved in separate processes in the brain. In the developing mouse kidney, *TM4SF10* is expressed transiently in podocyte precursors, and expression diminishes as the cell junctions of these precursors transition from an occluding-type junction typical of columnar epithelia to the specialized adherens junctions of slit diaphragms (36). Slit diaphragms are organized around the transmembrane proteins of the nephrin family (84, 85). Nephrin and the related protein, Neph1, participate in homo- and heterotypic intercellular adhesion along interdigitated podocyte cell extensions (86, 87). Slit diaphragm development results from signaling through Nephrin family proteins cytoplasmic domains which in turn effect changes in the underlying actin cytoskeleton, cell process extension, and cell process interdigitation, ultimately generating the podocyte side of the filtration barrier in the glomerulus. Following phosphorylation of Nephrin and Neph1 by the Src family kinase Fyn, Nck, PI3K, and Grb-2 bind specific phosphotyrosines and signal to reorganize the actin cytoskeleton (88–95). Interestingly, our unpublished data (L. Bruggeman and JSS) indicates that *TM4SF10* expression regulates the activity of Fyn and Nephrin maturation in podocytes, suggesting that *TM4SF10* may regulate the ability of nephrin to control actin dynamics and cell process extension, possibly by influencing Fyn phosphorylation of Neprhin. Consistent with this hypothesis is the finding that overexpression of *TM4SF10* phenocopies inhibition of neurite outgrowth by RhoA overexpression following NGF stimulation of PC12 cells (96). Similarly, unpublished data from one of our laboratories (JSS) showed that the overexpression of *TM4SF10* in MDCK cells blocks Fyn-dependent cell process extension. These results indicate that *TM4SF10* may have a general role in maintaining a columnar epithelia phenotype indirectly by preventing the formation of actin-dependent cell extensions. To date, interactions between VAB-9 and homologues of Nephrin and Fyn in *C. elegans* have not been investigated, although FRK-1, a non-receptor FER-like kinase, is required for embryonic morphogenesis (97–99). It will be of future interest to explore the interactions between TM4SF10, Fyn, and Nephrin and determine whether such interactions are unique to vertebrate podocytes or whether they are conserved in all epithelial cell types that express these proteins.

Further insight into the function of *vab-9* comes from an analysis of the MAGUK protein ZOO-1. ZOO-1 is the *C. elegans* ortholog of the tight junction membrane-associated guanylate kinase (MAGUK) protein ZO-1 (100). ZOO-1 is a predicted 129 kDa protein with three PDZ motifs in the N terminus, an SH3 domain, a guanylate kinase domain (predicted to be inactive), and a C-terminal ZU-5 domain, unique to this class of MAGUKS and netrins (47). Several data indicate that *zoo-1* closely interacts with *vab-9*. First, ZOO-1 localizes to cell junctions in the epidermis and, as VAB-9 requires HMR-1 (cadherin), but not HMP-1 or HMP-2, for junctional localization. Second, *zoo-1* localization requires VAB-9. Third, similar to *vab-9* mutants, loss of *zoo-1* function enhances the morphogenetic phenotypes of a weak *hmp-1* loss of function allele. Fourth, loss of *zoo-1* and *vab-9* function affects the organization of filamentous actin in the epidermis of elongating embryos. Fifth, loss of *zoo-1* function does not enhance the phenotypes of *vab-9* mutants, suggesting that they act in a common pathway. Sixth, both *vab-9* and *zoo-1* interact with mutations in genes of the actomyosin contractile machinery. Loss of *zoo-1* activity enhances *mel-11* (myosin light chain phosphatase) mutants and suppresses *let-502* (ROCK) mutants, whereas *vab-9* mutations suppress *mel-11* alleles (JSS, unpublished data). Surprisingly, *zoo-1* mutants alone have virtually no effect on morphogenetic phenotypes, although very mild abnormalities of F-actin accumulation at junctions were noted (100). Together these results suggest a regulatory pathway in which HMR-1 regulates VAB-9 and *vab-9*, in turn, regulates a subset of epidermal morphogenetic processes; a subset of these processes are in turn mediated by *zoo-1* and *mel-11* in both dependent and independent pathways. Since our own unpublished data suggests that VAB-9 may be required for MEL-11 subcellular localization in the epidermis, it should be interesting to determine the nature of the interactions between these protein classes in regulating F-actin organization at epidermal cell junctions and within epidermal epithelia, particularly since ZO-1 has been shown to influence the paracellular barrier through the regulation of perijunctional actomyosin (101).

7. Claudin-Like Genes *clc-1* to *clc-5*

Claudin family proteins exist in all metazoans. In humans and mice, there are as many as 24 different claudins. To date, five claudin-like proteins have been identified in *C. elegans* by virtue of sequence identity to mammalian claudins or to identified *C. elegans* claudins (81). Not surprisingly, sequence homology was relatively low, with CLC-1, CLC-2, and CLC-3 sharing 25%,

23%, and 26% identity with claudin-6, 5, and 4, respectively. CLC-4 shares just 36% identity with CLC-1. CLC-5 is the paralog of CLC-2, probably resulting from a gene duplication, being located just 3′ of *clc-2* in the genome. Despite this location, CLC-2 and CLC-5 are only 31% identical. Interestingly, BLAST searches identify similarities between several *C. elegans* claudins and the *nsy* family, suggesting possible overlap in function between the adhesive/charge/paracellular barrier properties of classic claudins and the channel regulatory properties of the gamma subunits of the voltage-gated calcium channels (102). Phylogenetic analysis also suggests a close association between *nsy-4*-like genes and the *clc* genes. Of the five putative *C. elegans* claudins, two have been further analyzed to determine the extent of their function as classical claudins. GFP fusion proteins were constructed for both CLC-1 and CLC-2. CLC-1 is localized at the apical region of the lateral cell membranes in all four regions of the *C. elegans* pharynx, the procorpus, metacorpus, isthmus, and terminal bulb. In the isthmus region, six cellular junctions connect three pharyngeal myoepithelial and three marginal cells. Freeze substitution electron microscopy reveals a thick subapical junction – as a classic adherens junction – and a thinner apical region with a more narrow paracellular space. It is possible that CLC-1 and the cell junction protein AJM-1 localize to this region and adherens junction proteins localize to the thicker basal junction. In the epidermis and intestine, AJM-1 and DLG-1 localize basal to the adherens junction. It may be possible that in the pharynx, the localization of the two junctions is reversed; however, that has not been demonstrated. It remains for immuno-TEM with different diameter gold particles linked to secondary antibodies or PALM microscopy of fluorescently labeled/tagged proteins to resolve this issue. An alternative possibility is that the narrow apical region discovered by freeze substitution is the locale for one or more CLC proteins and this may define a separate lateral region more typical of tight junctions in mammals, rather than the septate junctions of invertebrates. CLC-1 expression was also reported in the vulva, spermatheca, and pore cell of the excretory system, but was not assigned to a specific junctional complex. CLC-2::GFP shows expression in the hypodermal seam cells of adults. Following RNAi of *clc-1* and *clc-2*, a 10,000 MW TRITC-dextran is able to infiltrate to the interior of the RNAi worms in the pharynx and body cavity for *clc-1* and into the body cavity for *clc-2*, suggesting that both of these claudins maintain epithelial barriers. *clc-1* and *clc-2* expression only accounts for a subset of epithelial tissues with cell junctions in nematodes. Further studies should indicate whether *clc-3–5* are expressed in epithelia lacking CLC-1 and CLC-2 and whether inactivation of these claudins in other tissues has similar effects on the epithelial barrier function of such tissues or whether there are alternative phenotypes. It will be interesting to see whether all *C. elegans* claudins are localized to junctional regions with similar TEM appearances, since in vertebrates, specific claudins are associated with specific cell junction types. It will also be interesting to determine whether the CLC proteins can generate tight junction strands when expressed in L cells, as mammalian claudins. Expression of VAB-9 in L cells does not result in the formation of tight junction strands (JSS unpublished data).

8. *nsy-4* and Neuronal Cell Fate Specification

nsy-4 (*Neuronal SYmmetry* gene 4) was identified for its role in specifying the distinct fate of one AWC olfactory neuron; specifically, *nsy-4* is required for one AWC neuron to express the G protein-coupled odorant receptor gene *str-2* and detects the odor 2-butanone (102). Without *nsy-4* activity, both AWC neurons fail to express *str-2* and detect the odor 2,3-pentadione (102). These alternate fates, based on the expression of *str-2*, are referred to as AWC^{on} and AWC^{off}, respectively. In *C. elegans*, NSY-4 is most similar to uncharacterized genes K10D6.2 and C24H10.1 as well as claudin genes *clc-1 to clc-5* and R04F11.1. Outside *C. elegans*, NSY-4 is most similar to *Drosophila Stargazin*, TARPS, and gamma subunits of voltage-gated calcium channels. Surprisingly, *nsy-4* rescue by human claudin-1 expression was weak, but was stronger than stargazin/γ2 expression, while rescue by human γ7 channel was complicated by additional phenotypes. NSY-4 is thought to function as a gamma channel based on the genetic data, which indicate that *nsy-4* represses the function of calcium channel genes *unc-2* and *unc-36*. Together these findings suggest that NSY-4 may have additional functions in AWC specification beyond regulation of gamma channels. Additional functions may involve claudin-like adhesive functions. Genetic and expression studies demonstrate that NSY-4 levels correlate with the AWC^{on} fate: *nsy-4* expression in a single AWC is sufficient for the AWC^{on} fate, while overexpression of *nsy-4* in both AWC cells results in an increased frequency of animals with AWC^{on} fates in both AWC cells. NSY-4 regulates the choice of cell fate by influencing the activity of a signaling pathway consisting of calcium channels encoded by *unc-2* and *unc-36*, CAMKII encoded by *unc-43*, a Toll-interleukin 1 repeat protein encoded by *tir-1*, a MAPKKK encoded by *nsy-1*, a MAPKK encoded by *sek-1*, and a homeodomain protein encoded by *nsy-7* (103–107). The innexin protein, NSY-5, acts in parallel with NSY-4 to regulate the activity of this pathway, so that the overexpression of *nsy-4* can partially compensate for the loss of *nsy-5* and vice versa (108). While both proteins are localized at cell membranes (NSY-4 broadly and NSY-5 in puncta at junctions), neither protein affects the other's localization. Results from *nsy-5* mosaic analysis and targeted expression using cell-type-specific promoters indicate that transient *nsy-5* gap junction networks are involved in specifying cell fate and that the expression of AWC^{on} fates involves signaling across the midline as well as feedback inhibition of AWC^{on} fates from surrounding cells (108). Similarly, *nsy-4* is proposed to regulate communication between AWC cells in contacting axons across the midline.

Normally, the determination of which cell will express the AWC^{on} fate is stochastic, suggesting that some small fluctuation in activity upstream or parallel to NSY-4 tips the choice of signaling resulting in *str-2* expression in favor of one cell over another. Feedback signaling to reinforce this decision resulting in one, and only one, cell expressing the AWC^{on} fate is supported by genetic mosaic experiments. Such experiments show that the overexpression of NSY-4 in one AWC cell suppresses the expression of the AWC^{on} fate in the neighboring wild-type AWC cell. This signaling is reminiscent of the familiar stochastic cell fate specification of anchor cell and ventral uterine fates, which involves lateral signaling via the Delta-Notch ligand receptor system encoded by *lag-1/2* and *lin-12*, respectively (109–114). During the AC/VU decision, the LAG ligand signals through the LIN-12 receptor to specify anchor cell and ventral uterine fates and by analogy during AWC specification, it seems likely that NSY-4 is one half of a signaling pair and that a currently unidentified protein may function as an NSY-4 ligand/receptor. Since claudin family members are known to form homotypic and heterotypic pairs and since the *nsy-4* homolog *Stargazin* has been shown to mediate both homotypic and heterotypic cell adhesion when expressed in L cells, we speculate that NSY-4 interacts with NSY-4 and/or an unidentified claudin superfamily protein on neighboring AWC (and perhaps other) opposing cell membranes (33). Binding to NSY-4 then influences *nsy-4*-dependent signaling by refining forward and feedback signaling. Thus, NSY-4-related proteins, claudin family members, or VAB-9 may participate in heterotypic interactions with NSY-4 and mediate axon interaction and communication, participating in the specification of AWC cell fate and possibly the fates of other neurons. It is interesting to note that VAB-9 is expressed in the nerve ring, although the specific cells have not been identified (67). The expression pattern of other *nsy-4*-related genes has not yet been determined; while *clc-1* and *clc-2* are not expressed in the nerve ring, the expression patterns of *clc-3*, *clc-4*, and *clc-5* are currently unknown. Given the combinatorial possibilities, multiple distinct axonal signaling events could be mediated by these proteins. Specific cell–cell interactions mediated by distinct claudin pairs may activate multiple signaling events within the same or among closely related cells.

9. When Is a Claudin a Claudin?

The superfamily of proteins that fall into the general category of similarity to claudins has grown, begging the question of what exactly are the essential properties of a claudin, as opposed to a similar, but fundamentally distinct four-pass integral membrane protein. Most claudins generate a charge-selective pore in the paracellular space that regulates electrical properties of a cell or groups of cells due to specific charge permeability. Most claudins mediate adhesion and this adhesion can be homotypic or heterotypic between various claudin types. The formation of paired strands within the membrane by claudins reduces the paracellular space to zero and the number and arrangement of the strands correlates with the tightness of the epithelium. Most claudins appear able to traffic to cell membranes and generate paired strands in the absence of other cell adhesion molecules, but require targeting proteins to designate the proper membrane location. As other transmembrane proteins, claudins transduce extracellular information, such as the binding of ligands, small molecules, and viruses, to the nucleus to change gene expression or to transduce signals to junctional proteins (including themselves) to modulate the nature of the paracellular space. Claudin superfamily proteins are likely to have at least a majority of these properties; however, due to the various organisms, organs, tissues, and developmental and physiological systems in which claudin-like proteins have been shown to function, along with the dramatic divergence of the family, it is highly likely that the functional repertoire of claudin-like proteins will continue to expand. The claudin family proteins in *C. elegans* described in this review support this hypothesis, since they are involved in diverse roles such as maintaining tissue integrity, epithelial morphogenesis, regulating cell junction dynamics through filamentous actin, signal transduction, and cell fate specification. Our understanding of the roles for all the members of this family in *C. elegans* is incomplete, suggesting that future research will reveal novel functions.

Acknowledgments

We thank David Hall for sharing unpublished information and for the immunoelectron micrograph of the spermatheca shown in Fig. 5a. Drawings of the spermatheca in Fig. 5b, c are reprinted from Wormatlas with the permission of CSH press. This work was supported by NIH grant GM058038 to J.H.

References

1. Furuse, M., et al., *A single gene product, claudin-1 or -2, reconstitutes tight junction strands and recruits occludin in fibroblasts.* J Cell Biol, 1998. **143**(2): p. 391–401.
2. Furuse, M., et al., *Claudin-1 and -2: novel integral membrane proteins localizing at tight junctions with no sequence similarity to occludin.* J Cell Biol, 1998. **141**(7): p. 1539–50.
3. Morita, K., et al., *Endothelial claudin: claudin-5/TMVCF constitutes tight junction strands in endothelial cells.* J Cell Biol, 1999. **147**(1): p. 185–94.
4. Nitta, T., et al., *Size-selective loosening of the blood-brain barrier in claudin-5-deficient mice.* J Cell Biol, 2003. **161**(3): p. 653–60.
5. Gow, A., et al., *Deafness in Claudin 11-null mice reveals the critical contribution of basal cell tight junctions to stria vascularis function.* J Neurosci, 2004. **24**(32): p. 7051–62.
6. Furuse, M., et al., *Claudin-based tight junctions are crucial for the mammalian epidermal barrier: a lesson from claudin-1-deficient mice.* J Cell Biol, 2002. **156**(6): p. 1099–111.
7. Wilcox, E.R., et al., *Mutations in the gene encoding tight junction claudin-14 cause autosomal recessive deafness DFNB29.* Cell, 2001. **104**(1): p. 165–72.
8. Tatum, R., et al., *Renal salt wasting and chronic dehydration in claudin-7-deficient mice.* Am J Physiol Renal Physiol, **298**(1): p. F24–34.
9. Kollmar, R., et al., *Expression and phylogeny of claudins in vertebrate primordia.* Proc Natl Acad Sci USA, 2001. **98**(18): p. 10196–201.
10. Krause, G., et al., *Structure and function of claudins.* Biochim Biophys Acta, 2008. **1778**(3): p. 631–45.
11. Morita, K., et al., *Claudin multigene family encoding four-transmembrane domain protein components of tight junction strands.* Proc Natl Acad Sci USA, 1999. **96**(2): p. 511–6.
12. Blasig, I.E., et al., *On the self-association potential of transmembrane tight junction proteins.* Cell Mol Life Sci, 2006. **63**(4): p. 505–14.
13. Coyne, C.B., et al., *Role of claudin interactions in airway tight junctional permeability.* Am J Physiol Lung Cell Mol Physiol, 2003. **285**(5): p. L1166–78.
14. Mitic, L.L., V.M. Unger, and J.M. Anderson, *Expression, solubilization, and biochemical characterization of the tight junction transmembrane protein claudin-4.* Protein Sci, 2003. **12**(2): p. 218–27.
15. Van Itallie, C.M. and J.M. Anderson, *Claudins and epithelial paracellular transport.* Annu Rev Physiol, 2006. **68**: p. 403–29.
16. Kiuchi-Saishin, Y., et al., *Differential expression patterns of claudins, tight junction membrane proteins, in mouse nephron segments.* J Am Soc Nephrol, 2002. **13**(4): p. 875–86.
17. Simon, D.B., et al., *Paracellin-1, a renal tight junction protein required for paracellular Mg2+ resorption.* Science, 1999. **285**(5424): p. 103–6.
18. Holmes, J.L., et al., *Claudin profiling in the mouse during postnatal intestinal development and along the gastrointestinal tract reveals complex expression patterns.* Gene Expr Patterns, 2006. **6**(6): p. 581–8.
19. Colegio, O.R., et al., *Claudins create charge-selective channels in the paracellular pathway between epithelial cells.* Am J Physiol Cell Physiol, 2002. **283**(1): p. C142–7.
20. Van Itallie, C.M., et al., *Two splice variants of claudin-10 in the kidney create paracellular pores with different ion selectivities.* Am J Physiol Renal Physiol, 2006. **291**(6): p. F1288–99.
21. Evans, M.J., et al., *Claudin-1 is a hepatitis C virus co-receptor required for a late step in entry.* Nature, 2007. **446**(7137): p. 801–5.
22. Meertens, L., et al., *The tight junction proteins claudin-1, -6, and -9 are entry cofactors for hepatitis C virus.* J Virol, 2008. **82**(7): p. 3555–60.
23. Zheng, A., et al., *Claudin-6 and claudin-9 function as additional coreceptors for hepatitis C virus.* J Virol, 2007. **81**(22): p. 12465–71.
24. Hamazaki, Y., et al., *Multi-PDZ domain protein 1 (MUPP1) is concentrated at tight junctions through its possible interaction with claudin-1 and junctional adhesion molecule.* J Biol Chem, 2002. **277**(1): p. 455–61.
25. Itoh, M., et al., *Direct binding of three tight junction-associated MAGUKs, ZO-1, ZO-2, and ZO-3, with the COOH termini of claudins.* J Cell Biol, 1999. **147**(6): p. 1351–63.
26. Muller, D., et al., *Unusual clinical presentation and possible rescue of a novel claudin-16 mutation.* J Clin Endocrinol Metab, 2006. **91**(8): p. 3076–9.
27. D'Souza, T., R. Agarwal, and P.J. Morin, *Phosphorylation of claudin-3 at threonine 192 by cAMP-dependent protein kinase regulates tight junction barrier function in ovarian cancer cells.* J Biol Chem, 2005. **280**(28): p. 26233–40.

28. Tanaka, M., R. Kamata, and R. Sakai, *EphA2 phosphorylates the cytoplasmic tail of Claudin-4 and mediates paracellular permeability.* J Biol Chem, 2005. **280**(51): p. 42375–82.
29. Tatum, R., et al., *WNK4 phosphorylates ser(206) of claudin-7 and promotes paracellular Cl(-) permeability.* FEBS Lett, 2007. **581**(20): p. 3887–91.
30. Lane, N.J. and H.I. Skaer, *Intercellular junctions in insect tissues.* Adv Insect Physiol, 1980. **15**: p. 35–213.
31. Hua, V.B., et al., *Sequence and phylogenetic analyses of 4 TMS junctional proteins of animals: connexins, innexins, claudins and occludins.* J Membr Biol, 2003. **194**(1): p. 59–76.
32. Wu, V.M., et al., *Sinuous is a Drosophila claudin required for septate junction organization and epithelial tube size control.* J Cell Biol, 2004. **164**(2): p. 313–23.
33. Price, M.G., et al., *The alpha-amino-3-hydroxyl-5-methyl-4-isoxazolepropionate receptor trafficking regulator "stargazin" is related to the claudin family of proteins by Its ability to mediate cell-cell adhesion.* J Biol Chem, 2005. **280**(20): p. 19711–20.
34. Ihrie, R.A., et al., *Perp is a p63-regulated gene essential for epithelial integrity.* Cell, 2005. **120**(6): p. 843–56.
35. Attardi, L.D., et al., *PERP, an apoptosis-associated target of p53, is a novel member of the PMP-22/gas3 family.* Genes Dev, 2000. **14**(6): p. 704–18.
36. Bruggeman, L.A., S. Martinka, and J.S. Simske, *Expression of TM4SF10, a Claudin/EMP/PMP22 family cell junction protein, during mouse kidney development and podocyte differentiation.* Dev Dyn, 2007. **236**(2): p. 596–605.
37. Lane, N.J., *Tight junctions in invertebrates*, in *Tight Junctions*, M.a.A. Cereijido, J., Editor. 2001, CRC Press: New York. p. 39–59.
38. Spiegel, E. and L. Howard, *Development of cell junctions in sea-urchin embryos.* J Cell Sci, 1983. **62**: p. 27–48.
39. Tepass, U. and V. Hartenstein, *The development of cellular junctions in the Drosophila embryo.* Dev Biol, 1994. **161**(2): p. 563–96.
40. Noirot-Timothée, C. and C. Noirot, *Septate and sclariform junctions in arthropods.* Int Rev Cytol, 1980. **63**: p. 97–141.
41. Gilula, N.B., D. Branton, and P. Satir, *The septate junction: a structural basis for intercellular coupling.* Proc Natl Acad Sci USA, 1970. **67**(1): p. 213–20.
42. Satir, P. and N.B. Gilula, *The fine structure of membranes and intercellular communication in insects.* Annu Rev Entomol, 1973. **18**: p. 143–66.
43. Davidson, L.A., *A freeze fracture and thin section study of intestinal cell membranes and intercellular junctions of a nematode, Ascaris.* Tissue Cell, 1983. **15**(1): p. 27–37.
44. Lane, N.J. and L.S. Swales, *Stages in the assembly of pleated and smooth septate junctions in developing insect embryos.* J Cell Sci, 1982. **56**: p. 245–62.
45. Behr, M., D. Riedel, and R. Schuh, *The claudin-like megatrachea is essential in septate junctions for the epithelial barrier function in Drosophila.* Dev Cell, 2003. **5**(4): p. 611–20.
46. Genova, J.L. and R.G. Fehon, *Neuroglian, Gliotactin, and the Na+/K+ ATPase are essential for septate junction function in Drosophila.* J Cell Biol, 2003. **161**(5): p. 979–89.
47. Wu, V.M., et al., *Drosophila Varicose, a member of a new subgroup of basolateral MAGUKs, is required for septate junctions and tracheal morphogenesis.* Development, 2007. **134**(5): p. 999–1009.
48. Suzuki, A. and S. Ohno, *The PAR-aPKC system: lessons in polarity.* J Cell Sci, 2006. **119**(Pt 6): p. 979–87.
49. Fehon, R.G., I.A. Dawson, and S. Artavanis-Tsakonas, *A Drosophila homologue of membrane-skeleton protein 4.1 is associated with septate junctions and is encoded by the coracle gene.* Development, 1994. **120**(3): p. 545–57.
50. Baumgartner, S., et al., *A Drosophila neurexin is required for septate junction and blood-nerve barrier formation and function.* Cell, 1996. **87**(6): p. 1059–68.
51. Woods, D.F., J.W. Wu, and P.J. Bryant, *Localization of proteins to the apico-lateral junctions of Drosophila epithelia.* Dev Genet, 1997. **20**(2): p. 111–8.
52. Schulte, J., U. Tepass, and V.J. Auld, *Gliotactin, a novel marker of tricellular junctions, is necessary for septate junction development in Drosophila.* J Cell Biol, 2003. **161**(5): p. 991–1000.
53. Banerjee, S. and M.A. Bhat, *Neuron-glial interactions in blood-brain barrier formation.* Annu Rev Neurosci, 2007. **30**: p. 235–58.
54. Banerjee, S., et al., *Axonal ensheathment and septate junction formation in the peripheral nervous system of Drosophila.* J Neurosci, 2006. **26**(12): p. 3319–29.
55. Bhat, M.A., et al., *Axon-glia interactions and the domain organization of myelinated axons requires neurexin IV/Caspr/Paranodin.* Neuron, 2001. **30**(2): p. 369–83.

56. Stork, T., et al., *Organization and function of the blood-brain barrier in Drosophila.* J Neurosci, 2008. **28**(3): p. 587–97.
57. Bellen, H.J., et al., *Neurexin IV, caspr and paranodin – novel members of the neurexin family: encounters of axons and glia.* Trends Neurosci, 1998. **21**(10): p. 444–9.
58. Schafer, D.P. and M.N. Rasband, *Glial regulation of the axonal membrane at nodes of Ranvier.* Curr Opin Neurobiol, 2006. **16**(5): p. 508–14.
59. Blair, J.E., et al., *The evolutionary position of nematodes.* BMC Evol Biol, 2002. **2**: p. 7.
60. Knust, E. and O. Bossinger, *Composition and formation of intercellular junctions in epithelial cells.* Science, 2002. **298**(5600): p. 1955–9.
61. Costa, M., et al., *A putative catenin-cadherin system mediates morphogenesis of the Caenorhabditis elegans embryo.* J Cell Biol, 1998. **141**(1): p. 297–308.
62. Pettitt, J., et al., *The Caenorhabditis elegans p120 catenin homologue, JAC-1, modulates cadherin-catenin function during epidermal morphogenesis.* J Cell Biol, 2003. **162**(1): p. 15–22.
63. Bossinger, O., et al., *Zonula adherens formation in Caenorhabditis elegans requires dlg-1, the homologue of the Drosophila gene discs large.* Dev Biol, 2001. **230**(1): p. 29–42.
64. Firestein, B.L. and C. Rongo, *DLG-1 is a MAGUK similar to SAP97 and is required for adherens junction formation.* Mol Biol Cell, 2001. **12**(11): p. 3465–75.
65. Koppen, M., et al., *Cooperative regulation of AJM-1 controls junctional integrity in Caenorhabditis elegans epithelia.* Nat Cell Biol, 2001. **3**(11): p. 983–91.
66. McMahon, L., et al., *Assembly of C. elegans apical junctions involves positioning and compaction by LET-413 and protein aggregation by the MAGUK protein DLG-1.* J Cell Sci, 2001. **114**(Pt 12): p. 2265–77.
67. Simske, J.S., et al., *The cell junction protein VAB-9 regulates adhesion and epidermal morphology in C. elegans.* Nat Cell Biol, 2003. **5**(7): p. 619–25.
68. Aono, S., et al., *PAR-3 is required for epithelial cell polarity in the distal spermatheca of C. elegans.* Development, 2004. **131**(12): p. 2865–74.
69. Michaux, G., et al., *CHE-14, a protein with a sterol-sensing domain, is required for apical sorting in C. elegans ectodermal epithelial cells.* Curr Biol, 2000. **10**(18): p. 1098–107.
70. Shibata, Y., et al., *EAT-20, a novel transmembrane protein with EGF motifs, is required for efficient feeding in Caenorhabditis elegans.* Genetics, 2000. **154**(2): p. 635–46.
71. Bilder, D. and N. Perrimon, *Localization of apical epithelial determinants by the basolateral PDZ protein Scribble.* Nature, 2000. **403**(6770): p. 676–80.
72. Legouis, R., et al., *LET-413 is a basolateral protein required for the assembly of adherens junctions in Caenorhabditis elegans.* Nat Cell Biol, 2000. **2**(7): p. 415–22.
73. Bossinger, O., et al., *The apical disposition of the Caenorhabditis elegans intestinal terminal web is maintained by LET-413.* Dev Biol, 2004. **268**(2): p. 448–56.
74. McCarter, J., et al., *Soma-germ cell interactions in Caenorhabditis elegans: multiple events of hermaphrodite germline development require the somatic sheath and spermathecal lineages.* Dev Biol, 1997. **181**(2): p. 121–43.
75. Schedl, T., *Developmental genetics of the germ line* in *C. elegans II*, T.B. D. Riddle, B. Meyer and J. Priess, Editor. 1997, Cold Spring Harbor Laboratory Press: Cold Spring Harbor. p. 241–270.
76. Hall, D.H., et al., *Ultrastructural features of the adult hermaphrodite gonad of Caenorhabditis elegans: relations between the germ line and soma.* Dev Biol, 1999. **212**(1): p. 101–23.
77. Hall, D.H., Altun, Z.F., *Reproductive System*, in *C. elegans Atlas*, C.S.H.L. Press, Editor. 2008: Cold Spring Harbor. p. 243–291.
78. Piekny, A.J., et al., *The Caenorhabditis elegans nonmuscle myosin genes nmy-1 and nmy-2 function as redundant components of the let-502/Rho-binding kinase and mel-11/myosin phosphatase pathway during embryonic morphogenesis.* Development, 2003. **130**(23): p. 5695–704.
79. Piekny, A.J. and P.E. Mains, *Rho-binding kinase (LET-502) and myosin phosphatase (MEL-11) regulate cytokinesis in the early Caenorhabditis elegans embryo.* J Cell Sci, 2002. **115**(Pt 11): p. 2271–82.
80. Wissmann, A., J. Ingles, and P.E. Mains, *The Caenorhabditis elegans mel-11 myosin phosphatase regulatory subunit affects tissue contraction in the somatic gonad and the embryonic epidermis and genetically interacts with the Rac signaling pathway.* Dev Biol, 1999. **209**(1): p. 111–27.
81. Asano, A., et al., *Claudins in Caenorhabditis elegans: their distribution and barrier function in the epithelium.* Curr Biol, 2003. **13**(12): p. 1042–6.
82. Christophe-Hobertus, C., et al., *TM4SF10 gene sequencing in XLMR patients identifies*

common polymorphisms but no disease-associated mutation. BMC Med Genet, 2004. **5**: p. 22.

83. Christophe-Hobertus, C., et al., *Identification of the gene encoding Brain Cell Membrane Protein 1 (BCMP1), a putative four-transmembrane protein distantly related to the Peripheral Myelin Protein 22/Epithelial Membrane Proteins and the Claudins.* BMC Genomics, 2001. **2**(1): p. 3.
84. Holzman, L.B., et al., *Nephrin localizes to the slit pore of the glomerular epithelial cell.* Kidney Int, 1999. **56**(4): p. 1481–91.
85. Huber, T.B. and T. Benzing, *The slit diaphragm: a signaling platform to regulate podocyte function.* Curr Opin Nephrol Hypertens, 2005. **14**(3): p. 211–6.
86. Barletta, G.M., et al., *Nephrin and Neph1 co-localize at the podocyte foot process intercellular junction and form cis hetero-oligomers.* J Biol Chem, 2003. **278**(21): p. 19266–71.
87. Khoshnoodi, J., et al., *Nephrin promotes cell-cell adhesion through homophilic interactions.* Am J Pathol, 2003. **163**(6): p. 2337–46.
88. Verma, R., et al., *Nephrin ectodomain engagement results in Src kinase activation, nephrin phosphorylation, Nck recruitment, and actin polymerization.* J Clin Invest, 2006. **116**(5): p. 1346–59.
89. Verma, R., et al., *Fyn binds to and phosphorylates the kidney slit diaphragm component Nephrin.* J Biol Chem, 2003. **278**(23): p. 20716–23.
90. Garg, P., et al., *Neph1 cooperates with nephrin to transduce a signal that induces actin polymerization.* Mol Cell Biol, 2007. **27**(24): p. 8698–712.
91. Lahdenpera, J., et al., *Clustering-induced tyrosine phosphorylation of nephrin by Src family kinases.* Kidney Int, 2003. **64**(2): p. 404–13.
92. Li, H., et al., *SRC-family kinase Fyn phosphorylates the cytoplasmic domain of nephrin and modulates its interaction with podocin.* J Am Soc Nephrol, 2004. **15**(12): p. 3006–15.
93. Liu, X.L., et al., *Characterization of the interactions of the nephrin intracellular domain.* FEBS J, 2005. **272**(1): p. 228–43.
94. Harita, Y., et al., *Neph1, a component of the kidney slit diaphragm, is tyrosine-phosphorylated by the Src family tyrosine kinase and modulates intracellular signaling by binding to Grb2.* J Biol Chem, 2008. **283**(14): p. 9177–86.
95. Zhu, J., et al., *Nephrin mediates actin reorganization via phosphoinositide 3-kinase in podocytes.* Kidney Int, 2008. **73**(5): p. 556–66.
96. Laketa, V., et al., *High-content microscopy identifies new neurite outgrowth regulators.* Mol Biol Cell, 2007. **18**(1): p. 242–52.
97. Shen, K., R.D. Fetter, and C.I. Bargmann, *Synaptic specificity is generated by the synaptic guidepost protein SYG-2 and its receptor, SYG-1.* Cell, 2004. **116**(6): p. 869–81.
98. Shen, K. and C.I. Bargmann, *The immunoglobulin superfamily protein SYG-1 determines the location of specific synapses in C. elegans.* Cell, 2003. **112**(5): p. 619–30.
99. Putzke, A.P., et al., *Essential kinase-independent role of a Fer-like non-receptor tyrosine kinase in Caenorhabditis elegans morphogenesis.* Development, 2005. **132**(14): p. 3185–95.
100. Lockwood, C., R. Zaidel-Bar, and J. Hardin, *The C. elegans zonula occludens ortholog cooperates with the cadherin complex to recruit actin during morphogenesis.* Curr Biol, 2008. **18**(17): p. 1333–7.
101. Van Itallie, C.M., et al., *ZO-1 stabilizes the tight junction solute barrier through coupling to the perijunctional cytoskeleton.* Mol Biol Cell, 2009. **20**(17): p. 3930–40.
102. Vanhoven, M.K., et al., *The claudin superfamily protein nsy-4 biases lateral signaling to generate left-right asymmetry in C. elegans olfactory neurons.* Neuron, 2006. **51**(3): p. 291–302.
103. Lesch, B.J., et al., *Transcriptional regulation and stabilization of left-right neuronal identity in C. elegans.* Genes Dev, 2009. **23**(3): p. 345–58.
104. Sagasti, A., et al., *The CaMKII UNC-43 activates the MAPKKK NSY-1 to execute a lateral signaling decision required for asymmetric olfactory neuron fates.* Cell, 2001. **105**(2): p. 221–32.
105. Tanaka-Hino, M., et al., *SEK-1 MAPKK mediates Ca2+ signaling to determine neuronal asymmetric development in Caenorhabditis elegans.* EMBO Rep, 2002. **3**(1): p. 56–62.
106. Wes, P.D. and C.I. Bargmann, *C. elegans odour discrimination requires asymmetric diversity in olfactory neurons.* Nature, 2001. **410**(6829): p. 698–701.
107. Chuang, C.F. and C.I. Bargmann, *A Toll-interleukin 1 repeat protein at the synapse specifies asymmetric odorant receptor expression via ASK1 MAPKKK signaling.* Genes Dev, 2005. **19**(2): p. 270–81.
108. Chuang, C.F., et al., *An innexin-dependent cell network establishes left-right neuronal asymmetry in C. elegans.* Cell, 2007. **129**(4): p. 787–99.

109. Wilkinson, H.A., K. Fitzgerald, and I. Greenwald, *Reciprocal changes in expression of the receptor lin-12 and its ligand lag-2 prior to commitment in a C. elegans cell fate decision.* Cell, 1994. **79**(7): p. 1187–98.

110. Seydoux, G. and I. Greenwald, *Cell autonomy of lin-12 function in a cell fate decision in C. elegans.* Cell, 1989. **57**(7): p. 1237–45.

111. Chen, N. and I. Greenwald, *The lateral signal for LIN-12/Notch in C. elegans vulval development comprises redundant secreted and transmembrane DSL proteins.* Dev Cell, 2004. **6**(2): p. 183–92.

112. Henderson, S.T., et al., *lag-2 may encode a signaling ligand for the GLP-1 and LIN-12 receptors of C. elegans.* Development, 1994. **120**(10): p. 2913–24.

113. Tax, F.E., J.J. Yeargers, and J.H. Thomas, *Sequence of C. elegans lag-2 reveals a cell-signalling domain shared with Delta and Serrate of Drosophila.* Nature, 1994. **368**(6467): p. 150–4.

114. Lambie, E.J. and J. Kimble, *Two homologous regulatory genes, lin-12 and glp-1, have overlapping functions.* Development, 1991. **112**(1): p. 231–40.

115. Tamura, K., et al., *MEGA4: Molecular Evolutionary Genetics Analysis (MEGA) software version 4.0.* Mol Biol Evol, 2007. **24**(8): p. 1596–9.

Chapter 12

In Vivo Imaging of Tight Junctions Using *Claudin–EGFP* Transgenic Medaka

Tatsuo Miyamoto, Mikio Furuse, and Makoto Furutani-Seiki

Abstract

Tight junctions (TJs) function as a physiological barrier between epithelial and endothelial sheets by restricting the diffusion of fluid through the intercellular space. Recent morphological studies associated with TJs have revealed that the TJ is a dynamic rather than a static structure; indeed, several in vitro studies indicate that proper TJ function requires dynamic TJ behavior. Direct observation of the dynamic behavior of TJs is necessary to understand the essential roles of TJs in physiological contexts, such as during embryogenesis and metastasis. Here we describe a protocol for the generation of transgenic medaka (*Oryzias latipes*) that express *claudin–EGFP*. This fluorescent fusion protein enables real-time imaging of TJs in the living embryo. *Claudin–EGFP* transgenic medaka will be a useful tool to screen for mutations and for small molecules affecting cell–cell adhesion.

Key words: Transgenic medaka, *Oryzias latipes*, In vivo imaging, Tight junction

1. Introduction

The tight junction (TJ) is a type of cell–cell adhesion between epithelial and endothelial cells that provides a barrier to the diffusion of solutes through the intercellular space (1). The *claudins* are TJ adhesion molecules with four transmembrane domains and 24 members have been identified in each mammalian genome. They form TJ strands in cell membranes and contribute to the barrier function of TJs (2, 3). *Claudin-16* and *claudin-19* are responsible genes for familial hypomagnesemia; therefore, claudin-based TJs play an additional role as ion channels by regulating paracellular transport (4, 5). Recent studies have indicated that TJs are not only simple barriers but also multifunctional membrane structures. However, the molecular mechanisms that underlie the multiple physiological functions of TJs remain unclear.

Kursad Turksen (ed.), *Claudins: Methods and Protocols*, Methods in Molecular Biology, vol. 762, DOI 10.1007/978-1-61779-185-7_12, © Springer Science+Business Media, LLC 2011

Application of GFP (green fluorescent protein) technology to TJ studies in mammalian cultured cell lines has revealed that TJs exhibit a dynamic behavior in the cell membrane (6, 7). These recent reports suggest that TJ function depends on the dynamic behavior of the TJ. Despite of accumulating reports on TJ dynamics in vitro cell culture, behavior of TJs in 3D tissue in vivo needs to be examined. A transgenic medaka line that expresses Claudin fused with enhanced green fluorescent protein (EGFP) represents a powerful tool for visualizing TJs and for exploring TJ dynamics in the transparent medaka embryo (8). Here we present a detailed protocol for the generation of a transgenic medaka line that enables visualization of TJs in vivo via fluorescent proteins.

2. Materials

2.1. Preparation of the Claudin–EGFP Plasmid DNA for Injection into Medaka Embryos

1. The plasmid for microinjection includes the red sea bream *β-actin* promoter to provide ubiquitous expression in the medaka embryo (9) (see Note 1).
2. *Claudin–EGFP* plasmid DNA, isolated to injection grade purity using the Qiagen EndoFree plasmid MaxiPrep kit (Qiagen, Chatsworth, USA) (see Note 2).

2.2. Microinjection of Plasmid DNA into Medaka Embryos (see Note 3)

1. Microinjection needles: Relatively short needle with thick wall is used to inject through the tough chorion of medaka eggs. Filament-containing borosilicate thick wall glass capillaries (GC100F-10, Harvard apparatus) are pulled with a puller (Narishige PN-3, Narishige, Tokyo).
2. Hatching enzyme for removal of the chorion: Homogenize medaka embryos just before hatching (8–9 days after fertilization) in Ca^{2+} and Mg^{2+}-free PBS (1.47 mM KH_2PO_4, 8.06 mM Na_2HPO_4, 137 mM NaCl, and 2.68 mM KCl) on ice and store at 4°C overnight. Centrifuge the homogenate at 20,400 g for 10 min at 4°C. Dispense 100 ml aliquots of the supernatant into 1.5-ml centrifuge tubes and store at –80°C (see Note 4).
3. Embryo medium: 200 ml 50× stock solution, 1 ml 1% methylene blue in H_2O/10 L RO water. 50× Stock solution of embryo medium: NaCl 14.7 g, KCl 0.6 g, $CaCl_2{\cdot}2H_2O$ 2.4 g, $MgSO_4{\cdot}7H_2O$ 4.0 g/1 L RO water.
4. 1× BSS (balanced salt solution) medium for developing medaka eggs: Add 25 ml 20× BSS and 15 ml HEPES pH 7.0 fill up to 500 ml and filter sterilized before use. 20× BSS: 130 g NaCl, 8 g KCl, 4 g $MgSO_4{\cdot}7H_2O$, 4 g $CaCl_2{\cdot}2H_2O$, and 10 mg Phenol Red in 1 L MilliQ water and autoclave; 500 mM HEPES in MilliQ water autoclaved.

5. Agarose injection plate: Pour melt 2% agarose (LO3, Takara, Japan) in water into a 10-cm plastic dish to a depth of 3 mm. After the agarose has hardened, use a sterile razor blade to make several U-shaped grooves 1 mm wide and 1 mm deep for aligning embryos.
6. Injector and microneedle holder (Narishige IM-300, Narishige, Tokyo).
7. The Kyoto-Cab inbred medaka strain maintained in a closed circulation system (Aqua, Tokyo, Japan) with the water condition (pH 6.8–7.5, 200–450 mS/cm, 26–28°C, $NH_4 < 0.2$ mg/L, $NO_2 < 0.05$ mg/L, and $NO_3 < 20$ mg/L) and the day/night light cycle (12/12 h).
8. Collect fertilized medaka eggs immediately after spawning (at the onset of light). Transfer one-cell stage embryos (fertilized eggs) to 4°C to slow down the developmental process.

2.3. Whole-Mount Observation of Fish

1. 3% Methylcellulose in 1× BSS.
2. Epifluorescent microscope (Axio Plan2 compound microscope, Zeiss, Germany) or confocal microscope (TCS SP2, Leica Microsystems, Germany).

3. Methods

The dynamic nature of TJ behavior is considered to be fundamental to TJ function. To gain insights into morphology and physiology of TJs in vivo, it is essential to generate an animal model in which TJs can be visualized in a living animal. Medaka is suitable as a model organism for in vivo TJ studies because of the following advantages (10): (a) transparent embryos that develop outside mother's body facilitate in vivo imaging at the cellular and subcellular level, (b) a simple procedure for generating transgenic strains (11), (c) excellent genetic analysis using mutants generated by the phenotype-driven or gene-driven approaches (12), (d) easy maintenance of the strains and production of embryos in a small laboratory (sexual maturation at about 2 months after fertilization), (e) inbred strains that allow tissue transplantations, and (f) the whole genome draft sequence that enables efficient molecular cloning (13). Here we describe a protocol to generate a *claudin–EGFP* transgenic medaka as schematized in Fig. 1.

3.1. Preparation of Claudin–EGFP Plasmid DNA for Microinjection into Medaka Embryos

1. The key to successfully visualize TJs in vivo is the design of *claudin–EGFP* fusion gene. Hence 200 pg of mRNA of *claudin–EGFP* was microinjected into medaka eggs to check that TJs are properly labeled with the claudin–EGFP fusion protein, prior to the construction of a *claudin–EGFP* transgenic

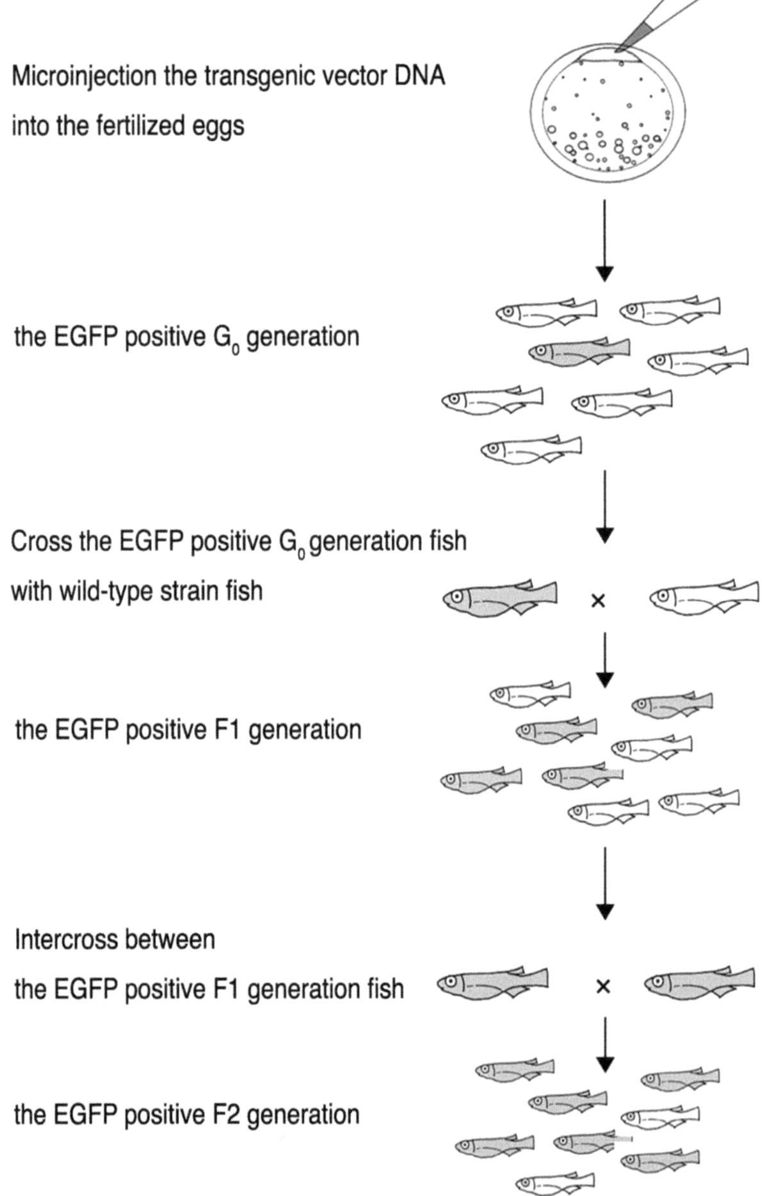

Fig. 1. Outline of the generation of *claudin–EGFP* transgenic medaka that allow in vivo TJ imaging.

vector (see below for microinjection protocol) (see Note 5). Indeed, in the case of *claudin-7*, only its C-terminal *EGFP* tagging allowed labeling of TJs in medaka embryos (8).

2. Insert the *claudin–EGFP* fragment downstream of the red sea bream *β-actin* promoter in the transgene plasmid (8). Check the *claudin–EGFP* junction sequence with the *EGFP* primers (*EGFP* N-terminal primer: 5′-CGTCGCCGTCCAGC TCGACCAG-3′, *EGFP* C-terminal primer: 5′-CATGGTC CTGCTGCTGGAGTTCGTG-3′).

3. Transform the DH5α strain of *Escherichia coli* with the *claudin–EGFP* plasmid and carry out midi-prep plasmid preparation using the Qiagen EndoFree plasmid MidiPrep kit (Qiagen, USA).

3.2. Microinjection of Claudin–EGFP Plasmid DNA into Medaka Embryos and Generation of a Transgenic Line

1. Back-fill a microinjection needle with the injection solution of 10 ng/μl *claudin–EGFP* plasmid DNA in circular form (for *claudin–EGFP* mRNA, use an injection solution of 200 ng/μl) (see Note 6).
2. Set the microinjection needle in the pipette holder of the air pressure injector and apply air pressure to the needle.
3. Collect fertilized eggs from the belly of medaka and transfer eggs to a Petri dish filled with embryo medium. To uncluster eggs, tangle and cut attachment filaments by holding the attachment filaments with two forceps.
4. Select fertilized eggs under a stereoscopic microscope and transfer to the agarose injection plate with a glass pipette. Align the eggs into the U-shape grooves with forceps.
5. Bring the tip of the microinjection needle close to the eggs under low magnification of the microscope. Break the tip of the microinjection needle by touching to the chorion of an embryo. Adjust the holding pressure of the injector to be relatively high, to ensure that no reflux occurs due to the high pressure upon puncture through the tough chorion of the egg.
6. Just before injection, the animal pole (the cytoplasmic region) of the egg is oriented perpendicular to the microinjection needle (see Note 7). Inject *claudin–EGFP* plasmid DNA solution into a cytoplasm of 1-cell stage embryo with a pulse of air pressure under high magnification (100×) (see Note 8).
7. Transfer the injected eggs from the agarose injection plate to a Petri dish containing with the embryo medium. Let the injected eggs develop at 28°C in an incubator (see Note 9).
8. Select the EGFP-positive embryos at 2 dpf (days postfertilization) using a fluorescence microscope (see below; observation of live medaka embryos). Grow embryos that have proper green fluorescence to sexual maturity (referred as the G_0 generation) (see Note 10).
9. Each of the G_0 generation fish is crossed with a wild-type fish. Select the G_0 fish that successfully produced F1 embryos with the proper pattern of EGFP expression (see Note 11). At least 100 F1 embryos should be checked for each cross, since G_0 fish harbor the transgene in a mosaic manner.
10. Select the EGFP-positive F1 embryos from each G_0 fish and grow up to Adult. F1 fish from the same G_0 fish are crossed to generate F2 fish to check, if the F2 fish that have the same insertion of the transgene develop normally and fertile to establish the transgenic strain.

3.3. Observation of Live Claudin–EGFP Transgenic Medaka Embryos

1. To observe *Claudin–EGFP* localization in medaka embryos at high resolution (Fig. 2), its chorion should be removed. Gently roll embryos on sandpaper (p2000 grit size, waterproof) using the forefinger, applying minimal pressure, and keeping finger parallel to the surface of the sand paper for around 45–60 s to remove some of the outer surface hairs of the chorion.
2. Incubate the embryos in Proteinase K (10 mg/ml in H_2O) for 1 h at 27°C.
3. Recover Proteinase K (store at −20°C for reuse) and wash embryos with the embryo medium five times.
4. Incubate the embryos in the hatching enzyme solution at 27°C. When the chorion is coming out from the embryos, immediately pick up individual embryo and transfer to 1× BSS in a 3.5-cm Petri dish by a glass pipette. The hatching enzyme is frozen stored at −20°C for reuse.
5. Thaw a small amount of 3% low gelling temperature agarose (in 1× BSS) by heating to 45°C and maintain at this temperature.
6. Transfer one dechorionated embryo to the cap depression of a 1.5-ml tube minimizing carrying over medium with the embryo. Immediately uptake the agarose and embryo from the cap depression and transfer to a 3.5-cm glass base dish. For imaging using an inverted microscope, embryos are orientated body facedown and placed close to the bottom glass. For imaging using an upright microscope, the thickness of agarose is minimized.
7. Transfer the 3.5-cm Petri dish into a 14-cm diameter Petri dish containing ice and water (water filled to roughly one-third total depth of large dish). While holding the chamber down firmly on the bottom of the 14-cm Petri dish, use a hair loop to gently orientate the embryo as desired in the molten agarose and hold the embryo until agarose solidifies.

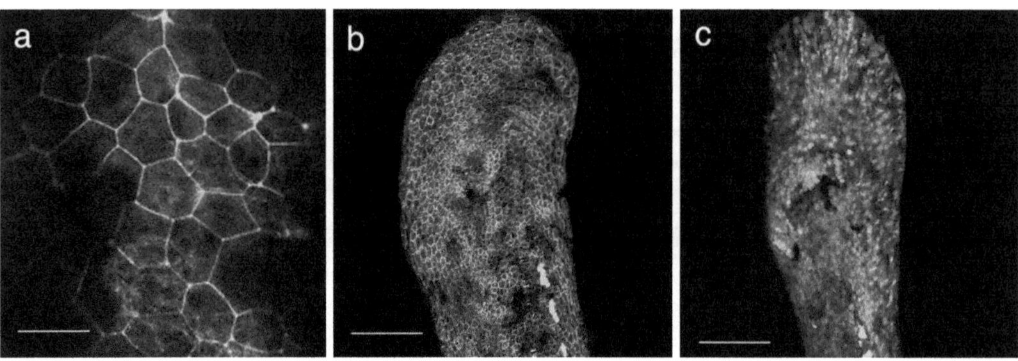

Fig. 2. Live confocal imaging of F1 *claudin-7–EGFP* transgenic embryos. (**a**) The epidermis of the *claudin-7–EGFP* transgenic medaka at 6 dpf. (**b**) The tail of the *claudin-7–EGFP* transgenic medaka at 6 dpf. (**c**) The tail of transgenic medaka expressing *EGFP* driven by the *β-actin* promoter. *Scale bars* 80 μm (**a**), 800 μm (**b**, **c**).

4. Notes

1. The red sea bream *β-actin* promoter is a strong promoter, strong enough to ubiquitously label TJs in all medaka embryo tissues. However, using the *β-actin* promoter might lead to overexpression of the claudin–EGFP fusion protein. Therefore, the use of endogenous promoters of medaka *claudins* is recommended for transgenic vectors.
2. Since endotoxin derived from *E. coli* affects development of medaka embryos, the plasmid DNA needs to be endotoxin-free.
3. Hatching enzyme solution can be reused several times by storing it in −80°C.
4. More detailed procedures of experiments using medaka fish (fish maintenance, egg production, and transgenesis) are available (14).
5. Prior to injection, about ten microinjection needles should be prepared because there is insufficient time to prepare them during injection.
6. Avoid air bubbles in the back-filled DNA solution by using Microloader, microloader (Eppendorf 930001007, Germany).
7. The cytoplasm of 1-cell stage embryo can be identified as the site free of oil drops (Fig. 1). Addition of phenol red in the injection solution assists proper injection of the DNA solution.
8. If the microinjection needle becomes clogged, apply air pressure, break the needle tip further, or replace the needle with a new one.
9. Dead embryos are stained blue with methylene blue.
10. Approximately 10% of the injected embryos exhibited EGFP signals in the G_0 generation (8).
11. Approximately 0.7% of the G_0 generation transmitted the *claudin-7–EGFP* transgene to the G_1 generation (8). Therefore, it is recommended to inject 1,000 eggs and grow them up.

Acknowledgment

We wish to thank all members of the Shoichiro Tsukita laboratory for helpful discussions.

References

1. Tsukita S, Furuse M, Itoh M (2001) Multifunctional strands in tight junctions. Nat Rev Mol Cell Biol 2: 285–93.
2. Furuse M, Fujita K, Hiiragi T, Fujimoto K, Tsukita S (1998) Claudin-1 and -2: novel integral membrane proteins localizing at tight junctions with no sequence similarity to occluding. J Cell Biol 141: 1539–50.
3. Furuse M, Sasaki H, Fujimoto K, Tsukita S (1998) A single gene product, claudin-1 or -2, reconstitutes tight junction strands and recruits occludin in fibroblasts. J Cell Biol 143: 391–401.
4. Simon DB, Lu Y, Choate KA, Velazquez H, Al-Sabban E, Praga M, Casari G, Bettinelli A, Colussi G, Rodriguez-Soriano J, McCredie D, Milford D, Sanjad S, Lifton RP (1999) Paracellin-1, a renal tight junction protein required for paracellular Mg^{2+} resorption. Science 283: 103–6.
5. Konrad M, Schaller A, Seelow D, Pandey AV, Waldegger S, Lesslauer A, Vitzthum H, Suzuki Y, Luk JM, Becker C, Schlingmann KP, Schmid M, Rodriguez-Soriano J, Ariceta G, Cano F, Enriquez R, Juppner H, Bakkaloglu SA, Hediger MA, Gallati S, Neuhauss SC, Nurnberg P, Weber S (2006) Mutations in the tight-junction gene claudin 19 (CLDN19) are associated with renal magnesium wasting, renal failure, and severe ocular involvement. Am J Hum Genet 79(5): 949–57.
6. Sasaki H, Matsui C, Furuse K, Mimori-Kiyosue Y, Furuse M, Tsukita S (2003) Dynamic behavior of paired claudin strands within apposing plasma membranes. Proc Natl Acad Sci USA 100: 3971–6.
7. Matsuda M, Kubo A, Furuse M, Tsukita S (2004) A peculiar internalization of claudins, tight junction-specific adhesion molecules, during the intercellular movement of epithelial cells. J Cell Sci 117: 1247–57.
8. Miyamoto T, Momoi A, Kato K, Tsukita S, Furuse M, Furutani-Seiki M (2009) Generation of transgenic medaka expressing claudin7-EGFP for imaging tight junction in living medaka embryos. Cell Tissue Res 335: 465–471.
9. Kato K, Takagi M, Tamaru Y, Akiyama S, Konishi T, Murata O, Kumai H (2007) Construction of an expression vector containing a beta-actin promoter region for gene transfer by microinjection in red sea bream Pagrus major. Fish Sci 73: 440–445.
10. Furutani-Seiki M, Wittbodt J (2004) Medaka and Zebrafish, an evolutionary twin study. Mech Dev 121: 629–637.
11. Ozato K, Kondoh H, Inohara H, Iwamatsu T, Wakamatsu Y, Okada TS (1986) Production of transgenic fish: introduction and expression of chicken delta-crystallin gene in medaka embryos. Cell Differ 19: 237–244.
12. Furutani-Seiki M, Sasado T, Morinaga C, Suwa H, Niwa K, Yoda H, Deguchi T, Hirose Y, Yasuoka A, Henrich T, Watanabe T, Iwanami N, Kitagawa D, Saito K, Asaka S, Osakada M, Kunimatsu S, Momoi A, Elmasri H, Winkler C, Ramialison M, Loosli F, Quiring R, Carl M, Grabher C, Winkler S, Del Bene F, Shinomiya A, Kota Y, Yamanaka T, Okamoto Y, Takahashi K, Todo T, Abe K, Takahama Y, Tanaka M, Mitani H, Katada T, Nishina H, Nakajima N, Wittbrodt J, Kondoh H (2004) A systematic genome-wide screen for mutations affecting organogenesis in Medaka, *Oryzias latipes*. Mech Dev 121: 647–658.
13. Kasahara M, Naruse K, Sasaki S, Nakatani Y, Qu W, Ahsan B, Yamada T, Nagayasu Y, Doi K, Kasai Y, Jindo T, Kobayashi D, Shimada A, Toyoda A, Kuroki Y, Fujiyama A, Sasaki T, Shimizu A, Asakawa S, Shimizu N, Hashimoto S, Yang J, Lee Y, Matsushima, Sugano S, Sakaizumi M, Narita T, Ohishi K, Haga S, Ohta F, Nomoto H, Nogata K, Morishita T, Endo T, Shin IT, Takeda H, Morishita S, Kohara Y (2007) The medaka draft genome and insights into vertebrate genome evolution. Nature 7145: 714–719.
14. Porazinski S, Wang H, Furutani-Seiki M (In press) Embryology and genetic methods-medaka. Methods in Molecular Biology.

Chapter 13

Claudins in a Primary Cultured Puffer Fish (*Tetraodon nigroviridis*) Gill Epithelium

Phuong Bui and Scott P. Kelly

Abstract

A primary cultured gill epithelium from the model organism *Tetraodon nigroviridis* (spotted green puffer fish) has been developed for the study of claudin tight junction (TJ) proteins and their potential role in the regulation of paracellular permeability across the gills of fishes. The cultured preparation is composed of polygonal epithelial cells that exhibit TJ protein immunoreactivity around the periphery and develop a surface morphology of concentric apical microridges. There is an absence of cells exhibiting intense Na^{+}-K^{+}-ATPase immunoreactivity and taken together, these characteristics indicate that the epithelium is composed of gill pavement cells only. In *Tetraodon*, 52 genes encoding for claudin isoforms (*Tncldn*) have been identified and 32 of these genes are expressed in whole gill tissue. Of these genes, 12 are responsive to alterations in environmental salinity in vivo (*Tncldn3a*, *-3c*, *-6*, *-8d*, *-10d*, *-10e*, *-11a*, *-23b*, *-27a*, *-27c*, *-32a*, and *-33b*). All claudin isoforms found in whole gill tissue can be found in cultured pavement cell gill epithelia with the exception of *Tncldn6*, *-10d*, and *-10e*. The cultured preparation is suitable for studying the "molecular machinery" of TJ proteins in fish gill pavement cells.

Key words: Tight junction, Osmoregulation, Pavement cells, Paracellular permeability

1. Introduction

Claudins have been well characterized in mammals (for review see refs. 1, 2); however, less is known about these proteins in other organisms. The most comprehensive examination of claudins in a nonmammalian vertebrate to date has been conducted using the piscine genomic model *Fugu* (= *Takifugu*) *rubripes* (3). In *Fugu*, the claudin superfamily has undergone expansion, likely due to both whole genome as well as tandem gene duplication events (3). This has resulted in the presence of genes encoding for as many as 56 claudin isoforms in this fish species (3). Furthermore, both in silico evidence and recently published reports (see (4–8)) support

Kursad Turksen (ed.), *Claudins: Methods and Protocols*, Methods in Molecular Biology, vol. 762,
DOI 10.1007/978-1-61779-185-7_13, © Springer Science+Business Media, LLC 2011

the notion that claudin superfamily expansion is a characteristic of fishes in general. A number of studies have considered a role for claudins and other tight junction proteins (e.g., occludin) in the regulation of paracellular permeability across ionoregulatory epithelia of fishes. These epithelia include the gill (4–7, 9–11), kidney (4, 6, 9, 10), and intestine (4, 9, 10, 12, 13) as well as the skin (4, 6). However, a particular emphasis has been placed on the TJ "machinery" of the gill epithelium since this tissue plays a leading role in the overall maintenance of homeostasis in fishes.

The fish gill epithelium is a physiologically complex heterogeneous tissue that overlies a rich vasculature (14). The gills of fishes are simultaneously involved in the processes of respiration, osmoregulation, acid/base balance, and waste nitrogen excretion (14). Fish gills are in direct contact with surrounding water and account for over 50% of the total surface area of the animal. These observations seem at odds with the arguably discrete nature of the tissue, which in modern bony fishes (e.g., teleosts) is housed in symmetrically arranged cephalic opercular chambers and is covered by a bony operculum. However, the fish gill also exhibits an exquisitely elaborate architecture. This allows for a concurrently large surface area of exposure, across which physiological processes take place, and a conservation of apparent mass. Despite these observations, and as is the case with all epithelia, the permeability of a large surface area of exposure must be tightly regulated in order for homeostasis to be achieved. In the case of the fish gill, maintaining a blood-to-water barrier poses a significant problem due to the significant electrochemical and concentration gradients that exist across this tissue. Furthermore, these gradients can rapidly alter in association with changing environmental conditions. In this regard, tight junction (TJ) heterogeneity is generally accepted to play an important role in the normal physiology of fish gills, especially with regard to salt and water balance (14). However, cellular heterogeneity as well as architectural complexity (as outlined above) does not make this biologically complex epithelium an easy tissue to study in vivo.

In an attempt to overcome the above outlined issues, primary cultured model gill epithelia have been developed from a number of fish species (15–17). Depending on the specific techniques used, primary cultured "reconstructed" gill preparations can be composed of either a homogeneous population of epithelial pavement cells (i.e., pavement cell epithelia) or contain both pavement cells and mitochondria-rich cells (15–19). In a natural setting, both these cell types come into direct contact with the external environment. Pavement cells cover the vast majority of the gill surface in what could be best described as a "cellular carpet" while mitochondria-rich cells (which account for ~2–10% of the gill epithelium cell population) are focal points of transcellular ionomotive activity. Regardless of composition, cultured gill epithelia possess permeability characteristics that mimic the in vivo properties of the fish gill very closely (for review see ref. 20). Cultured gill epithelia do not contain capillary endothelial cells, allowing for the examination of gill epithelial cell properties in isolation. Furthermore, the simplified architecture of cultured gill epithelia (i.e., a flat epithelial surface) permits the relatively uncomplicated determination of electrophysiological and radiotracer endpoints of epithelial permeability (21, 22). However, a restriction in the currently available cultured gill models is that they have not been developed from fish species that are broadly accepted as model vertebrates (e.g., puffer fish or zebra fish). This does not detract from the value of already available models but does set limitations on easy access to information on the "molecular machinery" of the TJ complex. To address this issue, a model gill epithelium has been developed for the spotted green puffer fish (*Tetraodon nigroviridis*) that, with continued effort, should permit experimental dissection of the specific roles played by TJ proteins in the regulation of gill permeability. The genome of this organism has recently been characterized (23), and the identification of claudin genes in *Tetraodon* has greatly benefited from the comprehensive characterization of claudins in the closely related Tetraodontiforme *Fugu rubripes* (see ref. 3). In addition to these advantages, *T. nigroviridis* could also be considered an appropriate model of choice for such endeavors because of its widespread availability, easy culture in a laboratory setting and most importantly, *T. nigroviridis* is a euryhaline fish species capable of acclimating to both freshwater and seawater environments (4, 6). In contrast, *Fugu rubripes* exhibits a limited ability to acclimate to freshwater (24).

2. Materials

2.1. Primary Culture of Puffer Fish Gill Epithelia

1. Sterile phosphate-buffered saline (PBS, pH 7.7): 137 mM NaCl, 8.1 mM Na_2HPO_4, 1.5 mM KH_2PO_4, and 2.7 mM KCl. Store at room temperature or at 4°C for subsequent preparation of chilled solutions (see below).
2. "WASH" solution: 200 IU/ml penicillin, 200 μg/ml streptomycin, 0.4 mg/ml gentamicin, and 2.5 μg/ml Fungizone® Amphotericin B in 15-ml sterile PBS (pH 7.7). Penicillin–streptomycin liquid, 5,000 Upenicillin/ml, 5,000 μg streptomycin/ml (GIBCO®, Invitrogen Canada, Inc.); gentamicin reagent solution, 50 mg/ml, liquid (GIBCO®, Invitrogen Canada Inc.); and Fungizone® Antimycotic, 250 μg amphotericin B/ml (GIBCO®, Invitrogen Canada, Inc.). This solution is prepared fresh for use on the same day using chilled PBS (pH 7.7) and held at 4°C.
3. Trypsin–EDTA solution: 0.05% trypsin in PBS (pH 7.7) with 5.5 mM EDTA. Add 5 ml 0.5% trypsin, 5.3 mM EDTA·4Na,

10×, liquid (GIBCO®, Invitrogen Canada, Inc.), and 0.5 ml sterilized 2% EDTA (2Na·$2H_2O$) in PBS (pH 7.7) to 44.5 ml sterile PBS (pH 7.7). Store at –20°C until use. After defrosting, aliquots can be store at 4°C for several days.

4. Fetal bovine serum (FBS, GIBCO®, Invitrogen Canada, Inc.): Divide into 6 ml aliquots and store at –20°C until use.
5. "STOP" solution: Add 2 ml FBS to 18 ml sterile PBS (pH 7.7, 4°C) to make a 10% FBS solution. Make fresh on day of use with chilled PBS (pH 7.7) and hold at 4°C.
6. "RINSE" solution: Add 0.5 ml FBS to 19.5 ml sterile PBS (pH 7.7, 4°C) to make a 2.5% FBS solution. Make fresh on day of use with chilled PBS (pH 7.7) and hold at 4°C.
7. Leibovitz's L-15 medium containing L-glutamine (Invitrogen Canada, Inc.) supplemented with 6% FBS, 100 IU/ml penicillin, 100 μg/ml streptomycin, and 200 μg/ml gentamicin (GIBCO®, Invitrogen Canada, Inc.). Supplemented media is prepared fresh on day of use and temperature equilibrated prior to addition to cells/epithelia (see Note 1).
8. Rat tail collagen type I: 5 mg/ml (Sigma-Aldrich Canada Ltd.). Make a working solution of 50 μg/ml by diluting stock collagen solution with sterile deionized water. Store working solution at 4°C.

2.2. Immunocytochemical Characterization of Cultured Puffer Fish Gill Epithelia

1. Phosphate-buffered saline (PBS, pH 7.7): 137 mM NaCl, 8.1 mM Na_2HPO_4, 1.5 mM KH_2PO_4, and 2.7 mM KCl. Store at 4°C.
2. 3% Paraformaldehyde (PFA, Fisher Scientific Company, USA), diluted in PBS (pH 7.7). Aliquots of 1.25 ml 3% PFA are stored at –30°C until use.
3. Methanol (Reagent Grade, BioShop® Canada, Inc.). Store at –30°C.
4. 0.01% Triton-X (Sigma-Aldrich Canada Ltd.), diluted in PBS (pH 7.7), store at 4°C.
5. Antibody diluting buffer (ADB): 10% Goat serum (Invitrogen Canada, Inc.), 3% Albumin (Bovine Serum Albumin, Biotechnology Grade, BioShop® Canada, Inc.), and 0.05% Triton-X (Sigma-Aldrich Canada Ltd.) in PBS (pH 7.7). Aliquots (1.5 ml) of ADB are stored at –30°C until use.
6. Primary antibodies: Mouse monoclonal anti-Na^+, K^+-ATPase α-subunit antibody: α5, 1:10 dilution in ADB; (Developmental Studies Hybridoma Bank, Iowa City, IA, USA). Rabbit polyclonal anti-zonula occludens 1 (ZO-1) antibody: anti-ZO-1, 1:50 dilution in ADB (Invitrogen Canada, Inc.). Primary antibody dilutions prepared from stock solutions on day of use.
7. Secondary antibodies: Flourescein isothiocyanate (FITC)-labeled goat anti-mouse: 1:500 dilution in ADB (Jackson

ImmunoResearch Laboratories, Inc., USA). Tetramethyl rhodamine isothiocyanate (TRITC)-labeled goat anti-rabbit: 1:500 dilution in ADB (Jackson ImmunoResearch Laboratories, Inc., USA). Secondary antibody dilutions are prepared from stock solutions on day of use.

8. Mounting medium: Molecular Probes ProLong® Antifade kit (Invitrogen Canada, Inc.) containing 4′, 6-diamidino-2-phenylindole (DAPI, Sigma-Aldrich Canada Ltd.). DAPI is diluted to 5 μg/ml in deionized water and stored in single use aliquots at −30°C. Just prior to use, mix ProLong® antifade reagent with the ProLong® mounting medium according to manufacturer's instructions and add 8 μl of defrosted DAPI solution.

2.3. Electron Microscope Examination of Cultured Puffer Fish Gill Epithelia

2.3.1. Scanning Electron Microscopy

1. Phosphate buffer (PB, pH 7.2): 0.1 M Na_2HPO_4, 0.1 M NaH_2PO_4. Store at 4°C.
2. 2.5% Glutaraldehyde (Sigma Chem Co) in PB (pH 7.2). Make fresh on day of use and hold at 4°C.
3. Acetone (BioShop® Canada, Inc.).
4. Tetramethylsilane (TMS, NMR Grade, Sigma-Aldrich).

2.3.2. Transmission Electron Microscopy

1. Phosphate buffer (PB, pH 7.2): 0.1 M Na_2HPO_4, 0.1 M NaH_2PO_4. Store at 4°C.
2. 2.5% Glutaraldehyde (Sigma Chem Co) in PB (pH 7.2). Make fresh on day of use and hold at 4°C.
3. 1% OsO_4 (EM Science, USA) in PB (pH 7.2). Store at −20°C.
4. Spurr resin (EM Science, USA).
5. 2% Uranyl acetate in methanol, store at 4°C.
6. Reynold's lead citrate: 0.08 M $Pb(NO_3)_2$, 0.12 M $Na_3C_6H_5O_7$, 16% 1 M NaOH.

2.4. RNA Extraction, cDNA Synthesis, and RT-PCR Analysis of Cultured Puffer Fish Gill Epithelia

1. TRIzol® reagent (Invitrogen Canada Inc.).
2. Chloroform (ACS, Reagent Grade, BioShop® Canada, Inc.).
3. Ethanol (Caledon Laboratories Ltd, Canada).
4. DEPC water (BioShop® Canada, Inc.). A 0.1% DEPC solution is prepared in deionized water. Solution is autoclaved before use.
5. Deoxyribonuclease (DNase) I, Amplification Grade – 1 U/μl and 10× DNase I Reaction buffer (Invitrogen Canada Inc.).
6. Oligo $(dT)_{12-18}$ primers – 0.5 μg/μl (Invitrogen Canada Inc.).
7. Superscript III Reverse Transcriptase – 200 U/μl (Invitrogen Canada, Inc.).
8. DTT – 0.1 M and 5× First-Strand Buffer (Invitrogen Canada Inc.).
9. dNTP – 10 mM (Invitrogen Canada, Inc.).

10. 10× PCR buffer (Invitrogen Canada, Inc.).
11. $MgCl_2$ – 50 mM (Invitrogen Canada, Inc.).
12. Taq DNA Polymerase – 5 U/μl (Invitrogen Canada, Inc.).

3. Methods

3.1. Preparation of Primary Cultured Puffer Fish Gill Epithelium

1. Gill arches are carefully removed from a spotted green puffer fish (*T. nigroviridis*) and immediately immersed in 5 ml of cold "WASH" solution in a small Petri dish. Gill arches are held in this solution for 10 min at room temperature during which time gill filaments are separated from cartilaginous branchial arches. Each row of filaments is then cut into four to five small pieces.
2. Following the above procedure, gill tissue is transferred to a 50-ml conical centrifuge tube for a second 10 min incubation in 5 ml chilled "WASH" solution. This incubation takes place at 4°C with occasional manual agitation. Following the second incubation, the "WASH" solution is removed by aspiration and replaced by 5 ml chilled "WASH" solution for a third and final incubation under identical conditions (i.e., 10 min at 4°C).
3. Following the incubations in "WASH" solution, epithelial cells are isolated from gill tissue using 0.05% trypsin with 5.5 mM EDTA. Four rounds of trypsin isolation are conducted (10 min each time at 4°C with occasional manual agitation). In each round of trypsination, gill tissue is immersed in 5 ml of trypsin solution and as the 10 min incubation time elapses, gill tissue is mechanically agitated to aid cell isolation. Mechanical agitation is conducted by running tissue up and down a sterile plastic pipette ~40 times. Following the latter three rounds of trypsination, cells are collected into "STOP" solution after filtration through a 100 μm cell strainer (see Note 2).
4. Cells suspended in "STOP" solution are collected as a pellet by centrifugation (10 min at 500×*g*, 4°C). The "STOP" solution is removed by aspiration and the cell pellet is resuspended in 20 ml of "RINSE" solution. The suspended cells are then centrifuged a second time (10 min at 500×*g*, 4°C).
5. Following centrifugation, "RINSE" solution is removed by aspiration and the remaining cell pellet is gently resuspended in 5 ml of Leibovitz's L-15 culture medium supplemented with 6% FBS, 100 IU/ml penicillin, 100 μg/ml streptomycin, and 200 μg/ml gentamicin.
6. Immediately following cell resuspension an aliquot is removed for the determination of cell yield. It is important to make sure that this aliquot is taken from a well-mixed cell suspension to ensure accuracy in determining cell yield. Cells are counted using a haemocytometer (see Note 3).

7. Once the cell yield has been determined, cells are seeded into cell culture plates at a density of 1×10^6 cells/cm^2. Prior to seeding, the wells in the culture plates are collagen coated by adding 400 μl of a 50 μg/ml collagen solution to each well (i.e., 10 μg collagen/cm^2). Cell culture plates with a growth area of 2 cm^2 (i.e., 24-well cell culture plates) are typically used. The total volume in the well should be ~1 ml (see Note 4).
8. Cells are held in a water-jacketed incubator in an air atmosphere at 27°C. After 24 h incubation, non-adherent cells are washed out and fresh media replaced (see Note 5). At this stage, cells normally exhibit ~80% confluence (see Fig. 1a). Cell are then incubated for a further 48 h during which time they form an epithelial monolayer that covers the entire growth area of the cell culture well (see Fig. 1b).

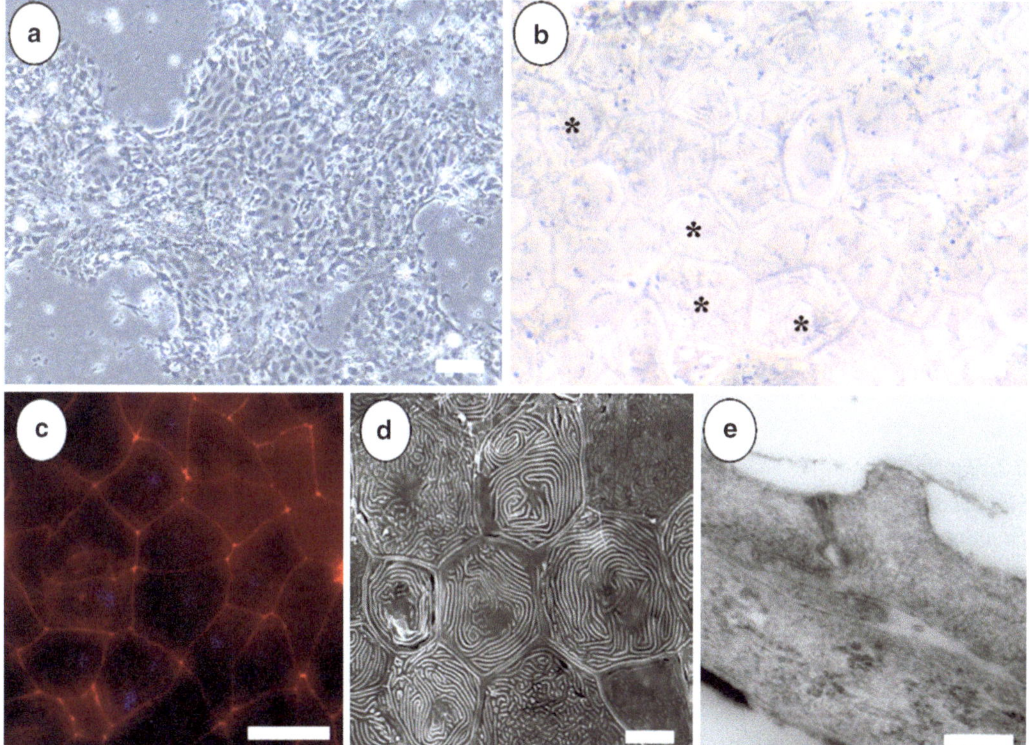

Fig. 1. Morphology of *Tetraodon* gill epithelia in primary culture. In panel (**a**), cells can been seen 24 h after seeding in culture wells. At this stage cells are usually ~80% confluent. Once cells become confluent and form an epithelium as illustrated in (**b**), they are polygonal and exhibit concentric apical microridges (*asterisks*). In panel (**c**), coimmunolocalization of the tight junction protein zona occludens 1 (ZO-1, *red*) with Na^+–K^+-ATPase reveals an absence of Na^+–K^+-ATPase-immunoreactive cells (NKA-ir, would normally appear in *green*). NKA-ir is routinely used as a marker for mitochondria-rich cells in fish gills. In panel (**d**), prominent concentric apical microridges are particularly easy to observe using scanning electron microscopy. This morphology is typical of gill pavement cells. Therefore, these morphological features as well as the absence of NKA-ir cells in cultured puffer fish gill epithelia strongly indicate that epithelia are composed exclusively of gill pavement cells. Panel (**e**) shows that in regions of apical cell-to-cell contact, cultured pavement cells exhibit typical TJ morphology. *Scale bars* = (**a**) 100 μm, (**b–d**) 10 μm, (**e**) 200 nm.

3.2. Immunocytochemistry of Primary Cultured Puffer Fish Gill Epithelium

1. Media is aspirated from cell culture wells and cultured gill epithelia are rinsed twice with 1 ml PBS (pH 7.7). Cultured gill epithelia are then fixed for 20 min in 3% PFA at room temperature. Following fixation, epithelia are rinsed twice with 1 ml cold PBS (pH 7.7) and 1 ml ice cold methanol is added to cell culture wells. Epithelia are then incubated in this solution at −30°C for 5 min.
2. Following methanol treatment, epithelia are washed (2 × 5 min) in ice cold PBS (pH 7.7).
3. Cultured gill epithelia are then permeabilized by incubating in 0.01% Triton-X for 10 min at room temperature. The 0.01% Triton-X solution is then removed by aspiration and 1 ml ADB is added to cultured epithelia for a 1 h incubation period at room temperature.
4. Following ADB incubation, epithelia are incubated overnight with primary antibody solutions for Na^+-K^+-ATPase and ZO-1 (see Note 6).
5. After incubating epithelia overnight with primary antibodies, epithelia are washed (3 × 5 min) with 1 ml PBS (pH 7.7) and then probed with secondary antibodies (FITC-labeled goat anti-mouse and TRITC-labeled goat anti-rabbit) for 1 h at room temperature.
6. Secondary antibodies are washed (3 × 5 min) from cultured preparations using PBS (pH 7.7). Epithelia are then prepared for examination by adding mounting medium containing DAPI. Epithelia are examined, and digital images captured, using a Nikon Inverted Microscope Eclipse Ti (Nikon Instruments, Inc., Melville, NY, USA) (see Fig. 1c).

3.3. Electron Microscope Examination of Cultured Puffer Fish Gill Epithelia

3.3.1. Scanning Electron Microscopy

1. Media is aspirated from cell culture inserts (see Note 7) and cultured gill epithelia are rinsed twice with 1 ml PBS (pH 7.7). Cultured gill epithelia are then fixed for 2 h in 2.5% glutaraldehyde in PB (pH 7.2) at 4°C. Following fixation, epithelia are rinsed (2 × 10 min each) with 1 ml PB (pH 7.2).
2. The permeable membrane at the base of a cell culture insert is then excised using a scalpel blade and placed in a scintillation vial containing PB (pH 7.2) for further processing. Epithelia are dehydrated through an ascending series of acetone (30, 50, 70, 80, 85, 90, 95, and 100%, 2 × 15 min at each concentration). Epithelia are then treated with TMS (2 × 10 min) and air dried at 37°C for ~10–20 s.
3. Epithelia are mounted on stubs and sputter coated (Hummer VI Au/Pd 40/60) for 2 min. All observations are made using a Hitachi S-520 scanning electron microscope and images are captured digitally using a passive image capture system (Hitachi, Quartz PCI Version 6) (see Fig. 1d).

3.3.2. Transmission Electron Microscopy

1. Media is aspirated from cell culture inserts and cultured gill epithelia are rinsed twice with 1 ml PBS (pH 7.7). Cultured gill epithelia are then fixed for 2 h using 2.5% glutaraldehyde in PB (pH 7.2) at 4°C. Following fixation, epithelia are rinsed twice (10 min each) with 1 ml PB (pH 7.2).
2. Epithelia are postfixed overnight at 4°C with 1% OsO_4 and rinsed (2 × 10 min) with PB (pH 7.2).
3. The bottom membrane of each insert is excised using a sharp scalpel blade and placed in a scintillation vial for further processing. Samples are dehydrated through an ascending series of ethanol (50, 70, and 85%, 15 min each; 95%, 2 × 15 min; and 100%, 3 × 15 min). Epithelia are then embedded in spurr resin. Sections are cut using an ultramicrotome (Sorvall Mt2b) and mounted on copper slot grids. Sections are stained with 2% uranyl acetate and Reynold's lead citrate. Samples are viewed using a Phillips EM201 transmission electron microscope (see Fig. 1e).

3.4. RNA Extraction and cDNA Synthesis

1. For total RNA extraction, media is aspirated and 500 μl of TRIzol® reagent is added to each well (see Note 8). Total RNA extraction is performed according to the manufacturer's instructions. RNA concentration is assessed by spectrophotometry and integrity by the A_{260}/A_{280} ratio.
2. An aliquot of total RNA (2 μg) is treated with DNase I, Amplification Grade for 15 min at room temperature followed by 10 min incubation at 65°C.
3. First-strand cDNA is synthesized by using 2 μg of DNase I-treated total RNA as per the manufacturer's instructions for SuperScript™ III Reverse Transcriptase. Oligo $(dT)_{12-18}$ is used as a primer.

3.5. RT-PCR Profiling of Claudin Isoforms in Cultured Puffer Gill Epithelia

1. To identify *Tetraodon* claudin isoforms, an in silico approach was used. Briefly, *Fugu rubripes* claudin sequences were retrieved from the National Center for Biotechnology Information database and used in conjunction with Genoscope's BLAT program (http://www.genoscope.cns.fr/externe/tetranew/) to search the *Tetraodon* genome for *Tetraodon* claudins exhibiting the highest degree of sequence identity. Sequences with the highest BLAT score (i.e., highest degree of sequence identity between *Fugu* and *Tetraodon*) were then submitted to a BLAST search against the nonredundant nucleotide database (thus cross-referencing *Fugu* and putative *Tetraodon* claudin orthologs). Upon confirmation of a putative *Tetraodon* claudin gene (*Tncldn*) using BLASTn, coding sequence (CDS) alignment using GeneDoc permit primer design based on highly conserved regions of the *Tetraodon* CDS. In *Tetraodon*, 52 genes encoding for claudin

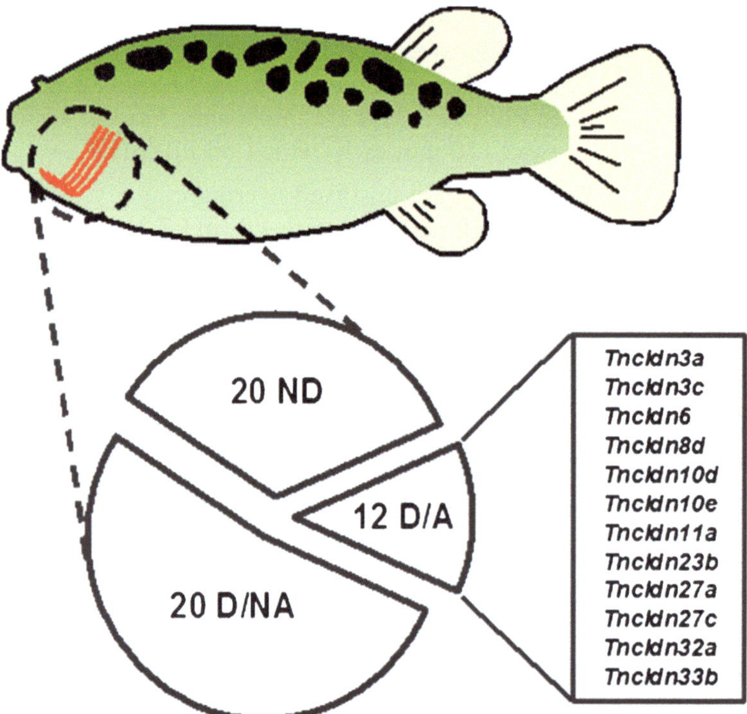

Fig. 2. In the spotted green puffer fish (*Tetraodon nigroviridis*), 52 genes encoding for claudin isoforms have been identified. In whole gill tissue, 20 genes encoding for *Tetraodon* claudin (*Tncldn*) isoforms are absent (not detected, *ND*). Of the remaining 32 genes, 12 have been found to exhibit significant alterations in mRNA abundance (detected/alter, *D/A*) in response to changes in environmental salinity (i.e., freshwater versus seawater acclimation) (see refs. 4, 6, 7). The remaining 20 *Tncldn* isoforms detected in the gills exhibit no alterations (detected/no alteration, *D/NA*) in response to environmental salinity change. The 12 salinity responsive *Tncldn* isoforms are listed on the right of the panel and are as follows; *Tncldn3a*, *-3c*, *-6*, *-8d*, *-10d*, *-10e*, *-11a*, *-23b*, *-27a*, *-27c*, *-32a*, and *-33b*.

isoforms have been identified and of these, 20 are absent from gill tissue, 20 are present but do not respond to alterations in environmental salinity, and 12 are present and respond to alterations in environmental salinity (see Fig. 2). Primer sequences used for RT-PCR analysis of *Tncldn* isoforms in gill tissue and cultured gill epithelia are outlined in Table 1.

2. PCR amplification of *Tncldn* isoforms is conducted in a PCR reaction mixture with a total volume of 25 μl. All components of the reaction mixture are obtained from Invitrogen Canada Ltd. and are as follows: 2.5 μl 10× PCR buffer, 0.75 μl $MgCl_2$, 0.3 μl dNTP, 0.5 μl 10 mM primer (forward), 0.5 μl 10 mM primer (reverse), 0.1 μl Taq DNA Polymerase, 18.35 μl sterile H_2O, and 2.0 μl cDNA template.

3. PCR amplification of *Tncldns* is performed under the following conditions: one cycle at 95°C (4 min); 40 cycles of denaturation at 95°C (30 s), annealing at 50–61°C (45 s), extension at 72°C (30 s); and one cycle of final extension at 72°C (5 min). Table 1 details specific annealing temperatures.

Table 1
Primer sets and annealing temperatures for *Tetraodon* claudin (*Tncldn*) isoforms present in gill tissue

Gene	Primer sequence	Amplicon length (bp)	Annealing temperature (°C)
Tncldn1	FOR: 5′-TCC TGC TGA GCG TCA TCT C-3′ REV: 5′-TTC CGA ACT CGT ACC TGG C-3′	225	55
Tncldn3a	FOR: 5′-TTG CCC TCT CCC AAG ACC-3′ REV: 5′-TCA AAC ATA GTC CTT CCT CTC CAG-3′	440	52
Tncldn3c	FOR: 5′-ATG TCG ATG GGT ATG GAG ATT G-3′ REV: 5′-CGA GGA CAG CCA GTA TGA TG-3′	275	51
Tncldn5a	FOR: 5′-TGG AAG GTG TCG GCT TTC-3′ REV: 5′-ATC TCC CTC TTC TTG GAC GC-3′	392	51
Tncldn5b	FOR: 5′-TCG CTG CGT GTT TGG AGC-3′ REV: 5′-CCT TCA CCA CCT CGT CCT TG-3′	339	56
Tncldn6	FOR: 5′-CCT GGT GTT GGG ATG TTT AGG-3′ REV: 5′-AAG ACA CAG CGA CTA TGG TGG-3′	152	56
Tncldn7a	FOR: 5′-GTG CTC CAT CAT TGC CTG -3′ REV: 5′-TCA CAC ATA CTC TTT GCT GCT G-3′	232	53
Tncldn8a	FOR: 5′-GCC AGA GGT TTG ATG TGC-3′ REV: 5′-CTG AGA ATC CAC CAC CAG AG-3′	234	52
Tncldn8b	FOR: 5′-GTC TCT TGT TGA CCC TCA TTG G-3′ REV: 5′-GGG ATA AGG ACG CAG ATG C-3′	379	54
Tncldn8c	FOR: 5′-TGA CGG CAT TCA TCG GAG-3′ REV: 5′-AGC AGG TTG ACT GGG TGG-3′	514	52
Tncldn8d	FOR: 5′-AGC AAA CCA CCT CAA ACC TAC-3′ REV: 5′-TTC ATC CAT AAA CCC TCC CAG-3′	275	54
Tncldn10d	FOR: 5′-ACC ACC TCC AAC TAC TAC TCC AAC-3′ REV: 5′-TTC CTG AGA AGT ATC CCA GCA C-3′	433	55
Tncldn10e	FOR: 5′-GGG ATT CTT GAT GAC AGT TCT G-3′ REV: 5′-TCA CCT TTC TCT TCG TCC AC-3′	682	53
Tncldn11a	FOR: 5′-ACG ACT GGG TGA ACA TTT GC-3′ REV: 5′-GTG TCC GAG GAG CAG CAG-3′	463	56
Tncldn12	FOR: 5′-ACC GTC AGT GTG GTT CCT G-3′ REV: 5′-ACT GTT CCT GTG TCA GCA CC-3′	575	52
Tncldn13	FOR: 5′-GTG GCT GGC TTT GGT TTG-3′ REV: 5′-GGC TGT TTG TGA AGG CAG TC-3′	628	54
Tncldn19	FOR: 5′-ATC GGA ATC ATC TCC ACC AC-3′ REV: 5′-CTT TAG AGC AGG AGC AGC AC-3′	511	54

(continued)

Table 1 (continued)

Gene	Primer sequence	Amplicon length (bp)	Annealing temperature (°C)
Tncldn23a	FOR: 5′-ATC CAG TCC GAC GCC TTC-3′ REV: 5′-TGG TCG TCA GGA AGT GGG-3′	310	56
Tncldn23b	FOR: 5′-TCC TCA CCT CGT TGT TCC TC-3′ REV: 5′-CAG ATT GAG CAT CGT GGC TG-3′	337	53
Tncldn27a	FOR: 5′-CGA GGA CGA GCG ATC TAA AG-3′ REV: 5′-GTT GTT GAT GAT GGT GTT GGC-3′	115	52
Tncldn27b	FOR: 5′-CCT GTG GAT GAA CTG TGT AAT GC-3′ REV: 5′-GGG TTC TGA AAG TCC GTG ATG-3′	303	54
Tncldn27c	FOR: 5′-GGA GGG TCA CTG CCT TTG-3′ REV: 5′-AGA TGT GGC TCT TGT AGG G-3′	527	50
Tncldn27d	FOR: 5′-GGC TCC TCC GTC TAC ATC-3′ REV: 5′-GGT TGT GTC CAC TGA GAC TG-3′	305	51
Tncldn28a	FOR: 5′-CTG GGA AGA CAA ATG CTA GGC-3′ REV: 5′-CCT GAT GAT GGT GTT TGC G-3′	426	57
Tncldn28b	FOR: 5′-CCT GGG GAG CAT CAT CAT C-3′ REV: 5′-GTC CTG GAT AAT GGT GTT GGT G-3′	385	57
Tncldn28c	FOR: 5′-TCG GCA CCA TCC TCA TCT G-3′ REV: 5′-CAC ACA GTT CAT CCA CAG ACC C-3′	110	57
Tncldn30a	FOR: 5′-CAT TGG CTG GAT CGG AGC-3′ REV: 5′-GCA GTT GGC ACA GAG GAG TC-3′	511	57
Tncldn30b	FOR: 5′-GAA GGC GAG GGT TTC AGT G-3′ REV: 5′-TCC TGG ATG ATG CTG CTT G-3′	99	56
Tncldn30c	FOR: 5′-ATG GTG TCT ACG GGG TTT CA-3′ REV: 5′-TCA GAC GTA GCC CTT TCC G-3′	645	61
Tncldn31	FOR: 5′-TGC TCT CCA TCT TGA TGT GC-3′ REV: 5′-CAT AAC CTT TGA CCT GTG AGG C-3′	390	55
Tncldn32a	FOR: 5′-GAA AGT GTC CTC CTT CAC CG-3′ REV: 5′-TTG GCG TCT TCG TTC TGC-3′	252	56
Tncldn33b	FOR: 5′-TGC TGG GAA TGC TGG TGG-3′ REV: 5′-CCT GAT ACC TTC GCT GCT GC-3′	342	57
β-actin	FOR: 5′-CCT CCG GTC GTA CCA CTG GTA T-3′ REV: 5′-CAA CGG AAG GTC TCA TTG CCG ATC GTG-3′	342	55

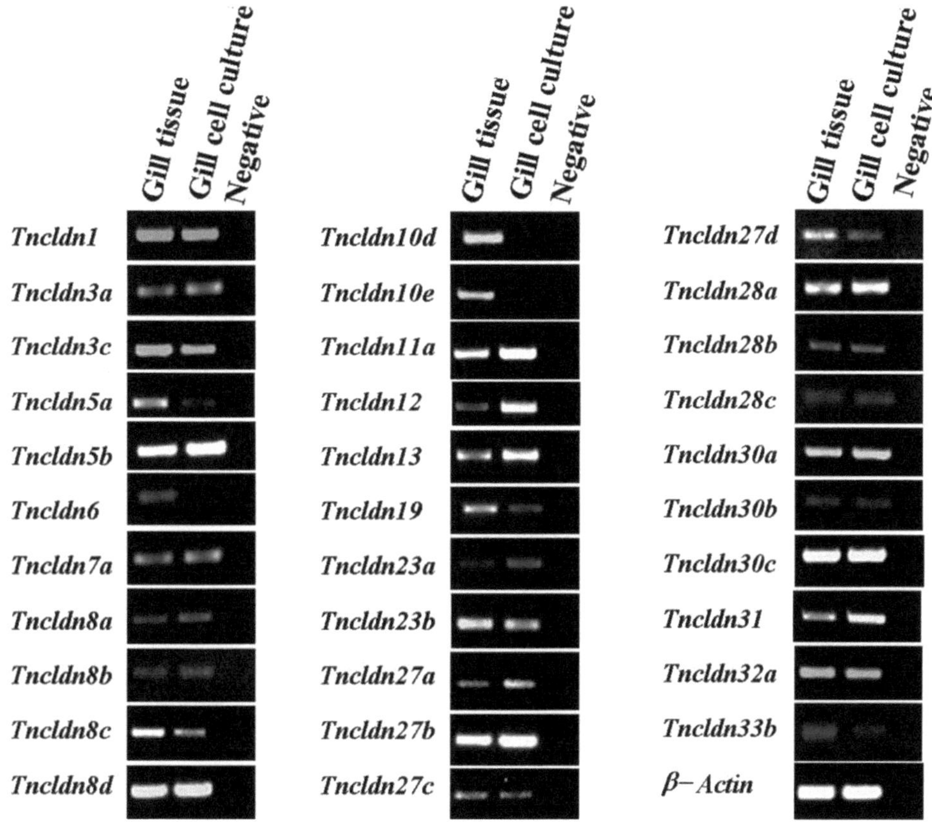

Fig. 3. A comparison of claudin isoform expression in puffer fish whole gill tissue and cultured puffer fish gill epithelia. In whole gill tissue, genes encoding for 32 claudin isoforms are present. Previous studies have demonstrated that of these 32 isoforms, 12 respond significantly to alterations in environmental salinity (i.e., freshwater versus seawater acclimation, see refs. 4, 6, 7). In cultured gill epithelia, 29 of the 32 genes encoding for claudin isoforms are present, with *Tncldn6*, *-10d*, and *-10e* being absent. Therefore, these claudin isoforms seem unlikely to be associated with gill pavement cells. However, *Tncldn6*, *-10d*, and *-10e* are responsive to salinity change (see ref. 7) and as such it seems likely that these isoforms may be associated with another gill cell type that is affected by salinity (e.g., mitochondria-rich cells). Negative controls contained PCR-grade water in the place of cDNA and β-actin served as a loading control.

4. β-Actin is amplified as an internal control and water is used in the place of cDNA as a negative control.
5. Amplicons are resolved using a 1.5% agarose gel (150 V for 1 h) and ethidium bromide is used for visualization. Gel images are captured using Quantity One® software (Bio-Rad Laboratories, Inc.) (see Fig. 3).

4. Notes

1. The optimal pH of L-15 media for the culture of *Tetraodon* gill cells was determined to be 7.45–7.50 at room temperature. Therefore, media is pH adjusted with sterile 1 M NaOH prior to use.
2. It has been found that by discarding the material/cells isolated during the first round of trypsination, subsequent attachment of harvested cells in the culture wells is enhanced. Therefore, although four rounds of trypsination are conducted to isolate epithelial cells from gill tissue, only the final three rounds of trypsinated cells are collected in "STOP" solution for subsequent use.
3. During the period of cell counting using a haemocytometer, the bulk cell suspension should be held at 4°C. It is important to determine cell numbers efficiently so that the seeding of the cells resuspended in media can take place as soon as possible (i.e., less than 30 min post-suspension). If left longer, or not held at 4°C, cell "clumping" has been observed to occur on occasions. Once cells have "clumped" they do not attach to culture plates very well.
4. Add media to culture wells prior to the addition/seeding of cells. When cells are seeded, they should be added in a manner which ensures that they are evenly distributed across the growth surface (e.g., do not swirl cells into the culture wells which cause them to congregate in the center of the growth surface, etc.). The final volume in each well is ~1 ml on the day of seeding.
5. Prior to any media change, equilibrate fresh L-15 media to 27°C. Add 2 ml of fresh media to each culture well.
6. Each well in a 24-well cell culture plate requires approximately 500 μl of diluted primary antibody. To prevent evaporation, the lid should be on and the plate placed in a moist chamber during incubation.
7. For electron microscopes studies, epithelia are cultured on cell culture inserts (0.9-cm^2 growth area, 0.4-μm pore size, 1.6×10^6 pores/cm^2 pore density; Falcon BD, Mississauga, ON, Canada). Inserts can be removed after the fixation of epithelia and easily mounted on stubs and/or sectioned for EM observation. Only epithelia that have developed a plateau in transepithelial resistance (TER) are used. The culture of epithelia on permeable cell culture inserts is part of ongoing studies that, when complete, will allow endpoints of paracellular permeability to be measured across cultured puffer fish gill epithelia.

8. On some occasions when using 24-well culture plates (growth area of 2 cm^2) the amount of total RNA extracted from an individual well has been insufficient for cDNA synthesis using the protocols outlined in our methodology. Therefore, to ensure that adequate RNA is obtained for subsequent analysis using the methodology described, TRIzol® reagent from two culture wells should be pooled for RNA extraction (i.e., 2×500 μl extracting RNA from 2×2 cm^2 area of cells).

Acknowledgments

The authors would like to extend their gratitude to Helen Chasiotis and Mazdak Bagherie-Lachidan for invaluable advice and assistance. This work was funded by an NSERC Discovery Grant, NSERC Discovery Accelerator Supplement, Canadian Foundation for Innovation New Opportunities Grant and an Ontario Early Researcher Award to SPK. The monoclonal α5 antibody developed by D.M. Fambrough was obtained from the Developmental Studies Hybridoma Bank (The University of Iowa, Department of Biological Sciences, Iowa City, IA, 52242, USA).

References

1. Van Itallie CM, Anderson JM (2006) Claudins and epithelial paracellular transport. Annu Rev Physiol 68: 403–429
2. Gonzalez-Mariscal L, Betanzos A, Nava P et al (2003) Tight junction proteins. Prog Biophys Mol Biol 81: 1–44
3. Loh YH, Christoffels A, Brenner S et al (2004) Extensive expansion of the claudin gene family in the teleost fish, *Fugu rubripes*. Genome Res 14: 1248–1257
4. Bagherie-Lachidan M, Wright SI, Kelly SP (2008) Claudin-3 tight junction proteins in *Tetraodon nigroviridis*: Cloning, tissue specific expression and a role in hydromineral balance. Am J Physiol Integrative Comp Physiol 294: R1638–R1647
5. Tipsmark CK, Kiilerich P, Nilsen TO et al (2008) Branchial expression patterns of claudin isoforms in Atlantic salmon during seawater acclimation and smoltification. Am J Physiol Integr Comp Physiol 294: R1563–R1574
6. Bagherie-Lachidan M, Wright SI, Kelly SP (2009) Claudin-8 and -27 tight junction proteins in puffer fish *Tetraodon nigroviridis* acclimated to freshwater and seawater. J Comp Physiol B 179: 419–431
7. Bui P, Bagherie-Lachidan M, Kelly SP (2010) Cortisol differentially alters claudin isoform mRNA abundance in a cultured gill epithelium from puffer fish (*Tetraodon nigroviridis*) Mol Cell Endocrinol 317: 120–126
8. Clelland ES, Kelly SP (2010) Tight junction proteins in zebrafish ovarian follicles: stage specific mRNA abundance and response to 17β-estradiol, human chorionic gonadotropin, and maturation inducing hormone. Biol Reprod
9. Chasiotis H, Kelly SP (2008) Occludin immunolocalization and protein expression in goldfish. J Exp Biol 211: 1524–1594
10. Chasiotis H, Effendi J, Kelly SP (2009) Occludin expression in epithelia of goldfish acclimated to ion poor water. J Comp Physiol B 179: 145–154
11. Chasiotis H, Wood CM, Kelly SP (2010) Cortisol reduces paracellular permeability and increases occludin abundance in cultured trout gill epithelia. Mol Cell Endocrinol
12. Clelland ES, Bui P, Bagherie Lachidan M et al (2010) Spatial and salinity-induced alterations in claudin-3 isoform mRNA along the gastrointestinal tract of the puffer fish *Tetraodon*

nigroviridis. Comp Biochem Physiol A Mol Integr Physiol 155: 154–163

13. Tipsmark CK, Sørensen KJ, Hulgard K et al (2010) Claudin-15 and -25b expression in the intestinal tract of Atlantic salmon in response to seawater acclimation, smoltification and hormone treatment. Comp Biochem Physiol A Mol Integr Physiol 155: 361–370
14. Evans DH, Piermarini PM, Choe KP (2005) The multifunctional fish gill: dominant site of gas exchange, osmoregulation, acid-base regulation, and excretion of nitrogenous waste. Physiol Rev 85: 97–177
15. Avella M, Berhaut J, Payan P (1994) Primary culture of gill epithelial cells from the sea bass *Dicentrarchus labrax*. In Vitro Cell Dev Biol 30A: 41–49
16. Wood CM, Pärt P (1997) Cultured branchial epithelia from freshwater fish gills. J Exp Biol 200: 1047–1059
17. Kelly SP, Wood CM (2002a) Cultured gill epithelia from freshwater tilapia (*Oreochromis niloticus*): Effect of cortisol and homologous serum supplements from stressed and unstressed fish. J Membrane Biol 190: 29–42
18. Fletcher M, Kelly SP, Pärt P et al (2000) Transport properties of cultured branchial epithelia from freshwater rainbow trout: A novel preparation with mitochondria-rich cells. J Exp Biol 203: 1523–1537
19. Kelly SP, Fletcher M, Pärt P et al (2000) Procedures for the preparation and culture of "reconstructed" rainbow trout branchial epithelia. Methods Cell Sci 22: 153–163
20. Wood CM, Kelly SP, Zhou B et al (2002) Cultured gill epithelia as models for the freshwater fish gill. BBA–Biomembranes 1566: 72–83
21. Wood CM, Gilmour KM, Pärt P (1998) Passive and active transport properties of a gill model, the cultured branchial epithelium of the freshwater rainbow trout (*Oncorhynchus mykiss*). Comp Biochem Physiol 119A: 87–96
22. Kelly SP, Wood CM (2001) Effect of cortisol on the physiology of cultured pavement cell epithelia from freshwater trout gills. Am J Physiol Regul Integr Comp Physiol 281: R811–R820
23. Jaillon O, Aury JM, Brunet F et al (2004). Genome duplication in the teleost fish *Tetraodon nigroviridis* reveals the early vertebrate proto-karyotype. Nature 431: 946–957
24. Lee KM, Kaneko T, Aida K (2005) Low-salinity tolerance of juvenile fugu *Takifugu rubripes*. Fisheries Sci 71: 1324–1331

Chapter 14

Manipulating Claudin Expression in Avian Embryos

Michelle M. Collins and Aimee K. Ryan

Abstract

Since the discovery of Claudin-1 and -2 by Tsukita and colleagues in the late 1990s [Furuse et al. J Cell Biol 141:1539–50,1998], claudin family members have been found to have critical roles in maintaining the integrity of epithelial and endothelial tight junctions [Furuse and Moriwaki Ann N Y Acad Sci 1165:58–61, 2009; Morita et al. Proc Natl Acad Sci USA 96:511–6, 1999; Tsukita and Furuse Ann N Y Acad Sci 915:129–35, 2000; Turksen and Troy J Cell Sci 117:2435–47, 2004]. The properties of distinct claudin family members in tight junction permeability and specificity have been extensively studied in vitro using cell culture models. In vivo, claudin family members are dynamically regulated during embryogenesis and alterations in their expression patterns can have detrimental effects on the formation and physiological function of the tissues in which they are expressed. The chick embryo provides an excellent system to dissect the roles of specific family members in vivo and to explore the effects of modulating claudin expression during the epithelial-to-mesenchymal and mesenchymal-to-epithelial transitions that are associated with tissue morphogenesis and differentiation. We are using the chick embryo to understand the roles of the claudin family of tight junction proteins during gastrulation and left–right patterning during embryogenesis. Here, we describe methodologies for manipulating claudin gene expression in specific target tissues during chick embryogenesis.

Key words: Claudins, Retroviral misexpression, Chick embryo, Ex ovo culture, Electroporation

1. Introduction

The avian embryo is a powerful model system that is used by developmental biologists to study the molecular mechanisms underlying morphogenesis (6). Chick embryos are staged according to well-standardized morphological criteria known as Hamburger-Hamilton (HH) stages (7), and it is relatively easy to obtain large numbers of developmentally synchronized embryos for experimental manipulation. Fertilized eggs can be purchased commercially, are available year-round, and can be stored for several days at 12–20°C

Kursad Turksen (ed.), *Claudins: Methods and Protocols*, Methods in Molecular Biology, vol. 762,
DOI 10.1007/978-1-61779-185-7_14, © Springer Science+Business Media, LLC 2011

prior to use. Once the eggs are placed into a 38°C incubator, embryonic development proceeds relatively quickly: embryos initiate gastrulation by 20 h, neurulation by 26–28 h, and formation and looping of the heart tube is complete by 48 h (7). Chick embryos can be cultured ex ovo during the first 3–4 days of development and this facilitates the manipulation of very young chick embryos. However, to examine phenotypic outcomes at later stages of development, it is best to manipulate gene expression in ovo. Importantly, gain-of-function and loss-of-function gene expression studies can be performed using similar techniques thereby permitting a comprehensive assessment of any given gene's function.

The chicken genome is approximately one-third the size of mammalian genomes, with a haploid content of 1.2×10^9 bp distributed over 38 autosomes and 2 sex chromosomes. In birds, the females have the heterogametic sex chromosomes (ZW), while males are homogametic (ZZ). Yet, despite these variations in organization, the chick genome demonstrates remarkable evolutionary conservation with mammals (NCBI Chicken Genome Resources). Twenty-four claudin family members have been described in mammals (1–5, 8, 9). Presently, 18 claudins have been annotated in the chick genome and all are highly similar to their murine homologues with amino acid identity ranging from 49% (claudin-22) to 85% (claudin-11) (Collins et al. unpublished observation). In addition, we have found that the expression patterns of several claudin family members are well-conserved between chick and mouse embryos, suggesting that they are likely to have conserved functions during vertebrate embryogenesis (Collins and Ryan, unpublished observations). In this chapter, we present two methods for manipulating claudin expression in the chick embryo: microinjection of retroviral particles and electroporation of plasmid DNA. We also include immunohistochemistry and in situ hybridization (ISH) protocols to identify those cells and tissues that have been targeted to ectopically overexpress claudin molecules.

2. Materials

2.1. Preparation of DNA

1. High-quality plasmid DNA should be prepared using a midiprep kit (e.g., QIAGEN Plasmid Midi Kit) or by cesium-chloride banding (10).
2. Filter-sterilized 0.1% Fast Green Dye (Sigma) in water.

2.2. Preparation of Embryos for Ex Ovo Plasmid Manipulation

1. Fertilized eggs purchased from a local supplier. If desired, pathogen-free eggs can be obtained from Charles River Laboratories.

2. Simple saline: 7.19 g NaCl/L water
3. Agar-albumen plates: Thin albumen can be collected from unfertilized eggs purchased at a supermarket. Crack egg into a petri dish and use a plastic transfer pipet to remove 4–5 ml of thin albumen per egg. Warm at 55°C. Boil 0.72 g agar in 120 ml simple saline. Cool to 55°C and add 120 ml pre-warmed thin albumen and 240 μl penicillin/streptomycin (5,000 U penicillin/5 mg streptomycin per ml). Mix well. Pipet 2.5 ml into 35 mm × 10 mm plastic plates, let set and store in humidified container at 4°C.
4. Dulbecco's Modified Eagle's Medium (DMEM; Gibco/BRL).
5. Filter paper supports: Fold a 70 mm circle of #2 Whatman filter paper (Cat. No. 1002-070) into quarters. Make two overlapping holes using a hole puncher. The hole size will depend on the stage of the embryo and duration of ex ovo culture. Trim to generate four 2 cm by 3 cm embryo supports. Wrap a stack of 20–25 supports in aluminum foil and autoclave to sterilize.
6. 10× Tyrode's Solution for electroporation: 80 g NaCl, 2 g KCl, 2.71 g $CaCl_2$, 0.5 g NaH_2PO_4, 2 g $MgCl_2$, and 10 g glucose in 1 L of water. Autoclave and dilute in water to 1× just prior to use.
7. Forceps.
8. Scissors.

2.3. Preparation of Retroviral Particles for Microinjection

1. DF-1 chick fibroblast cells (CRL12203; ATCC).
2. Lipofectamine™2000 (Invitrogen).
3. Dulbecco's Modified Eagle's Medium with 10% fetal bovine serum (Gibco/BRL).
4. Beckman SW28 rotor (or equivalent) and centrifuge tubes.

2.4. Injection and Electroporation Equipment Set-Up

1. 1 mm × 90 mm Narishige capillary tubes.
2. Vertical pipette puller.
3. Microinjector (e.g., Narishige IM300) connected to a nitrogen tank.
4. Square wave electroporator (e.g., BTX ECM830 or Protech CUY-21).
5. Electrodes: Paddle- or rod-shaped electrodes may be homemade or purchased. We use homemade platinum rod electrodes (11) and purchased paddle electrodes (CUY-701P2E; Protech). Choice of electrodes is determined by the target tissue (see below).

2.5. Analysis of Electroporated Embryos by Immunohistochemistry

1. 4% Paraformaldehyde (PFA) in PBS, pH 8. PFA can be prepared in advance and stored aliquoted at –20°C.
2. Ethanol.
3. Xylene.
4. Paraffin: melt at 60°C.
5. Peel-a-way disposable tissue embedding molds (Cat. No. 18986; Polysciences, Inc.).
6. Normal goat serum (Vector Laboratories).
7. 0.3% Triton X-100 in PBS.
8. Primary antibodies: Mouse anti-RCAS gag monoclonal antibody (AMV 3C2; Developmental Studies Hybridoma Bank) can be used to identify retrovirally infected cells and specific claudin antibodies to confirm protein expression. Supplier will depend on particular claudin family member being studied. We have purchased antibodies from Zymed, Spring Biosciences and Abnova for use on chick tissues.
9. Elite Vectastain ABC kit (Cat. No. PK-6100; Vector Laboratories).
10. Biotinylated anti-mouse IgG and anti-rabbit IgG antibody (Cat. No. BA-1400; Vector Laboratories) or biotinylated anti-rabbit IgG (Cat. No. BA-1000; Vector Laboratories).
11. Peroxidase substrate DAB kit (Cat. No. SK-4100 Vector Laboratories or Cat. No. 34002 Pierce).
12. Hematoxylin staining solutions (if desired for counterstain).

2.6. Analysis of Electroporated Embryos by Whole-Mount In Situ Hybridization

1. PBT: PBS with 0.1% Tween-20.
2. 6% Hydrogen peroxide: prepare fresh from 30% stock solution diluted in water.
3. Proteinase K solution: 1 μg Proteinase K in 1 ml PBS. Prepare from a 10 mg/ml stock solution. Store stock solution in 10 μl aliquots at –20°C. Discard after use.
4. Glycine solution: Place 0.1 g glycine into a 50 ml tube. Add 50 ml PBT immediately prior to use.
5. Postfix solution: 8 μl glutaraldehyde in 1 ml 4% paraformaldehyde in PBT.
6. Hybridization mix: 50% formamide, 5× SSC pH 5, 50 μg/ml yeast tRNA, 1% SDS, and 50 μg/ml heparin.
7. Digoxigenin-labeled antisense riboprobe: Synthesize an antisense riboprobe from linearized DNA template using the appropriate RNA polymerase and DIG RNA labeling mix (Cat. No. 11277073910; Roche). Precipitate riboprobe and resuspend in 0.1 M RNase-free DTT. Store at –80°C. Prior to use resuspend at 0.1–1 μg/ml in hybridization mix.

8. Wash solution 1: 50% formamide, 5× SSC pH 4.5, 1% SDS.
9. Wash solution 2: 50% formamide, 2× SSC pH 4.5.
10. TBST: 80 g NaCl, 2 g KCl, 250 ml Tris–HCl pH 7.5, 10 ml Tween-20 per 1 L water.
11. Levamisole (Sigma).
12. Heat-treated sheep serum (Gibco/BRL): Heat sheep serum at 70°C for 30 min. Store in 5 ml aliquots at –20°C.
13. Embryo powder: Collect 2–4 dozen day 4 chick embryos. Wash two times in ice-cold PBS. Add 2 volumes ice-cold acetone and homogenize. Incubate on ice 30 min. Centrifuge to pellet tissue. Dry tissue overnight at RT. Grind into a fine powder using a mortar and pestle and store at 4°C.
14. Antibody mix: For 5 ml of antibody mix, mix 7.5 mg embryo powder and 1.25 ml TBST in a microcentrifuge tube and heat at 65°C for 30 min. Chill on ice. Add 12.5 μl sheep serum and 1 μl anti-digoxigenin-alkaline phosphatase-conjugated antibody (Cat. No. 11093274910 Roche). Incubate with rocking at 4°C for 60 min. Microcentrifuge at 13,000 × *g* at 4°C for 10 min. Transfer supernatant to 15 ml conical tube and add 3.75 ml TBST and 37.5 μl heat-treated sheep serum. Store at 4°C prior to use.
15. NTMT: 0.1 M NaCl, 0.1 M Tris pH 9.5, 0.05 M $MgCl_2$, 0.1% Tween-20.
16. 4-Nitro blue tetrazolium chloride (NBT; Cat. No. 11383213001 Roche).
17. 5-Bromo-4-chloro-3-indolyl-phosphate (BCIP; Cat. No. 11383221001 Roche).

3. Methods

The two principle methods for gain-of-function and loss-of-function gene studies in chick embryos are injection of retroviral particles (12–14) or the introduction of eukaryotic expression vectors by electroporation (6, 11) (Fig. 1). Both approaches permit the targeted introduction of an expression cassette into a tissue of interest. They can be used to overexpress a protein within a tissue, to ectopically misexpress a protein in a tissue in which it is not normally expressed or to antagonize the function of an endogenous protein by overexpressing a siRNA specifically targeted to the gene of interest or by using a dominant negative isoform of the protein (15). Several expression vectors that utilize a strong ubiquitous promoter to drive expression of the downstream gene are available for use in chick embryos, including

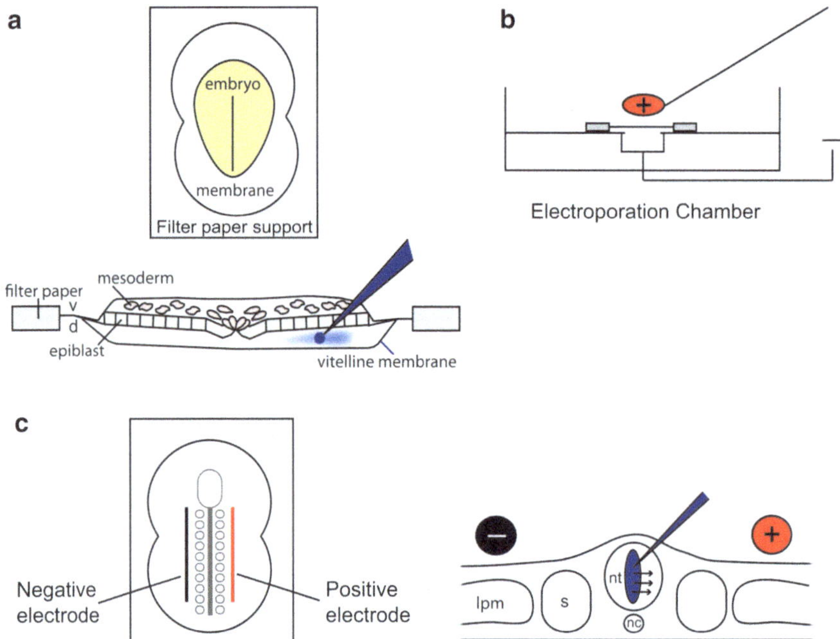

Fig. 1. Injection site and electroporation orientation for expression of retroviral particles or plasmid DNA. (**a**) Gastrulation to neurulation staged embryos are collected on filter paper supports, with the ventral (v) surface facing upwards and the dorsal (d) surface downwards. Schematic diagrams of an HH stage 4 chick embryo within a filter paper support (*top*) and a transverse section through the embryo (*bottom*) are shown. Plasmid DNA (*blue*) can be injected into the space between the epiblast of the embryo and the vitelline membrane (*as shown*) or retroviral particles can be injected directly into epiblast cells (*not shown*). (**b**) Embryos on the filter paper support are transferred to an electroporation plate with the dorsal side of the embryo facing downwards and an electrical current is passed across the embryo. For the orientation shown, DNA will be electroporated into cells of the epiblast. (**c**) On the left side is a diagram of a dorsal view of an older embryo showing electrode placement for neural tube and/or somite electroporation. The positive electrode is shown in *red* and the negative electrode in *black*. On the *right side* is a schematic of a transverse section through this embryo showing injection of DNA into the neural tube, placement of elctrodes and direction of DNA movement (*arrows*) in the electric field. Abbreviations: *lpm* lateral plate mesoderm, *nc* notochord, *nt* neural tube, *s* somite.

pcDNA3.1, pCMV5, pCAX, and pMES (11). In addition, more than one expression construct can be introduced into a single cell. Electroporation of a eukaryotic expression vector is sufficient for short-term experiments (24–36 h) but is limited by the fact that the plasmid is maintained as an episome and consequently the number of copies per cell decreases with each round of cell division.

In contrast, the choice of retroviral vectors is more limited and most laboratories use the replication competent retroviral vector RCAS, subtypes A–E (12, 14). However, injection of the retrovirus is preferable for long-term experiments since the viral DNA stably integrates into the genomic DNA and therefore all subsequent daughter cells will also express the protein of interest (13). Because this virus is replication competent, infected cells will generate viral particles that can infect neighboring cells and so broad domains of over-/mis-expression are possible. If a more limited domain of expression is desired, a replication incompetent retroviral vector should be used. Avian retroviruses do not require specific containment during their propagation and use since they do not normally infect nonavian hosts (14). One limitation of the retroviral approach is that the vector does not accommodate cDNA inserts greater than 2 kb. However, this is not a concern for manipulating claudin gene expression given that most claudins are less than 250 amino acids and encoded by cDNAs less than 750 bp.

Several methods exist to track the cells overexpressing the protein of interest in the embryo prior to stopping the experiment. The simplest way to do this is to use a cDNA that encodes a fluorescently tagged form of the protein, i.e., a green fluorescent protein (GFP)-fusion protein and monitor fluorescence using a fluorescent dissecting microscope during the incubation period. Given that a GFP moiety at the N terminus may affect insertion of any claudin into the cell membrane and a GFP moiety at the C terminus could hinder interactions of the claudin with other proteins in the tight junction cytoplasmic plaque, we prefer to coinject/electroporate a separate vector expressing GFP or RFP with our claudin expression vector (Fig. 2). Alternatively, the claudin cDNA can be cloned upstream of an IRES/GFP to generate a bicistronic message (e.g., pMES vector) and eliminate any effects on the functionality of the claudin protein. A bicistronic message can also be cloned into the retroviral vector. Alternatively, a GFP/RFP expression vector could be co-electroporated with the retroviral vector.

Our lab is examining the role of claudin family members during gastrulation and in the establishment of the left–right axis. Since these events occur during the earliest stages of embryonic development in embryos < 60 h, we perform most of our experiments using ex ovo cultured chick embryos using the method established by Chapman et al. (16). We have used electroporation to overexpress claudin family members throughout the epiblast in gastrulation stage embryos (Collins and Ryan, unpublished observations) and retroviral expression vectors to misexpress full-length and truncated variants of Claudin-1 and -3 in more restricted regions of the embryo (17). For our studies pertaining to left–right axis development, we usually manipulate only one side of the embryo allowing the unmanipulated side to serve as a negative control. At the end of the experiment, cells ectopically overexpressing the claudin protein can be identified by in situ hybridization or immunohistochemistry using reagents that detect the claudin molecule itself or RNA/proteins synthesized by the retrovirus (Fig. 2).

3.1. Preparation of DNA

1. cDNAs encoding the claudin gene of interest should be cloned downstream of a strong ubiquitous promoter in a eukaryotic expression vector. For the retroviral vectors, cDNAs

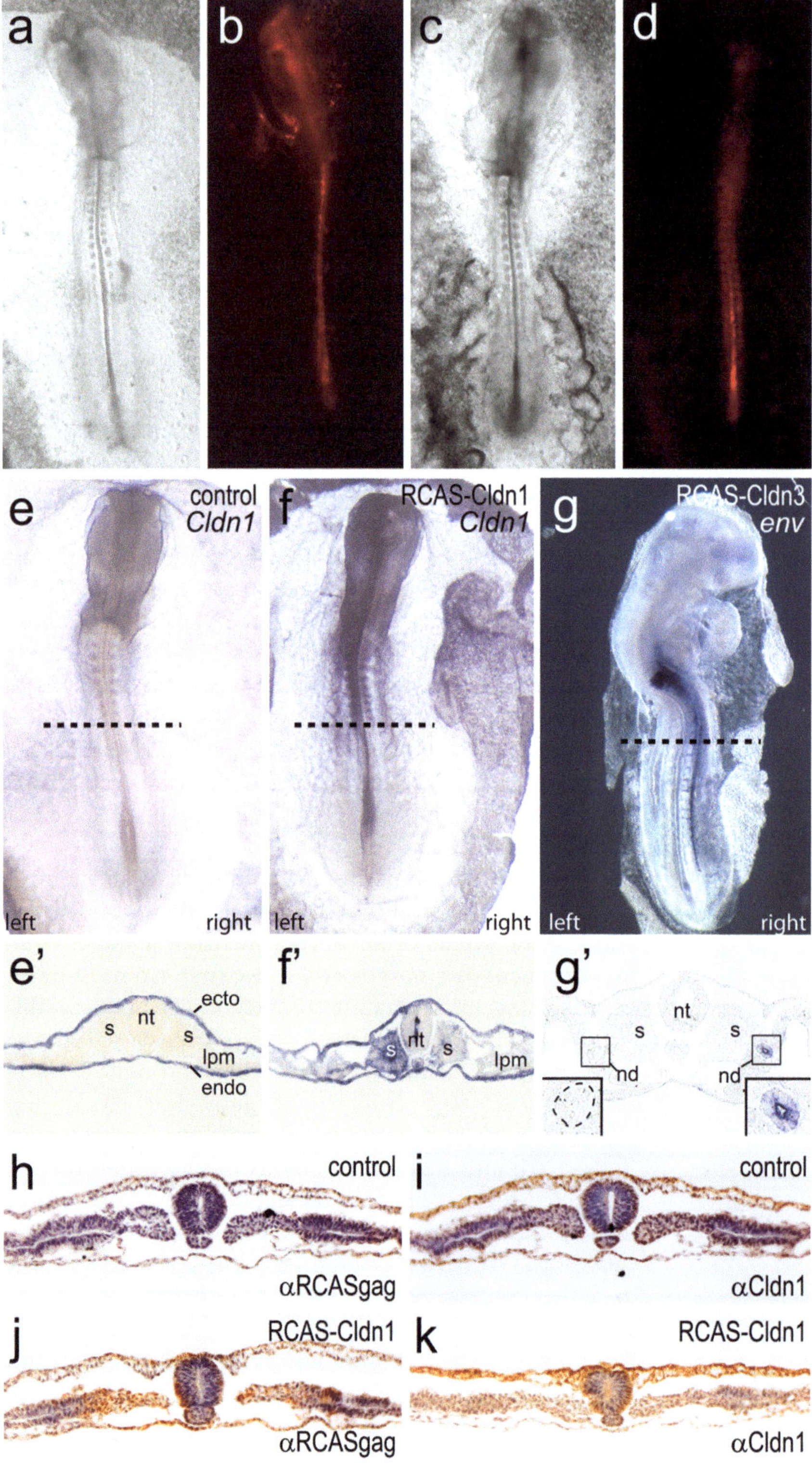
a
b
c
d
e
control
Cldn1
f
RCAS-Cldn1
Cldn1
g
RCAS-Cldn3
env
left
right
left
right
left
right
e'
ecto
nt
s
s
lpm
endo
f'
nt
s
s
lpm
g'
nt
s
s
nd
nd
h
control
αRCASgag
i
control
αCldn1
j
RCAS-Cldn1
αRCASgag
k
RCAS-Cldn1
αCldn1

are first cloned into the SLAX adaptor plasmid and then subcloned into the RCAS vector (12, 13).

2. For electroporation, supercoiled plasmid DNA should be concentrated to 3–5 μg/μl in sterile water. Add 0.5 μl 0.1% Fast Green dye to 5 μl plasmid DNA to aid in visualization of DNA during injection and electroporation. The empty parent vector should be prepared in parallel to use as a negative control.
3. For preparation of retroviral particles, RCAS-Claudin expression vector DNA should be prepared at 0.2–1 μg/μl.

3.2. Preparation of Retroviral Particles

1. Retroviral particles are prepared using a slightly modified protocol from that described in ref. (12).
2. DF-1 chick fibroblasts should be cultured in high-glucose DMEM with 10% FBS.
3. Transfect two wells of a six-well plate that contain DF-1 cells at 90% confluency with supercoiled RCAS-Claudin expression vector using Lipofectamine™2000 transfection reagent according to manufacturer's instructions.
4. Passage cells for 7–10 days, gradually expanding the cells to five 15-cm tissue culture dishes.
5. To obtain high concentrations of retroviral particles for microinjection, it is essential to ensure that 100% of the cells are infected. During expansion, plate an aliquot of the cells onto a glass coverslip and culture for 24 h. Fix cells for 15 min in 4% PFA/PBS. Perform immunohistochemical

Fig. 2. Identification of tissues overexpressing claudin proteins. (**a–d**) Brightfield (**a**, **c**) and fluorescent (**b**, **d**) images of embryos injected and electroporated at HH stage 4 with pcDNA-Claudin-1 and a mCherry reporter plasmid and cultured on agar-albumen plates until HH 12. Different expression patterns can be observed by monitoring mCherry fluorescence. The embryo on the left (**a, b**) exhibits mCherry expression in the neural tube, while the embryo on the right (**c**, **d**) shows mCherry expression in the neural tube and somites. (**e**, **f**) Whole mount in situ hybridization analysis of *Claudin-1* expression in an uninjected control embryo (**e, e′**) and an embryo that was injected with RCAS-Cldn1 retroviral particles on its left side at HH stage 4 (f, f′) using an *Claudin-1* riboprobe (*Cldn1*). (**e′**, **f′**) Transverse sections through embryos shown above (*at level of dotted line*). *Claudin-1* expression is observed in the ectoderm and endoderm of uninjected control embryos (**e′**) as previously described (17). In the RCAS-Cldn1-injected embryo, ectopic *Claudin-1* is detected in the neural tube, the left somite. Low levels of *Claudin-1* are also seen in the right somite. (**g**, **g′**) A HH stage 17 embryo that was injected at HH stage 10 with RCAS-Cldn3 retroviral particles into the right intermediate mesoderm adjacent to somite 10, which will give rise to the nephric duct. Whole mount in situ hybridization with an RCAS *env* antisense riboprobe at HH stage 17 shows that the entire right nephric duct was successfully targeted. A transverse section through the embryo (**g′**) confirms the specificity of the infection. Higher magnification views of the left and right nephric ducts are shown in the insets. (**h–k**) Immunohistochemical analysis of transverse sections through the embryo an uninjected control embryo (**h, i**) and an RCAS-Cldn1-injected embryo (**j, k**). An antibody against the RCAS gag protein (αRCASgag; **h**, **j**) or Claudin-1 (αCldn1; **i, k**) were used. In the control embryo, no staining is observed for the RCAS gag protein and *Claudin-1* is detected primarily in the epiblast. In the RCAS-Cldn1-infected embryo (**j, k**), RCAS gag and Claudin-1 are observed in most tissues. Protein is broadly expressed in shows that *Claudin-1* mRNA is misexpressed in of the RCAS-Claudin-1-injected embryo, but not the control injected embryo (F′). Abbreviations: *ecto* ectoderm, *endo* endoderm, *lpm* lateral plate mesoderm, *nt* neural tube, *s* somite.

analysis with the AMV-3C2 antibody to confirm that 100% of the cells have been infected.

6. Replace culture medium with DMEM with 2% FBS. After 24 h, collect the medium and replace with fresh medium.
7. Filter medium through a 0.45 μm sterile cellulose acetate filter. To pellet the retroviral particles, centrifuge the filtered medium at 22,000 rpm (64,000 × *g*) at 4°C for 3 h using a SW28 swinging bucket rotor. For best results, allow the rotor to stop without the brake. Carefully pour off supernatant. Cover opening of the tubes with Parafilm™ and store at 4°C overnight. Repeat this process twice after 20–24 and 40–48 h. Reuse the same centrifuge tubes for the Day 2 and 3 centrifugations.
8. After final spin, carefully pour off the supernatant and wipe the sides of the tubes to remove most of the medium. Then place tubes right-side up in a rack allowing remaining medium (~200 μl) to settle to bottom. Cover tubes with Parafilm™ and let stand at 4°C overnight to loosen the pellet of retroviral particles. Gently resuspend virus particles in a final volume of ~200 μl. Prepare 5–10 μl aliquots, freeze in liquid nitrogen, and store at −80°C. Discard aliquot after use (see Note 1).
9. To determine titer of retrovirus stock, inoculate cultured chick fibroblasts with 10^{-4} to 10^{-8} serial dilutions. Culture for 48 h. Perform immunohistochemical analysis as described below. Count the number of stained cells/colonies to determine titer. Minimum titers should be between 5×10^5 and 1×10^6 infectious particles/μl.

3.3. Preparation of Embryos for Injections and Electroporations

3.3.1. For Ex Ovo Culture and Manipulation

1. Ex ovo chick embryo cultures are performed essentially as described by Chapman et al. (16).
2. Place fertilized eggs with the narrowed tapered end facing downwards in a humidified incubator at 38°C until they reach the desired stage of development. The length of incubation can be estimated according to Hamburger and Hamilton (1) but will vary slightly for individual incubators. For older stages of embryogenesis, a rotating incubator should be used.
3. To collect embryos, crack egg into a 10 cm glass petri dish. The embryo will usually be located on the top surface. Gently remove the albumen from the surface around the embryo using Kimwipes™ (see Note 2), being careful to avoid the embryo itself.
4. Lower the filter paper support over the embryo and ensure that it has stuck to the surface by tapping the corners down with forceps. The embryo should be centered in the open area of the support, with the anterior–posterior axis directed parallel to the long side of the support. Cut around the filter paper and gently remove the filter with the attached embryo from the yolk, pulling away almost parallel to the yolk's surface.

5. Place yolk-side up on agar-albumen plates. The ventral side of the embryo will now be on top. Remove the yolk from the surface of the embryo by carefully pipetting simple saline across the embryo (see Note 3).

3.3.2. For In Ovo Culture and Manipulation

1. Incubate eggs on their sides in a humidified incubator at 38°C until they reach the desired stage of development. Let stand at room temperature for 20 min.
2. Spray surface of the eggs with 70% ethanol and allow to dry.
3. Poke a hole in the tapered end of the egg and remove 5 ml of albumen using an 18-gauge needle and 10 ml syringe.
4. Place a piece of clear cellophane tape on the center of the egg's surface and carefully cut a hole through the shell to reveal the embryo. If necessary enlarge the hole so that you have access to the embryo. The tape will prevent small pieces of shell from falling into the egg.
5. Place a piece of clear packing tape loosely over the hole to prevent the embryo from drying.
6. Following manipulation the opening must be tightly sealed with packing tape to prevent the embryos from drying.

3.4. Microinjection of Retroviral Particles

1. Injection needles can be made using glass capillary tubes (1 mm × 500 mm) drawn out to a fine tapered point using a vertical needle puller (see Notes 4 and 5).
2. Backfill the needle with the retroviral particle/0.1% Fast Green mix (see Note 6). The presence of the Fast Green allows the site of injection to be visualized.
3. Place the opened egg or the ex ovo cultured embryo on the agar-albumen plate under a dissecting microscope.
4. Gently insert the needle into the tissue of interest and inject retrovirus using a microinjector apparatus. Multiple sites of injection may be required to ensure broad expression. In older embryos, injections in specific tissue (e.g., the neural tube or somites) can be achieved (Fig. 1).
5. Seal the opening in the shell tightly with packing tape and return egg to the incubator. Dishes with ex ovo cultured embryos can be placed into a 15-cm petri dish that contains a piece of wet 3MM on the paper. This ensures that the embryos remain moist.

3.5. Microinjection and Electroporation of Plasmid DNA

1. The protocol for microinjection of plasmid DNA is essentially identical to that for microinjection of retroviral particles (Subheading 3.4, steps 1–4). If multiple plasmids are being injected together, the concentration of each should be 3–5 μg/μl in the injection mix. The retroviral vector can be

directly injected and electroporated into chick embryos (18) as an alternative to preparation and injection of retroviral particles to permit infection of a much broader area in the target cells/tissue.

2. Backfill the needle with DNA/0.1% Fast Green mix.
3. Plasmid DNA can be injected directly into a specific target tissue such as a somite or limb bud (19) (Fig. 1a) or into the lumen of a tissue, such as the neural tube (Fig. 1c). Injecting DNA into the lumen of a tissue will restrict movement of the DNA mix and confine it to a limited region for electroporation. Gross misexpression of plasmid DNA throughout the epiblast of gastrulation-stage embryos can be achieved by injecting the space between the dorsal side of the embryo and the vitelline membrane (Fig. 1a) (see Note 7).
4. Embryos should be electroporated immediately following DNA injection to ensure entry of the DNA into the cells.
5. The choice of paddle- versus rod-shaped electrodes is determined by the stage of the embryo and the tissue that is being injected. For broad expression of the construct in gastrulation-staged embryos, we use the paddle electrode set-up (see step 5; Fig. 1b). As described in step 2, DNA is injected in the space between the vitelline membrane and the embryo, the positive electrode should sit above the embryo and negative below to ensure DNA is electroporated into the epiblast cells. For electroporation of tissues such as the neural tube, DNA should be injected into the lumen of the neural tube and platinum rod electrodes should be used (see step 6; Fig. 1c). Negatively charged DNA will move from the cathode to the anode; therefore, the direction of electroporation will result in only one side of the neural tube expressing the gene of interest, while the other side acts as a negative control.
6. If using paddle electrodes for electroporation of ex ovo cultured embryos, transfer the injected embryo to an electroporation plate. The well containing the square electrode on the bottom of the dish should be filled with DMEM or Tyrode's solution (see Note 8). Place the embryo across the square well and add a drop of DMEM or Tyrode's solution to the top of the embryo. Lower the positive electrode down to meet the liquid without depressing the embryo. For gastrulation to neurulation stage embryos, five 5 V/50 ms pulses at 500 ms intervals can be used (see Note 9). After electroporation is completed, transfer the embryos back to the agar-albumen dishes. Paddle electrodes can also be used in ovo when both sides of the tissue are accessible, e.g., limb buds.
7. If electroporating with rods, electroporation can be done while the embryo is on the agar-albumen plate. Maintaining a

constant distance (4–6 mm) between electrodes is critical for consistency. Electrodes should be lowered gently onto the embryo flanking the target tissue. DMEM (~5–10 μl) should be placed on the embryo before and after electroporation, to moisten and cool the embryo, respectively (see Note 9). Using the BTX ECM830 electroporation set-up, we found that 5–7 20 V/20 ms pulses at 1 s intervals resulted in optimal electroporation.

8. After electroporation is completed, embryos should be returned to a humidified incubator at 38°C without turning. Eggs with in ovo electroporated embryos should be well sealed to prevent drying. Eight 35-mm dishes with ex ovo cultured embryos can be placed into a 15 cm petri dish with wet 3MM paper on the bottom to maintain humidity
9. If culturing embryos ex ovo for more than 48 h, transfer embryos to a fresh agar-albumen plate after 36–40 h.

3.6. Analysis of Electroporated/ Injected Embryos

3.6.1. Immunohistochemistry

1. Dissect embryos from the egg and place directly in cold PBS. Rinse embryos three times in fresh PBS. Transfer to 4% paraformaldehyde (PFA) and fix overnight at 4°C (see Note 10). Ex ovo cultured embryos are more fragile than in ovo cultured embryos and it may be easier to remove them from their support after they are fixed.
2. Wash embryos three times in PBS. Dehydrate through successive washes of 50% ethanol:PBS, 70% ethanol, 95% ethanol, twice in 100% ethanol, and twice in xylene. Wash times will depend on the stage of the embryo. Typically, 15 min washes are used for embryos less than 2 days and 30–45 min washes are used for older embryos. Xylene wash times should be extended until the embryo becomes clear.
3. Infuse embryos in melted paraffin in an oven (60°C) with a vacuum for 1–2 h. Transfer to fresh melted paraffin in a disposable Peel-A-Way™ mold and orient. For young embryos, we use a dissecting microscope to facilitate orientation.
4. Mount paraffin blocks onto appropriate support and collect 5–10 μM sections using a microtome. Transfer sections onto Superfrost®Plus microscope slides and dry overnight at ~37°C on a slide warmer.
5. Remove paraffin using two 5–10 min xylene washes and rehydrate sections through a descending series of ethanol:PBS washes for 5–10 min. Once rehydrated, wash slides three times in PBS. If necessary, perform antigen retrieval using boiling citrate buffer to expose the antigen (see Note 11).
6. Carefully add 200 μl of 10% normal goat serum in 0.3% Triton X-100/PBS to block nonspecific antibody binding to the tissue sections. Cover slides with Parafilm™ pre-cut to the size of the slide to ensure that sections do not dry out.

Incubate slides for 1 h at room temperature in a humidified chamber.

7. Remove Parafilm™ and allow excess solution to drain from slide. Replace solution with 100–200 μl of the primary antibody diluted in 5% normal goat serum in 0.3% Triton X-100/PBS. Cover and incubate for 1 h at room temperature or overnight at 4°C.
8. Remove excess primary antibody by washing slides three times for 10 min in PBS.
9. Incubate slides with 150–200 μl of the biotinylated secondary antibody at 1:500 in 0.3% Triton X in PBS for 1 h at room temperature. Cover slides with Parafilm™ to ensure that sections do not dry out.
10. Wash slides three times for 10 min with PBS to remove excess secondary antibody.
11. Incubate slides with ABC solution (Vector Laboratories) for 30 min, and wash slides in PBS.
12. For color development, incubate slides with DAB reagent for 2–10 min until brown color develops. Sections may be counterstained with hematoxylin, dehydrated, and cover-slipped with Permount™.

3.6.2. Whole-Mount In Situ Hybridization

1. The whole-mount in situ hybridization protocol that we use is modified from Wilkinson (1993). Collect embryos as described in Subheading 3.6, step 1 and fix overnight in 4% PFA at 4°C.
2. Wash embryos in PBT (PBS/0.1% Tween-20) and dehydrate embryos through successive PBT/methanol (75%, 50%, 25%, methanol) washes. Vials with embryos should be placed on a nutator or shaking platform during all washes. Embryos can be stored in methanol at -20°C.
3. Rehydrate embryos through successive PBT/methanol washes, and wash twice in PBT.
4. Bleach embryos in 6% hydrogen peroxide in PBT until white. This is not required for younger embryos, and older embryos will require up to an hour to bleach thoroughly.
5. Wash bleached embryos three times in PBT. During this time, make sure that glycine is weighed out and ready to be resuspended (for use in step 7) and that the glutaraldehyde/paraformaldehyde postfix solution is on ice (for use in step 8).
6. Replace PBT with 1 μg/ml proteinase K in PBT. Incubate at room temperature for 2–20 min, depending on the stage of the embryos. Generally, we treat HH stage 4–7 embryos for 3 min, HH stage 8–14 embryos for 5 min and embryos older than HH stage 25 for 15–20 min. The ex ovo cultured embryos are more fragile and consequently we limit the

proteinase K treatment to 1.5 min. The timing for this step is very critical.

7. To stop the proteinase K treatment, rinse embryos twice in glycine solution.
8. Wash embryos twice with PBT.
9. Replace solution with glutaraldehyde/PFA postfix solution and incubate on ice for 20 min.
10. Wash embryos twice with PBT.
11. Rinse embryos with warmed hybridization mix. Replace with fresh hybridization mix and incubate at 65°C for 1 h (see Note 12).
12. Remove solution and replace with appropriate antisense claudin riboprobe or RCAS probe overnight at 65–70°C (see Note 13).
13. Carefully remove probe (see Note 13) and wash embryos four times with Wash Solution 1 and four times with Wash Solution 2. These washes should be carried out at 65–70°C.
14. Wash embryos three times for 10 min at room temperature with TBST/2 mM levamisole.
15. Block embryos in 10% sheep serum in TBST for 1 h at room temperature.
16. Incubate embryos in antibody mix overnight at 4°C, rotating.
17. Wash embryos three times 5 min with TBST/2 mM levamisole.
18. Continue washing embryos for 5–6 h in TBST/2 mM levamisole, changing solution every 30–60 min. These washes may be extended overnight at 4°C.
19. Wash embryos three times in NTMT. For color development, incubate embryos in NTMT with BCIP/NBT (2.5 μl of each per ml NTMT). Wrap vials in aluminum foil to protect from light (see Note 14). Once color has developed, rinse with PBS and fix in 4% PFA.

4. Notes

1. Thawing and refreezing aliquots of retroviral particles can lead to a significant decrease in the titer of the stock. Therefore, we generally discard the aliquot. However, it can be stored at −80°C and used in place of transfection step to infect DF-1 cells to generate new viral stocks.
2. The surface of the yolk should be quite dry to allow the filter support to stick, however, over-drying will cause the yolk to tear.

3. The membrane surrounding the embryo must adhere to the filter paper support for embryo survival during incubation. To remove the yolk from the surface of the embryo, gently add simple saline dropwise to the surface and tilt the plate so that liquid can be drawn away from the embryo to avoid the membrane loosening from the corners of the support. A detailed description of the method for ex ovo culture of chick embryos can be found in ref. (16).
4. Needles will usually be drawn to a closed tip. Use forceps to gently break the end of the needle before using. Examine under a dissecting microscope to ensure that the opening is not too large.
5. To control for needle size between experiments, expel a single injection bolus and compare with a known size. Adjust the opening of the needle to increase this size.
6. Pressure should be balanced before filling to avoid DNA mix being expelled or drawn back away from the tip of the needle.
7. Generally, three or four injections will be required to fill the vitelline space. Alternatively, DNA can be injected just above the embryo and the direction of the electrodes can be flipped. This will allow the introduction of DNA into the epiblast and newly forming mesodermal cells.
8. DMEM can be used for later stages of embryogenesis. Tyrode's solution is recommended when electroporating pre-gastrulation to gastrulation stage embryos (20).
9. Electroporation conditions need to be determined empirically since they are dependent on the electroporation set-up, target tissue, and embryonic stage. Electroporation conditions should be optimized to allow for the greatest delivery of the plasmid DNA with the least amount of damage to the embryo.
10. Fixation conditions should be varied according to optimized conditions for the antibody. Certain antibodies may also work better on cryosections.
11. Antigen retrieval is performed using heat and 10 mM sodium citrate buffer, pH 6.5. Place slides into a Coplin jar and fill with citrate buffer. Microwave at 70% power for 3 min and twice at 40% power for 3 min. Ensure that sections are always covered with boiling solution. Let cool approximately 20 min before blocking.
12. Following prehybridization step, embryos can be stored in prehybridization solution at –20°C until use. They should then be reincubated at 65°C for 1 h prior to replacing prehybridization solution.

13. Probe and antibody mix can be saved and reused. Probes should be stored at −20°C and antibody mix at 4°C. The specificity of both mixes will improve with repeated use.
14. Color development should be monitored every 15–30 min for the first hour, and every hour thereafter. Embryos in coloring solution may be stored at 4°C overnight or at RT if color development proceeds longer than 1 day.

Acknowledgments

The authors thank Dr. I. Gupta for discussions and comments on the manuscript and A. Simard and Y. Zhang for technical assistance. This work was supported by funding from NSERC and CIHR. AKR is a member of the Research Institute of the McGill University Health Center, which is supported in part by the FRSQ.

References

1. Furuse, M., Fujita, K., Hiiragi, T., Fujimoto, K., and Tsukita, S. (1998) Claudin-1 and -2: novel integral membrane proteins localizing at tight junctions with no sequence similarity to occludin. *J Cell Biol* **141**, 1539–50.
2. Furuse, M., and Moriwaki, K. (2009) The role of claudin-based tight junctions in morphogenesis. *Ann N Y Acad Sci* **1165**, 58–61.
3. Morita, K., Furuse, M., Fujimoto, K., and Tsukita, S. (1999) Claudin multigene family encoding four-transmembrane domain protein components of tight junction strands. *Proc Natl Acad Sci USA* **96**, 511–6.
4. Tsukita, S., and Furuse, M. (2000) The structure and function of claudins, cell adhesion molecules at tight junctions. *Ann N Y Acad Sci* **915**, 129–35.
5. Turksen, K., and Troy, T. C. (2004) Barriers built on claudins. *J Cell Sci* **117,** 2435–47.
6. Stern, C. D. (2005) The chick; a great model system becomes even greater. *Dev Cell* **8**, 9–17.
7. Hamburger, V., and Hamilton, H. L. (1992) A series of normal stages in the development of the chick embryo. 1951. *Dev Dyn* **195**, 42.
8. Angelow, S., Ahlstrom, R., and Yu, A. S. (2008) Biology of claudins. *Am J Physiol Renal Physiol* **295**, F867–76.
9. Lal-Nag, M., and Morin, P. J. (2009) The Claudins. *Genome Biol* **10**, 235.
10. Sambrook, J., Fritsch, E. F., and Maniatis, T. (1989) Molecular Cloning: A laboratory manual. Cold Spring Harbor Laboratory Press.
11. Krull, C. E. (2004) A primer on using in ovo electroporation to analyze gene function. *Dev Dyn* **229**, 433–9.
12. Morgan, B. A., and Fekete, D. M. (1996) Manipulating gene expression with replication-competent retroviruses. *Methods Cell Biol* **51**, 185–218.
13. Logan, C., and Francis-West, P. (2008) Gene transfer in avian embryos using replication-competent retroviruses. *Methods Mol Biol* **461**, 363–76.
14. Hughes, S. H. (2004) The RCAS vector system. *Folia Biol (Praha)* **50,** 107–19.
15. Sauka-Spengler, T., and Barembaum, M. (2008) Gain- and loss-of-function approaches in the chick embryo. *Methods Cell Biol* **87**, 237–56.
16. Chapman, S. C., Collignon, J., Schoenwolf, G. C., and Lumsden, A. (2001) Improved method for chick whole-embryo culture using a filter paper carrier. *Dev Dyn* **220**, 284–9.
17. Simard, A., Di Pietro, E., Young, C. R., Plaza, S., and Ryan, A. K. (2006) Alterations in heart looping induced by overexpression of the tight junction protein Claudin-1 are dependent on its C-terminal cytoplasmic tail. *Mech Dev* **123**, 210–27.

18. Hermann, P. M., and Logan, C. C. (2003) Electroporation of proviral RCAS DNA alters gene expression in the embryonic chick hindbrain. *Biotechniques* **35**, 942–6, 48–9.
19. Iimura, T., and Pourquie, O. (2008) Manipulation and electroporation of the avian segmental plate and somites in vitro. *Methods Cell Biol* **87**, 257–70.
20. Voiculescu, O., Papanayotou, C., and Stern, C. D. (2008) Spatially and temporally controlled electroporation of early chick embryos. *Nat Protoc* **3**, 419–26.

Chapter 15

Identification of Claudins by Western Blot and Immunofluorescence in Different Cell Lines and Tissues

Lorenza González-Mariscal, Erika Garay, and Miguel Quirós

Abstract

Claudins are integral proteins of the TJ. Each epithelia in the organism expresses a unique set of claudins that determines the degree of sealing of the paracellular pathway and the ionic selectivity of the tissue. TJs are dynamic structures whose organization and composition change in response to alterations in the environment as well as under physiological and pathological conditions. Changes in claudin expression and subcellular distribution can be analyzed in western blot and immunofluorescence experiments, employing a wide array of available specific antibodies against claudins. In this chapter, we describe in detail protocols used for western blot and immunofluorescence detection of claudins in epithelial cell lines and in various tissue samples.

Key words: Claudin, Western blot, Immunofluorescence, Tight junctions, Epithelia

1. Introduction

Tight junctions (TJs) in epithelial cells localize at the most apical segment of the lateral membrane. TJ are integrated by a complex array of integral and peripheral proteins, among which claudins play a crucial role as back bone constituents of the filaments observed by freeze-fracture electron microscopy. The filaments that encircle epithelial cells below the apical microvilli are formed by lines of transmembrane particles with a 10 nm diameter that fuse when fixed with glutaraldehyde (1, 2). These particles are composed of claudin multimers, and their formation can be triggered upon claudin transfection, even in cells that lack TJs such as fibroblasts (3).

Kursad Turksen (ed.), *Claudins: Methods and Protocols*, Methods in Molecular Biology, vol. 762, DOI 10.1007/978-1-61779-185-7_15, © Springer Science+Business Media, LLC 2011

Claudins are tetraspan proteins that orient their amino and carboxyl terminal ends toward the cytosol, and expose two loops to the extracellular space. The first loop which is bigger than the second, contains several charged residues responsible for the ionic selectivity of the TJ (4–8), while the second contains aromatic and hydrophilic residues, conserved in most claudins, that are critical for the *trans*-interaction of claudins (9). Each particular tissue of the body of multicellular organisms exhibits a unique set of claudins that regulates by charge and size the passage of ions and molecules through the paracellular pathway.

Here we describe protocols employed to analyze the expression of claudins by western blot and immunofluorescence microscopy in diverse cell lines and tissues.

2. Materials

2.1. Cell Lysis

1. Phosphate buffer saline (PBS) from Gibco.
2. RIPA buffer: 40 mM Tris–HCl, pH 7.6, 150 mM NaCl, 2 mM EDTA pH 8, 10% glycerol, 1% Triton X-100, 0.5% sodium deoxycholate, and 0.2% SDS. Store at 4°C (see Note 1).
3. Gentle lysis buffer: 20 mM Tris–HCl, pH 7.6, 50 mM NaCl, 2 mM EDTA, and 1% Triton X-100.
4. Protease inhibitors: phenyl methane sulfonyl-fluoride (PMSF) 100 mM stock dissolved in isopropanol and protease inhibitor cocktail Complete™ (Roche).
5. Lowry protein assay (BioRad) or BCA protein assay reagent (Pierce).
6. Laemmli sample buffer (5×): 312.5 mM Tris–HCl, 10% SDS, 50% glycerol, 25% 2-mercaptoethanol, bromophenol blue 0.5%, pH 6.8. Store at −20°C in 1 ml aliquots.

2.2. SDS–Polyacrylamide Gel Electrophoresis

1. Separating gel buffer: 1.5 M Tris–HCl, pH 8.8. Store at 4°C.
2. Stacking gel buffer: 1 M Tris–HCl, pH 6.8. Store at 4°C.
3. Acrylamide solution: 30% (w/v) aqueous acrylamide/bisacrylamide (37.5:1) in water. Store at 4°C (see Note 2).
4. *N*,*N*,*N*,*N*′-Tetramethyl-ethylenediamine (TEMED): Store at 4°C.
5. Amonium persulfate (APS): prepare a 10% (w/v) solution in water and immediately freeze and store in 500 μl aliquots at −20°C.
6. Water from a Millipore MQ system with a resistivity of 18.2 MΩ cm.

7. SDS solution: 10% (w/v) solution. Store at room temperature (see Note 3).
8. SDS–polyacrylamide gel electrophoresis (SDS–PAGE) running buffer (10×): 250 mM Tris–HCl, 1.92 M glycine, pH 8.8, 1% (w/v) SDS (see Note 4). Store at room temperature.
9. Molecular weight markers: Dual Precision Plus Western blot standards (BioRad) or equivalent.

2.3. Detection of Claudins by Western Blotting

1. Transfer buffer: 48 mM Tris, 39 mM glycine, 20% v/v methanol, and 0.037% w/v SDS. Store at 4°C and use only once.
2. PVDF membrane (see Note 5).
3. Tris-buffer saline with Tween 20 (TBS-Tween): Prepare stock solutions of 2 M NaCl and 500 mM Tris–HCl, pH 7.5. Store at 4°C. With them prepare a solution containing 10 mM Tris–HCl, pH 7.5, 100 mM NaCl, and 0.2% v/v Tween 20. Store at 4°C.
4. Blocking buffer: 5% (w/v) non-fat dry milk, 3% (w/v) bovine serum albumin in TBS-Tween 0.1% (see Note 6).
5. Primary antibody dilution buffer: Blocking buffer.
6. Primary antibodies: We have employed the Invitrogen rabbit polyclonal antibodies against claudins: 1 (51–9000); 2 (51–6100); 3 (34–1700); 5 (34–1600); 7 (34–9100); and 16 (34–5400), as well as the mouse monoclonal antibody against claudin 4 (32–9400). Information on other commercial antibodies against claudins is found in Tables 1 and 2 (see Note 7). In addition, as loading control we employ antibodies against actin or tubulin.
7. Secondary antibodies: HRP goat anti-mouse or anti-rabbit IgG.
8. Chemiluminescence detection kit: ECL + Plus (GE Healthcare), Immobilon Western (Millipore) or equivalent.
9. Chemidoc Bio-Rad system can be employed to detect the chemioluminescence. However, if the signal is not too strong, we recommend the use of the autoradiography film Hyperfilm ECL (GE Healthcare).

2.4. Stripping and Reprobing Blots for Detection of Claudins

1. Stripping buffer: sodium dithionite 2% in TBS-Tween. Store at 4°C. Light sensitive.
2. Wash buffer: TBS-Tween.

2.5. Immunfluorescence for Detection of Claudins

1. Glass coverslips cut in 8 mm × 8 mm squares with a diamond tip pen.
2. 2-Methylbutane (Sigma Aldrich, St Louis, MO) (see Note 8).
3. Liquid nitrogen.

Table 1
Type of antibodies against claudins available from different manufacturers

Claudin/ manufacturer	Santa Cruz	Thermo scientific	Invitrogen	Sigma Aldrich	Abcam	Cell signaling	Abgent	Assay design	Bioworld technology	Genway biotech	Genetex	IBL	Life span biosciences	Origene technologies	Prosci	R&D	United States Biological
1	M, gP, rP	rP	M, rP	M, rP	M, rP	rP	rP	M	rP	rP	rP	rP	M, rP		rP	M	M, rP
2	rP	rP	M, rP	M, rP	M, rP, mP		M, rP		rP	rP	rP	rP	M, rP				M, rP
3	gP, rP	rP	rP	rP	rP			M	rP	rP	rP	rP	rP		rP	M	rP
4	M, gP	rP	M	M	M, rP		M	M	rP	rP	rP	rP	M, rP		M, rP	M	M, rP
5	gP, rP	rP	M, mP, rP		rP		M		rP	rP	rP	rP	M, rP				M, rP
6	gP				rP				rP	rP		rP	M, rP		rP		
7	gP, rP	rP		M, rP	M, rP				rP	rP		rP	M, rP		rP		M, rP
8	gP, rP	rP		rP	rP				rP	rP	rP	rP				M	rP
9	gP	rP		rP	rP					rP	rP		rP				
10	gP, rP	rP	M, rP	rP	rP					rP		rP	rP			M	rP
11	gP, rP	rP	rP	rP	rP				rP	rP		rP	M, rP			M	
12	gP	rP			rP					rP			rP			M	M, rP

13		rP			rP			rP	rP		rP			
14	gP	chP, gP, rP	rP		gP, rP			gP, rP	rP		rP	rP		gP, rP
15	gP, rP			rP	mP			rP		rP	M, rP			M, rP
16			rP	rP	rP			rP			rP			
17	rP	rP		rP	rP			rP	rP		rP		M	
18	gP		rP	rP				rP			rP			rP
19					rP									
20					rP	M	M	rP			rP			
23	rP	rP		rP	rP			rP	rP		rP			

M mouse monoclonal, *chP* chicken polyclonal, *gP* goat polyclonal, *mP* mouse polyclonal, *rP* rabbit polyclonal

Table 2
Commercial antibodies that recognize regions of Claudins different from the c-terminal segment

Claudin	First loop[a]	Second loop[a]	Amino terminal	PY	Middle[d]	Isoform
1	Santa Cruz, Abcam, Abgent	Abgent				
2			Abcam, Abgent	Abgent[b] (Y195, Y224)		
3	Santa Cruz			Abcam[c] (Y219)		
6				Abcam[c] (Y219)	Abcam	
7				Abcam[c] (Y210)		
10						10-b R&D
11			Santa Cruz			
14					Invitrogen	
18					Invitrogen	

[a] See Note 7
[b] See Note 20
[c] See Note 21
[d] Undisclosed epitope in the middle part of claudins

4. Tissue freezing-mounting media: Jung tissue freezing medium or equivalent.
5. Gelatin coated or charged and precleaned (Fisherbrand® ProbeOn™ Plus) microscope slides.
6. Three Coplin jars.
7. 2% *p*-Formaldehyde fixation solution: add 0.25 g *p*-formaldehyde to 9 ml of MiniQ water, add 20 μl 2 M NaOH and stir gently on a heating block at ~60°C, until the *p*-formaldehyde is dissolved. Add 2.5 ml of 5× PBS and 125 μl $MgCl_2$, and allow the mixture to cool to room temperature. Adjust the pH to 7.4 with 1 M HCl and the final volume to 12.5 ml. Filter the solution through a 0.45 μm membrane filter to remove any particulate matter. Make the *p*-formaldehyde solution fresh prior to use (see Note 9).
8. Methanol fixation: 100% methanol, ice cold.
9. Ethanol fixation: 70% ethanol, ice cold.
10. TX-100 permeabilization buffer: 0.5% Triton X-100 in PBS.
11. Acetone permeabilization: 100% acetone, ice cold.
12. Hydrophobic pen (PAP pen, Sigma Aldrich).

13. Cell lines blocking solution: 0.5% bovine serum albumin IgG free in PBS.
14. Tissue blocking solution: 1% bovine serum albumin IgG free in PBS.
15. Primary antibodies:

    Rabbit polyclonals against claudins: 1 (51–9000, Invitrogen); 2 (51–6100, Invitrogen); 3 (34–1700, Invitrogen); 5 (34–1600, Invitrogen); 7 (34–9100, Invitrogen), 11 (sc-25711, Santa Cruz Biotechnology, Inc.), and 16 (34–5400, Invitrogen).

    Mouse monoclonal against claudin 4 (32–9400, Invitrogen). Information on other commercial antibodies against claudins is found in Tables 1 and 2 (see Note 7).
16. Fluorescently labeled secondary antibodies: Alexa 488-conjugated donkey Ig anti-mouse and Alexa 594-conjugated donkey Ig anti-rabbit (Molecular Probes).
17. Antifade mounting solution: Vecta Shield (Vector Laboratories, Burlingame, CA).
18. Nikon Diaphot 200 fluorescence microscope (Nikon, Tokyo, Japan) and Leica SP2 confocal microscope (Leica, Wetzlar, Germany).

3. Methods

3.1. Cell Lysis and Preparation of Samples for SDS–PAGE

3.1.1. Cell Lines (MDCK, MDA-MB231, MCF10A)

1. Aspirate the medium bathing the confluent culture of cells seeded in 60 mm plates. Add 2 ml of ice-cold PBS per plate and repeat the procedure two more times.
2. Aspirate the PBS and immediately add 500 μl of RIPA buffer with protease inhibitors (see Note 1).
3. Scrape the cells from the dish with a rubber policeman and transfer the viscous cell suspension to a 1.5 ml tube. Store for 15 min on ice.
4. Sonicate the lysate two times for 30 s each at low intensity in an ultrasonic processor.
5. Quantitate the proteins in the cell lysate, with the Lowry or BCA protein assay (see Note 10).
6. Dilute the lysate samples in Laemmli sample buffer. For this purpose, employ the 5× stock of Laemmli sample buffer to reach a final 1× concentration in the lysate samples.
7. Heat the samples for 10 min at 90–96°C in a heat block (see Note 11).
8. After cooling the samples briefly on ice, they are ready for loading onto SDS–PAGE or can be stored at −70°C.

3.1.2. Tissues (Skin and Placenta)

Skin

1. Employ a regular razor blade to shave the mice hair off from 2×2 cm area of the animal back skin.
2. Dissect a portion of the epidermis with dissection scissors and transfer it to a 1.5 ml tube.
3. Wash the dissected epidermis two times with ice-cold PBS containing 1 mM PMSF.
4. Aspirate the PBS and immerse the tubes containing the tissue in liquid nitrogen. If needed, samples can next be stored at –70°C.
5. Ground the frozen tissue with a prechilled mortar and pestle, until a homogeneous powder is obtained. Transfer with a spatula the tissue powder to prechilled 1.5 ml tubes.
6. Place tissue powder in an eppendorf tube until it reaches the 100 μl mark. Then add 1 ml of RIPA buffer with protease inhibitors (see Note 1). Place under gentle rotation for 15 min at 4°C.
7. Sonicate the lysate three times for 30 s each at high intensity in an ultrasonic processor.
8. Spun the lysate for 15 min at 4°C at 16,000×*g*. Recover the supernatant and transfer it to a new and prechilled 1.5 ml tube (see Note 12).
9. Continue as described in Subheading 3.1.1 from steps 5 to 8.

Placenta

1. Immediately after delivery, separate the placenta from the decidual tissue, and rinse it in ice-cold PBS.
2. Cut 1 cm^2 biopsies, transfer the samples to 15 ml Falcon tubes, and wash again with ice-cold PBS containing 1 mM PMSF.
3. Continue as described in Subheading 3.1.2 from steps 4 to 8.
4. Continue as described in Subheading 3.1.1 from steps 5 to 8.
5. If Triton X-100 soluble and insoluble fractions wish to be obtained, treat the powered tissue described in Subheading 3.1.2, step 5 with the Gentle lysis buffer containing the protease inhibitors, for 30 min at 4°C under continuous agitation. Continue with Subheading 3.1.2, steps 7 and 8, and designate the resulting supernatant as the Triton X-100 soluble fraction. To obtain the Triton X-100 insoluble fraction, resuspend the pellet in RIPA buffer and continue with Subheading 3.1.2, steps 7 and 8. Discard the pellet.
6. Continue as described in Subheading 3.1.1 from steps 5 to 8.

3.2. SDS–PAGE

1. The following instructions are specific for the use of the Bio-Rad Mini-PROTEAN 3 gel system (Bio-Rad, Hercules, CA, USA). Start by cleaning the glass plates with a common detergent and then wash them extensively with deionized water.

Dry and store the clean plates until usage. Before assembly of the gel system clean the glasses with 70% ethanol and air dry.

2. Since the molecular weight of claudins is ~20 kDa, we strongly recommend preparing 15% separating gels (1.5 mm thick). To prepare two gels mix 7.5 ml acrylamide solution, 3.8 ml separating gel buffer, 3.5 ml water, 150 μl SDS solution, and 150 μl APS. Add 6 μl TEMED and mix carefully. Immediately pour the gel by filling the space between the glass plates up to 1.5 cm below the top of the smaller glass plate. Carefully overlay with water.
3. Leave the gel to polymerize for 20 min, then prepare the stacking gel.
4. To prepare two stacking gels mix 1.3 ml acrylamide solution, 1 ml stacking gel buffer, 5.5 ml water, 80 μl SDS solution, and 80 μl APS. Pour off the water from the separating gel and remove residual water with blotting paper. Add 8 μl TEMED to the stacking gel solution and mix carefully. Pour the stacking gel up to the top of the smaller glass plate and insert the comb.
5. After 30 min the gel should be fully polymerized.
6. Dilute 200 ml of 10× running buffer with 1,800 ml of MilliQ water and mix well.
7. Carefully remove the comb from the stacking gels before assembling the gels in the inner electrophoresis chamber, and place the chamber into the buffer tank. Fill the inner chamber completely with running buffer and ensure that the chamber is not leaky. Then fill the outer chamber making sure that the bottom of the gel is well immersed into the buffer. Eliminate the air bubbles trapped in the outer chamber at the bottom of the gel with the help of a syringe with a bent needle filled with running buffer.
8. Before loading the samples wash out the wells with running buffer applied with a 200 μl pipette. Ensure that the wells are not blocked by gel slices.
9. Load 20 μg of protein per well in a volume ranging from 10 to 30 μl. Load the molecular weight marker to one well.
10. Cover the gel chamber with the lid and connect it to the power supply. Run the gel at 90–100 V and stop the current when the blue dye front reaches the bottom of the gel.

3.3. Western Blot Analysis of Claudins

1. After separation by SDS–PAGE, proteins are transferred to a PVDF membrane in a Trans-Blot SD Semi-Dry Transfer Cell (Bio-Rad, Hercules, CA) (see Note 13).
2. Cut a sheet of PVDF membrane slightly larger than the size of the separating gel (8.5 × 6 cm) and leave it for 1 min to moisten in methanol (see Note 5). Then transfer the PVDF membrane

into blotting buffer and shake gently to wash away the methanol. Leave the membrane in the buffer for 10 min. Subsequently, immerse two sheets of gel blotting paper (ExtraThick blot paper, 2.5 mm, BioRad) in blotting buffer. Disconnect the SDS–PAGE unit from the power supply, disassemble and separate the glass plates, remove the stacking gel, and place the separating gel in blotting buffer for 10 min.

3. Place one sheet of blotting paper on the anode plate of the blotter and remove the air bubbles trapped between the anode and the blotting paper by carefully rubbing with a pipette. Add 2 ml of blotting buffer onto the gel blotting paper, place the PVDF membrane into this buffer, and cover with 2 ml of blotting buffer. Next place the gel onto the PVDF membrane and cover with another 2 ml of blotting buffer. Place the second sheet of blotting paper on the top of the gel. Complete the blotting stack by mounting the cathode plate.
4. Connect the blotting apparatus to a power supply and carry out the transfer for 30 min at 400 mA.
5. Once the transfer is complete, switch off and disconnect the system from the power supply. Carefully disassemble the blotting stacks.
6. Place the membrane in a plastic container adequate for the size of the membrane and cover it with 5 ml of blocking buffer (see Note 6). Incubate in a rocking platform for 1 h at room temperature.
7. Remove the blocking solution and cut the PVDF membrane in two at the level of the 37 kDa molecular weight marker. Incubate the PVDF membrane segment containing the higher molecular weight markers with the antibody against actin or tubulin in blocking solution for 1 h at room temperature or overnight at 4°C. Incubate the segment of the PVDF membrane containing the lower molecular weight markers with the antibodies against the chosen claudin in blocking solution for a minimum of 1 h at room temperature or overnight at 4°C. We recommend the following dilutions and incubation times for claudin antibodies: (a) Tissues (1) Skin: claudin-1, 1:1,500, 1 h incubation; claudin-3, 1:250, overnight; claudin-4, 1:166, overnight; claudin-5, 1:250, overnight. (2) Placenta: claudin-1, 1:500, overnight; claudin-3. 1:125, overnight; claudin-4: 1:166, overnight; claudin-5, 1:500, overnight; claudin-15, 1:166, overnight; claudin-16, 1:125, overnight. (b) Cell lines (1) MDA-MB231 and MCF-10A: Claudin-1, 1:500, 1 h; claudin-3, 1:125 claudin-4, 1:166, overnight; claudin 7, 1:125, overnight. (2) MDCK: claudin-1, 1:1,500, 1 h; claudin-1, 1:800, overnight; claudin-3, 1:500, overnight; cluadin-4, 1:500, overnight; claudin-5, 1:250, overnight.
8. Wash the membranes six times with TBS-Tween for 5 min.

9. Incubate with secondary antibody in blocking solution for 1 h at room temperature.
10. Wash the membranes as before. Remove the blot from the TBS-Tween and allow the buffer to drain from the blot almost completely, but do not allow the blot to dry up. Place the membrane on a tray with the gel side facing up, and distribute the chemiluminescence developing reagent, previously prepared according to the manufacturer's instructions, over the entire membrane by gentle rocking.
11. Obtain several different exposures of the membrane either on standard chemiluminescence film or in a digitized format with a chemiluminescence detection apparatus (see Note 14). The chemiluminescence signal can next be quantified in a Chemi Doc System using the Quantity One software (see Note 15).

3.4. Stripping and Reprobing the Blots for Claudins

1. Once a positive result has been obtained with the antibody specific against a given claudin, the membrane can be stripped of the signal and reprobed with another anti claudin antibody.
2. Cover the membrane with the stripping buffer (5 ml per blot) and incubate for 2 h at room temperature in a rocking platform. Remember that the stripping buffer is light sensitive.
3. Wash the stripped blot three times, for 5 min each with TBS-Tween and continue as described in Subheading 3.3, steps 6–11.

3.5. Preparation of Samples for Immunofluorescence

3.5.1. Cell Lines (MDCK, MDA-MB231, MCF10A)

1. Cells are cultured on sterile glass coverslips placed inside petri dishes (six coverslips of 8×8 mm/35 mm diameter petri dish).
2. Cell are washed twice with ice-cold PBS.
3. Cells are fixed with 2% *p*-formaldehyde fixation solution for 30 min at 4°C or 100% methanol for 20 min at –20°C (see Note 16).
4. Cells are washed three times with ice-cold PBS.
5. Cells fixed with 2% *p*-formaldehyde are permeabilized by incubation in TX-100 permeabilization buffer, for 10 min at room temperature. Methanol fixed cells do not require permeabilization.
6. Cells are washed three times with ice-cold PBS.
7. Prepare a humid chamber by linking together with masking tape as a hinge, the lids of two clean but not sterile 24-multiwell dishes. On the inside of this chamber, place on one side a humid tissue paper and in the other a Parafilm layer. On the Parafilm layer, place neatly spaced 30 μl drops of cell line blocking solution, and on top of each drop gently place, with

the help of fine tip tweezers, a glass coverslip, taking care that the cells on the coverslip face the blocking solution. Close the humid chamber.

8. Incubated for 1 h at room temperature.
9. Prepare the dilution of claudin antibodies in cell line blocking solution (MDCK, MDA-MB231, and MCF-10A): Claudin-1, 1:50; claudin-3, 1:50; and claudin-4, 1:50.
10. In the humid chamber, place another line of neatly spaced 30 μl drops of claudins antibody solution, below the previous line of drops of blocking solution. With care, remove with the tweezers the coverslips from the blocking solution, drain the excess liquid on a filter paper, and transfer the coverslips to the antibody drops. Close the humid chamber and incubated overnight at 4°C.
11. Open the humid chamber and transfer each coverslip to an individual well in a 24-multiwell dish containing 1 ml PBS. Make sure that the cells are placed facing upward. Wash five times with PBS.
12. Dilute the secondary antibodies in blocking solution according to the manufacturer's instructions.
13. Prepare another humid chamber with a line of drops of the secondary antibodies and transfer the coverslips from the wells to the antibody drops, making sure that the cells face the antibody solution. Close the humid chamber and cover it with aluminum foil to protect the fluorescent antibodies from light exposure. Incubate for 1 h at room temperature.
14. Repeat step 11.
15. Mount the coverslips onto glass microscope slides using 2–3 μl Vecta Shield antifade mounting solution. With nail polish, seal the borders of the coverslip to avoid cell drying.
16. Store the slides in the dark at –20°C.

3.5.2. Tissues (Skin, Placenta, Uterus, and Testis)

1. Place ~15 ml of ice-cold 2-methylbutane in a cylindrical metal container of ~6 cm height and 4.5 cm diameter (e.g., metal container of Complete™ packing).
2. Immerse for 2 min tissue samples, no bigger than 0.5 × 0.5 cm, in ice-cold 2-methylbutane (see Note 8).
3. Transfer the metal container with the tissue samples in 2-methylbutane, to liquid nitrogen for 5 min. The samples can next processed immediately for cryo-sectioning or placed in a plastic container and stored at –70°C in a Revco.
4. Make a small flat bottom mold with aluminum foil. Place a drop of tissue freezing mounting media in the bottom of the mold, move the mold to the inside of a cryostat at –20°C, and place one piece of tissue over it. Fill the mold with tissue

freezing mounting media. Be careful to exclude large bubbles. Leave the mold inside the cryostat at –20°C for 5 min. Alternatively, mount the tissue in the aluminum mold over a platform of dry ice pellets within a styrofoam container.

5. Remove the aluminum foil and proceed to cryosection or store the tissue samples in tightly wrapped aluminum foil envelopes, maintained within a plastic container at –70°C in a Revco.
6. Cut 6–8 μm sections with a Leica Cryostat (Leica CM 1510-S) and place them onto electrocharged or gelatin-coated slides (see Note 17). Store overnight at –70°C in a Revco.
7. Place the slides with the frozen sections on a Coplin jar with 70% ethanol at –20°C for 30 min. Transfer the slides to another Coplin jar with 100% acetone for 3 min.
8. Transfer the slides three times to Coplin jars with ice-cold PBS.
9. Move the slides to a Coplin jar with TX-100 permeabilization buffer, for 10 min at room temperature and repeat step 8.
10. Remove the slides one by one from the Coplin jars, drain them, and dry the surface not containing the tissue sections. Draw a circle around the tissue sections with a hydrophobic pen. Place 40 μl of tissue blocking solution inside the circle. Leave to quench for 1 h at room temperature (see Note 18).
11. Prepare the dilution of claudin antibodies in cell line blocking solution (a) Rat uterus: claudin-1, 1:33; claudin-3, 1:12.5; claudin-5, 1:12.5; claudin-7, 1:100. (b) Human placenta: claudin-1, 1:25; claudin-3, 1:12.5; claudin-4, 1:167; claudin-5 1:125, claudin-16, 1:85. (c) Mouse skin: claudin-1, 1:150.
12. Drain the blocking solution and place 40 μl of the claudin antibodies dilution inside the hydrophobic pen circle. Incubate overnight at room temperature.
13. Transfer the slides five times to Coplin jars with ice-cold PBS
14. Dilute the secondary antibodies in blocking solution according to the manufacturer's instructions.
15. Remove the slides one by one from the Coplin jars, drain them, and dry the surface not containing the tissue section. Place 40 μl of secondary antibody solution inside the hydrophobic pen circle. Leave for 1 h at room temperature.
16. Repeat step 13.
17. Remove the slides one by one from the Coplin jars, drain them, and dry the surface not containing the tissue section.
18. Cover the slides with 6–15 μl of Vecta Shield antifade mounting solution and place a coverslip on top. With nail polish seal the borders of the coverslip to avoid tissue drying.
19. Store the slides in the dark at –20°C.

3.6. Interpretation of Results

3.6.1. Interpretation of Claudin Western Blots

The main problem with claudin western blots relies in the correct identification of the claudin band as some claudin antibodies, in a nonspecific manner, recognize other low molecular weight proteins. In addition, some claudin antibodies recognize more than one claudin (see Note 19). To correctly identify the band of claudins, we suggest employing any of the following strategies: (1) Load in the SDS–PAGE an additional lane with a sample from a cell line where the specific claudin can be easily identified (Fig. 1a). (2) Employ two antibodies from different manufacturers or research groups against the same claudin, and identify the common band in both blots (Fig. 1b). (3) When the specific antigenic peptide is available from the antibody manufacturer, test if its preincubation with the claudin antibody inhibits the appearance of a particular band ~20 kDa. (4) Do several exposure times of the claudin blots because sometimes the first band to appear ~20 kDa is not the specific claudin band (Fig. 1c).

Each claudin particle visualized by freeze-fracture electron microscopy is proposed to be composed of claudin multimers. Therefore, the detection by western blot, with claudin-specific antibodies, of bands with molecular weights higher than 20 kDa has sometimes been interpreted as the result of claudin oligomerization. In fact, the presence of 8% of the phospholipid detergent, perfluoro-octanoic acid (PFO), in the extraction buffer, followed by denaturing 13% SDS–PAGE leads to the

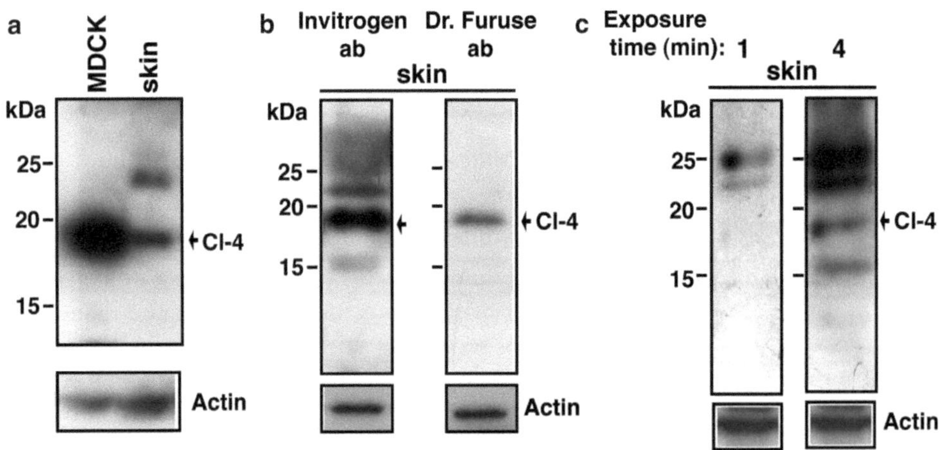

Fig. 1. Identification of claudin-specific bands in a western Blot. (**a**) An additional lane with a sample from MDCK cells was included to identify the specific claudin-4 band present in the mouse skin sample. (**b**) Antibodies against claudin-4 from Invitrogen or generously provided by Dr M. Furuse (Kobe University, Japan) were tested in different lanes containing mouse skin samples with the purpose of identifying a common band in both lanes. (**c**) 1 and 4 min exposure times were employed in this blot where claudin-4 was detected in mouse skin samples. Observe that the first bands to appear are above 20 kDa and do not correspond to claudin-4.

appearance of claudin-4 monomer (20.4 kDa) and several oligomers: dimers (41.7 kDa), trimers (41.7 kDa), tetramers (77.6 kDa), pentamers (89.1 kDa), and hexamers (107.2 kDa) (Mitic et al., Vol 12, 1994). This indicates that PFO permits maintenance of oligomeric claudin species and that SDS is unable to completely disrupt a complex previously exposed to PFO. In contrast, no claudin oligomers have been identified in claudin western blots of cell or tissues extracted with RIPA or gentle lysis buffers.

3.6.2. Interpretation of Claudin Immunofluorescence

TJs localize at the uppermost portion of the lateral membrane, and proteins such as ZO-1, ZO-2, and occludin concentrate precisely at this point. In contrast, claudin staining at the lateral membrane sometimes goes below that observed for other TJ proteins. Thus, while in MDCK cells, Claudin-1 expression is concentrated at the TJ (Fig. 2a), other claudins distribute along the whole basolateral membrane. Such is the case for example of claudin-7 in rat uterus (10) (Fig. 2b) and in rabbit renal tubules (11) (Fig. 2c). Some claudins change their localization according to the physiological state of the organism. For example, in rat uterus, claudin-3 redistributes from the TJ region to the whole basolateral surface as the animal progresses from diestrus to proestrus (10) (Fig. 2d). To observe claudin staining along the basolateral membrane of epithelial cell lines, confocal immunofluorescence should be done on yz sections.

The study of the subcellular distribution of claudins is highly recommended when the amount of claudin mRNA or protein detected either by microarray assays or western blot analysis, is observed to change as a result of physiological, pathological, or experimental conditions. Care should be taken on the interpretation of results, since an increase in the amount of claudin expression does not automatically imply the appearance of more complex and sealed TJs, and in fact the opposite might be true. For example, in colon carcinoma and metastasis, claudin-1 overexpression is accompanied by a decreased expression of the protein at the cell borders, and an enhanced nonjunctional claudin-1 staining, localized largely in the nucleus and cytoplasm (12). Furthermore, claudins-3 and -4 are overexpressed in numerous carcinomas (13–16), and in the case of claudin-4 it has been observed that Tyr208, which localizes within the PDZ binding motif of the molecule, is a target of ephrin receptor EphA2, a kinase that belongs to a family of receptors frequently overexpressed in cancerous tissues (for review see ref. 17). The phosphorylation of Tyr208 diminishes claudin-4 interaction with the PDZ containing protein ZO-1, and as a result claudin-4 can no longer integrate in an efficient manner to the cell borders (18).

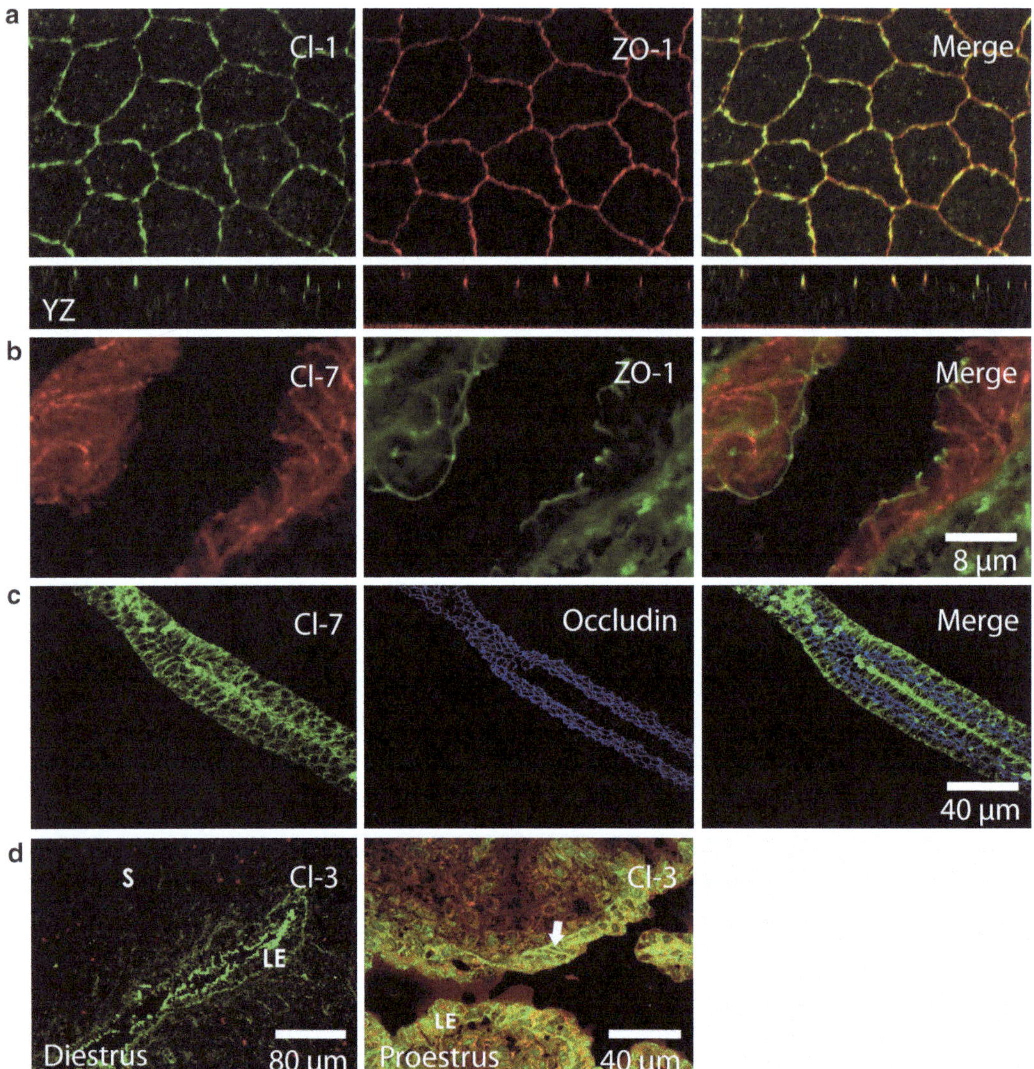

Fig. 2. Immunofluorescence detection of claudins. (**a**) Claudin-1 staining at the lateral membrane of MDCK cells is restricted to the TJ region. Observe the colocalization of claudin-1 (*green*) and ZO-1 (*red*) in the yz sections. (**b**) In rat uterus (from Mendoza-Rodriguez et al. Cell Tissue Res. 2005, 319:315–329, (10) ©2005 with permission from Springer) and in (**c**) isolated rabbit renal tubules (from Gonzalez-Mariscal et al. Nephrol. Dial. Transplant, 2006, 21:2391–2398, (11) ©2006 with permission from Oxford University Press), claudin-7 distributes along the whole basolateral membrane, whereas ZO-1 and occludin localization is restricted to the TJ. (**d**) Claudin-3 localization in the uterus changes from the TJ region to the basolateral membrane, as the animal progresses from diestrus to proestrus (From Mendoza-Rodriguez et al. Cell Tissue Res. 2005, 319:315–329, (11) ©2005 with permission from Springer). *LE* luminal epithelium, *S* stroma.

4. Notes

1. Immediately before using the RIPA buffer for cell lysis, add 1 mM PMSF and the protease inhibitor cocktail Complete™ (for each 60 mm diameter cell culture dish add 15 µl of the solution prepared by diluting 1 Complete™ pellet in 1 ml of water).

2. Acrylamide is a neurotoxin when unpolymerized and, hence, care should be taken to avoid exposure or incorporation.
3. Storage of the 10% SDS stock solution at room temperature prevents SDS precipitation.
4. pH electrodes can be damaged when they are used with SDS-containing solutions, because dodecyl sulfates precipitate in the porous plug and on the electrode surface. So, measure the pH before adding SDS.
5. PVDF membranes are preferable to nitrocellulose membranes because of their better physical stability and the lower propensity of proteins to run through the membrane. PVDF membranes should be wetted with 100% methanol just before use. Once wet they should never be allowed to dry during the blotting and detection process.
6. Bovine serum albumin for the western blot blocking buffer does not need to be IgG free.
7. Antibodies against epitopes on the first and second extracellular loops of claudins can be employed with live cells or in fixed cells without permeabilization. They can be used to analyze the arrival of claudins to the plasma membrane and in functional assays (19).
8. 2-Methylbutane is flammable.
9. Prepare the *p*-formaldehyde solution in the hood to avoid exposure to *p*-formaldehyde vapors. Never allow the solution to boil.
10. Alternatively at this step you can freeze at –70°C a small sample of the cell lysate for later quantification of proteins.
11. Take care not to heat above 100°C, to avoid unexpected opening of the tubes and potential loss of samples. To be on the safe side, we recommend to use safe lock reaction tubes.
12. In the skin samples, a whitish, low-density material, floating on top of the cell lysate supernatant is found. Be sure to include this material in the recovered supernatant as it is rich in claudins. In other tissues this material is also present albeit at a lower amount.
13. We favor the use of a semi-dry blotting system because it is easier, faster and requires less buffer volumes than tank blotting. Small proteins such as claudins (~20 kDa) are transferred with better reproducibility in this system.
14. There must be taken several different exposure times of the claudin blots since sometimes the first bands to appear around ~20 kDa are not the specific ones.
15. To have an adequate chemioluminescence quantification, the claudin signal in each sample should be normalized to that of actin or tubulin (claudin/actin or tubulin).

16. *p*-Formaldehyde fixation is usually chosen as the best way for preserving cell morphology. However, *p*-formaldehyde can reduce or destroy the antigenicity of some proteins and hence 100% methanol fixation should be tried. Claudin-3 is better detected with the Invitrogen antibody, if the cells are fixed with methanol.
17. To coat slides with gelatin, heat 500 ml of distilled water to 60°C and completely dissolve 1.5 g of gelatin, Type A, with a magnetic stirrer. Add 0.25 g of chromium potassium sulfate and stir. The solution should turn pale blue. Dip racks of clean slides in the warm gelatin solution, drain the slides onto tissue paper, and then stand the slides on ends to air dry overnight on a dust-free container or cover with foil.
18. Treatment with 0.005% Evans blue for 10 min at room temperature can be used, in addition, to quench the green unspecific tissue autofluorescence and to provide a red staining of the tissue to facilitate its observation.
19. Rabbit polyclonal against the carboxyl segment of claudin-1 from Invitrogen with Cat. No. 71-7800 strongly crossreacts with claudin-3; rabbit polyclonal against the first loop of claudin-1 from Santa Cruz Biotechnologies, Inc. with Cat. No. sc-28668, crossreacts with claudin-2, and rabbit polyclonal against the first loop of claudin-3 from Santa Cruz Biotechnologies, Inc. with Cat. No. sc-28666, crossreacts with claudins 4, 6, and 9.
20. Tyrosine 195 in claudin 2 constitutes a c-Src phosphorylation site.
21. Residues Y219 in claudins 3 and 6, and Y210 in claudin-7 localize within PDZ binding motifs. Previously, it was demonstrated that phosphorylation of an equivalent residue in claudin-4 (Tyr208) diminishes the interaction of the claudin with ZO-1, and as a result the affected claudin can no longer integrate to the cell borders (18). This tyrosine located at position −1 is conserved in claudins 1–10, 14, 15, and 17–20 as an Eph phosphorylation site.

Acknowledgments

This work was supported by grant 98448 from the Mexican Council for Science and Technology (Consejo Nacional de Ciencia y Tecnología (CONACYT)). E.G. and M.Q. were recipients of doctoral fellowships from CONACYT (203572 and 209822).

References

1. Staehelin, L. A. (1973) Further observations on the fine structure of freeze-cleaved tight junctions. *J. Cell Sci. 13*, 763–786.
2. van Deurs, B. and Luft, J. H. (1979) Effects of glutaraldehyde fixation on the structure of tight junctions: a quantitative freeze-fracture analysis. *J. Ultrastruct. Res. 68*, 160–172.
3. Furuse, M., Fujita, K., Hiiragi, T., Fujimoto, K., and Tsukita, S. (1998) Claudin-1 and -2: novel integral membrane proteins localizing at tight junctions with no sequence similarity to occludin. *J. Cell Biol. 141*, 1539–1550.
4. Colegio, O. R., Van Itallie, C., Rahner, C., and Anderson, J. M. (2003) Claudin extracellular domains determine paracellular charge selectivity and resistance but not tight junction fibril architecture. *Am. J. Physiol. Cell Physiol. 284*, C1346–C1354.
5. Colegio, O. R., Van Itallie, C. M., McCrea, H. J., Rahner, C., and Anderson, J. M. (2002) Claudins create charge-selective channels in the paracellular pathway between epithelial cells. *Am. J. Physiol. Cell Physiol. 283*, C142–C147.
6. Yu, A. S., Enck, A. H., Lencer, W. I., and Schneeberger, E. E. (2003) Claudin-8 expression in Madin-Darby canine kidney cells augments the paracellular barrier to cation permeation. *J. Biol. Chem. 278*, 17350–17359.
7. Van Itallie, C. M., Rogan, S., Yu, A., Vidal, L. S., Holmes, J., and Anderson, J. M. (2006) Two splice variants of claudin-10 in the kidney create paracellular pores with different ion selectivities. *Am. J. Physiol. Renal Physiol. 291*, F1288–F1299.
8. Yu, A. S. (2009) Molecular basis for cation selectivity in claudin-2-based pores. *Ann. N. Y. Acad. Sci. 1165*, 53–57.
9. Piontek, J., Winkler, L., Wolburg, H., Muller, S. L., Zuleger, N., Piehl, C., Wiesner, B., Krause, G., and Blasig, I. E. (2008) Formation of tight junction: determinants of homophilic interaction between classic claudins. *FASEB J. 22*, 146–158.
10. Mendoza-Rodriguez, C. A., Gonzalez-Mariscal, L., and Cerbon, M. (2005) Changes in the distribution of ZO-1, occludin, and claudins in the rat uterine epithelium during the estrous cycle. *Cell Tissue Res. 319*, 315–330.
11. Gonzalez-Mariscal, L., Namorado, M. C., Martin, D., Sierra, G., and Reyes, J. L. (2006) The tight junction proteins claudin-7 and -8 display a different subcellular localization at Henle's loops and collecting ducts of rabbit kidney. *Nephrol. Dial. Transplant. 21*, 2391–2398.
12. Dhawan, P., Singh, A. B., Deane, N. G., No, Y., Shiou, S. R., Schmidt, C., Neff, J., Washington, M. K., and Beauchamp, R. D. (2005) Claudin-1 regulates cellular transformation and metastatic behavior in colon cancer. *J. Clin. Invest 115*, 1765–1776.
13. Lodi, C., Szabo, E., Holczbauer, A., Batmunkh, E., Szijarto, A., Kupcsulik, P., Kovalszky, I., Paku, S., Illyes, G., Kiss, A., and Schaff, Z. (2006) Claudin-4 differentiates biliary tract cancers from hepatocellular carcinomas. *Mod. Pathol. 19*, 460–469.
14. Kominsky, S. L., Vali, M., Korz, D., Gabig, T. G., Weitzman, S. A., Argani, P., and Sukumar, S. (2004) Clostridium perfringens enterotoxin elicits rapid and specific cytolysis of breast carcinoma cells mediated through tight junction proteins claudin 3 and 4. *Am. J. Pathol. 164*, 1627–1633.
15. de Oliveira, S. S., de Oliveira, I. M., De Souza, W., and Morgado-Diaz, J. A. (2005) Claudins upregulation in human colorectal cancer. *FEBS Lett. 579*, 6179–6185.
16. Rangel, L. B., Agarwal, R., D'Souza, T., Pizer, E. S., Alo, P. L., Lancaster, W. D., Gregoire, L., Schwartz, D. R., Cho, K. R., and Morin, P. J. (2003) Tight junction proteins claudin-3 and claudin-4 are frequently overexpressed in ovarian cancer but not in ovarian cystadenomas. *Clin. Cancer Res. 9*, 2567–2575.
17. Surawska, H., Ma, P. C., and Salgia, R. (2004) The role of ephrins and Eph receptors in cancer. *Cytokine Growth Factor Rev. 15*, 419–433.
18. Tanaka, M., Kamata, R., and Sakai, R. (2005) EphA2 phosphorylates the cytoplasmic tail of Claudin-4 and mediates paracellular permeability. *J. Biol. Chem. 280*, 42375–42382.
19. Kausalya, P. J., Amasheh, S., Gunzel, D., Wurps, H., Muller, D., Fromm, M., and Hunziker, W. (2006) Disease-associated mutations affect intracellular traffic and paracellular Mg^{2+} transport function of Claudin-16. *J. Clin. Invest. 116*, 878–891.

Chapter 16

Expression and Function of Claudins in Hepatocytes

Takashi Kojima and Norimasa Sawada

Abstract

Tight junctions of hepatocytes play crucial roles in the barrier to keep bile in bile canaliculi away from the blood circulation, which we call the blood–biliary barrier. Tight junction proteins of hepatocytes are regulated by various cytokines and growth factors via distinct signal transduction pathways. To investigate changes in expression and function of tight junction proteins including claudins via signal transduction pathways in hepatocytes during EMT induced by TGF-β, we examined effects of TGF-β on expression and localization of the integral tight junction proteins, claudin-1, -2, and occludin, as well as the fence function by using primary cultures of adult rat hepatocytes.

Key words: Claudins, Occludin, Hepatocytes, Fence function

1. Introduction

The polarization of hepatocytes involves the formation of functionally distinct sinusoidal (basolateral) and bile canalicular (apical) plasma membrane domains that are separated by tight junctions (1). Tight junctions are formed by not only the integral membrane proteins claudins, occludin, JAMs, and tricellulin but also many peripheral membrane proteins (2–5). The claudin family, consisting of 24 members, is solely responsible for forming tight junction strands and shows tissue- and cell-specific expression of individual members (2).

In murine livers, claudin-1, -2, -3, -5, -7, -8, -12, and -14 are detected together with occludin, JAM-A, CAR and tricellulin, and claudin-1, -2, and -3 are expressed in the bile canalicular region of hepatocytes (2, 5–8). In the rat liver, claudin-2 shows a lobular gradient increasing from periportal to pericentral hepatocytes, whereas claudin-1 and -3 are evenly expressed in the whole liver lobule (9, 10). In the human liver, occludin, JAM-A, ZO-1,

Kursad Turksen (ed.), *Claudins: Methods and Protocols*, Methods in Molecular Biology, vol. 762,
DOI 10.1007/978-1-61779-185-7_16, © Springer Science+Business Media, LLC 2011

ZO-2, claudin-1, -2, -3, -7, -8, -12, -14, and tricellulin are detected together with well-developed tight junction structures, and claudin-2 shows a lobular gradient increasing from periportal to pericentral hepatocytes as in the livers of rat and mouse, whereas claudin-1 is expressed in the whole liver lobule (1, 11).

Tight junction proteins of hepatocytes are regulated by various cytokines and growth factors via distinct signal transduction pathways (11). By using primary cultures of adult rat hepatocytes at day 10 after plating, in which epithelial cell polarity is well-maintained by tight junctions, we examined effects of TGF-β on expression and localization of the integral tight junction proteins, claudin-1, -2, and occludin, as well as the fence function (12).

2. Materials

2.1. Isolation and Cell Culture

1. Ca^{2+}, Mg^{2+}-free Hanks' balanced salt solution (HBSS) (Gibco/BRL, Gaithersburg, MD) supplemented with 0.5 mM EGTA (Sigma, St. Louis, MO), 0.5 mg/l insulin from bovine pancreas (Sigma), and 100 U/ml penicillin and 100 μg/ml streptomycin (Clonetics Corp. San Diego, CA).
2. HBSS containing 40 mg collagenase (Yakult Co., Tokyo, Japan).
3. Percoll (GE Healthcare UK, Ltd., Buckinghamshire, UK).
4. L-15 medium (Gibco/BRL) supplemented with 0.2% bovine serum albumin (BSA, Sigma), 20 mM HEPES (Sigma), 0.5 mg/l insulin (Sigma), 10^{-7} M dexamethasone (Sigma), 1 g/l galactose (Sigma), 30 mg/l proline (Sigma), 20 mM $NaHCO_3$, 10 ng/ml epidermal growth factor (EGF, Collaborative Res., Lexington, MA), and 100 U/ml penicillin and 100 μg/ml streptomycin (Clonetics Corp.).
5. Rat tail collagen (500 μg dried tendon/ml in 0.1% acetic acid). Store at 4°C.
6. 35 and 60 mm culture dishes (Corning Glass Works, Corning, NY).
7. 2% Dimethylsulfoxide (DMSO, Aldrich Chemical Co., Inc., Milwaukee, WI).
8. 10^{-7} M glucagon (Glucagon S, Yamanouchi, Tokyo, Japan).
9. Murine TGF-β (PEPROTECH, London, UK).
10. Inhibitors: PD98059, SB203580, LY294002, and GF-109203X (Calbiochem-Novabiochem Corporation, San Diego, CA).

2.2. Reverse Transcription-Polymerase Chain Reaction for Claudins

1. TRIzol reagent (Gibco/BRL).
2. SuperScript® III First-strand SuperMix (Invitrogen).
3. Premix Taq DNA polymerase (Takara, Inc., Shiga, Japan).
4. Ethidium bromide (Sigma).

2.3. Western Blotting for Claudins and Signal Transduction Molecules

1. Sample buffer (1 mM $NaHCO_3$ and 2 mM phenylmethylsulfonyl fluoride, Sigma).
2. Protein assay kit (Pierce Chemical Co., Rockford, IL).
3. 4/20% gradient SDS–polyacrylamide gels (Daiichi Pure Chemicals Co., Tokyo, Japan).
4. Running buffer: 25 mM Tris–HCl, 192 mM glycine, 0.1% sodium dodecyl sulfate (SDS).
5. Nitrocellulose membranes (Immobilon-P, Millipore Corporation, Bedford, MA).
6. Transfer buffer: 25 mM Tris–HCl, 192 mM glycine, and 10% methanol.
7. TBS: 25 mM Tris–HCl, pH 8.0, 125 mM NaCl.
8. Blocking buffer: 0.1% Tween 20 and 4% skim milk in TBS.
9. TBS-T: 0.05% Tween 20 in TBS.
10. Ponceau S (Sigma).
11. Primary antibodies: occludin, clauidn-1, -2 (Zymed Laboratories, San Francisco, CA), phospho-MAPK, phospho-Akt, Akt, phosphor-p38MAPK (Cell Signaling, Beverly, MA), ERK1/2 (Promega Corporation, Madison, WI), c-H-Ras (Santa Cruz Biotechnology, Santa Cruz, CA), phospho-Smad-2 (Upstate biotechnology, Lake Placid, NY), and actin (Sigma).
12. Secondary antibodies: horseradish peroxidase-conjugated anti-mouse IgG or anti-rabbit IgG (DAKO, A/S, Denmark).
13. Enhanced chemiluminescence western blotting system (GE Healthcare UK, Ltd., Buckinghamshire, UK).
14. WB stripping solution (NACALAI TESQUE, INC, Kyoto, Japan).

2.4. Confocal Immunofluorecence for Claudins

1. Glass-bottom microwell plates (Mat Tek Corp., Ashland, MA).
2. –20°C cold acetone and ethanol (1:1) (Sigma).
3. Phosphate-buffered saline (PBS): Prepare 10× stock with 1.37 M NaCl, 27 mM KCL, 100 mM Na_2HPO_4, 18 mM KH_2PO_4 (pH 7.4). Store at room temperature. Prepare working solution by the dilution of one part with nine parts water.

4. Primary antibodies: occludin, clauidn-1, -2 (Zymed Laboratories).
5. Secondary antibodies: Alexa 488-conjugated anti-mouse IgG or anti-rabbit IgG (Molecular Probes, Inc., Eugene, OR).
6. DABCO: 2% 1,4-diazabicyclo(2,2,2)octan and 50% glycerol in PBS.

2.5. Freeze-Fracture Replicas for Tight Junction Strands

1. 2.5% Glutaraldehyde/0.1 M PBS (pH 7.3).
2. 40% Glycerin solution (Sigma).
3. 10% Sodium hypochlorite solution.
4. Teflon dish.

2.6. Measurement for Fence Functions

1. P buffer (pH 7.4): 10 nM HEPES (Sigma), 1 mM sodium pyruvate (sigma), 10 mM glucose (Sigma), 3 mM $CaCl_2$ (Sigma), and 145 mM NaCl (Sigma).
2. BODIPY-FL-sphingomyelin (Molecular Probes).
3. Bovine serum albumin (BSA, Sigma).
4. Glass-bottom microwell plates (Mat Tek Corp.).

3. Methods

3.1. Isolation and Cell Culture

1. Adult male Sprague-Dawley rats (weight range, 300–400 g) were used to isolate hepatocytes by the two-step liver perfusion method (13) with some modification (14).
2. Briefly, the liver was perfused in situ through the portal vein with 150 ml of Ca^{2+}, Mg^{2+}-free Hanks' balanced salt solution (HBSS) supplemented with 0.5 mM EGTA, 0.5 mg/l insulin, and antibiotics (see Note 1).
3. After the initial perfusion, the liver was perfused with 200 ml of HBSS containing 40 mg collagenase for 10 min (see Note 2).
4. The isolated cells were purified by Percoll isodensity centrifugation (15) (see Note 3).
5. Viability of the cells, as judged by the trypan blue exclusion test, was more than 90% in the experiments.
6. The cells were suspended in L-15 medium with 0.2% BSA, 20 mM HEPES, 0.5 mg/l insulin, 10^{-7} M dexamethasone, 1 g/l galactose, 30 mg/l proline, 20 mM $NaHCO_3$, 10 ng/mL EGF, and antibiotics (modified L-15 medium).
7. The isolated hepatocytes were plated at a density of 5.5×10^5 cells/cm^2 on 35 and 60 mm culture dishes, which were coated with rat tail collagen (500 μg dried tendon/ml in 0.1% acetic acid), and placed in a 5% CO_2 to 95% air incubator at 37°C (see Note 4).

8. The medium was replaced with fresh medium every other day. After 96 h of culture, 2% DMSO and 10^{-7} M glucagon were added to the modified L-15 medium. Then the cells were cultured until 10 days after plating (14) (see Note 5).
9. The cells were treated with 0.01–20 ng/ml murine TGF-β for 24 h. Some cells were pretreated with 50 μM PD98059 (MAPK inhibitor), 20 μM SB203580 (p38 MAPK inhibitor), 10 μM LY294002 (PI3K inhibitor), and 10 μM GF-109203X (PKC inhibitor) at 30 min before treatment with 10 ng/mL TGF-β for 24 h (see Note 6).

3.2. RT-PCR for Claudins

1. Total RNA was extracted from the cells using TRIzol reagent. For reverse transcription-polymerase chain reaction (RT-PCR), 1 μg of total RNA was reverse transcribed into cDNA using the manufacturer's recommended conditions (SuperScript® III First-strand SuperMix).
2. The RT product was amplified using Premix Taq DNA polymerase.
3. In RT-PCR, primers used to detect SIP1, Snail, E-cadherin, occludin, claudin-1, claudin-2, and G3PDH (12) (Fig. 1). Conditions applied for PCR were: 96°C for 30 s, 30 cycles of 96°C for 15 s, 55°C for 30 s, 72°C for 1 min, and 72°C for 7 min.
4. Samples were separated by electrophoresis in ethidium bromide-impregnated 1% agarose gels.

3.3. Western Blotting for Claudins and Signal Transduction Molecules (Fig. 2)

1. For western blotting of total cell lysates, the dishes were washed with PBS twice, and 300 μl of sample buffer was added to 60 mm dishes. The cells were scraped and collected in microcentrifuge tubes and then sonicated for 10 s.
2. Protein concentrations in the extracts were measured using a protein assay kit and calculated as standard of albumin value.
3. Briefly, 15 μg of protein of each sample per lane was applied and separated by electrophoresis (25 mA, 3 h) on 4/20% gradient SDS–polyacrylamide gels using running buffer.
4. After electrophoretic transfer (10 V, 60 min) to immobilon-P membranes by semi-dry transfer cell (Bio-Rad, Hercules, CA) and transfer buffer, the membranes were saturated with a blocking buffer for 30 min at room temperature. The quality of the transfer was controlled by Ponceau S staining of the membranes.
5. The membranes were incubated with primary antibodies (1:1,000 in TBS) at room temperature for 1 h and washed with three times for 5 min with TBS-T. Membranes were incubated with secondary antibodies (1:1,000 in TBS) at room temperature for 1 h, and washed three times for 5 min with TBS-T, and

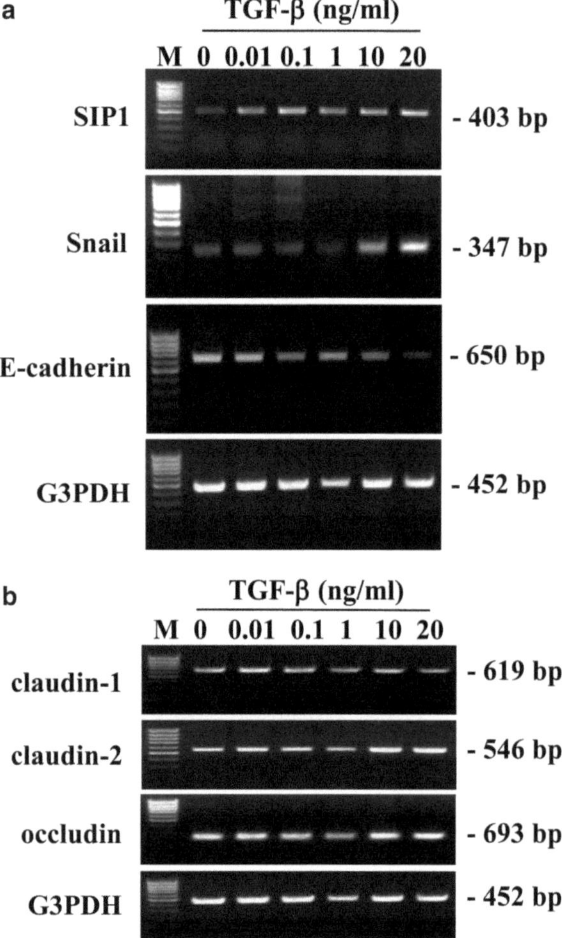

Fig. 1. (**a**) RT-PCR for SIP1, Snail, and E-cadherin in cultured rat hepatocytes after treatment with 0.01–20 ng/ml TGF-β. Upregulation of SIP1 and Snail and downregulation of E-cadherin are observed. (**b**) RT-PCR for claudin-1, -2, and occludin in cultured rat hepatocytes after treatment with 0.01–20 ng/ml TGF-β. Downregulation of claudin-1 and upregulation of claudin-2 are observed.

detection was carried out using an enhanced chemiluminescence (ECL) western blotting system (see Notes 7 and 8).

6. The membranes are removed from the ECL reagents, blotted with Kim-Wipes, placed between the acetate sheets and then placed in an X-ray film cassette with film (Kodak).

3.4. Confocal Immunofluorecence for Claudins

1. For immunocytochemistry, cells grown on glass-bottom microwell plates were fixed with cold acetone and ethanol (1:1) at –20°C for 10 min (see Note 9) (Fig. 3).
2. After the plates were rinsed three times for 5 min with PBS, they were incubated with primary antibodies (1:100 in PBS)

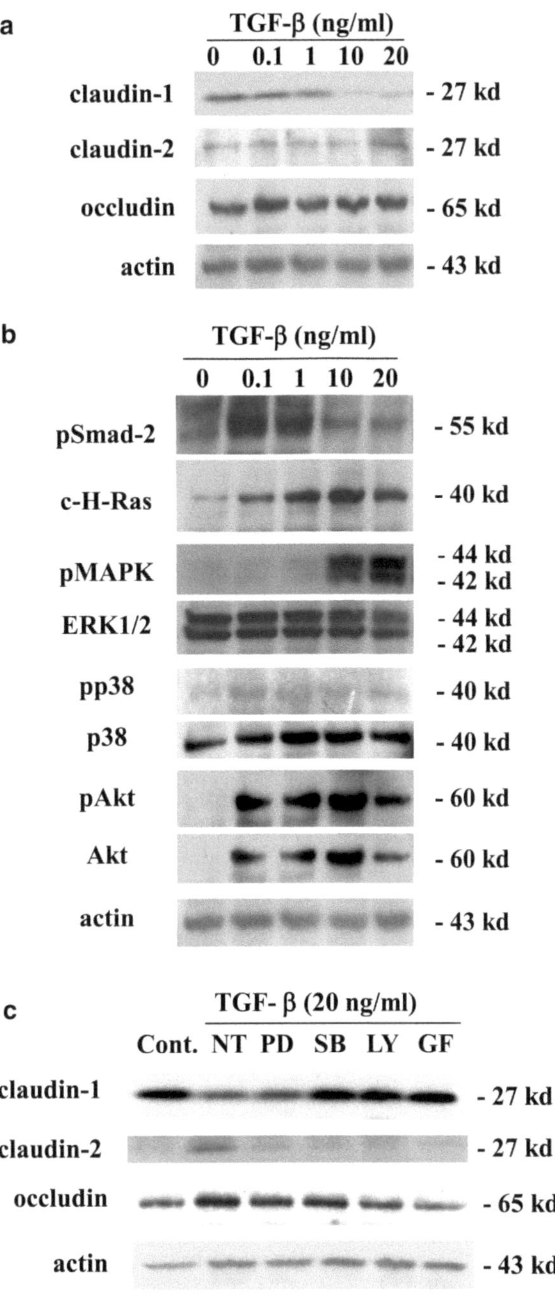

Fig. 2. (**a**) Western blotting for claudin-1, -2, and occludin in cultured rat hepatocytes after treatment with 0.01–20 ng/ml TGF-β. Downregulation of claudin-1 and upregulation of claudin-2 and occludin are observed. (**b**) Western blotting for pSmad-2, c-H-Ras, pMAPK, ERK1/2, pp38 MAPK, p38 MAPK, pAkt and Akt in rat hepatocytes after treatment with 0.01–20 ng/ml TGF-β. Expression of all signal transduction pathways is observed. (**c**) Western blotting for claudin-1, -2, and occludin in rat hepatocytes treated with inhibitors before treatment with 20 ng/ml TGF-β. *C* control, *NT* nontreatment of inhibitors, *PD* PD98059, *SB* SB203580, *LY* LY294002, *GF* GF109203X. The changes of claudin-1, -2, and occludin are prevented by inhibitors of distinct signal transduction pathways.

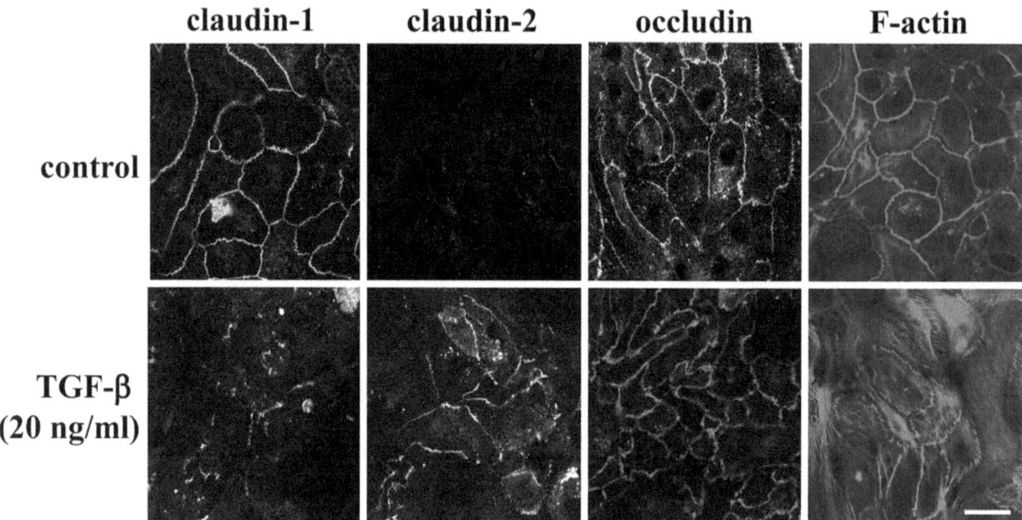

Fig. 3. Immunocytochemistry for claudin-1, -2, occludin, and F-actin in rat hepatocytes after treatment with 20 ng/ml TGF-β. Downregulation of claudin-1 and upregulation of claudin-2 are observed at cell borders. F-actin is changed from lines to zipper-like structures at cell borders. Bar: 20 μm in (**b**).

at room temperature for 1 h and were rinsed three times for 5 min with PBS (see Note 10). The plates were incubated with secondary antibodies (1:200 in PBS) at room temperature for 1 h and were rinsed three times for 5 min with PBS. The samples are mounted with DABCO using slide glasses (see Note 11).

3. The samples were examined using a laser-scanning confocal microscope (MRC 1024; Bio-Rad).

3.5. Freeze-Fracture Replicas for Tight Junction Strands (Fig. 4)

1. Cells were cultured on 60-mm dishes. Dishes were washed with PBS, and the cells were scraped from the dishes and collected in microcentrifuge tubes.
2. They were fixed then with 2.5% glutaraldehyde/0.1 M PBS (pH 7.3). After fixation, they were immersed in 40% glycerin solution (see Note 12).
3. The specimens were mounted on a copper stage, cooled in liquid nitrogen, and fractured at –170 to –180°C in a JFD-7000 freeze-fracture device (JEOL Ltd., Tokyo, Japan). Platium-carbon replicas were made without etching.
4. After thawing, the replicas were floated on filtered 10% sodium hypochlorite solution for 30 min in a Teflon dish (see Note 13). Replicas were washed in distilled water, mounted on copper grids.
5. The grids were examined at an acceleration voltage of 80 kV with a JEM-1210 transmission electron microscope (JEOL Ltd.).

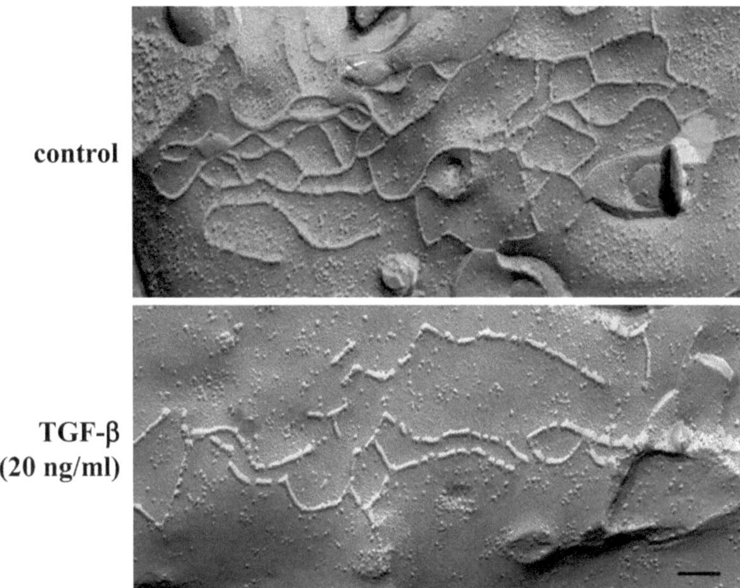

Fig. 4. Freeze-fracture replicas of rat hepatocytes after treatment with 20 ng/ml TGF-β. In treatment with TGF-β, the numbers of tight junction strands are decreased and the some strands are changed from lines to fragments. Bar: 50 nm.

3.6. Measurement for Fence Functions (Fig. 5)

1. For measurement of the tight junctional fence function, we used diffusion of BODIPY-sphingomyelin (16) with some modification (17).
2. Sphingomyelin/BSA complexes (5 nM) were prepared in P buffer using BODIPY-FL-sphingomyelin and defatted BSA (16). Store at –20°C.
3. Cells plated on glass-bottom microwell plates were loaded with BODIPY-sphingomyelin/BSA complex for 1 min on ice, after which they were rinsed with cold modified L-15 medium at 4°C and mounted in modified L-15 medium on a glass slide.
4. The samples were analyzed by confocal laser scanning microscopy. All pictures shown were generated within the first 5 min of analysis (see Note 14).

4. Notes

1. In the initial perfusion in situ through the portal vein, the enough washouts of the blood components are very important.
2. During the perfusion, HBSS-containing collagenase have to be keeping at 37°C using water bath.

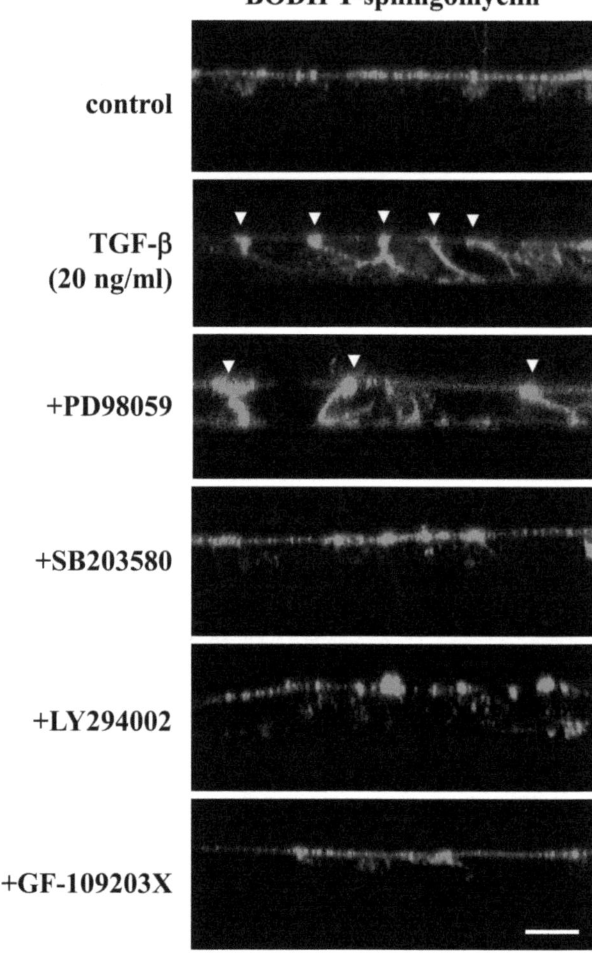

Fig. 5. Fence function as examined by diffusion of labeled BODIPY-sphingomyelin of rat hepatocytes treated with inhibitors before treatment with 20 ng/ml TGF-β. In treatment with TGF-β, the probe strongly labels the basolateral surfaces and appears to penetrate the cells (*arrowheads*), while the probe is effectively retained in the apical domain of the control. The diffusion of the probe is inhibited by SB203580, LY294002, and GF109203X but not PD98059 (*arrowheads*). Bar: 20 μm.

3. Although the standard ratio of the medium containing the isolated cells and Percoll solution is 1:1, it is necessary to change the ratio by the cell survival activity.
4. The coating with rat tail collagen is completely dried up under the UV light over night.
5. 2% DMSO has weak toxicity against the rat hepatocytes. When 2% DMSO exposures to the rat hepatocytes, the high cell-density is very important.
6. Please forget that the inhibitors add the medium with TGF-β after pretreatment with the inhibitors.

7. In phospho-antibodies, TBS should be used during all process. In claudin-antibodies, PBS can be also used.
8. It is possible to perform the stripping using WB stripping solution and reprobing blots using claudin-antibodies but not phospho-antibodies.
9. The monolayer cells are easily removed from glass-bottom microwell plates. Please perform slow, careful fixation.
10. For economy, only 40–50 μl of diluted antibody per sample needs to be used.
11. Air bubbles are undesirable in the mounting solution.
12. The samples replaced with 40% glycerin solution can be store at 4°C for several months.
13. If the cells are remained, the samples should be incubated with the sodium hypochlorite solution overnight.
14. The preparation is very important to perform Z-section analysis using confocal laser scanning microscopy within 5 min after mounting.

Acknowledgments

This work was supported by Grants-in-Aid from the National Project "Knowledge Cluster Initiative" (2nd stage, "Sapporo Biocluster Bio-S"), Program for developing the supporting system for upgrading the education and research, the Ministry of Education, Culture, Sports Science, and Technology, and the Ministry of Health, Labor and Welfare of Japan.

References

1. Kojima, T., Sawada, N., Yamaguchi, H., Fort, A. G., and Spray, D. C. (2009) Gap and tight junctions in liver: composition, regulation, and function. In: Arias, I. M., et al (ed) The Liver: Biology and Pathobiology, 5th edn. Lippincott Williams&Wilkins, Philadelphia, pp201–220.
2. Tsukita, S., Furuse, M., Itoh, M. (2001) Multifunctional strands in tight junctions. *Nat Rev Mol Cell Biol.* **2**, 285–293.
3. Sawada, N., Murata, M., Kikuchi, K., Tobioka, H., Kojima, T., Chiba, H. (2003) Tight junctions and human disease. *Med Electron Microsc.* **36**, 147–156.
4. Schneeberger, E. E., Lynch, R. D. (2004) The tight junction: a multifunctional complex. *Am J Physiol Cell Physiol.* **286**, C1213–C1228.
5. Ikenouchi, J., Furuse, M., Furuse, K., Sasaki, H., Tsukita, S., Tsukita, S. (2005) Tricellulin constitutes a novel barrier at tricellular contacts of epithelial cells. *J Cell Biol.* **171**, 939–945.
6. Fechner, H., Haack, A., Wang, H., Wang, X., Eizema, K., Pauschinger, M., Schoemaker, R. G., van Veghel, R., Houtsmuller, A. B., Schultheiss H. P., Lamers, J. M. J., Poller, W. (1999) Expression of coxsackie adenovirus receptor and alpha$_v$-integrin does not correlate with adenovector targeting in vivo indicating anatomical vector barriers. *Gene Ther* **6**, 1520–1535.
7. Kojima, T., Yamamoto, T., Murata, M., Chiba, H., Kokai, Y., Sawada, N. (2003) Regulation of the blood-billiary barrier: interaction

between gap and tight junctions in hepatocytes. *Med Electron Microsc* **36**, 157–164.

8. Wilcox, E. R., Burton, Q. L., Naz, S., Riazuddin, S., Smith, T. N., Ploplis, B., Belyantseva, I., Ben-Yosef, T., Liburd, N. A., Morell, R. J., Kachar, B., Wu, D. K., Griffith, A. J., Riazuddin, S., Friedman, T. B. (2001) Mutations in the gene encoding tight junction claudin-14 cause autosomal recessive deafness DFNB29. *Cell* **104**, 165–172.
9. Rahner, C., Mitic, L. L., Anderson, J. M. (2001) Heterogeneity in expression and subcellular localization of claudins 2, 3, 4, and 5 in the rat liver, pancreas, and gut. *Gastroenterology* **120**, 411–422.
10. Yamamoto, T., Kojima, T., Murata, M., Takano, K., Go, M., Hatakeyama, N., Chiba, H., Sawada, N. (2005) p38 MAP-kinase regulates function of gap and tight junctions during regeneration of rat hepatocytes. *J Hepatol* **42**, 707–718.
11. Kojima, T., Murata, M., Yamamoto, T., Lan, M., Imamura, M., Son, S., Takano, K., Yamaguchi, H., Ito, T., Tanaka, S., Chiba, H., Hirata, K., Sawada, N. (2009) Tight junction proteins and signal transduction pathways in hepatocytes. *Histol Histopathol* **24**, 1463–1472.
12. Kojima, T., Takano, K., Yamamoto, T., Imamura, M., Murata, M., Son, S., Yamaguchi, H., Osanai, M., Chiba, H., Himi, T., Sawada, N. (2008) Transforming growth factor-β induces epithelial to mesenchymal transition by down-regulation of claudin-1 expression and the fence function in adult rat hepatocytes. *Liver Int* **28**, 534–545, 2008.
13. Seglen, P. O. (1976) Preparation of isolated rat liver cells. *Methods Cell Biol* **13**, 29–83.
14. Kojima, T., Yamamoto, M., Mochizuki, C., Mitaka, T., Sawada, N., Mochizuki, Y. (1997) Different changes in expression and function of connexin 26 and connexin 32 during DNA synthesis and redifferentiation in primary rat hepatocytes using a DMSO culture system. *Hepatology* **26**, 585–597.
15. Kreamer, B. L., Staecker, J. L., Sawada, N., Sattler, G. L., Hsia, M. T., Pitot, H. C. (1986) Use of a low-speed, iso-density percoll centrifugation method to increase the viability of isolated rat hepatocyte preparations. *In Vitro Cell Dev Biol* **22**, 201–211.
16. Balda, M. S., Whitney, J. A., Flores, C., Gonzalez, S., Cereijido, M., Matter, K. (1996) Functional dissociation of paracellular permeability and transepithelial electrical resistance and disruption of the apical-basolateral intramembrane diffusion barrier by expression of a mutant tight junction membrane protein. *J Cell Biol* **134**, 1031–1049.
17. Kojima, T., Kokai, Y., Chiba, H., Yamamoto, M., Mochizuki, Y., Sawada, N. (2001) Cx32 but not Cx26 is associated with tight junctions in primary cultures of rat hepatocytes. *Exp Cell Res* **263**, 193–201.

Chapter 17

Analysis of Changes in the Expression Pattern of Claudins Using Salivary Acinar Cells in Primary Culture

Junko Fujita-Yoshigaki

Abstract

Primary saliva is produced from blood plasma in the acini of salivary glands and is modified by ion adsorption and secretion as the saliva passes through the ducts. In rodents, acinar cells of salivary glands express claudin-3 but not claudin-4, whereas duct cells express both claudins-3 and -4. The distinct claudin expression patterns may reflect differences in the permeability of tight junctions between acinar and duct cells. To analyze the role of claudins in salivary glands, we established a system for the primary culture of parotid acinar cells, where the expression patterns of claudins are remarkably changed. Real-time RT–PCR and immunoblot analyses reveal that the expression levels of claudins-4 and -6 increased, whereas claudins-3 and -10 decreased. We found that the signal to induce those changes is triggered during cell isolation and is mediated by Src and p38 MAP kinase. Here, we introduce the methods used to determine the signal pathway that induces the change in claudin expression.

Key words: Claudin, Tight junction, Salivary gland, Parotid acinar cells, Primary culture, Signal pathway, Src, p38 MAP kinase, Dedifferentiation

1. Introduction

Claudins show specific patterns of expression in various epithelial and endothelial tissues (1, 2). Each type of claudin is thought to create unique barrier properties in terms of charge selectivity, and combinations of different claudin family members expressed in cells determine the overall barrier properties of the tight junctions (TJ) (3, 4). Thus, the expression pattern of claudins should be closely related to the function and characteristics of epithelial tissues.

Kursad Turksen (ed.), *Claudins: Methods and Protocols*, Methods in Molecular Biology, vol. 762, DOI 10.1007/978-1-61779-185-7_17,

Salivary glands mainly consist of two types of epithelium, acini and ducts, which have different roles in the production of saliva. Primary saliva is produced in acini and is then modified as it passes through the duct system (5). For primary saliva secretion, Ca^{2+} elevation induced by cholinergic stimulation activates Ca^{2+}-dependent apical Cl^- channels and basolateral K^+ channels. The efflux of K^+ and Cl^- produces a transepithelial potential difference between the apical and basolateral spaces, which is a driving force of Na^+ and water diffusion from blood plasma to the lumen across TJ. As a result, isotopic primary saliva is produced. This model requires a high permeability of acinar TJ for Na^+ and water. After production, primary saliva is modified during its passage through the ducts by ion absorption and secretion with no further secretion or absorption of water, resulting in the saliva finally becoming hypotonic. To make saliva hypotonic, TJ in the ducts are considered to have a low permeability of water. Differences in the permeability of TJ between acinar and duct cells may be due to expression patterns of claudins.

The distribution of claudins has been reported in rodents and humans, but the expression patterns differ among animals and tissues. In rat salivary glands, claudin-3 expression was detected both in acinar and in duct cells, while claudin-4 was only expressed in duct cells (6). The differences in claudin expression are likely to correlate with differences in their functions. In addition, changes in the expression pattern of claudins during the development of submandibular glands has been reported (7). Each claudin appears in distinct types of cells at distinct developmental stages and some of them disappear after that. Such changes in the expression of claudins may be necessary to regulate the microenvironment to differentiate the salivary glands properly.

We have established a system for the primary culture of parotid acinar cells to study the regulatory mechanism(s) of saliva production (8). It is well known that retaining the function of salivary acinar cells in vitro is difficult. The acinar cells in our culture system maintained their capacity for signal-dependent exocytosis, but gradually lost their function and acquired duct-like characteristics. At the same time, the expression pattern of claudins changed remarkably (9), and there was a correlation between changes in claudin expression and the degree of differentiation of acinar cells. We found that the signal to induce that change in expression is triggered during cell isolation and is mediated by Src and p38 MAP kinase (p38 MAPK) (10). It is likely that the signal is a response to cellular stresses such as tissue injuries. We present here the methods used to analyze the signal pathway that induces the change in the expression pattern of claudins by parotid acinar cells in primary culture.

2. Materials

2.1. Isolation of Acinar Cells from Rat Parotid Glands

1. HBSS-BSA: Hanks' balanced salt solution (Gibco/BRL, Bethesda, MD) buffered with 20 mM HEPES–NaOH (pH 7.4) containing 0.5% bovine serum albumin (HBSS–0.5% BSA) or 4% BSA (HBSS–4% BSA) (see Note 1).
2. Enzymes for cell dispersion: collagenase A (Worshington Biochemical Co., Freehold, NJ, USA) and hyaluronidase (Roche Diagnostics).
3. Nylon mesh (pore size is about 1 mm).

2.2. Culture of Isolated Parotid Acinar Cells

1. Waymouth's medium (Gibco/BRL) containing ITS-X supplement (Gibco/BRL 51500-056), 1 μM hydrocortisone, 10 nM cystatin (Sigma C0408), 100 U/ml penicillin, 0.1 mg/ml streptomycin, and 10% rat serum. Rat serum was prepared from blood taken from 8-week-old rats under anesthesia via the abdominal aorta (see Note 2).
2. Collagen I-coated 35-mm dishes (Iwaki Co. Ltd., Tokyo, Japan) and collagen I-coated cell culture inserts (effective diameter of the membrane is 10.5 mm) in 12-well plates (BD Falcon).

2.3. Analysis of Expression of Claudin mRNAs by Real-Time RT–PCR

1. TRIzol reagent (Invitrogen, Carlsbad, CA, USA).
2. RNase-free DNase kit (Qiagen) and RNeasy MinElute Cleanup kit are purchased from Qiagen.
3. QuantiTect® SYBR® RT-PCR kit (Qiagen).
4. PCR tube: Low tube strips and Flat cap strips (Bio-Rad Laboratories, Hercules, CA, USA).
5. DNA Engine Opticon™ System (MJ Research, Waltham, MA, USA).

2.4. Analysis of Protein Expression of Claudins by Immunoblotting

1. Homogenizing buffer: 20 mM HEPES–NaOH (pH 7.4), 150 mM NaCl, 1 mM EDTA, 1 mM EGTA, 1 mM PMSF, 10 mM NaF, 1 mM Na_3VO_4, 50 mM β-glycerophosphate (see Note 3).
2. Complete Protease Inhibitor Cocktail, EDTA-free (Roche Diagnostics).
3. SDS–PAGE: Multigel II Mini (15–20% gradient gel, 13-well) (Cosmo Bio, Tokyo, Japan).
4. TBS-T: 0.05% Tween 20 in Tris-buffered saline.
5. For blotting, Hybond-LFP, Blocking agent and ECF Western blotting kit including alkaline phosphatase-conjugated secondary antibodies are purchased from GE Healthcare.

6. Antibodies: rabbit polyclonal antibodies to claudins-1, -2, -3, -7, -8, -10, and occludin and mouse monoclonal antibody to claudin-4 (Zymed Laboratories, San Francisco, CA, USA), and goat polyclonal antibody to claudin-6 (Santa Cruz Biotechnology, Santa Cruz, CA, USA).
7. Image analyzer: Typhoon Trio (GE Healthcare).

2.5. Immuno-fluorescence Microscopy for the Distribution of Claudins in the Culture

1. Collagen I-coated glass base dish (Iwaki, Tokyo, Japan). The diameter of the glass is 12 mm.
2. Fixative: 10% Formalin (3.7% formaldehyde) in phosphate-buffered saline (PBS), freshly prepared before use.
3. Blocking solution: 1% BSA (Sigma A7888) and 0.05% preimmune goat IgG (Sigma I5256) in PBS.
4. Primary antibodies: The same antibodies to claudin-3, claudin-4, and occludin that are used in the immunoblot analysis can also be used for immunofluorescence microscopy: rabbit polyclonal antibodies to ZO-1 (Zymed Laboratories) and amylase (Sigma-Aldrich).
5. Second antibodies: Alexa Fluor 488 goat anti-mouse IgG (A-11029), Alexa Fluor 568 goat anti-mouse IgG (A-11031), Alexa Fluor 488 goat anti-rabbit IgG (A-11034), and Alexa Fluor 568 goat anti-rabbit IgG (A-11036) (Invitrogen).
6. Nuclear labeling dye: TO-PRO-3 iodide (Invitrogen).
7. Mounting medium: Prolong Gold Antifade Reagent (Invitrogen).

2.6. Analysis of the Signal Pathway Regulating Claudin Expression and Dedifferentiation

1. All kinase inhibitors were from Calbiochem (San Diego, CA, USA). Inhibitors for Src kinase (PP1, PP2, PP3, and Src inhibitor-1) are dissolved in dimethylsulfoxide (DMSO) to a concentration of 20 mM as stock solutions. Inhibitors for p38 MAPK (SB203580, SB202190, and PD169316) are dissolved in DMSO to a concentration of 5 mM as stock solutions. They are stored in –20°C.
2. SDS–PAGE: Mutigel II Mini (10%, 13 well).
3. Antibodies: Mouse monoclonal anti-p38 MAPK antibody (BD Biosciences) and rabbit polyclonal anti-phospho-p38 MAPK (T180/Y182) antibody (R&D Systems, Minneapolis, MN, USA).
4. Blocking solution for anti-phospho-p38 MAPK (T180/Y182) antibody: 3% BSA in TBS-T.

3. Methods

To analyze changes in claudin expression, cells that form functional TJ are required. Because the maintenance of salivary acinar cells in vitro is very difficult, the methods for isolation and culture of parotid acinar cells are important for this analysis.

3.1. Isolation of Acinar Cells from Rat Parotid Glands

1. Parotid glands are excised from two male Sprague-Dawley rats (6 weeks old, 150–200 g each) anesthetized with sodium pentobarbital (Kyoritsu Seiyaku Co., Tokyo, Japan), and put into HBSS–0.5% BSA (5 ml) immediately.
2. After the glands are minced finely with a knife, they are suspended in new HBSS–0.5% BSA (5 ml) and are centrifuged at 700 rpm (90 × *g*) for 1 min to remove fats.
3. The precipitate is incubated in collagenase A (0.35 U/ml) and hyaluronidase (75 μg/ml) in 10 ml HBSS–0.5% BSA (see Note 4), at 37°C under 100% O_2 for 60 min, with rotation at 220 rpm by using a rotary shaker (innova 3100, New Brunswick, Edison, NJ, USA). After 30 min, the glands are further minced with small scissors in the incubation buffer.
4. After 60 min, dispersed acinar cells are filtered through a nylon mesh (four layers) and are loaded onto HBSS–4% BSA. After centrifugation of the filtrate at 700 rpm (90 × *g*) for 2 min, precipitated cells are suspended in HBSS–0.5% BSA and filtered again through eight-layers of nylon mesh (see Note 5).
5. The filtrate is centrifuged at 700 rpm (90 × *g*) for 1 min, and then is washed twice with Waymouth's medium (see Note 6).
6. The precipitate is resuspended in 2.5 ml Waymouth's medium containing 10% rat serum at a final concentration of about 3 mg/ml (see Note 7). When cells are isolated, about 25% of them are single cells, while the others form clusters of 2–10 cells.
7. Cell viability is checked by Trypan blue exclusion.

3.2. Culture of Isolated Parotid Acinar Cells

Isolated cells are diluted to 0.2 mg/ml in Waymouth's medium containing 10% rat serum and are cultured at 37°C in 5% CO_2 in collagen I-coated dishes (see Note 8). When TER is measured, cells are cultured in collagen I-coated cell culture inserts. The medium is changed at 1 day. Since cells tend to attach to each other, they form large clusters of more than 100 cells immediately after starting the culture. After culture for 24 h, the cells attach to the dishes as clusters, while most of the single cells float. At 48 h, most cells have spread as a monolayer (Fig. 1).

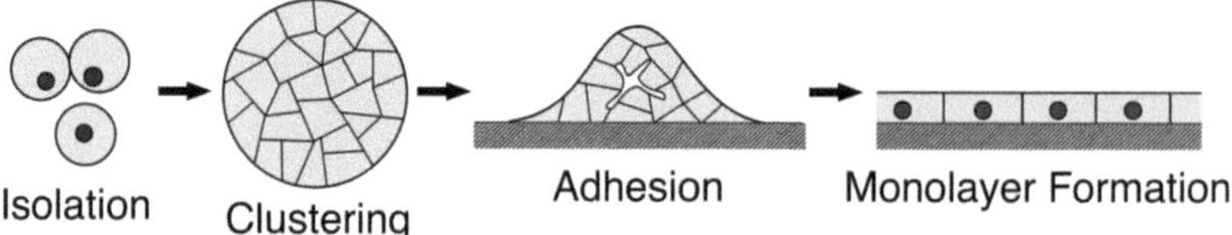

Fig. 1. Morphological changes in the organization of isolated parotid acinar cells during culture. Isolated cells immediately form clusters that consist of more than 100 cells and attach to dishes as clusters. The cells then expand on the dishes and form monolayers within 2–3 days. TJs are observed at the apical site of the monolayer, surface of the clusters, and lumen inside the clusters.

3.3. Analysis of Expression of Claudin mRNAs by Real-Time RT–PCR

1. Total RNAs are isolated from cells just after isolation and after culture for 1–3 days using the TRIzol reagent. After treatment with DNase I, RNA is purified using an RNeasy MinElute Cleanup kits. The amounts of RNA are quantified by measuring the absorbance at 260 nm.
2. Expression levels of mRNAs are determined using a QuantiTect® SYBR® RT–PCR kit and the DNA Engine Opticon™ System. One hundred nanograms of total RNA is used for each reaction. Primer pairs used for real-time RT–PCR are taken from ref. (11) and are shown in Table 1. Primer pairs for the amplification of rat claudin-3 were obtained from Qiagen. Relative RNA equivalents for each sample are obtained by normalizing to glyceraldehydes-3-phosphate dehydrogenase (GAPDH) levels (see Note 9). An example of the changes in claudin expression over time is shown relative to the values of expression levels just after isolation in Fig. 2.
3. PCR products are evaluated by melting curve analysis according to the manufacturer's instructions and by determining their sizes on 2.0% agarose gels.

3.4. Analysis of Protein Expression of Claudins by Immunoblotting

1. Cells are harvested and lysed in homogenizing buffer containing 1× Complete Protease Inhibitor Cocktail.
2. Proteins are separated by SDS–PAGE, and are transferred to Hybond-LFP membranes. The same amounts of proteins are applied to each lane of SDS–PAGE.
3. Membranes are blocked at room temperature for 50 min with 5% Blocking Agent in TBS-T.
4. The blocking reagent is discarded and the membranes are blotted with anti-occludin (1 μg/ml), anti-claudin-1 (1 μg/ml), anti-claudin-2 (1 μg/ml), anti-claudin-3 (0.25 μg/ml), anti-claudin-4 (0.05 μg/ml), anti-claudin-6 (2 μg/ml), anti-claudin-7 (1 μg/ml), anti-claudin-8 (1.25 μg/ml), or anti-claudin-10 (1.25 μg/ml) antibodies in TBS-T for 2 h.

Table 1
Primers used for real-time RT–PCR analysis

Target gene	Primer sequence	Product (bp)	Accession No.
Ocln (occludin)	5′-CTGTCTATGCTCGTCATCG-3′ 5′-CATTCCCGATCTAATGACGC-3′	294	NM_031329
Cldn1 (claudin-1)	5′-TGTAATTTCAGGTCTGGCGACA-3′ 5′-GGATAAGGCCGTGGTGTTGG-3′	222	NM_031699
Cldn2 (claudin-2)	5′-TTGGTAGCTGGAGTCATCCTC-3′ 5′-CTCTTGGCTTTGGGCTGTTG-3′	134	NM_001106846
Cldn4 (claudin-4)	5′-CGAGCCCTGATGGTCATCAG-3′ 5′-CGGAGTACTTGGCGGAGTAG-3′	358	NM_001012022
Cldn5 (claudin-5)	5′-CCGGTGTCTCAGAAGTACGA -3′ 5′-GGTGCTGAGTACTTGACTGG-3′	152	NM_031701
Cldn6 (claudin-6)	5′-CACTACCTGTGTGGAAGACAAG -3′ 5′-GTTGTAGAAGTCCTGGATGATGG-3′	136	NM_001102364
Cldn7 (claudin-7)	5′-TGGCAGGTCTTGCTGCTTTG-3′ 5′-TGCCCAGCCGATAAAGATGG-3′	128	NM_031702
Cldn8 (claudin-8)	5′-GGTGAAGCCCTCTACATAGG-3′ 5′-CGTGGAAACTCCTCTGAGTG-3′	148	NM_001037774
Cldn9 (claudin-9)	5′-CTGGTTGCTGAAGCTCTCAAG-3′ 5′-ACGGGAAGGGATGGAGTAG-3′	165	NM_001011889
Cldn10 (claudin-10)	5′-GTCAGGTCTGTGTTCCATG-3′ 5′-TGACACCGCCAATGATGC-3′	155	NM_001106058
Cldn11 (claudin-11)	5′-TTCTCCCCTGTATCCGAATG-3′ 5′-AAGCTCACGATGGTGATCTC-3′	157	NM_053457
Cldn12 (claudin-12)	5′-AGTGACTGCCTGATGTACGA-3′ 5′-TGATGTTAGGCACCGAAGAC-3′	169	NM_001100813
Cldn14 (claudin-14)	5′-GAAGTGCACTCGCTGTGCCA-3′ 5′-TCATGCCACTGGGCAGCAGA-3′	164	NM_001013429
Cldn15 (claudin-15)	5′-TGGCCCTCTCTGGCTATGTC-3′ 5′-CTGCAATGGCCAGCAGCTTG-3′	156	NM_001107135
Cldn16 (claudin-16)	5′-GACTCTCTGGAGGTGAGCAC-3′ 5′-TCATCAGTGCCCGAGTCACC-3′	145	NM_131905

(continued)

Table 1 (continued)

Target gene	Primer sequence	Product (bp)	Accession No.
Cldn17 (claudin-17)	5′-GCAGAAGCGGGAACTTGGAG-3′ 5′-TCCTCTGGCTGTCTTTTGGTG-3′	167	NM_001107112
Cldn18 (claudin-18)	5′-TCCAGCCATGCTGCAAGCAG-3′ 5′-AGTCATCTTGGCCTTGGCAG-3′	139	NM_001014096
Cldn19 (claudin-19)	5′-CCCTGGACGGTCATATCCAG-3′ 5′-AGAGAGCACCTCCGGAGATG-3′	162	NM_001008514
Cldn20 (claudin-20)	5′-GGAGGAGCCGTGTACATTGG-3′ 5′-CTTTGGCTGCCAGGTAGAGG-3′	144	NM_001109394
Cldn22 (claudin-22)	5′-CTCAGGCTCACAGAACGTTC-3′ 5′-GAGGCAGGATAAAACCCATC-3′	100	NM_001110143
Cldn23 (claudin-23)	5′-GTTGCTGCTCAATCTCGTCA-3′ 5′-TGAGTCCCGAAGTAGTTCCA-3′	174	NM_001033062
Gapdh	5′-TAAAGGGCATCCTGGGCTACACT-3′ 5′-TTACTCCTTGGAGGCCATGTAGG-3′	200	NM_017008

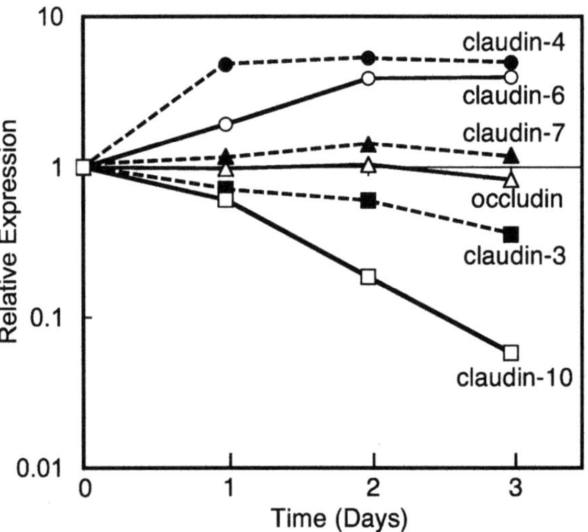

Fig. 2. Real-time RT–PCR analysis of changes in expression levels of claudin mRNA. After normalized to the expression level of GAPDH mRNA, all values are shown as values relative to cells just after isolation (time 0). Although the expression level of occludin mNRA was hardly changed for the 3 days in culture, claudins-4 and -6 increased and claudins-3 and -10 decreased.

5. The membranes are rinsed three times and washed three times (once for 10 min and twice for 5 min) with TBS-T.
6. The membranes are blotted with alkaline phosphatase-conjugated secondary antibodies (1:10,000) in TBS-T for 1 h.
7. After final rinsing and washing, immunoreactivity is determined using an ECF Western blotting kit and images are acquired using Typhoon Trio (excitation by 488-nm laser and detection with 520BP40 filter). Intensities of immunoreactivities are quantified using ImageQuantTL software.

3.5. Immunofluorescence Microscopy for the Distribution of Claudins in the Culture

1. Cells are cultured in collagen I-coated glass-base dishes or in collagen I-coated cell culture inserts (see Note 10).
2. After 3-day culture, the dishes are washed with Waymouth's medium (37°C), fixed with 10% formalin in PBS for 10 min, and permeabilized with 0.2% Triton X-100 in PBS for 10 min.
3. The cells are then washed with 4 ml PBS for 10 min three times.
4. Cells outside of the glass are removed using Kimwipes S-200 (Nippon Paper Crecia, Tokyo, Japan), and only the glass part of the dish is blocked and labeled with antibodies. The volume of each solution used for blocking and for labeling is 150 μl.
5. After blocking with 1% BSA and 0.05% goat IgG for 1 h, the cells are labeled using anti-amylase (1:250) (see Note 11), anti-occludin (10 μg/ml), anti-ZO-1 (10 μg/ml), anti-claudin-3 (5 μg/ml), or anti-claudin-4 (10 μg/ml) antibodies for 1 h in a humidified chamber.
6. After three washes with 4 ml PBS, cells are labeled using the appropriate Alexa Fluor-labeled secondary antibodies (1:50) and 1 μM TO-PRO-3 iodide for 1 h.
7. After three further washes, dishes are mounted using Antifade (200 μl) and covered with a cover slip.
8. Fluorescence images are acquired by using LSM-510 (Carl Zeiss) confocal microscopy.

3.6. Analysis of the Signal Pathway Regulating Claudin Expression and Dedifferentiation

To determine the signal pathway that induces changes in claudin expression, the effects of various kinase inhibitors on claudin expression are examined (see Note 12).

3.6.1. Effect of Inhibitors on Claudin Expression

1. After cell isolation from the glands, inhibitors are added to the medium at the beginning of culture (see Note 13). When the culture medium is changed after 24 h, the inhibitors are added to the new medium (see Note 14).
2. Cells are harvested after 3 days of culture using homogenizing buffer.

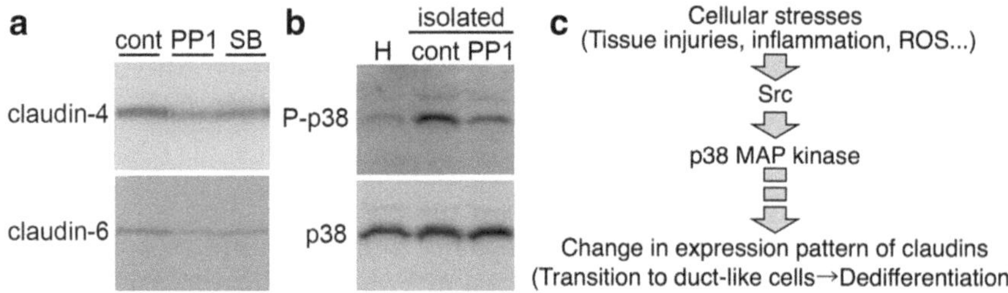

Fig. 3. Signal pathway that induces the change in expression pattern of claudins in parotid acinar cells. (**a**) Expression of claudins-4 and -6 at 3 days after isolation. In the absence of inhibitor (cont), both claudins are expressed, but their expression levels are suppressed by the addition of 10 μM PP1 (PP1) or 20 μM SB203510 (SB). (**b**) Activation of p38 MAPK induced during cell isolation. In the homogenate of intact glands (H), p38 MAPK is not activated (P-p38: phosphorylated p38 MAPK). After isolation (isolated), activation is observed in cells isolated in the absence of inhibitors (cont), but it is suppressed in the presence of 10 μM PP1 (PP1). This result indicates that the dispersion of cells by enzymatic digestion induced the activation of p38 MAPK, which is mediated by Src kinase activity. (**c**) A scheme of the signaling pathway that induces the change in claudin expression. Cellular stresses, such as tissue injuries, inflammation, or reactive oxygen species, induce the activation of p38 MAPK via Src kinase activity. The activation of p38 MAPK causes changes in the expression patterns of claudins, resulting in the transition to duct-like cells and dedifferentiation.

3. The expression levels of claudins are determined by immunoblot analysis. After the protein concentration of the extract is determined, the same amounts of proteins are applied to each lane of the SDS–PAGE. The immunoblotting protocol is the same as described in Subheading 3.4. An example of the immunoblot analysis on the effects of inhibitors is shown in Fig. 3a. From these results, Src and p38 MAPK are considered to be involved in the signaling.

3.6.2. Effect of Inhibitors on Activation of p38 MAPK During the Culture

When p38 MAPK is activated, it is phosphorylated at Thr180 and Tyr182. To examine the signal pathway that activates p38 MAPK, the amounts of the phosphorylated and total p38 MAPK are determined.

1. After cell isolation, inhibitors are added to the medium at the beginning of culture. When the culture medium is changed after 24 h, the inhibitors are added to the new medium.
2. Cells are harvested just after and at 2–72 h after isolation using the homogenizing buffer.
3. The immunoblotting protocol is the same as described in Subheading 3.4, except for the blocking solution used for the anti-phospho-p38 MAPK antibody. For the purpose, 3% BSA is used. Antibodies to the phosphorylated and total p38 MAPK are used at 0.5 and 0.1 μg/ml, respectively, and the signal intensities are measured using Typhoon Trio. Levels of activation are calculated as relative values to the total amounts of p38 MAPK in the same samples.

3.6.3. Effect of Inhibitors on the Activation of p38 MAPK During Cell Isolation

Since analysis of the activation of p38 MAPK revealed that it had already been activated just after isolation, the effects of Src inhibitors during isolation of the cells were examined.

1. To examine the effects of Src inhibitors during cell isolation, all solutions used for cell isolation (HBSS–0.5% BSA, HBSS–4% BSA, and Waymouth's medium) contained the various inhibitors.
2. Cells are harvested just after isolation using the homogenizing buffer.
3. The immunoblotting protocol is the same as described in Subheading 3.6.2. An example of the results is shown in Fig. 3b, which clearly shows that the Src kinase inhibitor PP1 suppresses the activation of p38 MAPK (see Note 15). The expected signal pathway is shown in Fig. 3c.

4. Notes

1. Because the supply of oxygen is important for the survival of isolated acinar cells, the HBSS is bubbled with oxygen for 15 min before the addition of BSA. After its addition, oxygen is blown across the surface of the medium. The glands in the medium are also maintained under oxygen during the cell dispersion.
2. Rat serum should be prepared from young adult rats. If the serum is prepared from older rats, blood lipids may disturb the attachment of the cells to the dishes. To avoid decreases in activity, serum is stored at –20°C and is add to the medium just before use.
3. Although phosphatase inhibitors are not necessary for immunoblot analysis of claudins, the homogenizing buffer-containing phosphatase inhibitors is always used to examine the phosphorylation of p38 MAPK in the same samples in Subheadings 3.6.2 and 3.6.3.
4. Trypsin is not used for cell dispersion, because in the presence of trypsin, almost all cells are dispersed and do not attach to the dishes.
5. For filtration of the dispersed cells, Cell strainer (BD Falcon) with 70- or 100-μm pores were also tested: although they can be used, cells isolated with those Cell strainers lost their amylase activity, which is an index of differentiation, more rapidly than those filtered through the nylon mesh.
6. Dispersed cells are usually contaminated with erythrocytes and fibroblasts after digestion with enzymes. To remove contaminating cells, the cells are simply centrifuged at 700 rpm

(90 × g) for 1 min at lease three times, after which almost all cells are acinar cells.

7. Since the cells tend to aggregate, counting the cell number is very difficult and the results are variable: cell concentrations are normalized by the concentration of proteins.
8. Collagen I-coated dishes are usually used. In many cases, extracellular matrix (ECM) influences the differentiation of epithelial tissues. Since laminin has been reported to induce the differentiation of human submandibular gland cells (12), the type of ECM used was expected to affect the cell function. Several kinds of ECM-coated dishes were used (collagen IV-, laminin-, fibronectin-coated and non-coated dishes), however, no difference was detected in differentiation degree or expression of claudins by the cells cultured in collagen I-coated dishes although all types of dishes can be used for culture.
9. It is difficult to quantify mRNA expression levels in the culture, because the expression patterns of many genes are markedly changed during culture. mRNA levels of occludin and claudins are usually normalized relative to the value of GAPDH expression. If another gene is used as a reference, the values will be changed. On the other hand, in immunoblotting analysis, the amounts of occludin and claudin are normalized to the amount of total proteins. As a result, there can be differences between the results of real-time RT–PCR and immunoblot analyses. It is difficult to say which method to normalize the values is the most reasonable. The mRNA expression levels of GAPDH and occludin are similar when the same amounts of total RNA are used. Moreover, the amount of occludin detected by the immunoblot analysis appears to be unchanged during the culture. Therefore, GAPDH is considered a suitable reference for mRNA expression.
10. When collagen I-coated culture inserts are used, the insert filters should be cut with scissors and put into the glass-based dishes for the labeling procedure. After labeling, the pieces are put on slide glasses with the cell sides up and are then covered with cover slips.
11. Because genes that are not expressed in intact acinar cells begin to be expressed, there was a possibility that cells other than acinar cells contaminate the cultures. To exclude that possibility and to confirm that changes in gene expression really occur in acinar-derived cells, the cultured cells were co-stained with an anti-amylase antibody, which detects the amylase localized in secretory granules. The presence of secretory granules is the best marker of acinar cells.

12. For analysis of the signaling pathways, we selected inhibitors that do not affect cell viability. If cell viability is decreased, one cannot determine whether the inhibitor suppresses the change in specific gene expression or disturbs all functions of the cells.
13. When inhibitors of Src kinase are added to the medium, 20-mM stock solutions are made and are diluted to 1:2,000 (final concentration of 10 μM) with the culture medium. The same volume of DMSO is added to the control medium (final 0.05%). For inhibitors of p38 MAPK, 5-mM stock solutions are made because the solubility of PD169316 is low. p38 MAPK inhibitors are diluted to 1:250 with the culture medium (final concentration of 20 μM). The same volume of DMSO is added to the control medium (final 0.4%). We confirmed that 0.4% DMSO has no effect on the changes in gene expression examined in this study. When lower concentrations of inhibitors are used, the stock solutions are diluted with DMSO and the same volumes are added to the culture medium.
14. The timing and periods of treatment with inhibitors are important. Inhibitors are usually added to the medium at the beginning of culture, and the cells are cultured in the presence of inhibitors for 3 days and are harvested at 3 days after isolation. The effects of inhibitors on claudin expression and dedifferentiation were evaluated at 3 days, except when time-dependent changes were examined.
15. We found that p38 MAPK had already been activated just after cell isolation, although it was not activated in the homogenate of the intact glands. Its activation is not induced by culture. Because incubation with Src inhibitors for only the first 24 h can suppress the changes in claudin expression, we consider that the signal to induce the change is triggered during the cell isolation and is finished within 1 day although the pattern of gene expression continues to change for 3 days.

References

1. Tsukita, S., Furuse, M., and Itoh, M. (2001) Multifunctional strands in tight junctions, *Nat Rev Mol Cell Biol* 2, 285–293.
2. Turksen, K., and Troy, T. C. (2004) Barriers built on claudins, *J Cell Sci* *117*, 2435–2447.
3. Furuse, M., and Tsukita, S. (2006) Claudins in occluding junctions of humans and flies, *Trends Cell Biol* *16*, 181–188.
4. Heiskala, M., Peterson, P. A., and Yang, Y. (2001) The roles of claudin superfamily proteins in paracellular transport, *Traffic* 2, 93–98.
5. Turner, R. J. (1993) Ion transport related to fluid secretion in salivary glands., in *Biology of the salivary glands* (Dobrosielski-Vergona, K., Ed.), pp.105–125, CRC Press, Inc., Boca Raton, FL.
6. Peppi, M., and Ghabriel, M. N. (2004) Tissue-specific expression of the tight junction proteins claudins and occludin in the rat salivary glands, *J Anat* *205*, 257–266.
7. Hashizume, A., Ueno, T., Furuse, M., Tsukita, S., Nakanishi, Y., and Hieda, Y. (2004) Expression patterns of claudin family of tight

junction membrane proteins in developing mouse submandibular gland, *Dev. Dyn.* ***231***, 425–431.

8. Fujita-Yoshigaki, J., Tagashira, A., Yoshigaki, T., Furuyama, S., and Sugiya, H. (2005) A primary culture of parotid acinar cells retaining capacity for agonists-induced amylase secretion and generation of new secretory granules, *Cell Tissue Res* ***320***, 455–464.
9. Qi, B., Fujita-Yoshigaki, J., Michikawa, H., Satoh, K., Katsumata, O., and Sugiya, H. (2007) Differences in claudin synthesis in primary cultures of acinar cells from rat salivary gland are correlated with the specific three-dimensional organization of the cells, *Cell Tissue Res* ***329***, 59–70.
10. Fujita-Yoshigaki, J., Matsuki-Fukushima, M., and Sugiya, H. (2008) Inhibition of Src and p38 MAP kinases suppresses the change of claudin expression induced on dedifferentiation of primary cultured parotid acinar cells, *Am J Physiol Cell Physiol* ***294***, C774-785.
11. Michikawa, H., Fujita-Yoshigaki, J., and Sugiya, H. (2008) Enhancement of barrier function by overexpression of claudin-4 in tight junctions of submandibular gland cells, *Cell Tissue Res* ***334***, 255–264.
12. Hoffman, M. P., Kibbey, M. C., Letterio, J. J., and Kleinman, H. K. (1996) Role of laminin-1 and TGF-beta 3 in acinar differentiation of a human submandibular gland cell line (HSG), *J. Cell Sci.* ***109***, 2013–2021.

Chapter 18

Development of Biological Tools to Study Claudins in the Male Reproductive Tract

Daniel G. Cyr, Évemie Dubé, Julie Dufresne, and Mary Gregory

Abstract

It is estimated that between 12 and 15% of couples are infertile. More than half of these are related to problems associated with male reproductive dysfunction. Of those, 40% occur from idiopathic or unexplained causes. While spermatozoa are formed in the testis, testicular spermatozoa are immature and cannot swim or fertilize. These critical functions are acquired as spermatozoa transit through the epididymis in the specific luminal environment created in part by the tight junctions of the blood-epididymis barrier. To understand the normal and pathological conditions attributable to human and animal epididymal function, we have needed to develop biological tools to characterize the physiological, cellular, and molecular functions of tight junctions and claudins (Cldns) in the epididymis. We have shown that by developing epididymal cell lines we have gained valuable insight into the functions of epididymal Cldns, the regulation of the Cldn1 gene and how these can be mistargeted in infertile men. Here we describe some of the techniques that have been used to address these critical aspects of epididymal Cldns.

Key words: Male infertility, Blood-epididymis barrier, Principal cell line, Immunogold labeling, Electron microscopy, Promoter analyses, Luciferase constructs

1. Introduction

In general, male infertility is classified into three major categories: non-testicular dysfunction (endocrine, genetic, and drug-induced), testicular disorders (Sertoli cell only syndrome, arrested spermatogenesis, cryptorchidism, etc.), and post-testicular disorders (obstruction, infection, anti-sperm antibodies, and coital disorders). In patients with normal spermogram and diagnosed as idiopathic infertile, it is generally thought that post-testicular factors contribute to or are entirely responsible for the patient's infertility, as the problem appears associated with sperm maturation (1, 2). Thus post-testicular infertility may be the result of

Kursad Turksen (ed.), *Claudins: Methods and Protocols*, Methods in Molecular Biology, vol. 762,
DOI 10.1007/978-1-61779-185-7_18,

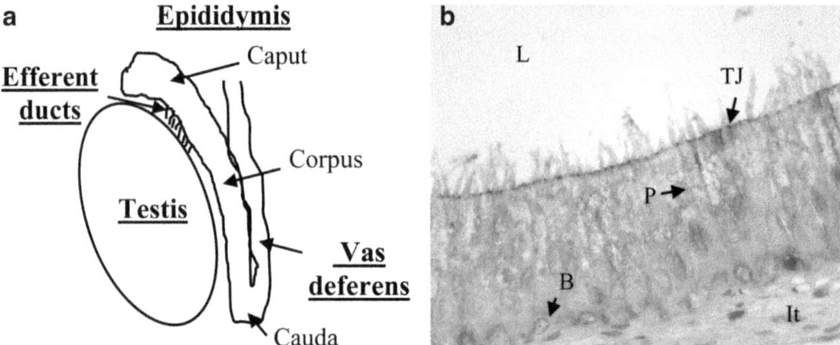

Fig. 1. Tights junctions in the human epididymis. (**a**) Schematic representation of the human epididymis. The epididymis, which is a long convoluted tubule connecting the efferent ducts to the vas deferens, is generally divided into three regions (caput, corpus, and cauda). (**b**) Epididymal tight junctions are apically located between adjacent principal cells as shown by CLDN1 immunostaining. *L* lumen, *It* interstitium, *P* principal cells, *B* basal cells, *TJ* tight junctions.

pathological dysfunction in the epididymis, which results in incomplete or dysfunctional sperm maturation.

The epididymis is the major component of the testicular excurrent duct system (Fig. 1). Testicular input to the tissue is conveyed via the efferent ducts, which anastomose to form a single, highly convoluted epididymal duct. Morphologically, the human epididymis can be subdivided into three distinct regions: the head (caput), the body (corpus), and the tail (cauda). The epididymis can be divided into two main compartments: the epithelium and the lumen. The luminal environment is maintained by active secretion and absorption by epithelial principal cells and the blood-epididymis barrier.

Our studies were the first to characterize the proteins that comprise epididymal tight junctions. We demonstrated that occludin is present in epididymal tight junctions during embryonic development following the development of the epididymis, but that tight junctions were heterogeneous and that occludin was not expressed in certain regions of the tissue (3). This led to studies which showed that Cldn1 was localized to epididymal tight junctions and that it was regulated by androgens, but only in the initial segment, indicating that Cldns may be regulated differently in different regions of the epididymis (4, 5).

A limitation to understanding the regulation of Cldns was the lack of an epididymal cell line. We have developed the first immortalized rat and human epididymal principal cell lines (6, 7). These cell lines express transcripts for hormone receptors, epididymal-specific genes, several connexins (gap junction), and Cldns (6, 7). These cell lines were used to demonstrate that SP1/SP3 transcription factors were crucial for Cldn1 expression (8).

To determine if epididymal function was altered in infertile patients, gene expression profiling was done on caput epididymides from nonobstructive azoospermic patients (6). The data revealed that the epididymis was altered in infertile patients and that mRNA levels of over 400 genes were altered in infertile men (9). Differentially regulated genes included several encoding proteins involved in sperm maturation, water, and ion channels. Furthermore, the intracellular targeting of Cldn10 and ZO-1 was altered, while Cldn1 cytoplasmic staining was increased. These data demonstrated that epididymal function and epididymal Cldns were altered in nonobstructive azoospermic infertile patients.

Using human epididymal cells lines and siRNA, we demonstrated that downregulation of a single Cldn could alter the formation of tight junctions, determined by measuring transepithelial resistance (TER) across the cells in culture (7). This indicated that a loss or mistargeting of a single Cldn, as observed in nonobstructive infertile patients (7), is sufficient to compromise the blood-epididymis barrier.

In this chapter, we present four methods to understand the regulation of Cldns in human epithelial cells. These are: (1) the development of epididymal cell lines, a technique applicable to other cell types; (2) immunogold labeling for electron microscopy, to identify the cellular localization of Cldns; (3) the use of siRNA to determine the role of specific Cldns, and (4) gene reporter assays to identify critical regions for the regulation of the Cldn genes.

2. Materials

2.1. Isolation of Rat or Human Epididymal Epithelial Cells

1. Male rats (40 days old) can be purchased from Charles River Laboratories (St. Constant, QC).
2. Surgical instruments (scissors, scalpel, and forceps), 70% ethanol, petri dishes.
3. Dulbecco's Modified Eagle's Medium (DMEM; Gibco, Burlington, ON) supplemented with 50 U/ml penicillin (Sigma), 50 μg/ml streptomycin (Sigma-Aldrich, Oakville, ON), 2 mM L-glutamine (Sigma-Aldrich), 5 nM testosterone, 10 μg/ml insulin (Sigma-Aldrich), 10 μg/ml transferrin (Sigma-Aldrich), 80 ng/ml hydrocortisone (Sigma-Aldrich), 10 ng/ml epidermal growth factor (Sigma-Aldrich), 10 ng/ml cAMP (Sigma-Aldrich), 2 ng/ml sodium selenium (Sigma-Aldrich), 200 ng/ml tocopherol (Sigma-Aldrich), 200 ng/ml retinol (Sigma-Aldrich), and fetal bovine serum (FBS; Sigma-Aldrich).
4. Collagenase type I (stock 200 mg/ml, Gibco) and RQ1 RNase-free DNase (stock 1U/μl, Promega, San Luis Obispo, CA)
5. 6- or 24-Well culture plates coated with mouse collagen IV (BD Biosciences).

2.2. Immortalization of Epididymal Epithelial Cells

1. A plasmid containing the SV40 LTAg and neomycin resistance genes (6).
2. Fugene 6 (Roche Diagnostics, Laval, QC).
3. DMEM (Gibco) with nutrients (Sigma-Aldrich) as described in Subheading 2.1.
4. 96-Well culture plates coated with mouse collagen IV (BD Biosciences, Mississauga, ON) or sterile cloning cylinders (Sigma-Aldrich).
5. Primers can be designed using the Oligo 6 (Molecular Biology Insights, Cascade, CO) software. A BLAST search is performed on NCBI PrimerBlast to ensure the specificity of the primers. Primers can be synthesized commercially by different companies.
6. Illustra RNAspin Mini Isolation kit (GE Healthcare, Baie d'Urfe, QC). All reagents are provided by the manufacturer except for the ethanol and β-mercaptoethanol (Fisher Scientific, Nepean, ON). These chemicals must be manipulated under a chemical fume hood.
7. Denature 500 ng (2 μl) of RNA by heating at 65°C for 10 min. Place the RNA on ice for 5 min. Add 18 μl of RT (reverse transcriptase) mix to the RNA [1× PCR buffer containing 1.5 mM $MgCl_2$ (GE Healthcare), 3.5 mM $MgCl_2$ (Sigma-Aldrich), 1 mM dNTPs (Invitrogen, Burlington, ON), 2.5 μM Oligo dT (R&D systems, Minneapolis, MN), 20 U RNAguard (Sigma-Aldrich), 100 U MMLV-RT (Sigma-Aldrich) diluted in autoclaved and 0.22 μm filtered water]. Incubate for 10 min at room temperature followed by 1 h at 42°C.
8. For PCR amplification, add 4.5 μl of cDNA to 20.5 μl of mix [1× PCR buffer containing 1.5 mM $MgCl_2$ (GE Healthcare), 0.5 mM $MgCl_2$ (Sigma-Aldrich), 0.2 mM dNTPs (Invitrogen), 0.5 μM of each of the primers, 1.25 U TAQ polymerase (GE Healthcare) diluted in autoclaved, and 0.22 μm filtered water].
9. Agarose gel: 1.5% (w/v) of ultrapure agarose (Invitrogen) diluted in 1× TBE (Tris-Borate EDTA buffer). To dissolve the agarose, heat the solution for several minutes in a microwave. Stop the microwave frequently and swirl the flask because the agarose solution can easily boil over. Let the molten agarose cool down to 60°C (5 min at room temperature). Add ethidium bromide (Fisher) to a final concentration of 0.5 μg/ml. Ethidium Bromide is mutagenic and should be handled with extreme caution. A stock solution can be prepared and stored at 4°C in the dark. Pour the gel into the gel casting tray, eliminate air bubbles, and insert the electrophoresis comb. Allow the gel to polymerize for 30 min. Use 1× TBE as the running buffer for electrophoresis.

10. 10× TBE: 108 g Tris Base (Fisher), 55 g Boric Acid (Fisher), 40 ml of 0.5 M EDTA (Fisher) pH 8.0. Complete with water to 1 L.
11. 6× Loading buffer: 30% glycerol, 0.25% bromophenol blue diluted in water. Bromophenol blue migrates at a rate equivalent to ~300 bp DNA.
12. DNA ladder (Invitrogen).

2.3. Immunogold Labeling

2.3.1. Tissue Preparation and Fixation of Tissues for Frozen Sections for Immunogold Labeling

1. 1% Glutararaldehyde in 0.1 M PBS, pH 7.4.
2. PBS, 0.15 M and 0.1 M, pH 7.4.
3. Sucrose solution, 1.5 M sucrose in PBS.
4. LR-White resin (Canemco, Inc., Gore, QC) for embedding.
5. Liquid nitrogen.

All chemicals are available from Sigma-Aldrich unless otherwise indicated. Ideally, tissues should be fixed with freshly prepared solution. Note that tissues may remain in fixation solution for several days. Place tissues in PBS (1×) the day prior to processing.

2.3.2. Materials for Immunogold Staining

1. Parafilm (Fisher) or White silicone grid mats (Canemco).
2. Copper or nickel grids (Canemco), 200-mesh, coated with 0.25% formvar in 1,2-dichloroethylene (Canemco).
3. BCO-Bovine serum albumin, 2% Casein, 0.5% Ovalbumin, 0.1 M, in PBS, pH 7.4. (Sigma-Aldrich).
4. 0.15% Glycine in PBS.
5. PBS – Phosphate-buffered saline, 0.1 and 0.15 M.
6. DPBS – Dulbecco's phosphate-buffered saline containing calcium chloride and magnesium chloride.
7. Uranyl oxalic acid (EMS, Hatfield, PA) – uranyl acetate (4%) plus oxalic acid (0.3M) in 1:1 v/v ratio, pH 7–7.5 with NH_4OH.
8. Methyl cellulose – 2% wt/vol in water plus 0.4% uranyl acetate (Anachemia, Lachine, QC).
9. Primary antibody (e.g., Claudin-1, Zymed/Invitrogen) diluted in BCO; usually a 1:5–1:10 dilution is appropriate.
10. Secondary antibody, conjugated with gold, diluted 1:20 in BCO.

All reagents are available from Sigma-Aldrich, unless otherwise indicated.

2.4. Study of Claudins Using RNA Interference

1. DMEM (Gibco) supplemented with antibiotics and nutrients (Subheading 2.1).
2. 0.05% Trypsin/EDTA (Gibco).

3. Hiperfect reagent (Qiagen, Mississauga, ON).
4. All siRNA are chemically synthesized commercially (e.g., Qiagen).

2.5. Study of Claudin Promoter Activity

2.5.1. Rapid Amplification of cDNAs Ends (5′ RACE)

1. Tissue samples or cells expressing the claudin gene of interest.
2. Illustra RNAspin Mini kit (GE Healthcare). (See Subheading 2.2, step 6).
3. SMARTer™ RACE cDNA Amplification kit (Clontech, Mountain View, CA).
4. Gene-specific primers (see Subheading 2.2, step 5).

2.5.2. Plasmid Constructs

1. Gel purification kit (QIAEX II, Qiagen).
2. TA cloning kit (TOPO® TA Cloning kit, Invitrogen).
3. Small scale plasmid extraction kit (llustra plasmidPrep Mini Spin kit, GE Healthcare).
4. Restriction enzymes (New England Biolabs, Pickering, ON).
5. PCR reagents and promoter-specific primers to amplify genomic DNA (see Subheading 2.2).
6. pGL3-Basic reporter vector (Promega).
7. Absolute ethanol and sodium acetate (3 M, pH 5.2) for precipitation (12).
8. T4 DNA ligase (Invitrogen).
9. Competent bacteria: strain JM109 or other *recA* host to avoid recombination.
10. Sequencing primers for pGL3-Basic.

2.5.3. Luciferase Assays

1. Control vectors: pGL3-Basic empty vector, pGL3-Control vector, and phRL-TK (Promega).
2. Medium scale plasmid extraction kit (Plasmid Midi kit, Qiagen).
3. Cell line: RCE-1 cells (6) or adherent cell line expressing the claudin endogenously.
4. Lipofectamine 2000 (Invitrogen).
5. Dual-Luciferase Assay kit (Promega).

3. Methods

The use of cell lines to study junctional proteins, and in particular the function and regulation of Cldns, remains an essential tool. In the case of the human epididymis, the lack of available cell lines and difficulty in obtaining tissue samples (there are no epididymal cancers) has required the development of cell lines to address Cldn function and regulation. While this protocol is specific for epididymal cells, it can be modified and used for the development of other epithelial cell lines.

3.1. Isolation of Rat or Human Epididymal Epithelial Cells

1. Epididymal tubules are obtained from rats or humans by surgery.
   (a) For rat tissue, place a euthanized male rat on a sterile work area and wipe the lower abdomen with 70% ethanol. An incision should be made in the lower abdomen near the scrotum to quickly extract the testis and the excurrent ducts. Once the tissue is excised, immerse the tissue in DMEM containing 2 mM glutamine, 50 U/ml penicillin, and 50 μg/ml streptomycin. Transfer the tissue to a sterile laminar flow hood. Place the tissues in a Petri dish. Separate the epididymis from the testis and eliminate all the fat along the length of the epididymis using a pair of fine scissors or a scalpel. The epididymis can be divided into the caput, corpus, and cauda.
   (b) Human epididymal tissues can be obtained following an orchidectomy. Immerse the tissues in DMEM-containing antibiotics and glutamine and transport them quickly from surgery to the laboratory.
2. Mince tubules into small tissue fragments (2–3 mm^3) in a Petri dish under sterile conditions.
3. Transfer the tubules to a flask and add 10 ml DMEM culture medium containing 2 mM glutamine, 50 U/ml penicillin, 50 μg/ml streptomycin, 2 mg/ml collagenase, and 20 U/ml DNase. Incubate in a shaking water bath for 50 min at 37°C at low speed.
4. Dissociate the tissue between digestions by gently pipetting the tissue up and down. Allow the tissue fragments to sediment to the bottom of the flask and replace the supernatant with fresh enzyme solution. Repeat step 3.
5. Transfer the cell suspension into a 15-ml tube and centrifuge at 34 × *g* for 3 min. Discard the supernatant and resuspend the pellet in DMEM culture medium supplemented with antibiotics and nutrients. At this step, only 1% FBS should be added to the culture medium to limit the growth of non-epithelial cells.
6. Place small tissue fragments in 6- or 24-well plates coated with mouse collagen IV and maintain the cultures in a humidified chamber at 32°C with 5% CO_2.
7. Once the cells are well adhered to the bottom of the wells (approximately 2–4 days), change the culture medium every 48 h for 12–18 days until 70–80% confluence is reached (Fig. 2a). The cells can subsequently be used for immortalization.

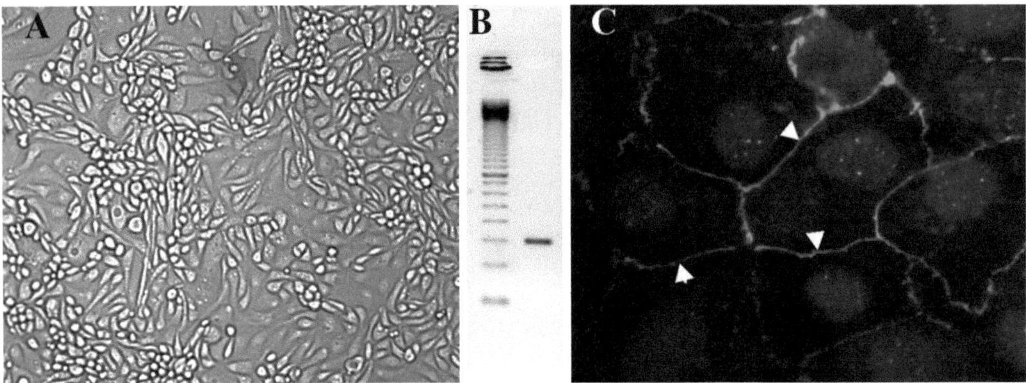

Fig. 2. Development of human epididymal cell lines. (**a**) Representative photomicrograph of human epididymal primary principal cells grown at confluence on collagen IV-coated plates. (**b**) The expression of SV40 LTAg was verified by RT–PCR in a human epididymal cell line immortalized with a plasmid encoding for the LTAg. *Left lane*, 100 bp DNA ladder; *right lane*, human epididymal cell line. (**c**) Immunolocalization of ZO-1 (also called TJP1) in a human epididymal cell line. ZO-1 (*arrowheads*) is localized to the plasma membrane. Magnification ×640.

3.2. Immortalization of Epididymal Epithelial Cells

1. Transfect the cells with a plasmid containing the SV40 LTAg and neomycin resistance genes using Fugene 6 according to the manufacturer's instructions.
2. Select stable transfectants by incubating the cells in DMEM medium (supplemented with antibiotics, nutrients, and 1% FBS) containing 200 μg/ml neomycin for 14 days.
3. After selection, maintain the cells in DMEM medium containing antibiotics, nutrients, and 5–10% FBS. The culture medium should be changed every 48 h.
4. Generate cell lines by isolating single cells in 96-well culture plates coated with collagen IV using serial dilution. Cloning cylinders can also be used.
5. The homogeneity of the cell lines obtained can be verified by FACS based on the complexity and size of the cells.
6. The presence of the SV40 LTAg can be verified by RT–PCR (Fig. 2b).
   (a) Isolate total cellular RNA using the Illustra RNAspin Mini Isolation kit according to the manufacturer's instructions.
   (b) Reverse transcribe 500 ng of total RNA using an oligo $(dT)_{16}$ primer.
   (c) Amplify the DNA sequence (forward primer, 5′-AATAG CAAAGCAAGC AAGAGT, reverse primer, 5′-AAAAT GGAAGATGGAGTAAA) using the following program: 94°C for 5 min, 35 cycles of 94°C for 30 s, 51.2°C for 30 s, 72°C for 1 min, and cool to 4°C. It is important to include a PCR amplification done with RNA only

(no RT enzyme) to confirm that the sample is free of contaminating genomic DNA (see Note 4).

(d) To visualize the PCR product, dilute the amplified DNA in 1× loading buffer and separated by electrophoresis on a 1.5% agarose gel containing ethidium bromide. The bands and DNA ladder (to assess the size of the amplicon) can then be visualized using a UV light or with a densitometer equipped with a UV light.

7. The expression of several tissue-specific markers should be checked to ensure that the cells are of the correct origin. For the epididymis, several antibodies are commercially available to verify the expression of cytokeratin, an epithelial marker, and vimentin, a mesenchymal marker, by immunofluorescence. In addition, several epididymal markers have been identified over the last decade and their expression can be verified by RT–PCR. It is important to use epididymal markers that are specifically expressed by epididymal epithelial cells.

8. The expression of junctional markers, such as Cldns, can also be verified at the mRNA and protein levels using RT–PCR, western blot, immunofluorescence or immunogold labeling, and electron microscopy (Fig. 2c).

3.3. Immunogold Labeling

To assess the ultrastructural localization of Cldns in the epididymis, immunogold labeling using weakly fixed tissues and cryosections for electron microscopy has provided a reliable technique in which the primary antibody can be recognized (Fig. 3, Cldn-1).

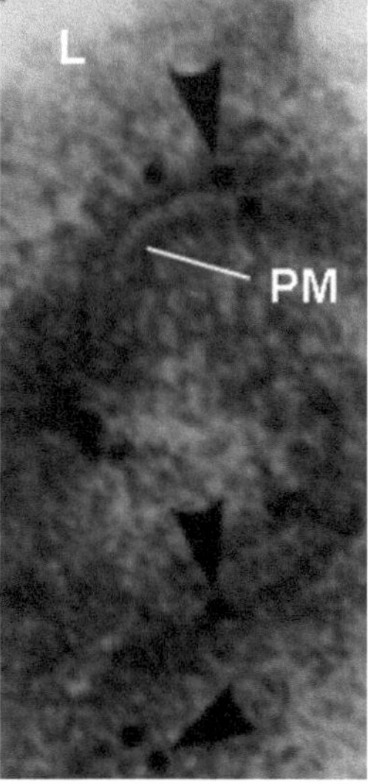

Fig. 3. Immunogold labeling of CLDN1 between adjacent principal cells in the apical area of the blood-epididymis barrier. Gold particles (*arrowheads*) indicate the position of Cldn1 in the tight junction. *L* lumen, *PM* plasma membrane.

3.3.1. Tissue Preparation

1. Tissues are fixed in 0.5% glutaraldehyde and 4% paraformaldehyde in 0.1 M phosphate buffer (pH 7.4). After fixation, the tissue is trimmed into small pieces (0.5 mm^3), immersed for 2 h in fixative at 4°C, washed two or three times in 0.15 M PBS (pH 7.4), and then treated with PBS containing 1.5 M sucrose. The tissue is then frozen in liquid nitrogen until sectioning. Ultrathin sections are cut and placed on either copper or nickel grids (200-mesh) coated with 0.25% Formvar in 1,2-- dichloroethylene. The degree of fixation of the tissue will affect the antigenicity of the primary antiserum. A higher percentage of glutaraldehyde will provide a better fixation but the antigenicity of the antiserum may be lost or significantly reduced.

3.3.2. Immunogold Labeling

1. Spread clean Parafilm into the bottom of a Petri dish (or other very smooth surface). Place 5–10 μl drops of each solution listed below in a line on Parafilm sheet. Make sure that the drops are well formed and sufficiently distinct and apart from each other. White silicone grid mats can also be used instead of Parafilm.
2. Gently grasp the edge of each grid with fine forceps and dip the grid into the first drop of solution. Incubate for the appropriate period of time and continue dipping each grid, one at a time, into sequential drops.
3. Carefully blot excess fluid from each grid onto filter paper (in between washes) by gently touching the edge of the grid on the filter paper.
4. Place each grid onto a drop of BCO (2% bovine serum albumin, 2% casein, and 0.5% ovalbumin) in PBS, tissue face down and allow the grid to float on top of the drop for 10 min.
5. Rinse each grid on a drop of glycine, 0.15% in PBS for 10 min, to quench any residual reactive aldehyde groups resulting from the fixation with glutaraldehyde.
6. Place the grids onto 10–15 μl drops of primary antibody, such as claudin-1, diluted 1:5 or 1:10 in BCO for 1 h. Cover the Petri dish to minimize evaporation and dust.
7. Wash the sections on six drops of DPBS (Dulbecco's Phosphate-Buffered Saline containing calcium chloride and magnesium chloride); 6 × 5 min washes.
8. Rinse each section on a drop of BCO for 10 min.
9. Incubate the sections on 20 μl drops of gold-labeled secondary antibody, (e.g., anti-rabbit IgG conjugated to 10 nm gold), diluted in BCO, usually 1:20 dilution, for 30 min. Secondary antibodies raised against IgGs of different species can be used with gold particles of different sizes to localize multiple antigens on the same grid.
10. Wash the sections on six drops of DPBS; 6 × 5 min washes.

11. Sections are then rinsed on drops of distilled water, 6×5 min washes. Gently blot excess water from each grid dry on filter paper.
12. The tissue is counterstained by placing sections on a drop of uranyl oxalic acid (for negative staining of sections to provide contrast): use 1:1 ratio of 4% uranyl acetate plus 0.3 M oxalic acid, pH 7–7.5 with NH_4OH. Incubate for *exactly* 5 min.
13. To rinse the grids, place them on top of distilled water drops for 90 s. Repeat once.
14. The grids are then placed on a drop of methyl cellulose (2% wt/vol in water plus 0.4% uranyl acetate), *on ice*, to protect the sections.

All steps are conducted at room temperature unless otherwise indicated.

3.4. Claudin Functions Using RNA Interference

Small-interfering RNAs (siRNA; 10) are double-stranded RNA molecules (20–25 nucleotides) that are involved in the RNA interference (RNAi) pathway. siRNA are designed to target a specific mRNA sequence by hybridizing to this sequence and activating its degradation by the RNA-inducing silencing complex (RISC). Once the target mRNA is degraded, the transcript cannot be translated into its corresponding protein. In 2001, synthetic siRNA were shown to induce RNA interference in mammalian cells (11). Using claudin-specific siRNA, it is possible to temporarily knockdown the expression of an endogenous claudin and study its role in either barrier function or paracellular transport.

1. Prior to transfection, cells are treated with trypsin/EDTA, centrifuged 7 min at 1,000×*g* and resuspended in DMEM supplemented with antibiotics and nutrients. Cells (1.4×10^5) are plated in a 24-well plate coated with collagen IV in a final volume of 0.5 ml culture medium.
2. Using the HiPerfect Reagent, cells are transfected with the siRNA of interest (final concentration 5 nM) according to the manufacturer's instructions (see Note 1).
3. At different time points (24, 48, 72, and 96 h), the efficiency and specificity of each siRNA of interest (see Note 2) can be assessed by real-time RT–PCR or western blots (see Note 3).

3.5. Claudin Promoter Activity

Specific Cldn mRNA and genomic sequence information can be found in the GenBank database of the National Center for Biotechnology Information (NCBI). Many free web tools such as Proscan and Transfac software can help identify conserved sequence motifs and transcription factor binding sites present in gene promoters. Such tools are helpful in the analysis of specific promoter sequences and help guide the experiments described in this section.

3.5.1. Determination of the Transcriptional Start Site Position

1. One of the ways that is used to identify the transcriptional start site is by Rapid Amplification of cDNA Ends (RACE). Total RNA is extracted from the tissue or cells using a commercial kit such as the Illustra RNAspin Mini kit. The RNA must be of excellent quality and free of contaminating DNA (see Note 4).
2. To perform a RACE reaction: amplify the 5 portion of the mRNA with the SMARTer™ RACE cDNA Amplification kit. Use at least two different gene-specific primers, in two separate reactions, whose positions ensure a difference in the size of the two products that is easily visible on an agarose gel (Subheading 2.2, step 9). Alternatively, set up a nested RACE reaction with the internal primer.
3. Purify the 5′ RACE products by gel electrophoresis, extract the amplified DNA, and clone them. A TA cloning kit can be use at this point, provided that a final step of 7–12 min at 72°C is included following the RACE reaction. After restriction enzyme digest analyses, the selected clones can be sequenced. Many commercial companies offer rapid and inexpensive sequencing reactions. It is important to obtain the sequences from a sufficient number of clones to confirm the position of the transcriptional start site of the mRNA.

3.5.2. Plasmid Construction for Gene Reporter Assays

Preparation of the Promoter Fragments

1. Produce sequential deletion constructs of the promoter by PCR amplification using one reverse primer, located downstream of the transcription start site, and different forward primers located at different distances from the 5′ promoter region. Insert restriction sites in the primers to ensure directional cloning into the pGL3-Basic reporter vector (see Note 5). For example, a *Nhe*I site can be inserted in the reverse primer and a *Kpn*I site in the forward primers (8). Use genomic DNA as template and a reaction volume that would be sufficient for subsequent purifications (between 100 and 200 μl).
2. Precipitate the PCR products with absolute ethanol and sodium acetate (12) and purify them by gel electrophoresis to remove PCR reagents.
3. Double-digest the concentrated PCR products using the appropriate restriction enzymes. Purify the promoter fragments by gel electrophoresis and estimate the amount of resulting DNA by measuring the absorbance at 260 nm.

Preparation of the Reporter Plasmids

1. Amplify the plasmid DNA, pGL3-Basic, in bacteria and subsequently extract the plasmid by using a mini prep kit or the alkaline-lysis method (12).
2. Digest the plasmid DNA (2 μg) using appropriate restriction enzymes.

3. Purify the plasmid by agarose gel electrophoresis, extract the DNA band from the gel, and estimate the amount of DNA recovered by measuring absorbance at 260 nm.
4. Proceed to a ligation reaction for each promoter fragment for insertion into the reporter vector. Use 50–60 ng of plasmid and at least 1 mol equivalent of each fragment per reaction in a final volume of 15 μl. Refer to Sambrook et al. (12) for more details.
5. Transform competent bacteria with each construct. Analyze the plasmid in isolated colonies by restriction enzyme digest and sequence the DNA insert in the clones using sequencing primers that are contained within the pGL3-Basic vector sequence.

3.5.3. Luciferase Assays

1. Prepare pure plasmid DNA for each construct to be tested for luciferase activity using a commercial midi prep kit. The following plasmids should also be prepared and use as controls: a negative control (the promotorless pGL3-Basic empty vector), a positive control (e.g., the pGL3-Control vector or the pRSV-L (8)), and a vector used as an internal control to correct for transfection efficiency (e.g., the phRL-TK vector).
2. Grow adherent cells, the RCE-1 cells (6) or a cell line that has been shown to transcribe the claudin gene of interest. Cells are cultured to 80% confluency at the time of transfection. Each transfection should be done in triplicate wells. For experiments conducted in 24-well plates, transfect the cells using Lipofectamine 2000 (2 μl, see Note 6) and 1 μg of each of the reporter construct combined with 100 ng of the internal control vector (phRL-TK).
3. Incubate the cells 24–48 h prior to lysis with the Passive Lysis Buffer included in the Dual-Luciferase Assay kit. Determine the firefly and *Renilla* luciferase activities using the Dual-Luciferase Assay kit and a luminometer. Express the relative luciferase activity as the ratio of firefly to *Renilla* luciferase activity. The luciferase activity ratio can be normalized to the protein concentration by quantifying the amount of total protein in the cellular lysate.

4. Notes

1. Optimize the siRNA transfection according to the manufacturer's instructions by checking the amount of siRNA necessary for efficient knockdown, the ratio of HiPerfect reagent to siRNA, and the cell density at transfection.

2. Different controls should be included in the experimental set up, such as untreated cells, cells treated with the transfection reagent, and cells transfected with a negative siRNA (a non-silencing siRNA that has no homology to any known mammalian gene). In addition, the siRNA can be tagged with a fluorochrome to ensure the efficiency of the transfection.
3. A housekeeping gene or protein (such as GAPDH or β-actin) should always be used to normalize the steady-state levels for loading differences or for differences in total RNA or protein concentrations between samples.
4. The purity of the RNA can be estimated by agarose/ethidium bromide gel electrophoresis. A PCR reaction can also be done to detect any contaminating genomic DNA. Use of a commercial kit including a DNase treatment is suggested, but the RNA could also be extracted by the guanidine thiocyanate method (13).
5. A minimal number of base pairs between the restriction site and the 5′ end of the primer is necessary to ensure optimal cleavage of the DNA. This number is specific for each restriction enzyme.
6. The transfection reagent as well as the reagent:DNA ratio should be adapted for the particular cell line used to achieve optimal transfection efficiency. The manufacturer's instructions should be followed with care.

References

1. McLachlan, R.I. (2002) Basis, diagnosis and treatment of immunological infertility in men. *J Reprod Immunol* **1–2**, 35–45.
2. Sullivan, R. (2004) Male fertility markers, myth or reality. *Anim Reprod Sci* **82–83**, 341–7.
3. Cyr, D.G., Hermo, L., Egenberger, N., Mertineit, C., Trasler, J., and Laird, D. (1999) Cellular immunolocalization of occludin during embryonic and postnatal development of the mouse testis and epididymis. *Endocrinology* **140**, 3815–25.
4. Gregory, M., Dufresne, J., Hermo, L. and Cyr, D.G. (2001) Claudin-1 is not restricted to tight junctions in the epididymis. *Endocrinology* **142**, 854–63.
5. Cyr, D.G., Finnson K.W., Dufresne J., and Gregory M. (2002) Cellular interactions and the blood-epididymal barrier. In B. Robaire and B. T. Hinton eds., The Epididymis: from Molecules to Clinical Practice. Plenum Press. New York. pp103–18.
6. Dufresne, J., St-Pierre, N., Viger, R.S., Hermo, L., and Cyr, D.G. (2005) Characterization of a novel rat epithelial epididymal cell line. *Endocrinology* **146**, 4710–20.
7. Dubé, E., Dufresne, J., Chan, P., Hermo, L. and Cyr, D.G. (2010) Assessing the role of claudins in maintaining the integrity of epididymal tight junctions using novel human epididymal cell lines. *Biol Reprod* **82**, 1119–28.
8. Dufresne, J., Cyr, D.G. (2007). Activation of an SP binding site is crucial for the expression of claudin 1 in rat epididymal principal cells. *Biol Reprod* **76**, 825–32.
9. Dubé, E., Hermo, L., Chan, P.T.K., and Cyr, D.G. (2008). Alterations in gene expression in the caput epididymis of non-obstructive azoospermic men. *Biol. Reprod* **78**, 342–51.
10. Hamilton, A., Baulcombe, D. (1999). A species of small antisense RNA in posttranscriptional gene silencing in plants. *Science* **286**, 950–52.

11. Elbashir, S., Harborth, J., Lendeckel, W., Yalcin, A., Weber, K. and Tuschl, T. (2001). Duplexes of 21-nucleotide RNAs mediate RNA interference in cultured mammalian cells. *Nature* **411**, 494–98.
12. Sambrook, J., Fritsch, E. F., and Maniatis, T. (1989). Molecular Cloning: A laboratory manual. Cold Spring Harbor Laboratory Press. Cold Spring Harbor, New York. 545p
13. Chomczynski, P. and Sacchi, N. (1987). Single-step method of RNA isolation by acid guanidinium thiocyanate-phenol-chloroform extraction. *Analytical Biochemistry* **162**, 156–59.

Chapter 19

Using Molecular Tracers to Assess the Integrity of the Intestinal Epithelial Barrier In Vivo

Julian A. Guttman

Abstract

The examination of the epithelial barrier has been a primary site of focused research for years. Despite the importance of this site to numerous intestinal diseases, the determination of the integrity of this barrier has been clouded by controversies as to the validity of certain techniques and the ease of use regarding others. To determine the barrier integrity in vivo, we have adapted a simple tracer-based microscopic assay that was initially used in other systems to the in vivo intestinal epithelium. This technique is widely adaptable to other tracer molecules that can be applied to the tissue and consequently generates images depicting barrier maintenance or functional breach.

Key words: Tight junction, Barrier, Epithelium, Tracer, Disruption, Intestine, Colon, Microscopy

1. Introduction

Tight junctions positioned at the apical region between adjacent epithelial cells govern the regulation of molecular movements between the luminal and adlumenal spaces in the intestine. Since the identification of these junctions, the evaluation of their barrier regulating properties have been a focus of intense research as intercellular disruption of this epithelial barrier has been proposed to be involved in diarrhea generation, as well as inflammatory responses. Consequently, research in the fields of enteric pathogenesis, inflammatory bowel diseases as well as other enteric diseases, have positioned the intestinal epithelial barrier at the forefront of host alterations to be examined.

Analysis of the intestinal epithelial barrier has been accomplished using various strategies such as immunolocalization of tight junction proteins, high-resolution electron microscopy, and

Kursad Turksen (ed.), *Claudins: Methods and Protocols*, Methods in Molecular Biology, vol. 762,
DOI 10.1007/978-1-61779-185-7_19, © Springer Science+Business Media, LLC 2011

epithelial resistance measurements either in the form of calculations of transepithelial resistance (TER) when examining cultured cells or Ussing chambers when studying the epithelial barrier in tissues. Despite these often-used approaches, their ease of use coupled with their independent validity with regard to barrier function when in the context of numerous additional host alterations can be challenging (1). Consequently, to visually assess the integrity of the epithelial barrier, we have adapted a molecular tracer procedure used to determine the barrier integrity in other systems (2, 3) to the in vivo murine intestinal system, to assess the epithelial barrier integrity of the colon.

2. Materials

2.1. Animals

1. C57/Bl6 mice were purchased from Charles River, Canada. Animals were housed at the university animal facility, provided food and water ad libitum. Animals were allowed to recuperate from shipping for at least 4 days prior to any experimental treatments.

2.2. Biotin Tracer Preparation

1. PBS containing 1 mM $CaCl_2$: 150 mM NaCl, 5 mM KCl, 0.8 mM KH_2PO_4, 3.2 mM Na_2HPO_4, 1 mM $CaCl_2$, pH 7.3. This solution can be made the day prior to use.
2. EZ-Link Sulfo-NHS-Biotin, molecular size 443 Da (Pierce Chemical Co, now part of Thermo Scientific) is prepared fresh for each experiment by dissolving the EZ-Link Sulfo-NHS-Biotin to 2 mg/ml in PBS containing 1 mM $CaCl_2$. This solution is kept on ice.

2.3. Tissue Preparation

1. Phosphate-Buffered Saline (PBS): 150 mM NaCl, 5 mM KCl, 0.8 mM KH_2PO_4, 3.2 mM Na_2HPO_4, pH 7.3. This solution should be autoclaved and can be stored at room temperature.
2. Paraformaldehyde (Canmeco, Quebec, Canada): 3% paraformaldehyde is prepared from a 16% stock using the individual components of PBS to result in 150 mM NaCl, 5 mM KCl, 0.8 mM KH_2PO_4, 3.2 mM Na_2HPO_4. This solution should then be brought to pH 7.3, can be prepared the day prior to use and stored at 4°C.

2.4. Making Blunt-Ended Needles

1. Use a pair of wire cutters to cut off the angled point of a 22-gage needle. This will also physically compress the needle.
2. Sand the tip of the needle with a series of sand paper, starting at 80 grit extending trough to 220 grit. This will result in the formation of a needle with a blunt end (see Note 1).

3. Methods

Tight junctions found at the epithelial barrier regulate the intercellular movement of material in the intestine. Although there are a number of strategies to detect barrier integrity, the addition of a molecular tracer to the luminal side of the barrier in conjunction with fluorescent molecular observation provides a straightforward, visual method to observe barrier integrity. This strategy can utilize conventional fluorescent microscopy or more advanced confocal imaging to conclude if the barrier remains intact, or has been breached, by simply observing the location of the tracer molecule within the tissue sample (Fig. 1).

3.1. Administration of Biotin Tracer in the Murine Colon

1. Mice are euthanized, the abdomen is opened and the abdominal organs are moved to expose the colon.
2. The most distal portion of the rectum is cut to free the distal colon from the animal.
3. Using forceps, gently pull the most distal part of the colon away from the animal. The colon and attached intestines will

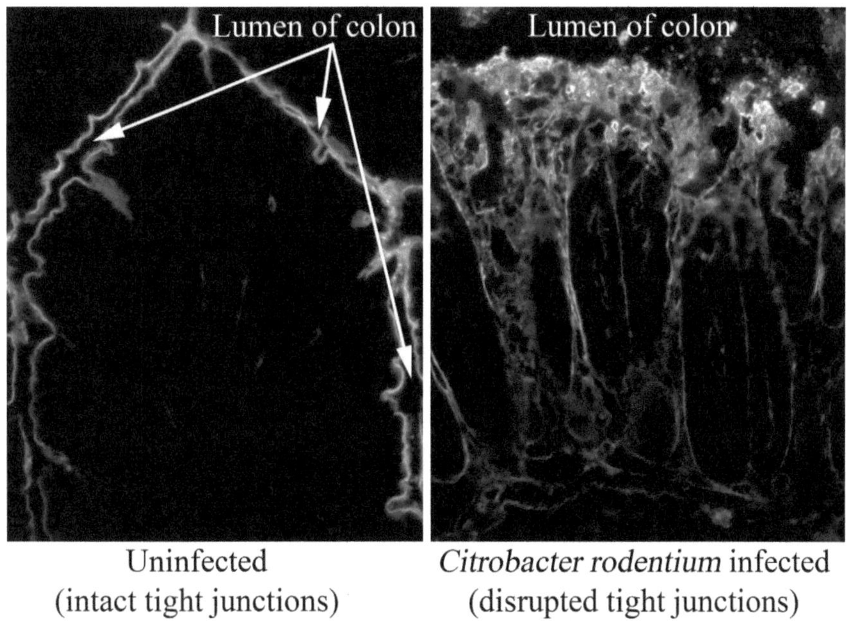

Fig. 1. Images of biotin tracer labeling in the colons of mice that have intact tight junctions or disrupted tight junctions. Uninfected and 7-day infected *Citrobacter rodentium* (used to disrupt the epithelial barrier) murine colonic tissue were treated with the biotin tracers and reacted with Alexa 488 streptavidin as described in this chapter. In the uninfected tissue that retains intact tight junctions, the biotin molecule is contained to the luminal aspect of the tissue and does not penetrate the epithelium. During *C. rodentium* infections (4), these bacteria disrupt the tight junctions in the epithelium at sites where they interact with the tissue (at the apex of the epithelium), the biotin tracer permeates the infected tissue and is even found deep in the epithelium.

linearize, while the rostral part of the intestinal system remains attached to the animal.

4. Orient the colon so that it is linear and somewhat isolated from the remainder of the animal.
5. Using a blunt-ended 22-gage needle, gently insert the needle into the distal (cut) region of the colon.
6. Clamp the exterior of the colon with vessel cannulation forceps (Fine Science Tools, North Vancouver, Canada) to grasp the needle through the colonic tissue (see Note 2).
7. Inject enough of the biotin tracer into the lumen of the colon (see Note 3). The tissue will expand but do not force in too much tracer so that the liquid burst out of the colon.
8. Allow the biotin to be in contact with the tissue for 3.5 min.
9. Once the time has elapsed release the needle and excise 1 cm of tissue located just distal to the location of the needle, thus the portion of the colonic tissue that had not been mechanically squeezed by the vessel cannulation forceps.

3.2. Tissue Fixation, Sectioning and Microscopy

1. Once the administration of the biotin tracer is completed, the 1 cm of the tissue just rostral to the location of the needle used for injection is excised and placed into room temperature 3% paraformaldehyde for 3 h.
2. The tissue is then washed three times using PBS at 10 min intervals, for a total of 30 min of washing (see Note 4).
3. Then the tissue is mounted onto a metal stub from the cryostat.
4. To mount the tissue, place a drop of Sakura Tissue-Tek OCT compound (VWR international, Canada) on the metal stub, then carefully place one end of the tissue into the OCT compound. While gently holding the other end of the tissue with forceps and the metal stub with larger forceps carefully place the base of the metal stub into liquid nitrogen. As the stub slowly freezes so will the OCT compound, thus anchoring the portion of the tissue that is in the OCT compound. Keep the base of the metal stub in the liquid nitrogen until the tissue completely freezes. You will be able to see a front of freezing climbing up the tissue as it freezes. Thus, you will have the base of the tissue embedded in OCT compound attached to the stub as well as tissue not in OCT compound, but frozen, that will be used for cryosectioning.
5. Once the tissue is frozen, place the metal stub with the attached tissue into the cryostat for about 20 min to allow the tissue to equilibrate to the cryostat temperature. The cryostat should be kept between −15 and −20°C.

6. Tissue sections should be cut at 5–7 μm thicknesses and put onto glass slides.
7. The tissue is now permeabilized using one of two techniques. (a) Plunge the slide(s) into –20°C acetone for 5 min, then place the slide on the bench at room temperature to air dry the tissue until it is completely dry or (b) plunge the slide(s) into room temperature PBS containing 0.2% Triton X-100 for 5 min then rinse the slides in PBS for three consecutive 10 min washes for a total of 30 min.
8. The samples are now treated with a 1:500 dilution of Alexa Fluor 488 streptavidin (Invitrogen Canada, now part of Life Technologies) for 30 min at room temperature (see Note 5). If the PBS/Triton X-100 permeabilization is used, as much of the liquid that remains on the slide should be removed without drying the tissue itself prior to the addition of the streptavidin.
9. Slides are then washed with 3–5 ml of PBS, the excess liquid is removed without drying out the tissue and coverslips mounted using either the hard setting ProLong® Gold antifade reagent with DAPI (Invitrogen Canada, now part of Life Technologies) or the wet-setting Vectashield (Vector Laboratories, Ontario, Canada) (see Note 6).

4. Notes

1. Instead of making a blunt-ended needle, 22-gage gavage needles can be purchased from Fine Science Tools (North Vancouver, Canada).
2. Your hand may get fatigued if you have a large number of mice to perform these experiments on, to alleviate some of the strain, following the plunger depression to inject the tracer into the colon, the syringe can be released from one hand while maintaining force on the forceps with the other hand around the needle. The tracer will not leak back into the syringe and it will allow you to switch hands to continually squeeze the colonic tissue around the needle.
3. The general administration strategy employed for the biotin tracer molecule can be easily adapted to other molecules such as fluorescent dextrans of various sizes.
4. Using a 10 ml Pasteur pipette for the tissue washes (following the fixation) allows the tissue to be washed without the tissue being sucked into the pipette.
5. The Alexa Fluor 488 streptavidin runs off of the sample when added to the tissue. To restrain the Alexa Fluor 488 streptavidin

to the tissue, a liquid blocking Super PAP pen (Invitrogen Canada, now part of Life Technologies) can be used to encircle the tissue. Simply add the Alexa Fluor 488 streptavidin to the center of the circle.

6. Both Life Technologies and Vector Laboratories sell hard setting and wet-setting mounting media. Additionally, both companies sell their mounting media containing DAPI or not containing DAPI. The selection of mounting media does not influence the biotin tracer results.

Acknowledgments

This work was supported through operating grants from the CIHR and NSERC. JAG is a CIHR New Investigator.

References

1. Guttman, J. A., and Finlay, B. B. (2009) Tight junctions as targets of infectious agents, *Biochim Biophys Acta 1788*, 832–841.
2. Furuse, M., Hata, M., Furuse, K., Yoshida, Y., Haratake, A., Sugitani, Y., Noda, T., Kubo, A., and Tsukita, S. (2002) Claudin-based tight junctions are crucial for the mammalian epidermal barrier: a lesson from claudin-1-deficient mice, *J Cell Biol 156*, 1099–1111.
3. Nitta, T., Hata, M., Gotoh, S., Seo, Y., Sasaki, H., Hashimoto, N., Furuse, M., and Tsukita, S. (2003) Size-selective loosening of the blood-brain barrier in claudin-5-deficient mice, *J Cell Biol 161*, 653–660.
4. Guttman, J. A., Li, Y., Wickham, M. E., Deng, W., Vogl, A. W., and Finlay, B. B. (2006) Attaching and effacing pathogen-induced tight junction disruption in vivo, *Cell Microbiol 8*, 634–645.

Chapter 20

Laboratory Methods in the Study of Endometrial Claudin-4

Paulo Serafini, André Monteiro da Rocha, Gary Daniel Smith, Eduardo Leme Alves da Motta, and Edmund Chada Baracat

Abstract

Immunohistochemistry is a suitable method for the detection of proteins from the Claudin family and several antibodies are commercially available for the detection of Claudin congeners. Immunodetection of Caludin-4 in the paraffin-embedded specimens might be a useful tool for studying the role of these proteins in the cyclic transformation of the endometrium and its role in the endometrial receptivity; furthermore, other components of the junctional zone involved in the transformational process of the endometrium can be detected by means of immunohistochemistry/immunofluorescence with several polyclonal or monoclonal antibodies. The aim of this chapter is to comprehensively overview the materials and methods to perform the endometrial biopsy and to detect Claudin-4 in paraffin-embedded samples of endometrium. Additionally, the interpretation of the results is addressed.

Key words: Claudin-4, Endometrium, Immunohistochemistry, Immunofluorescence

1. Introduction

Antibodies can be combined with some substances without any effects on antigen binding (1). This property allowed antibodies to be conjugated to fluorescent compounds, enabling the diagnosis of several diseases by fluorescent antibody techniques (2–4). Because plain antibody–antigen reactions are not detectable by light microscopy, the labeling of antibodies with fluorescent dyes has been key to the detection of antigens in cells and tissues by microscopy (2). Immunofluorescence is now a common tool for research of proteins and their involvement in development, physiology, and pathology (5–21). Antibodies can also be conjugated with other useful substances, such as enzymes that react with specific substrates and chromogens to generate a colored precipitate, thus

Kursad Turksen (ed.), *Claudins: Methods and Protocols*, Methods in Molecular Biology, vol. 762,
DOI 10.1007/978-1-61779-185-7_20,

enabling optical detection of antigen-bound antibodies (22). Immunohistochemistry is a suitable method for the detection of proteins from the Claudin family and several antibodies are commercially available for the detection of Claudin congeners (23–29). The aim of this chapter is to describe methods employed to detect the expression of claudin in human endometrium.

2. Materials

1. Vaginal speculum.
2. Antiseptic solution.
3. Pipelle catheter.
4. Buffered formalin: Dilute 37% formalin in deionized water (1:10) and then adjust pH with 1 M NaOH.
5. Hematoxilin.
6. Eosin.
7. Xylol.
8. Ethanol 95%.
9. Ethanol 85%.
10. Sodium citrate (10 mM)/citric acid buffer (10 mM): 2.941 g of sodium and citrate, 2.101 g of citric acid. Dissolve the following in 800 ml of deionized water. Adjust volume to 1 L with the addition of deionized water. Store solution at 4°C.
11. Polyclonal antibody against Claudin-4 – Santa Cruz Biotechnology, Inc., USA (reference C-18: sc-17664).
12. Goat anti-rabbit secondary antibody for enzymatic immunohistochemistry (reference sc-2040).
13. Phosphate buffer saline (PBS): 8 g of NaCl, 0.2 g of KCl, 1.44 g of Na_2HPO_4, and 0.24 g of KH_2PO_4. Dissolve the following in 800 ml distilled H_2O. Adjust pH to 7.4. Adjust volume to 1 L with the addition of distilled H_2O. Stock for 15 days at 4°C.
14. Hoescht 33342: Add 5 μg of Hoescht 33342 to 1 ml of PBS to produce a stock solution. Store stock solution at 4°C. Dilute 2 μl of the stock solution in 1,998 μl of PBS to produce the working solution.
15. Mounting media: Mix 75 ml of glycerol, 25 ml of PBS, and 100 mg of propil galate. Store at 4°C.
16. 0.3% H_2O_2: Dilute 1 ml of commercially available 30% H_2O_2 in 9 ml of deionized H_2O_2.
17. Biotin.

18. Diaminobenzidine.
19. Mayer's hematoxilin.
20. Synthetic resin for histology.

3. Methods

Immunohistochemical detection of Claudin-4 involves a series of steps as described below. Two options for visualization are provided, namely immunofluoresence and enzymatic immunohistochemistry.

3.1. Endometrial Biopsy

Endometrial biopsy is a simple procedure performed in an outpatient facility without sedation or local anesthesia. It should be performed in patients who are not pregnant and preferentially at least one or two cycles before the patient attempts to conceive. Samples of endometrium are preferentially obtained during the "window of implantation", from 19th to 23rd days of a 28-day menstrual cycle (see Note 1) as follows:

1. Patients should be in a lithotomy position.
2. Vaginal speculum is inserted to expose the cervix.
3. The exposed cervix should be cleansed with antiseptic solution and excess should be removed.
4. The Pipelle catheter should be "moulded" considering the cervical-uterine axis (see Note 2) (23).
5. The catheter is gently inserted with a delicate approach into the fundus, previously measured by sonography, after the patient takes a deep breath.
6. The catheter is moved forward gently for the estimated length of the uterine cavity.
7. The plunger of the catheter should be pulled to form vacuum during its movement to the distal portion of the uterus.
8. During this movement, the catheter opening should be facing the anterior wall of the uterus.
9. The catheter should be placed in the cranial portion of the uterine cavity with the catheter opening facing the posterior wall of the uterus.
10. The maneuver for endometrial sampling should be repeated (see Note 3).
11. Endometrial samples should each be at least 3 mm^3.
12. Immediately immerse the collected samples in 4% buffered formalin fixative solution (pH 7.0) (23, 30) (see Note 4).

3.2. Tissue and Slide Preparation

Endometrial samples should be subjected to standard histology preparation.

1. Fixed samples should be positioned in plastic processing or enclosure cassettes.
2. Samples should be submitted to a 12-h dehydration, diaphanization and paraffin exposure cycle (23).

One representative section (3–4 μm thick) should be obtained and stained with hematoxilin and eosin (30). Endometrial dating should be performed by a experienced pathologist according to Noyes' criteria (31) (see Note 5).

3.3. Immuno fluorescence

3.3.1. Section Deparaffination and Hydration

1. Specimen deparaffination should be performed prior to immunostaining.
2. Slides are immersed in two subsequent xylol baths of 5 min prior to hydration.
3. Specimen hydration should be accomplished by sequential immersion in 95 and 85% for 3 min.

3.3.2. Antigen Retrieval

1. Antigen retrieval should be performed by microwave heating.
2. Slides are placed in a heat-resistant bulk and immersed in sodium citrate (1 mM)/citric acid (1 mM) buffer.
3. Samples should be heated in microwave oven (1,300 W) for 9 min at 70% potency, twice (see Note 6).
4. Replace buffer between heating periods.
5. Slides should be allowed to cool for 20 min at room temperature.
6. Slides should be submitted to three 5-min washes with deionized water.

3.3.3. Blocking Nonspecific Reactions and Binding

Nonspecific reaction binding should be blocked before incubation with the primary antibody (see Note 7).

1. Serum should be diluted in PBS to a final blocking solution that is 10–15% serum.
2. Slides should be placed flat and level (i.e., on glass or plastic bars), tissue side up, in a humidified chamber.
3. 1 ml of blocking solution should be pipetted over the slide surface.
4. Slides should be allowed to sit undisturbed for 20 min.
5. Samples should be given three 5-min washes in PBS.

3.3.4. Primary Antibody Incubation

1. Primary antibody (see Note 8) against Claudin-4 should be freshly diluted in PBS (1:200) supplemented with 1.5% of the blocking solution.

2. Slides should be placed in a humidified chamber.
3. Primary antibody should be pipetted on top of specimens (approximately 250 μl per section).
4. The chamber must be sealed (i.e., with clingy plastic wrap or tape).
5. The tissue should be allowed to incubate with the primary antibody at room temperature for 60 min.
6. Primary antibody must be poured off from the slides and the tissue should be given three 5-min washes in PBS.

3.3.5. Secondary Antibody Incubation for Immunofluorescence

1. Secondary antibody conjugated to a fluorescein (see Note 9) should be freshly diluted 1:200 in PBS supplemented with 1.5% of blocking serum.
2. Each slide should placed in a dark humidified chamber (i.e., a chamber made of an opaque material or covered in aluminum foil).
3. Samples should be covered with 1 ml of secondary antibody solution.
4. Slides are incubated for 45 min.
5. Te secondary antibody solution should be poured off.
6. Immunolabeled slides of sections should then be covered with 1 ml of Hoescht 33342 solution (1 μg/ml) for 5 min to stain DNA.
7. The Hoescht stain must be removed with three 5-min rinses in PBS before cover-slipping (see Note 10).
8. The rinsed slides should be drained.
9. Fifteen microliters of mounting media should be pipetted onto each specimen section.
10. Cover slips should be placed gently to avoid the formation of air bubbles.
11. The surfaces of the cover-slipped slides should be dried before examining in fluorescence microscope.

3.4. Enzymatic Immunohistochemistry

Preparation of specimens for enzymatic immunohistochemistry shares many steps with the protocol for immunofluorescence. Namely, slide preparation, tissue deparaffination and hydration, unspecific reaction blocking, and primary antibody incubation are done as described above in Subheadings 3.3.1–3.3.4.

3.4.1. Blocking of Endogenous Peroxidase

Because endogenous peroxidase activity can interfere with immunohistochemistry results, creating a brownish background stain (see Note 11), it should be blocked.

1. Endogenous peroxidase activity can be quenched by incubation of sections in 0.3% H_2O_2 in PBS for 5 min before unspecific antigen/antibody reaction blocking.
2. Incubation in the H_2O_2 solution should be followed by three 5-min washes in PBS.

3.4.2. Secondary Antibody Incubation and Development for Immunohistochemistry

1. Biotin-conjugated secondary antibody should be diluted in PBS (1:200) and pipetted onto each slide to cover the specimen.
2. Incubation should be performed for 30 min in a humidified chamber at room temperature.
3. The slides should then be washed three times in PBS before commencing a 30-min incubation in a streptavidin-peroxidase (similarly applied).
4. Specimens should be washed three times in PBS of 5 min each.
5. Development should be conducted by submersion of slides into diaminobenzidine (DAB) solution (prepared fresh according to the manufacturer's instructions) for 5 min (see Note 12).
6. Reaction should be stopped by three washes in PBS.

3.4.3. Counterstaining and Slide Mounting

1. Slides should be counterstained with Mayer's hematoxylin.
2. Hold slides in hematoxylin stain for 3 min.
3. Slides should be washed by allowing the slide holding dish to sit in running tap water for 10 min.
4. Sides should be dehydrated by placing in sequential solutions of ethanol: once in 85% (3 min), once in 95% (3 min), and three times in absolute ethanol (3 min each).
5. Dehydrated specimens should be diaphanized in three consecutive xylol baths (3 min each).
6. Finally, the prepared specimens should be cover-slipped with synthetic resin mounting media prior to reading and interpretation.

3.5. Slide Reading and Interpretation

Slide reading should be performed by two independent observers to decrease subjective interference. Ensuring that slide readers are blinded to treatments or sample identity can remove potential bias. Positivity and membrane binding are key features for specimen categorization (32).

1. Claudin-4 scores should be computed as a multiplication of frequency × intensity.
2. Results should range between 0 and 12.

3. Frequency scores reflect the percentage of membrane that is immunopositive according to the following rubric:
   (a) 1 – 0% to 10%;
   (b) 2 – 10% to 50%;
   (c) 3 – 50% to 90%;
   (d) 4 – >90%.
4. Immunolabeling intensity scores are recorded as:
   (a) 1 – weak;
   (b) 2 – moderate;
   (c) 3 – strong.
5. Categories can be designated according to the final frequency × intensity score according to the following score ranges:
   (a) negative, 0–1.9;
   (b) weak, 2.0–4.9;
   (c) moderate, 5.0–8.9;
   (d) strong, 9.0–12.0 (30) (see Note 13).

4. Notes

1. Endometrial biopsy can be obtained during any phase of the menstrual cycle. However, it has been extensively documented that the molecular behavior of the endometrium is distinct during the period of receptivity to embryo implantation (23, 33–37). Thus if one's intention is to assess claudin-4 expression its relation to uterine receptivity, it should be done between the 19th and 23rd days of the menstrual cycle.
2. This is an important step to avoid cervical bleeding.
3. Performance of the endometrial biopsy with these movements might yield a complete sampling of the uterine cavity and a more complete examination.
4. Alternatively, the uterine samples can be collected following insertion of a Betocchi's hysteroscope that still permits execution in an office equipped with general anesthesia in case it becomes necessary, thought it should be possible to accomplish with local or without anesthesia. The major benefit of this alternative would be the ability to collect samples from specific endometrial sites, such as posterior fundus which is considered to be the implantation site by many (38–42).
5. Only samples obtained during the window of implantation (menstrual dates) with histologically confirmed timing should be used for subsequent analysis.

6. Slides should be placed in a microwavable container for antigen retrieval. Because antigen retrieval in household microwave ovens allows evaporation of retrieval buffers, the buffer level should be checked every 2 min to avoid overdrying of specimens.
7. Blocking is performed using normal serum from the species in which the secondary antibody was raised. For example, if the secondary antibody was raised in donkey or donkey cells, the blocking serum should be normal donkey serum.
8. Monoclonal and polyclonal antibodies against Claudin-4 can be raised in several species and they are available from several manufactures. Our laboratory obtains reliable results with rabbit polyclonal antibody from Santa Cruz Biotechnology, Inc., USA (reference C-18: sc-17664). There are several other antibodies produced by other manufactures that can be used if one is interested in investigating other members of the claudin family, such as claudin-4 and -5, among others.
9. Similar to the primary antibody, appropriate secondary antibody can be raised in different species and can be purchased from several manufactures. Our laboratory uses goat anti-rabbit secondary antibody for enzymatic immunohistochemistry (reference sc-2040) and donkey anti-rabbit secondary antibody conjugated with the isothiocyanate (reference sc-2365) derivative of the green fluorescent dye fluorescein (abbreviated FITC) for immunofluorescence.
10. Excess Hoescht 33342 concentration or inadequate post-stain washing of the slides can result in an undesirable bluish background stain.
11. Brown DAB background staining may be particularly undesirable for claudin-4 immunohistochemistry analysis due to the importance of the deposition of the immunocomplex on membranes.
12. Small variations in DAB exposure time can result in large changes in staining intensity. The time elapsed during DAB exposure must remain consistent between samples in order for the observations to be quantifiable.
13. Individuals' scores can be correlated with physiological factors of interest. For example, we demonstrated previously that strong CLDN4 immunolabeling was associated with failure to establish pregnancy after in vitro fertilization (IVF). That is, patients whose uterine endometrium specimens showed strong CLDN4 immunolabeling were 10.5 times less likely to achieve pregnancy than those with weak immunolabeling (30). However, the predictive value of claudin-4 immunoreactivity in relation to IVF results may be stronger, if results are combined with immunolabeling results for endometrial leukemia inhibitor factor (LIF).

References

1. Eagle H, Smith DE and Vickers P (1936). The Effect of Combination with Diazo Compounds on the Immunological Reactivity of Antibodies. *J Exp Med* **63**: p.617–643
2. Coons AH, Creech HJ and Jones RN (1941). Immunological properties of an antibody containing a fluorescent group. *Proc Soc Exp Biol Med* **47**: p.200–202
3. Sheldon WH (1953). Leptospiral antigen demonstrated by the fluorescent antibody technic in human muscle lesions of *Leptospirosis icterohemorrhagiae*. *Proc Soc Exp Biol Med* **84**: p.165–7
4. Teague RE (1960). Fluorescent antibody techniques in the diagnosis of communicable diseases. *J Ky Med Assoc* **58**: p.1356
5. Coons AH (1951). Fluorescent antibodies as histochemical tools. *Fed Proc* **10**: p.558–9
6. Coons AH (1957). The application of fluorescent antibodies to the study of naturally occurring antibodies. *Ann N Y Acad Sci* **69**: p.658–62
7. Coons AH (1959). The diagnostic application of fluorescent antibodies. *Schweiz Z Pathol Bakteriol* **22**: p.700–23
8. Anderson WA (1975). Application of immunological principles in dermatology. *J Natl Med Assoc* **67**: p.423–7, 454
9. Moon HD and McIvor BC (1960). Elastase in the exocrine pancreas: localization with fluorescent antibody. *J Immunol* **85**: p.78–80
10. Coons AH and Kaplan MH (1950). Localization of antigen in tissue cells; improvements in a method for the detection of antigen by means of fluorescent antibody. *J Exp Med* **91**: p.1–13
11. Yanai R, Ko JA, Nomi N, Morishige N, Chikama T, Hattori A, Hozumi K, Nomizu M and Nishida T (2009). Upregulation of ZO-1 in cultured human corneal epithelial cells by a peptide (PHSRN) corresponding to the second cell-binding site of fibronectin. *Invest Ophthalmol Vis Sci* **50**: p.2757–64
12. Alanne MH, Pummi K, Heape AM, Grenman R, Peltonen J and Peltonen S (2009). Tight junction proteins in human Schwann cell autotypic junctions. *J Histochem Cytochem* **57**: p.523–9
13. Ko JA, Yanai R, Morishige N, Takezawa T and Nishida T (2009). Upregulation of connexin43 expression in corneal fibroblasts by corneal epithelial cells. *Invest Ophthalmol Vis Sci* **50**: p.2054–60
14. Nighot PK, Moeser AJ, Ryan KA, Ghashghaei T and Blikslager AT (2009). ClC-2 is required for rapid restoration of epithelial tight junctions in ischemic-injured murine jejunum. *Exp Cell Res* **315**: p.110–8
15. Grumbach Y, Quynh NV, Chiron R and Urbach V (2009). LXA4 stimulates ZO-1 expression and transepithelial electrical resistance in human airway epithelial (16HBE14o-) cells. *Am J Physiol Lung Cell Mol Physiol* **296**: p.L101–8
16. Chen-Quay SC, Eiting KT, Li AW, Lamharzi N and Quay SC (2009). Identification of tight junction modulating lipids. *J Pharm Sci* **98**: p.606–19
17. Vajda S, Bartha K, Wilhelm I, Krizbai IA and Adam-Vizi V (2008). Identification of protease-activated receptor-4 (PAR-4) in puromycin-purified brain capillary endothelial cells cultured on Matrigel. *Neurochem Int* **52**: p.1234–9
18. Liu LB, Xue YX, Liu YH and Wang YB (2008). Bradykinin increases blood-tumor barrier permeability by down-regulating the expression levels of ZO-1, occludin, and claudin-5 and rearranging actin cytoskeleton. *J Neurosci Res* **86**: p.1153–68
19. Lourenco SV, Coutinho-Camillo CM, Buim ME, Uyekita SH and Soares FA (2007). Human salivary gland branching morphogenesis: morphological localization of claudins and its parallel relation with developmental stages revealed by expression of cytoskeleton and secretion markers. *Histochem Cell Biol* **128**: p.361–9
20. Inai T, Sengoku A, Guan X, Hirose E, Iida H and Shibata Y (2005). Heterogeneity in expression and subcellular localization of tight junction proteins, claudin-10 and -15, examined by RT-PCR and immunofluorescence microscopy. *Arch Histol Cytol* **68**: p.349–60
21. Heller F, Florian P, Bojarski C, Richter J, Christ M, Hillenbrand B, Mankertz J, Gitter AH, Burgel N, Fromm M, Zeitz M, Fuss I, Strober W and Schulzke JD (2005). Interleukin-13 is the key effector Th2 cytokine in ulcerative colitis that affects epithelial tight junctions, apoptosis, and cell restitution. *Gastroenterology* **129**: p.550–64
22. Ramos-Vara JA (2005). Technical aspects of immunohistochemistry. *Vet Pathol* **42**: p.405–26
23. Serafini P, Da Rocha AM, De Toledo Osorio CA, Smith GD, Hassun PA, da Silva IG, Da Motta EL and Baracat EC (2009). Protein profile of the luteal phase endometrium by tissue microarray assessment. *Gynecol Endocrinol* p.1–6

24. Pan XY, Li X, Weng ZP and Wang B (2009). Altered expression of claudin-3 and claudin-4 in ectopic endometrium of women with endometriosis. *Fertil Steril* **91**: p.1692–9
25. Mazaud-Guittot S, Meugnier E, Pesenti S, Wu X, Vidal H, Gow A and Le Magueresse-Battistoni B (2010). Claudin 11 deficiency in mice results in loss of the sertoli cell epithelial phenotype in the testis. *Biol Reprod* **82**: p.202–13
26. Morrow CM, Tyagi G, Simon L, Carnes K, Murphy KM, Cooke PS, Hofmann MC and Hess RA (2009). Claudin 5 expression in mouse seminiferous epithelium is dependent upon the transcription factor ets variant 5 and contributes to blood-testis barrier function. *Biol Reprod* **81**: p.871–9
27. Kaarteenaho-Wiik R and Soini Y (2009). Claudin-1, -2, -3, -4, -5, and -7 in usual interstitial pneumonia and sarcoidosis. *J Histochem Cytochem* **57**: p.187–95
28. Nemeth Z, Szasz AM, Tatrai P, Nemeth J, Gyorffy H, Somoracz A, Szijarto A, Kupcsulik P, Kiss A and Schaff Z (2009). Claudin-1, -2, -3, -4, -7, -8, and -10 protein expression in biliary tract cancers. *J Histochem Cytochem* **57**: p.113–21
29. Sahin U, Koslowski M, Dhaene K, Usener D, Brandenburg G, Seitz G, Huber C and Tureci O (2008). Claudin-18 splice variant 2 is a pan-cancer target suitable for therapeutic antibody development. *Clin Cancer Res* **14**: p.7624–34
30. Serafini PC, Silva ID, Smith GD, Motta EL, Rocha AM and Baracat EC (2009). Endometrial claudin-4 and leukemia inhibitory factor are associated with assisted reproduction outcome. *Reprod Biol Endocrinol* 7: p.30
31. Noyes RW and Haman JO (1953). Accuracy of endometrial dating; correlation of endometrial dating with basal body temperature and menses. *Fertil Steril* **4**: p.504–17
32. Soini Y (2004). Claudins 2, 3, 4, and 5 in Paget's disease and breast carcinoma. *Hum Pathol* **35**: p.1531–6
33. Schmidt A, Groth P, Haendler B, Hess-Stumpp H, Kratzschmar J, Seidel H, Thaele M and Weiss B (2005). Gene expression during the implantation window: microarray analysis of human endometrial samples. *Ernst Schering Res Found Workshop* p.139–57
34. Franchi A, Zaret J, Zhang X, Bocca S and Oehninger S (2008). Expression of immunomodulatory genes, their protein products and specific ligands/receptors during the window of implantation in the human endometrium. *Mol Hum Reprod* **14**: p.413–21
35. Van Vaerenbergh I, McIntire R, Van Lommel L, Devroey P, Giudice L and Bourgain C (2010). Gene expression during successful implantation in a natural cycle. *Fertil Steril* **93**: p.268.e15–8
36. Haouzi D, Mahmoud K, Fourar M, Bendhaou K, Dechaud H, De Vos J, Reme T, Dewailly D and Hamamah S (2009). Identification of new biomarkers of human endometrial receptivity in the natural cycle. *Hum Reprod* **24**: p.198–205
37. Savaris RF, Hamilton AE, Lessey BA and Giudice LC (2008). Endometrial gene expression in early pregnancy: lessons from human ectopic pregnancy. *Reprod Sci* **15**: p.797–816
38. Harper MJ (1992). The implantation window. *Baillieres Clin Obstet Gynaecol* **6**: p.351–71
39. Klentzeris LD (1997). The role of endometrium in implantation. *Hum Reprod* **12**: p.170–5
40. Tabibzadeh S (1998). Molecular control of the implantation window. *Hum Reprod Update* **4**: p.465–71
41. Valles CS and Dominguez F (2006). Embryo-endometrial interaction. *Chang Gung Med J* **29**: p.9–14
42. Achache H and Revel A (2006). Endometrial receptivity markers, the journey to successful embryo implantation. *Hum Reprod Update* **12**: p.731–46

Chapter 21

Role of Claudins in Oxidant-Induced Alveolar Epithelial Barrier Dysfunction

Yu Sun, Richard D. Minshall, and Guochang Hu

Abstract

Claudins are the most important components of the tight junctions at the interface of the basolateral and apical membranes of polarized epithelial and endothelial cells. They determine the barrier properties of cell–cell contact existing between two neighboring cells and regulate paracellular permeability. Although maintenance of barrier properties requires intact epithelial tight junctions, relatively little is known about the role of claudins expressed in the alveolar epithelium in the regulation of epithelial permeability in response to inflammatory stimuli and oxidative stress, or injury. The present study was undertaken to determine whether differential expression of tight junction claudins is a mechanism for regulation of oxidant-induced pulmonary epithelial hyperpermeability. Here, we show that claudin-2 plays an important role in the regulation of epithelial barrier function during oxidative stress.

Key words: Western blot, Immunostaining, Confocal microscopy, Epithelium, Claudin, Oxidants

1. Introduction

Claudins, ~23-kDa integral membrane proteins bearing four transmembrane domains, were identified as primary constituents of tight junctions (1). Genes expressing the claudins comprise a large family consisting of at least 24 members in mammals displaying organ- and tissue-specific patterns of expression (2–5). Although claudin-3, claudin-4, and claudin-5 were found to be three of the major claudins expressed in rat type II alveolar epithelial cells (6, 7), claudin-2 is also expressed in the human lung cell line A549 (8). Claudins play an important role in establishing "velcro-like" homophilic and heterophilic adhesive

Kursad Turksen (ed.), *Claudins: Methods and Protocols*, Methods in Molecular Biology, vol. 762,
DOI 10.1007/978-1-61779-185-7_21, © Springer Science+Business Media, LLC 2011

interactions between adjacent epithelial cells to create a tight barrier between the airspace and the underlying blood-perfused tissue (9). During acute lung injury, changes in claudin expression in the lungs have been shown to account for differences in alveolar barrier function relevant to pulmonary edema formation due to an increase in paracellular permeability. Downregulation of claudin-4 expression might lead to altered tight junction structure and be related to the impairment of epithelial function in acute lung injury. However, claudin-3 expression did not change with ventilator-induced lung injury (10). Claudin-2 has been shown to regulate the expression of cation-selective channels near the tight junctions of epithelial cell monolayers and thereby affect paracellular permeability (11, 12). Herein, we investigated the role of claudin-2, 3, 4, and 5 in the regulation of H_2O_2-induced epithelial hyperpermeability in lung A549 epithelial cell monolayers. We found that only claudin-2 expression dramatically decreased after exposure to H_2O_2, which caused gap formation in epithelial monolayers. Our results indicate the crucial role of claudin-2 in oxidant-induced increase in alveolar epithelial paracellular permeability (see Fig. 1). Commercial antibodies to the claudin proteins were used to detect changes in protein levels by Western blotting, and confocal microscopy of immunofluorescently stained cells was used to verify the relationship between the level of claudin expression and epithelial gap formation.

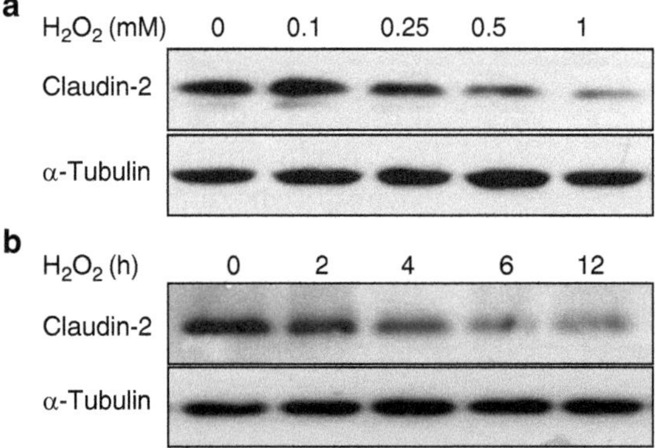

Fig. 1. Effects of H_2O_2 on claudin-2 expression in A549 epithelial cells. (**a**) H_2O_2 caused a decrease in claudin-2 expression in a concentration-dependent manner. A549 cells were exposed to different concentrations of H_2O_2 (0–1 mM) for 4 h. (**b**) Time course of H_2O_2-induced claudin-2 expression. Cells were stimulated with 0.5 mM H_2O_2 for the indicated times. All *blots* are representative of three separate experiments.

2. Materials

2.1. Cell Culture and Lysis

1. High glucose Dulbecco's modified Eagle medium (DMEM, Gibco/BRL, Bethesda, MD) supplemented with 10% fetal bovine serum (FBS, Serum Source International, Charlotte, NC), 50 U/mL of penicillin, and 50 μg/mL of streptomycin (Gibco).
2. 10× solution of trypsin/ethylenediaminetetraacetic acid (trypsin/EDTA, Gibco).
3. 30% (w/w) hydrogen peroxide solution (H_2O_2, Fisher Scientific) freshly diluted in DMEM in the dark and immediately used for treatment.
4. Modified RIPA lysis buffer for cell lysis: 50 mM Tris–HCl, pH 7.5, 150 mM sodium chloride (NaCl), 1 mM EDTA, 0.25% (w/v) sodium deoxycholate, 1% (w/v) Igepal CA-630, 0.1% (w/v) sodium dodecyl sulfate (SDS, Bio-Rad), and 1 mM sodium fluoride (NaF). Mix well and store in aliquots at 4°C. 1 mM sodium orthovanadate (Na_3VO_4), 1 mM phenylmethylsulfonyl fluoride (PMSF), and protease inhibitor cocktail is added immediately before use.
5. Disposable cell scrapers (Fisher brand).
6. BCA protein assay kit (Pierce).

2.2. SDS-Polyacrylamide Gel Electrophoresis

1. Cassettes 1.0 mm (Invitrogen).
2. Prepare the following stock solution:
   (a) 50% (w/v) acrylamide/BIS (29:1) solution. Store at room temperature. Keep solution A in a dark glass bottle.
   (b) Separating gel buffer:1 M Tris–HCl, pH 8.8. Store at room temperature.
   (c) 10% (w/v) SDS. Store at room temperature.
   (d) Stacking gel buffer: 0.375 M Tris–HCl, pH 6.8. Store at room temperature.
   (e) 5% (w/v) ammonium persulfate. Prepare a fresh solution on the day of use.
3. *N,N,N,N'*-tetramethylethylenediamine (TEMED, Bio-Rad).
4. Isopropanol (Fisher Scientific).
5. 10× Running buffer (1 L): 30.3 g (0.25 M) Tris Base, 144.0 g (1.92 M) glycine, and 10.0 g (1%, w/v) SDS. Bring to 1,000 mL with deionized H_2O. Do not adjust pH. Store at room temperature.
6. Precision plus protein standards (Bio-Rad).

2.3. Western Blotting for Claudin-2, Claudin-3, Claudin-4, and Claudin-5

1. Transfer buffer (4 L): 11.72 g glycine, 17.43 g Tris-Base, 1.5 g SDS, and 800 mL methanol. Bring to 4 L with deionized H_2O. Do not adjust pH.
2. Tris-buffered saline with Tween (TBS-T) (1 L): 10 mM Tris–HCl, pH 8.0, 0.15 M NaCl, and 1% (v/v) Tween-20. Bring to 1 L with deionized H_2O.
3. Blocking buffer: 5% (w/v) bovine serum albumin (BSA, Sigma–Aldrich) in TBS-T.
4. Trans-Blot transfer medium (Pure nitrocellulose membrane 0.2 μm) from Bio-Rad and Gel Blot paper (7 × 10 cm) from Whatman.
5. Primary antibody: mouse anti-claudin-2, rabbit anti-claudin-3, mouse anti-claudin-4, and mouse anti-claudin-5 (Invitrogen, Zymed).
6. Primary antibody dilution buffer: TBS-T supplemented with 2% (w/v) BSA.
7. Secondary antibody: goat anti-mouse or goat anti-rabbit IgG conjugated to horseradish peroxidase (Santa Cruz).
8. SuperSignal West Pico Chemiluminescent substrate from Thermo Scientific. It consists of two solutions – a luminol/enhancer solution and a stable peroxide solution.

2.4. Confocal Immunofluorescence Microscopy

1. 12-mm microscope cover glass from Fisher Scientific.
2. 2× HBSS: 1 bottle of HBSS powder (Sigma), 1.16 g Na-HEPEs, 1.32 g HEPEs, and 0.35 g sodium bicarbonate ($NaHCO_3$). Adjust pH to 7.4 and bring to 500 mL with deionized H_2O. Sterilize using a 0.22-μm filter and store at 4°C. Dilute 2× HBSS with an equal volume of deionized H_2O to make 1× HBSS.
3. 4% (v/v) paraformaldehyde: Dilute 16% (v/v) paraformaldehyde (Electron Microscopy Sciences) to 8% (v/v) with an equal volume of 2× HBSS. Then dilute 8% (v/v) to 4% (v/v) with an equal volume of 1× HBSS.
4. Quench solution: 100 mM glycine in 1× HBSS.
5. Permeabilization solution: 0.1% (v/v) Triton X-100 in HBSS.
6. Antibody dilution buffer: 5% (w/v) goat serum, 0.2% (w/v) BSA, and 0.1% (w/v) sodium azide in HBSS.
7. Secondary antibody: goat anti-rabbit or goat anti-mouse Alexa 546 (Molecular Probes).
8. Alexa Fluor 488 phalloidin (Molecular Probes).
9. Nuclear stain: 1 μg/mL 4,6-diamidino-2-phenylindole (DAPI, Molecular Probes) in 1× HBSS.
10. Mounting medium: ProLong anti-fade mounting medium (Molecular Probes).

3. Methods

Claudins show specific patterns of expression in different tissues. To obtain reliable and reproducible results, therefore, it is important to choose different claudin antibodies and adjust the concentration of antibodies according to the expression level of claudins. In addition, to investigate the role of claudins in the regulation of epithelial barrier integrity, double-labeled fluorescent staining of tight junction proteins and actin (F-actin) is often performed using a claudin antibody and Alexa Fluor-phalloidin (13). Alexa Fluor-phalloidin, a specific label for F-actin, labels polymerized actin observed in stress fibers and along the cortical cytoplasm adjacent to junctional complexes, which allows detection of small intercellular gaps between cells (14). Phalloidin binds specifically at the interface between F-actin subunits, locking adjacent subunits together. In control cultures, F-actin was distributed primarily at the periphery of the cytosol, forming dense peripheral bands. After stimulation with H_2O_2, the distribution of F-actin in the cytoplasm became more diffuse (see Fig. 2).

3.1. Western Blotting

3.1.1. Preparation of Samples

1. A549 cells (human lung adenocarcinoma epithelial cell line) are passaged with trypsin/EDTA when approaching confluence to provide new maintenance cultures on T-75 flasks and experimental cultures on six-well plates.
2. Plate A549 cells at the density of 1.6×10^5 cells/well in a six-well plate. The cells will approach confluence after 48 h. At this time, the cultures are rinsed with PBS and incubated for 18 h in DMEM without serum.
3. The cells are treated with H_2O_2 according to the protocol. After treatment, the cells are washed three times with ice-cold 1× HBSS. A volume of 300 μl of cold RIPA buffer is added to each well. The cells are lysed on ice for 15 min and then scraped into labeled microcentrifuge tubes.
4. Centrifuge cell lysates at $16{,}000 \times g$ for 20 min at 4°C. Supernatants are transferred to new tubes and the pellet is discarded.
5. The concentration of protein in the samples is determined using a BCA protein assay kit.
6. Take 20 μg protein from each sample and mix with 5× sample buffer. The tubes are closed and then boiled for 10 min. Allow the tubes to cool to room temperature and centrifuge for 10 s to bring down water condensation on the lid. The samples are ready for separation by SDS-polyacrylamide gel electrophoresis (SDS-PAGE).

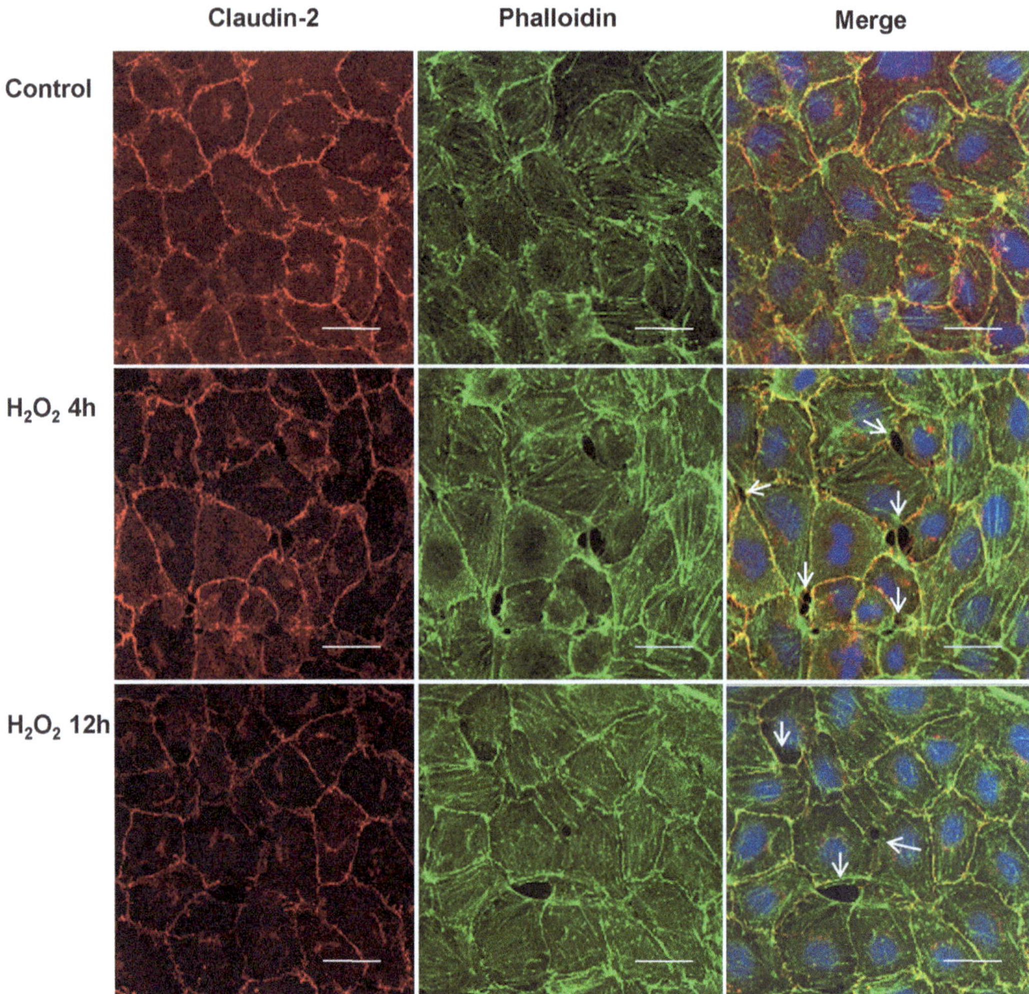

Fig. 2. Immunofluorescence confocal microscopy of claudin-2 and F-actin in A549 cells. Cells were grown to confluence and treated with H_2O_2 (0.5 mM) for the indicated times. After fixation and permeabilization, the cells were incubated with anti-claudin-2 primary antibody followed by Alexa 546-conjugated secondary antibody (*red*). The F-actin and nucleus were stained with Alexa 488 phalloidin (*green*) and DAPI (*blue*), respectively. In the second and third panels, note intercellular gap formation (*arrow*) in response to H_2O_2, and absence of claudin-2 in areas where gaps in the monolayer have formed. The first panel (*control*) shows no gap formation and claudin-2 loss in cell–cell contacts. Scale bars = 20 μm. Shown are representative results of three independent experiments.

3.1.2. SDS-PAGE

1. XCell SureLock Mini cell electrophoresis system is used.
2. Prepare 12% separating gel solution. Mix 6.0 mL of solution A, 9.4 mL of solution B, 250 μL of solution C, and 8.8 mL of H_2O. Add 625 μL of solution E and 6.25 μL of TEMED and mix well just before the gel solution is to be poured into the cassette. This amount of gel solution is enough to make four 1.0-mm thick gels. Immediately after the gel solution has been prepared and mixed, fill the cassette and leave space for the

stacking gel. Carefully overlay with ~400 μL of isopropanol. Allow the separating gel to polymerize for approximately 1 h.

3. Prepare 4% stacking gel solution. Mix 1.0 mL of solution A, 4.2 mL of solution D, 125 μL of solution C, and 6.3 mL of H_2O. Add 1.0 mL of solution E and 5.0 μL of TEMED and mix well just before the gel solution is to be poured. Remove the overlay solution and fill the cassettes with the stacking gel solution. Insert the comb without trapping air bubbles. Wait for about 30 min for the stacking gel to polymerize completely.
4. Prepare the 1× running buffer. Dilute 100 mL of the 10× running buffer with 900 mL of H_2O and mix well.
5. Carefully remove the comb from the stacking gel ensuring not to damage the wells, remove any spacer from the bottom of the gel cassette when the stacking gel has set. Place the cassette into the electrophoresis chamber and fill the chambers with running buffer. Load the samples and protein standard marker with a loading tip to the bottom of the well. Fill the tank with the electrophoresis buffer (running buffer).
6. Connect the chamber to the power supply and apply a voltage of 60 until the samples reach the separating gel. At this time, increase the voltage to 120.
7. When the dye front reaches the bottom of the gel, turn off the power supply.
8. Remove the cassettes from the chamber and the gels are now ready to transfer (see Notes 1).

3.1.3. Western Blotting for Claudins

1. To transfer the protein blots that have been separated by SDS-PAGE onto nitrocellulose membrane, prepare two pieces of fiber pads, two pieces of Whatman gel blot paper, and one piece of nitrocellulose membrane (6×8 cm) for each gel holder cassette.
2. Prepare a staining box of transfer buffer and pre-wet all the materials in it. Lay out a gel holder cassette in transfer buffer and then place the first sheet of fiber pad and one piece of Whatman gel blot paper in sequence onto the clear side (face anode). Then place the nitrocellulose membrane onto the Whatman paper.
3. Disassemble the gel unit. Remove and discard the stacking gel and excess separating gel and make sure the orientation of the separating gel (see Notes 2). Then lay the gel on the top of the membrane. It is important to make sure that no air bubbles are trapped between the membrane and the gel.
4. Finally, the second Whatman paper and fiber pad are carefully laid in sequence on the top of the gel (see Notes 3). Using a 15-mL tube as a roller, squeeze out any air bubbles in the resulting sandwich. Then close the cassette.

In summary, stack materials in the following order:

cassette (clear side): face to anode

sponge

Whatman paper

membrane

gel

Whatman paper

sponge

cassette (black side): face to cathode

5. Mini Trans-Blot cell transfer system is used. The cassette is placed into the transfer chamber with the black side facing black such that the membrane is between the gel and the anode. It is vitally important to ensure this orientation, otherwise the proteins will be lost from the gel into the buffer rather than transferred to the membrane.
6. Place cooling unit filled with ice into the transfer chamber to maintain a temperature no warmer than room temperature and activate a magnetic stir bar in the chamber.
7. The lid is put on chamber and connected with the power supply. Then the transfer can be accomplished at 100 V for 75 min.
8. Once the transfer is complete, the membranes are taken out of the gel holder cassette (see Notes 4). The blue protein standard molecular weight marker should be clearly visible. Mark the orientation of the membrane by cutting the lower right corner of it. Then, the membranes are incubated in blocking buffer for 1 h at room temperature on a rocking platform.
9. The blocking buffer is discarded and the primary antibody diluted in TBS-T/2% BSA. Incubate the membrane for 2 h at room temperature or overnight at 4°C in the sealed pouch on a rotating device.

   The dilution of primary claudin antibodies are as follows:

   Claudin-2: 1:1,000

   Claudin-3: 1:1,000–2,000

   Claudin-4: 1:800

   Claudin-5: 1:500
10. The primary antibody is then removed and the membrane washed three times for 5 min each with TBS-T on a rocking platform.
11. The secondary antibody 1:4,000–5,000 is freshly prepared in blocking buffer and the membrane incubated for 1 h at room temperature on a rocking platform.
12. The secondary antibody is discarded and the membrane washed three times for 10 min each with TBS-T on a rocking platform.

13. The working solution of SuperSignal West Pico Chemiluminescent Substrate is prepared by combining two solutions in a 1:1 ratio. The membrane is rinsed briefly with PBS and placed into the chemiluminescent solution. Ensure that the membranes are completely covered for about 1 min.
14. The blot is removed from the ECL reagents and wrapped in a transparent plastic sheet.
15. The wrapped membranes are then placed in an X-ray film autoradiography cassette and exposed to films for a suitable exposure time in a dark room (see Notes 5).

3.2. Confocal Immunofluorescence for Claudins

1. Prepare sterilized 12-mm glass coverslips by autoclaving in a glass beaker with pieces of tissue paper. Pass the coverslips through the flame and then place it in each of the wells of a 24-well plate.
2. Prepare A549 cells at a density of $2.0–2.5 \times 10^5$ cells/mL, plate 110 μL as drops onto each coverslip ($2.2–2.75 \times 10^4$ cells), and incubate 20 min for the cells to attach. Then add 500 μL of culture medium to each well. The cells are cultured for 48 h to confluence and changed to serum-free DMEM after rinsing in PBS. After 2–4 h of starvation, the cells are treated with H_2O_2 and then rinsed rapidly twice with ice-cold HBSS.
3. 400 μL of 4% paraformaldehyde solution is then added and the cells are incubated for 30 min at room temperature on the rocking platform.
4. The paraformaldehyde is discarded and the residual quenched by rinsing in 100 mM glycine three times for 5 min each, followed by further three washes with HBSS for 10 min each at room temperature on the rocking platform.
5. The cells are permeabilized by incubation in permeabilization solution for 30 min at room temperature.
6. The samples are incubated with primary antibody (1–1.5 μg/mL) in antibody dilution buffer overnight at 4°C or for 2 h at room temperature.
7. The primary antibody is removed and the sample washed three times for 10 min each with HBSS.
8. The cells are pre-incubated with permeabilization solution for 30 min.
9. The secondary antibody is prepared at 1:250 in antibody dilution buffer and added to the samples for 2 h at room temperature. From this step, the sample should be put under aluminum foil and kept in dark.
10. The samples are rinsed with HBSS and 1 μg/mL DAPI is added for 20 min to stain the DNA and identify nuclei.

11. Wash the samples three times for 10 min each with HBSS.
12. The samples are then ready to be mounted. Add one drop of mounting medium, avoiding bubbles to glass microscopy slide and invert coverslip carefully onto the drop (see Notes 6).
13. The samples are stored to dry at room temperature overnight in dark and stored at -20°C in dark for up to several months.
14. The slides are viewed by a laser-scanning confocal microscope (Zeiss LSM 510 META).

4. Notes

1. Once you have removed the gel from the unit and the cassette, perform the transfer immediately.
2. When inserting the gel knife between the two plates and pushing up and down on the knife handle to separate the plates, use caution to avoid excessive pressure toward the gel.
3. To ensure a snug fit, use an additional pad since pads lose their resiliency after many uses. Replace pads when they begin to lose resiliency and are discolored.
4. Wear gloves at all times during the entire blotting procedure to prevent contamination of gels and membranes. Do not touch the membrane or gel with bare hands.
5. For storing nitrocellulose membranes, air-dry the membrane and store the membrane in an air-tight plastic bag at room temperature or 4°C. Avoid storing nitrocellulose at -20°C or -80°C, as they will shatter.
6. Nail polish can be used to seal the coverslips to the slides. When the varnish is dry, the samples can be viewed immediately.

Acknowledgments

This work was supported by the American Heart Association Scientist Development Grant 0730331N and NIH NHLBI Grant 1R01HL104092 (GH) and NIH NHLBI Grants 5R01 HL071626 and P01 HL060678 (RDM).

References

1. Furuse, M., Fujita, K., Hiiragi, T., Fujimoto, K., Tsukita, Sh. (1998) Claudin-1 and -2: novel integral membrane proteins localizing at tight junctions with 534 Cell-to-cell contact and the extracellular matrix no sequence similarity to occludin. *J Cell Biol.* **141**, 1539–1550.
2. Furuse, M. and Tsukita, S. (2006) Claudins in occluding junctions of humans and flies, *Trends Cell Biol.* **16**, 181–188.
3. Krause, G., Winkler, L., Mueller, S.L., Haseloff, R.F., Piontek, J., Blasig. I.E. (2008) Structure and function of claudins. *Biochim Biophys Acta.* **1778**, 631–645.
4. Kiuchi-Saishin, Y., Gotoh, S., Furuse, M., Takasuga, A., Tano, Y., Tsukita, S. (2002) Differential expression patterns of claudins, tight junction membrane proteins, in mouse nephron segments. *J Am Soc Nephrol.* **13**, 875–886.
5. Rahner, C., Mitic, L.L., Anderson, J.M. (2001) Heterogeneity in expression and subcellular localization of claudins 2, 3, 4, and 5 in the rat liver, pancreas, and gut. *Gastroenterology.* **120**, 411–422.
6. Wang, F., Daugherty, B., Keise, L.L., Wei, Z., Foley, J.P., Savani, R.C., et al. (2003) Heterogeneity of claudin expression by alveolar epithelial cells. *Am J Respir Cell Mol Biol.* **29**, 62–70.
7. Koval, M. (2009) Tight junctions, but not too tight: fine control of lung permeability by claudins. *Am J Physiol.* **297**, L217–218.
8. Peter, Y., Comellas, A., Levantini, E., Ingenito, E.P., Shapiro, S.D. (2009) Epidermal growth factor receptor and claudin-2 participate in A549 permeability and remodeling: implications for non-small cell lung cancer tumor colonization. *Mol Carcinog.* **48**, 488–497.
9. Morin, P.J. (2005) Claudin proteins in human cancer: promising new targets for diagnosis and therapy. *Cancer Res.* **65**, 9603–9606.
10. Wray, C., Mao, Y., Pan, J., Chandrasena, A., Piasta, F., Frank, J.A. (2009) Claudin-4 augments alveolar epithelial barrier function and is induced in acute lung injury. *Am J Physiol.* **297**, L219–227.
11. Furuse, M., Furuse, K., Sasaki, H., Tsukita, S. (2001) Conversion of zonulae occludentes from tight to leaky strand type by introducing claudin-2 into Madin-Darby canine kidney I cells. *J Cell Biol.* **153**, 263–272.
12. Amasheh, S., Meiri, N., Gitter, A.H., Schöneberg, T., Mankertz, J., Schulzke, J.D., et al. (2002) Claudin-2 expression induces cation-selective channels in tight junctions of epithelial cells. *J Cell Sci.* **115**, 4969–4976.
13. Sun, Y., Hu, G., Zhang, X., Minshall, R.D. (2009) Phosphorylation of caveolin-1 regulates oxidant-induced pulmonary vascular permeability via paracellular and transcellular pathways. *Circ Res.* **105**, 676–685.
14. Waschke, J., Bruggeman, P., Baumgartner, W., Zillikens, D., Drenckhahn, D. (2005) Pemphigus foliaceus IgG causes dissociation of desmoglein 1-containing junctions without blocking desmoglein 1 transinteraction. *J Clin Invest.* **115**, 3157–3165.

Chapter 22

Tracing the Endocytosis of Claudin-5 in Brain Endothelial Cells

Svetlana M. Stamatovic, Richard F. Keep, and Anuska V. Andjelkovic

Abstract

Claudin-5 is a transmembrane tight junction protein highly expressed in brain endothelial cells, the site of the blood–brain barrier. The properties of the brain endothelial tight junction complex are considered to be dependent on claudin-5 cell–cell interaction, putting this protein in a position to play a major role in the maintenance of brain endothelial barrier integrity. Thus, alterations in claudin-5 function can lead to "opening" of the paracellular route and increased brain endothelial barrier permeability. Recent work from the authors's laboratory has established that caveolae-dependent internalization/recycling of claudin-5 is a mechanism underlying transient increases in brain endothelial paracellular permeability in the presence of pro-inflammatory mediators. The biochemical and microscopic techniques presented here were used to investigate trafficking of claudin-5 during those changes in paracellular permeability.

Key words: Claudin-5, Internalization, Caveole, Recycling, Green fluorescent protein – GFP, Blood–brain barrier

1. Introduction

Claudin-5 is an integral membrane tight junction (Tj) protein that belongs to the PMP22/EMP/MP20/claudin family of proteins of which there are 24 members (for review, see ref. 1–3). Claudin-5, like other claudins, has a four-transmembrane spanning pattern: two extracellular loops and two cytoplasmic termini: a very short internal N-terminal sequence (2–6 amino acids) and a longer internal C-terminal sequence (21–63 amino acids) (3). The first extracellular loop (49–52 amino acids) influences paracellular charge selectivity, while the second extracellular loop (16–33 amino acids) is the receptor for a bacterial toxin (3–6). The C-terminus possesses the binding site for the cytoplasmic proteins

Kursad Turksen (ed.), *Claudins: Methods and Protocols*, Methods in Molecular Biology, vol. 762, DOI 10.1007/978-1-61779-185-7_22, © Springer Science+Business Media, LLC 2011

ZO-1, ZO-2, ZO-3, MUPP1, PATJ through a PDZ motif (4, 6, 7). Claudin-5 function in the Tj complex is to limit paracellular ion movement selectively and this produces the high electrical resistance of the brain endothelial barrier (8, 9). There are strong indications that brain endothelial claudin-5 may form pores of variable size (~10 Å), which may be involved in transjunctional movement of water (8–11). Deletion of claudin-5 (claudin-5 knockout mice) results in a size-selective increase in blood–brain barrier (BBB) permeability, although it does not inhibit Tj formation (3, 12). Thus, claudin-5 is considered a major component maintaining BBB integrity.

Under conditions that increase BBB paracellular permeability, there is ongoing Tj complex remodeling. This particularly affects the adhesion trans-interaction of claudin-5. Morphological and biochemical analyses have shown that claudin-5 redistributes/relocates between the membrane and the cytosolic fractions (13). In this way, the Tj complex loses a major adhesion force responsible for occlusion of the paracellular pathway. Two specific events are indicated as being involved: phosphorylation of claudin-5 (mostly of threonine and serine residues) and/or endocytosis (14, 15). Several recent studies have indicated that phosphorylation of Thr207 could be directly involved in disturbing the interaction of claudin-5 with zonula occludens protein-1 (ZO-1), which could further facilitate the redistribution of claudin-5 (16, 17). In recent studies from our laboratory, we have highlighted potential mechanisms of claudin-5 redistribution, focusing on endocytosis, and have shown caveolae-mediated internalization and recycling of claudin-5 during the increases in brain endothelial permeability induced by a chemokine, monocyte chemoattractant protein-1 (CCL2) (15). Similar patterns of claudin redistribution have also been described in some other barrier systems, including the blood–retinal barrier and epithelial barriers (18, 19). Therefore, endocytosis/internalization of claudin-5 could be considered as a basic mechanism regulating claudin-5 kinetics in endothelial or epithelial cells and a critical one for enhancing paracellular permeability and establishing barrier properties.

In this chapter, we outline the methods that have been used in our laboratory to map claudin-5 trafficking during increases in brain endothelial barrier permeability. The same protocols with slight modification could be used on other cells and systems. The protocols are divided into three parts: the first highlights the engineering of GFP-tagged claudin-5, which facilitates tracking on claudin-5 internalization; the second outlines microscopic techniques for studying internalization of claudin-5; and the last part focuses on biochemical methods for characterizing the process of internalization/endocytosis.

2. Materials

1. bEnd.3 cells (ATCC).
2. Dulbecco's modified Eagle's medium (DMEM), inactivated fetal calf serum (FBS), HEPES, 200 mM glutamine, 100× antibiotic/antimycotic, and trypsin-EDTA 0.05% solution; store aliquots at −20 C (all from Invitrogen).
3. pAcGFP-C plasmid (Clontech) and whole-length claudin-5 cDNA (Open Laboratory).
4. Designed primers, PCR kit (Platinum Tag DNA polymerase, 10× PCR buffer; 50 mM $MgCl_2$), DNase RNase-free water, and 10 mM dNTP mix (all from Invitrogen).
5. Restriction enzymes *Sal* I and *Hin*dIII (both from Invitrogen).
6. NucleoSpin Extract II kit and In-Fusion Dry-Down PCR cloning kit (both from Clontech).
7. SOC media, *Luria–Bertani* (LB) medium, kanamyci*n*, *Escherichia coli* (Fusion Blue competent cells, all from Clontech), and LB agar/kanamycin/ X-gal IPTG plates.
8. Mini and Maxi plasmid purification kit (Invitrogen), glycogen, 2-propanol, and 70% ethanol.
9. Sterile phosphate-buffered saline (PBS) solution (1 mM KH_2PO_4, 155 mM NaCl, and 2.97 mM $Na_2HPO_4{\cdot}7H_2O$, pH 7.4).
10. Lipofectamine 2000, OptiMEM, and selection antibiotic G418 (all from Invitrogen).
11. Monocyte chemoattractant protein-1 (MCP-1/CCL2; PrepoTech), and lipopolysaccharide (LPS) from *E. coli* 0111:B4 (Sigma–Aldrich).
12. Tracers: Alexa596-cholera toxin and BODIPY-TR ceramide (both from Invitrogen).
13. Antibodies: anti-caveolin-1, -Rab4, -Rab5, -Rab9 antibodies (BD Bioscience); anti-claudin-5 (Invitrogen); anti-GFP antibodies (Clontech); anti-EAE1 and anti-β-actin antibodies (Abcam); Texas Red conjugated anti-mouse and anti-rabbit antibodies (Vector Laboratories); and horseradish peroxidase (HRP)-conjugated anti-mouse or rabbit antibodies (Vector Laboratories).
14. Bovine serum albumin (BSA), Triton-X100, polyoxyethylene-20-sorbitan monolaurate (Tween-20) (all from Sigma), and paraformaldehyde (Sigma–Aldrich).
15. Sulfo-NHS-SS-biotin (Pierce), EZview Red streptavidin affinity gel, and EZview™ Red protein A affinity gel (Sigma–Aldrich).

16. Tris–HCl, NaCl, sodium dodecyl sulfate (SDS), sodium deoxycholate, NH_4Cl, $MgCl_2$, $CaCl_2$, glutathione, NaOH, and protease inhibitor cocktails (all from Sigma–Aldrich).
17. 2-(4-Morpholino) ethanesulfonic acid (MES), ethylenediaminetetraacetic acid (EDTA), aprotinin, leupeptin, Na-orthovanadate, NaF, sucrose, KCl, magnesium acetate tetrahydrate (Mg-acetate), dithiothreitol (DTT), 4-(2-hydroxyethyl)piperazine-1-ethanesulfonic acid, *N*-(2-hydroxyethyl) piperazine-*N′*-(2-ethanesulfonic acid) (HEPES); soyabean trypsin inhibitor, CH_3CO_2H, Tris, penicillin G, 1,4-piperazinediethanesulfonic acid, piperazine-1,4-bis(2-ethanesulfonic acid), and piperazine-*N,N′*-bis(2-ethanesulfonic acid) (PIPES) (all from Sigma–Aldrich).
18. Laemmli sample buffer, Ponceau S (Sigma–Aldrich), nitrocellulose membrane, nonfat dry milk, chemiluminescent HRP substrate kit, and 15% Tris–HCl SDS-PAGE gels (all from Bio-Rad).

3. Methods

3.1. Engineering Green Fluorescent Protein-Tagged Claudin-5

To obtain functionally active claudin-5 in brain endothelial cells, the GFP tag has to be added at the N-terminus as the C-terminus contains a PDZ-binding domain essential for the association of claudin-5 with the Tj scaffolding protein ZO-1 (Fig. 1). As with native claudin-5, GFP-tagged claudin-5 is directed to the plasma membrane. The presence of the GFP tag enables real-time fluorescence imaging of internalization and recycling to the plasma membrane and facilitates co-localization studies (e.g., with endocytotic vesicles).

3.1.1. Generating the GFP-Claudin-5 Chimer

1. To generate claudin-5 cDNA with appropriate restriction sites for subcloning, 100 ng of existing plasmid containing whole-length claudin-5 and PCR primers are used. For PCR, the suggested primers are sense 5′-AAGGCCTCTGTCGACATGGGGTCTGCAGCGTTGGA-3′; antisense 5′-AGAATTCGCAAGCTTTTAGACATAGTTCTTCTTGTCGTAATC-3′, which have incorporated sites for the restriction enzymes *SalI* and *Hin*dIII. Platinum Tag DNA polymerase (Invitrogen) is suggested. PCR cycles for amplification are 95°C 1 min, 65°C min, 72°C 1 min. A total of 35 cycles with a hot start are strongly recommended.
2. After completing amplification, the PCR product is analyzed by electrophoresis on an agarose/EtBr gel to confirm the presence of a single DNA fragment and to estimate the concentration of the PCR product. If more than one DNA band

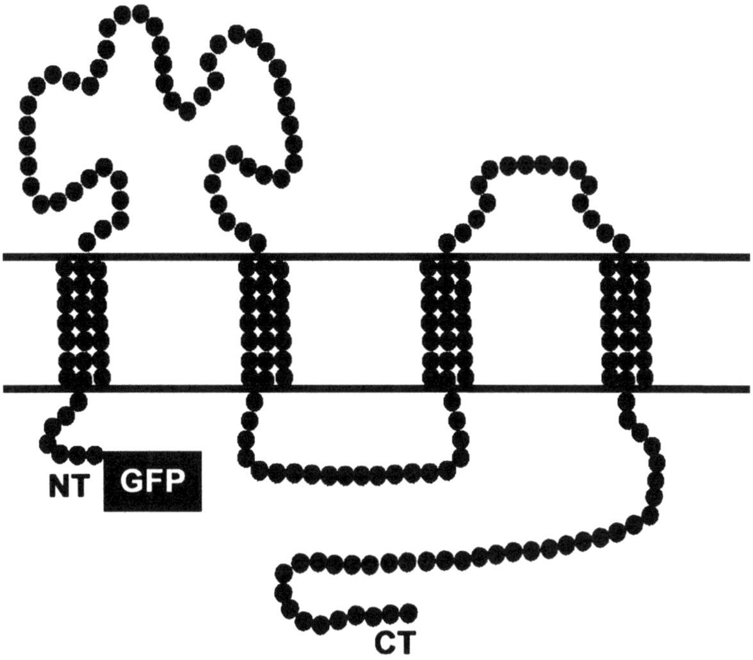

NT GFP

CT

Fig. 1. Schematic representation of green fluorescent protein-tagged claudin-5 (GFP-claudin-5). GFP tag has to be added at the N-terminus (NT) due to the fact that C-terminus (CT) contains a PDZ-binding domain essential for the association of claudin-5 with the tight junction scaffolding protein ZO-1.

is found, use a NucleoSpin Extract II kit (Clontech) to purify the PCR products. The amount of PCR products can be estimated by measuring against a known standard or molecular weight marker ladder run on the same gel.

3. Digest the pAcGFP-C vector with restriction enzymes *Sal* I and *Hin*dIII following the manufacturer's suggestions. After digestion, purify the linearized vector with NucleoSpin Extract II kit (Clontech). Verify the linearized pAcGFP-C vector by electrophoresis on an agarose/EtBr gel.
4. Ligate the linearized pAcGFP-C and claudin-5 cDNA using an In-Fusion Dry-Down PCR Cloning kit (Clontech) following the manufacturer's instructions. The molar ratio between the PCR insert and the vector should be 2:1. Incubate the reaction for 15 min at 37°C, followed by 15 min at 50°C, and then transfer the tubes to ice. Dilute the reaction with 40 μl TE buffer (10 mM Tris–HCl and 1 mM EDTA) and mix well. Verify the cloning reaction by electrophoresis on an agarose/EtBr gel.
5. Transform the Fusion Blue competent cells (>10^8 cfu/μg) with the diluted reaction from step 4 (5–50 μl of bacteria), using the heat shock method (30 min on ice, 45 s at 42°C, followed by 1 min on ice). Add 450 μl of SOC media

(Invitrogen) and shake extensively for 1 h at ~30 × *g* at 37°C in an incubator-shaker.

6. Take 50 μl of cells and add SOC media (Invitrogen) up to 100 μl. Mix and centrifuge at 6,000 rpm (500 × *g*) for 5 min, and then resuspend cells in 100 μl fresh SOC media. Spread the transformed Fusion Blue competent cells on a LB plate containing 30 μg/ml kanamycin and overlay with IPTG and X-gal (LB agar/kanamycin/X-gal IPTG plates). Incubate the plate for 24 h at 37°C.
7. Next day, pick up the white colonies and extract the DNA using standard miniprep protocols.
8. Verify the insertion of claudin-5 cDNA by digestion with *SalI* and *Hin*dIII restriction enzymes and electrophoresis on an agarose/EtBr gel.
9. Sequence the claudin-5-GFP.
10. Prepare large-scale GFP-claudin-5 construct using a Maxi plasmid purification kit.

3.1.2. Expression of GFP-Claudin-5 and Generation of Stable Cell Line

1. bEnd.3 cells are grown to 50–70% confluence in 35-mm dishes in growth media (DMEM, supplemented with 10% FBS, 2 mM glutamine, and 1× antibiotic/antimycotic).
2. Remove media and wash the cells 2× with sterile PBS (pH 7.4).
3. Overlay the cells with OptiMEM and leave them in an incubator at 37°C during the preparation of transfection solution.
4. Prepare the transfection solution which contains 3 μl Lipofectamine 2000 per 1 μg construct pAcGFP-C – claudin-5 in OptiMEM. As a control, use the pAcGFP-C vector only.
5. Replace the OptiMEM with transfection solution. Incubate the cells for 24 h at 37°C in a tissue-culture incubator.
6. Stop transfection by replacing the transfection solution with bEnd.3 growth media. Leave the cells to recover for 24 h.
7. To form a stable cell line, expose the transfected cells to growth media containing *G418* (start with 200 μg/ml, and after 3 days replace media with 50 μg/ml). Approximately 10–14 days are necessary to remove untransfected cells and purify the GFP-claudin-5 bEnd.3 cells.
8. Test claudin-5-GFP expression level by immunofluorescence and Western blot analysis (Fig. 2).

3.2. Internalization Assay: Microscopy

Using bEnd.3 cells stably expressing GFP-claudin-5 and tracers for different endocytotic vesicles, internalization of claudin-5 can be examined by live cell imaging (Subheading 3.2.2) and standard immunofluorescence on fixed tissue (Subheading 3.2.3). Such results can be quantified to assess the relative importance of different internalization pathways (Subheading 3.2.4).

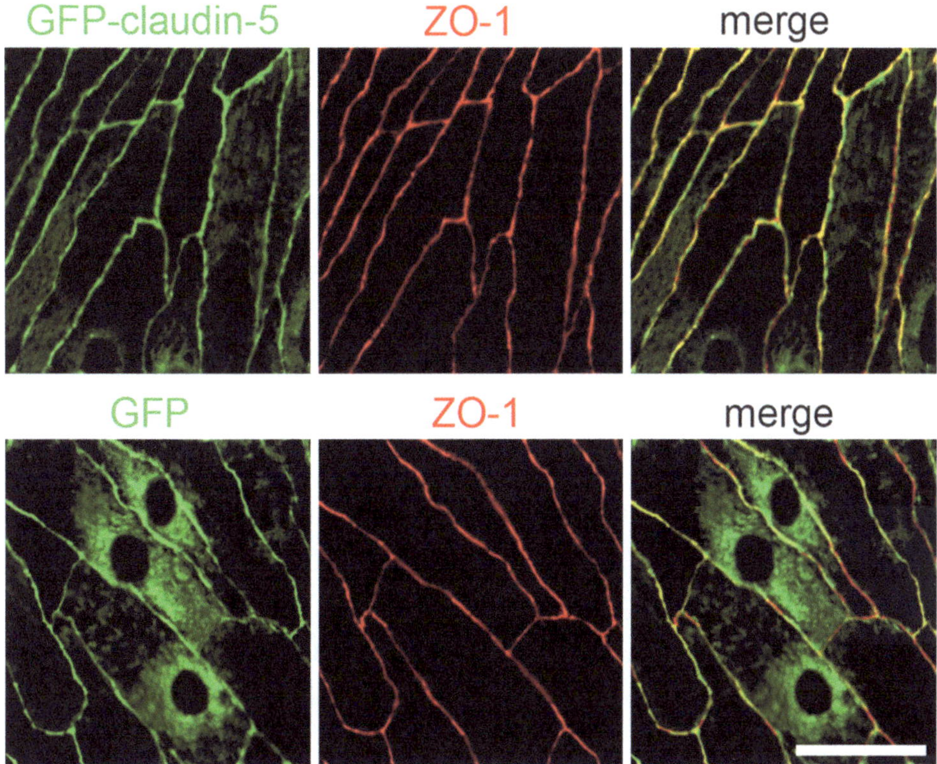

Fig. 2. Staining with anti-GFP and anti-ZO-1 antibodies, confirmed the correct localization of GFP-claudin-5 proteins in bEnd.3 cells, with co-localization in merged images. Scale bar 200 μm.

3.2.1. Tracer Assay

1. GFP-claudin-5 bEnd.3 cells are plated on a 35-mm FluoroDish (World Precision Instrument Inc). Based on our recent studies (15), claudin-5 localization and functional properties are achieved at day 7 after initial plating. Thus, cells should be used experimentally after this time period (see Subheading 3.3.1 step 1, method for evaluating maturity of barrier).
2. Remove growth media, wash the cells with PBS (pH 7.4) and overlay the cells with DMEM without phenol red.
3. Prepare tracers: for visualization of caveolae-dependent internalization, use either Alexa596-cholera toxin or BODIPY-TR ceramide (both from Invitrogen). Working solution for Alexa596-cholera toxin is 10 μg/ml, while that for BODIPY-TR ceramide is 5 μM, prepared in DMEM w/o phenol red.
4. Remove media and overlay the cells with one of the tracer's solution. Leave the cells for 30 min to allow tracer accumulation on the cell surface.
5. To initiate increased permeability and internalization of claudin-5, add either CCL2 (100 ng/ml) or LPS (5 μg/ml)

as a stimulus. The incubation time should be between 0 and 60 min. The samples can be observed by live cell imaging (Subheading 3.2.2) or processed for immunocytochemistry (Subheading 3.2.3) (see Notes 1 and 3).

3.2.2. Live Cell Image

Time-lapse microscopy is performed using a Leica DMIRB inverted microscope (Leica Microsystems, Germany, objective 40×). The stage is surrounded by a temperature hood to maintain a constant temperature of 37°C during experiments using a heater (fan and temperature controller). The time-lapse experiments are conducted for 0–60 min. The images are collected every 5 min using an Olympus DP-30 high-sensitivity grayscale CCD camera (Olympus, Japan) attached to the microscope and stored on a hard disk (see Note 2).

3.2.3. Immunofluorescence

1. Remove the experimental media at the required time points and fix the samples with 4% paraformaldehyde (in 0.1 M sodium phosphate buffer, pH 7.2) for 20 min at 20°C.
2. Wash the samples with PBS (1 mM KH_2PO_4, 155 mM NaCl, and 2.97 mM $Na_2HPO_4{\cdot}7H_2O$, pH 7.2) 3×5 min.
3. Incubate the samples with blocking solution containing 5% normal goat serum (Vector Laboratories) and 0.05% Tween-20 (polyoxyethylene-20-sorbitan monolaurate) in PBS (pH 7.2) for 30 min at room temperature.
4. Incubate the samples in primary antibody solution at 4°C for 18 h. The primary antibody solution contains 5% normal goat serum (Vector Laboratories), 0.05% Tween, PBS (pH 7.2), and selected primary antibody: anti-caveolin-1 (at working dilution 1:1,000), anti-early endosome marker EEA1 (working dilution 1:500), anti-Rab4 (working dilution 1:1,000), anti-Rab5 (working dilution 1:500), or anti-Rab9 (working dilution 1:500).
5. Wash the samples with PBS 3×10 min.
6. Add secondary antibody solution containing either anti-rabbit or anti-mouse conjugated with rhodamine (Vector Laboratories: working dilution 1:200), 3% normal goat serum, 0.005% Tween, and PBS (pH 7.2). Incubate the samples for 2 h at room temperature in the dark.
7. Wash the samples with PBS (pH 7.2) for 3×15 min.
8. Mount the samples with Vectashield mounting media (Vector laboratories) (see Notes 1 and 3).

3.2.4. Quantitative Immunoflouresence: Evaluation of Co-localization

1. Set up conditions on confocal microscopy (e.g., Zeiss LSM META 510 laser scanning microscope) to avoid interference between channels and saturation. The contrast, brightness, and pinhole should be kept at a constant value.

2. From selected area, obtain z-stock of consecutive optical sections.
3. Launch Image J (NIH Image) and upload co-localization finder plug-in function.
4. Split images on red and green channels or upload two images from two different channels.
5. Correct background using the following formula: corrected co-localization = measured co-localization – background co-localization/1 – background co-localization/100, and apply to ever-formed stock.
6. Go to "plugins," "co-localization finder" to obtain image with marked co-localization sides as well as the numerical value of Pearson's correlation coefficient (Pearson's Rr), overlap rate, % pixels). Pearson coefficient ranges between –1 (perfect negative correlation) and +1 (perfect positive correlations between two images). A coefficient of 0 means no correlation between two images (15, 20, 21).

3.3. Internalization Assays: Biochemistry

In addition to microscopy, a number of biochemical techniques are available to study claudin-5 internalization. These include biotinylation assays (Subheadings 3.3.1 and 3.3.2) and subcellular fractionation (Subheading 3.3.3). These techniques are complementary to the microscopy methods in the previous section and have advantages in terms of quantification of internalization.

3.3.1. Biotinylation Assay for Endocytosis of Claudin-5

1. bEnd.3-GFP-claudin-5 cells are grown to confluence in 24-mm Transwell culture systems (Corning). Three days after initial plating, the transendothelial electrical resistance should be first evaluated (TEER, EVOM^2, Word Precision Instrument Inc.) and, thereafter, every 2 days. bEnd.3 cells with a resistance of approximately 60 Ωcm^2 (between 7 and 10 days of initial plating) can be utilized in experiments. This resistance ensures that tight junction and barrier properties are satisfied.
2. Remove media from cells and wash 2× with PBS. Overlay the apical side of bEnd.3 cells with ice-cold PBS containing sulfosuccinimidyl 2-(biotinamido) ethyl-dithioproprionate (sulfo-NHS-SS-biotin; 0.5 mg/ml; Pierce). Gently rotate the culture plate for 60 min at 4°C to allow biotinylation of cell-surface proteins. Then wash the cells with PBS (1 mM $MgCl_2$ and 0.1 mM $CaCl_2$, pH 7.4) containing 50 mM NH_4Cl to quench free sulfo-NHS-SS-biotin.
3. To initiate internalization, incubate the cells with a stimulus that increases brain endothelial barrier permeability (e.g., CCL2 [100 ng/ml], or LPS [5 μg/ml]) in DMEM w/o phenol red for varying times (0–60 min) at 37°C.

4. Stop internalization after varying time points, wash the cells 2×20-min with ice-cold glutathione solution (50 mM glutathione, 75 mM NaCl, 75 mM NaOH, and 1% BSA) which will remove all cell-surface biotin groups.
5. Scrape cells from the surface with a disposable cell lifter and add 500 ml of lysis buffer (25 mM Tris–HCl, pH 7.4, with 150 mM NaCl, 0.1% SDS, 1% Triton X-100, and 1% deoxycholate, with a cocktail of protease inhibitors; Roche). Transfer cell lysate to 1.5 ml low-retention microcentrifuge tube, rotate the tubes for 1 h at 4°C, and then centrifuge at 12,000×*g* for 15 min. Collect supernatant and transfer to fresh 1.5-ml microcentrifuge tube.
6. Incubate the collected supernatant with EZview Red streptavidin affinity gel (Sigma) previously prepared according to the manufacturer's protocol. Briefly, transfer 20–50 μl of 50% slurry per sample to fresh microcentrifuge tubes and add 750 μl lysis buffer to equilibrate the beads. Vortex the tube for 30 s and spin in a microcentrifuge for 1 min at 5,000×*g*. Carefully remove supernatant and repeat equilibration one more time.
7. Transfer the experimental sample to equilibrated EZview red streptavidin affinity gel beads, briefly vortex, and incubate them for 2 h at 2–8°C with gentle rotation to precipitate biotin-labeled protein.
8. Centrifuge the tubes for 1 min at 5,000×*g* and aspirate supernatant. Wash the pellet three times by resuspending pellet in 1 ml lysis buffer and centrifuge for 1 min at 5,000×*g*.
9. After removing the final wash supernatant, shear the beads through a 27-G needle attached to a 1-ml syringe. Resuspend the gel beads in 50 ml of Laemmli sample buffer and prepare the sample for SDS-PAGE and Western blot analysis see Subheading 3.5, (see Note 4).

Suggested controls: To assure the quality of biotinylation and stripping procedures, two additional controls should be included: (a) *surface*, the level of surface biotinylated proteins and (b) *gs*, the level of biotin-labeled protein on the cell surface after quenching surface biotin. For this purpose, the bEnd.3 cells should be biotinylated and not exposed to the stimulus that alters brain endothelial barrier permeability. The *surface* samples are collected after biotinylation and removing excess biotin immediately preceding steps 5–8. For the *gs* samples, after biotinylation, wash to remove excess biotin, glutathione strip and proceed as indicated in steps 5–8. In addition, to evaluate the level of sequestrated/internalized pool of claudin-5, comparing with total biotin-labeled claudin-5, a sample should be collected after step 3 and processed through steps 5–8. This sample is denoted as *total*, which indicates the total amount of biotin-labeled claudin-5 (22).

3.3.2. Biotinylation Assay for Recycling of Claudin-5

1. bEnd.3-GFP-claudin-5 cells are grown to confluence in a 24-mm Transwell chamber system. Three days after initial plating, TEER should be measured. The cells are ready for experimentation when the TEER reaches approximately 60 Ω cm^2.
2. Remove growth media, wash the cells twice in PBS, and proceed with biotinylation (see Subheading 3.3.1, step 2).
3. Induce alterations in brain endothelial permeability and endocystosis by adding a stimulus such as CCL2 or LPS, as suggested in Subheading 3.3.1 step 3.
4. After 60 min of exposure, remove media with stimulus and replace with fresh DMEM w/o phenol red. Return the cells to 37°C for varying times (0–60 min) during which there may be recycling of claudin-5 to the membrane surface.
5. Recycling is stopped by removing the media and washing with ice-cold glutathione solution (50 mM glutathione, 75 mM NaCl, 75 mM NaOH, and 1% BSA) two times to remove all cell-surface biotin groups. The cells are further processed as suggested in Subheading 3.3.1, steps 5–8.

The level of recycling of claudin-5 is estimated from the content of intracellular claudin-5 after removing the internalization stimulus. The suggested controls in Subheading 3.3.1 should also be useful for this type of assay (Fig. 3).

3.3.3. Subcellular Fractional Analysis

Isolation of Caveolin-Rich Membranes

1. bEnd.3-GFP-claudin-5 cells are grown to confluence in 60-mm tissue-culture dishes. Seven days after initial plating, the cells are ready for experiments.
2. Remove the growth media and wash the cells 2× with PBS (pH 7.2) at room temperature.
3. Initiate internalization of claudin-5 with a stimulus that increases brain endothelial barrier permeability (e.g., CCL2 or LPS).

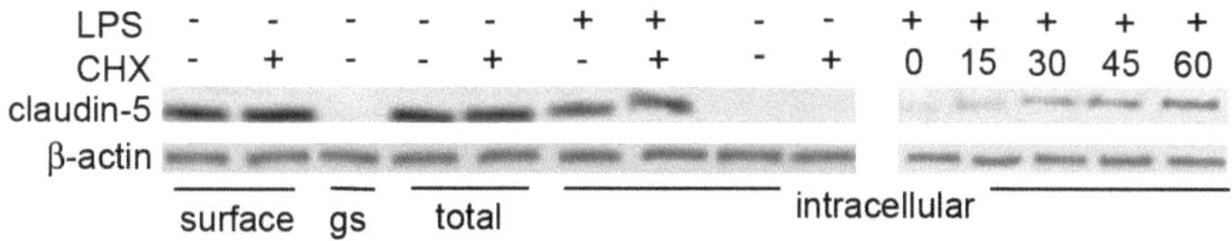

Fig. 3. Internalization of surface biotinylated claudin-5 proteins. Confluent bEnd.3, treated with and without cycloheximide (CHX), were surface biotinylated at 0°C and then exposed to LPS (5 μg/ml) for 60 min at 37°C to allow internalization. Any membrane bound biotin was removed by glutathione solution (glutathione stripping – gs). *Surface* indicated the biotinylated claudin-5 on cell surface, gs indicated glutathione stripping of surface biotin; *total* indicated total amount of biotinylated claudin-5, and *intracellular* indicated the portion of biotinylated-internalized claudin-5. Adjusted blot showing the time course (0–60 min) of internalized biotinylated claudin-5 during bEnd.3 exposure to LPS.

4. Stop the internalization by removing media and washing the cells with PBS (pH 7.2).
5. Overlay the cells with 1 ml ice-cold lysis buffer for 20 min. The lysis buffer is 1% (v/v) Triton-X 100 in MNE buffer, the latter containing 25 mM MES (pH 6.5), 150 mM NaCl, 5 mM EDTA, 1 μg/ml aprotinin, 1 μg g/ml leupeptin, 1 mM Na-orthovanadate, and 10 mM NaF.
6. Scrape cells from the culture plate and collect them in Dounce-type homogenizer. Homogenize the samples with 20 strokes on ice and transfer to fresh microcentrifuge tubes.
7. Centrifuge the samples for 10 min at 400 × *g* at 4°C to remove nuclei (pellet). Collect supernatant.
8. Dilute the supernatant 1:2 with 80% (w/v) sucrose in MNE buffer and place at the bottom of 12.5-ml ultracentrifuge tubes (Beckman).
9. Overlay the supernatant gently and slowly (with the pipette tip leaning on tube wall) with 6 ml of 35% sucrose in MNE followed by 3 ml of 5% sucrose in MNE. This will form the 5–40% discontinuous sucrose gradient.
10. Centrifuge the samples at 200,000 × *g* for 20 h at 4°C in a swinging bucket rotor (model SW41; Beckman Instruments) to allow separation of the low-density caveolae.
11. Collect fractions (0.92 ml) from top to bottom of the gradient. Collected samples can be used for SDS-PAGE, immunoprecipitation, and Western blot analysis (Fig. 4a). The enrichment of caveolin-1 in different fractions is determined by Western blot (23) (see Note 5).

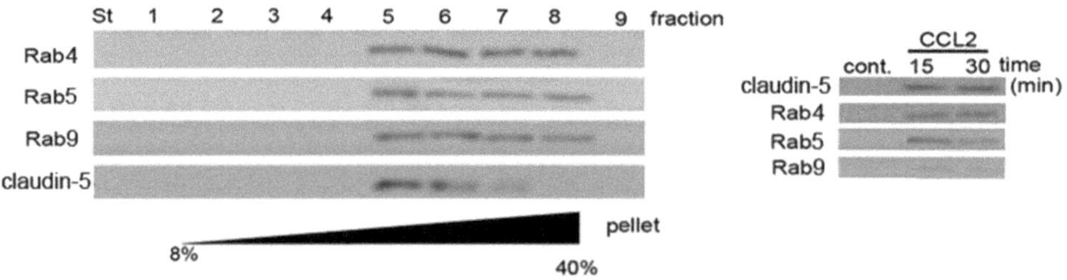

Fig. 4. After exposure of bEnd.3 cells to CCL2 (100 ng/ml), endosome-rich fractions were prepared as described Subheading "Isolation of Endosomes." Fractions were analyzed by Western blotting. Std-standard control line, Fraction 1 is the top of gradient. Fractions 1 + 2 homogenization buffer (8% sucrose); fraction 3, 25% sucrose/homogenization buffer interphase; fraction 4, 25% sucrose; fraction 5, 35%/25% interphase (early endosomes); fraction 6, 35% sucrose; fraction 7, 40%/35% interphase, fraction 8, 40% sucrose; fraction 9, pelleted material. In addition, immunoprecipitation and Western blot analysis were performed to analyze the endosome Rab4-, Rab5-, and Rab9-rich fraction. Anti-Rab4, -Rab5 and -Rab9 antibodies were from BD Bioscience. Data represent one of three successful experiments. Time points 15 and 30 min represent treatment of bEnd.3 cells for this indicated time. Control is untreated cells.

Isolation of Endosomes

1. bEnd.3-GFP-claudin-5 cells are grown to confluence in growth media (Subheading "Isolation of Caveolin-Rich Membranes") in 60-mm tissue-culture dishes. Seven days after initial plating, the cells are ready for experiments.
2. Wash the cells twice with PBS (pH 7.2).
3. Initiate internalization with a stimulus that increases barrier permeability (CCL2, LPS, step 3, Subheading "Isolation of Caveolin-Rich Membranes") and incubate the cells with the stimulus for 0–60 min at 37°C.
4. Stop internalization by removing experimental media and washing the cells with 5 ml PBS (pH 7.2).
5. Overlay the cells with 1 ml hypoosmotic buffer containing 15 mM KCl, 1.5 mM magnesium acetate, 1 mM DTT, and 10 mM-HEPES, pH 7.5. Scrape cells and collect them in microcentrifuge tube.
6. Place the cells in a "nitrogen bomb" for 20 min at 750 lb/in^2 (5.17 MPa; nitrogen cavitation lysis). Then transfer the samples to a Dounce-homogenizer and homogenize with 20 strokes to complete cell lysis.
7. Restore isotonicity by adding 0.1 volume hyperosmotic buffer 700 mM KCl, 40 mM magnesium acetate, 1 mM DTT, and 10 mM HEPES, pH 7.5.
8. Centrifuge the cells for 5 min at 800 × *g*.
9. Wash pellet with 2 ml of a buffer containing 85 mM KCl, 5.5 mM Mg acetate, 1 mM DTT, and 10 mM HEPES, pH 7.5.
10. Add 1 g/ml trypsin to the supernatant and incubate for 3 min at 37°C.
11. Stop the trypsinization with 1.5 μg/ml of soybean trypsin inhibitor.
12. Centrifuge the samples for 20 min at 145,000 × *g*.
13. Resuspend resulting crude membrane pellet in 1 ml of homogenization buffer (0.25 M sucrose, 1 mM EDTA, and 10 mM Tris, pH 8).
14. Load the samples on a discontinuous sucrose gradient. The sucrose gradient is made by gently layering the following (in exact order) sucrose solutions: 0.5 ml of 2.5 M sucrose in water, 4 ml of 40%, 4 ml of 30%, and 3 ml of 20% sucrose in gradient buffer (1 mM EDTA/10 mM acetic acid/10 mM Tris, pH 7.5).
15. Centrifuge the samples for 2 h at 100,000 × *g* in an SW 40 Beckman rotor.
16. Collect the endosomes (1 ml) from the 30%/20% sucrose gradient interface, a cloud-like phase in the lower part of the tube.

17. Resuspend collected endosome fraction (approximately 500 μl) in 1 ml of translocation buffer (110 mM KCl, 15 mM $MgCl_2$, 1 μg/ml penicillin G, and 20 mM PIPES, pH 7.1).
18. The isolated endosomes are now ready for further analysis including SDS-PAGE gel, Western blot, and immunoprecipitation ((24), Fig. 4b). The enrichment of the endosome fraction is determined by Western blotting for Rab4, Rab5, and Rab9, endosome markers (see Note 5).

3.4. Immuno-precipitation and Co-immunoprecipitation

Co-localization of claudin-5 and intracellular vesicles can be established by immunoprecipitation and co-immunoprecipitation. Samples from Subheading "Isolation of Caveolin-Rich Membranes" and "Isolation of Endosomes" can be further processed using the following protocol:

1. To collected samples, add 10 μl of an appropriate antibody dilution (use a working dilution of 1 μg/ml of antibody). Vortex briefly.
2. Incubate after thorough gentle mixing for 4 h at 2–8°C to allow antibody–antigen complexes to form.
3. Before using the EZview™ Red protein A affinity gel, carefully mix until it is completely and uniformly suspended.
4. Transfer 50 μl of the 50% slurry in a 1.5-ml microcentrifuge tube and place on ice.
5. Wash and equilibrate beads with the same buffer as in collected samples in Subheading 3.3.1, step 6, by adding 750 μl of buffer, vortexing for 5 s, and centrifuge in a microcentrifuge for 1 min at 8,000×*g*. Repeat washing 2×, remove supernatant, and place beads on the ice.
6. Transfer the samples/antibody from step 4 to equilibrated EZview beads Red protein G gel beads, vortex briefly, and incubate with thorough, gentle mixing for 2 h at 2–8°C.
7. Centrifuge the samples in a microcentrifuge for 1 min at 8,000×*g*. Set on ice. Remove supernatant carefully and set the tube with the bead pellet on ice.
8. Wash bead pellet by adding 750 μl of RIPA buffer (50 mM Tris, pH 8.0, 150 mM NaCl, 1% Triton X-100, 0.5% sodium deoxycholate, and 0.1% SDS). Vortex briefly and incubate with thorough, gentle mixing at 2–8°C for 5 min. Centrifuge for 1 min at 8,000×*g*. Remove supernatant.
9. Repeat step 8 twice more.
10. To elute the bead–antibody–antigen complex, add 25 μl of RIPA buffer to bead pellet. Vortex briefly.
11. Then add 25 ml of 2× Laemmli sample buffer. Vortex briefly.

12. Boil the samples for 5 min (100°C). Vortex and centrifuge for 1 min at 8,000 × *g* in a microcentrifuge to pellet the EZview Red protein G affinity gel beads. Use these samples for SDS-PAGE Western blot analysis.

3.5. Western Blotting

1. Prepare samples as described in Subheadings 3.1.2, 3.3.1, 3.3.2, "Isolation of Caveolin-Rich Membranes" and "Isolation of Endosomes." Add 25 μl of 2× concentrated Laemmli sample buffer and boil for 5 min.
2. Separate proteins on 15% SDS-polyacrylamide gel using a minigel apparatus and Tris/glycine/SDS buffer (50 mM Tris–HCl, pH 8.3, 196 mM glycine, and 0.1% SDS).
3. Transfer proteins to a nitrocellulose membrane at 50 V for 2 h with ice-cold Tris/glycine/methanol transfer buffer (25 mM Tris, 192 mM glycine, and 10% methanol). The transferred proteins are visualized by immersing the membrane in 0.2% Ponceau S (in acetic acid). Mark down the molecular weight standards with ballpoint pen and rinse off the Ponceau S staining with repeated washes with distilled water.
4. Wash the membrane 2 × 10 min with TBS/T (0.05 M Tris–HCl, 0.15 M NaCl, pH 7.5, and 0.1% Tween).
5. Block the membrane with 5% nonfat dry milk in TBS/T, pH 7.5, for 2 h with constant gentle shaking.
6. Wash the membrane 1 × 10 min with TBS/T and add the solution containing the primary antibody (e.g., claudin-5, working dilution 1:1,000) and 5% nonfat dry milk in TBS/T. Incubate overnight (18 h) at 4°C with gentle constant shaking.
7. Wash the membrane with TBS/T (pH 7.5), 5 × 10 min.
8. Incubate the membrane with secondary HRP-conjugated antibody dissolved in incubation buffer and 5% nonfat dry milk in TBS/T for 2 h at 20°C.
9. Visualize the reaction using a chemiluminescent HRP substrate kit (v/v Luminol and peroxides solution) and develop the film.
10. After developing procedures, wash the membrane with TBS/T, 5 × 10 min.
11. Add Restore Western blot stripping buffer (Pierce) and incubate the membrane for 15 min at room temperature. Wash the membrane once with TBS/T and use a chemiluminescent HRP substrate kit to check the quality of the stripping procedure.
12. Wash the membrane 3 × 10 min with TBS/T.
13. Immerse the membrane in blocking buffer solution (see step 5 above).

14. Proceed through steps 6–9, but this time use a primary antibody that will recognize internal load control proteins (e.g., β-actin or GADPH) and the corresponding secondary antibody.

4. Notes

1. The above protocols reflect our previous work on CCL2-induced claudin-5 internalization in brain endothelial cells, which is via caveolae (15). Internalization pathways may differ with cell type, with internalization stimulus, and with the Tj protein (e.g., occludin vs. claudin-5). This protocols may easily be adapted to examine those different scenarios. For example, lysine-fixable Texas Red-Dextran and Texas Red-Transferrin can be used as tracers to track macropinocytosis and clathrin-mediated endocytosis (15).
2. The protocols rely on GFP-claudin-5 undergoing internalization by the same pathways as native claudin-5. While evidence indicates that this is the case for brain endothelial cells (15), experiments comparing the distribution of GFP-claudin-5 and native claudin-5 (e.g., by immunofluorescence in fixed tissue) should be undertaken for other cell types.
3. The above microscopy techniques can also be used to examine recycling of claudin-5 to the cell membrane after removal of the internalization stimulus. Protein synthesis inhibitors can be used to examine whether the reappearance of claudin-5 at the cell membrane is due to recycling rather than newly synthesized protein.
4. Western blotting is used in concert with many of the techniques described above. It should also be noted that under certain circumstances, redistribution of Tj proteins from the plasma membrane may lead to degradation (with no potential recycling). Western blots of total claudin-5 should be undertaken in the presence and absence of internalization stimuli to examine whether degradation is occurring.

Acknowledgments

We expresses our appreciation to Dr Michael Wang for his enthusiasm for sharing with us the details of his protocols for GFP-tagging membrane proteins. In addition, we are grateful to the Molecular Imaging Laboratory in the Department of Cell and Developmental Biology at the University of Michigan for their support for the Live Cell Imaging technique. This work is supported by Grant (A.V.A) NS 044907 from the National Institutes of Health and Office of the Vice President for Research Faculty Grants and Awards Program, University of Michigan.

References

1. Bazzoni, G. (2006) Endothelial tight junctions: permeable barriers of the vessel wall. *Thromb. Haemost.*, **95**(1), 36–42.
2. Gonzalez-Mariscal L., Betanzos A., Nava P., Jaramillo B.E. (2003) Tight junction proteins. *Prog. Biophys. Mol. Biol.*, **81**(1), 1–44.
3. Heiskala M., Peterson P.A., Yang Y. (2001) The roles of claudin superfamily proteins in paracellular transport. *Traffic*, **2**(2), 93–98.
4. Furuse M., Sasaki H., Tsukita S. (1999) Manner of interaction of heterogeneous claudin species within and between tight junction strands. *J Cell Biol.*, **147**(4), 891–903.
5. Katahira J., Inoue N., Horiguchi Y., Matsuda M., Sugimoto N. (1997) Molecular cloning and functional characterization of the receptor for *Clostridium perfringens* enterotoxin. *J Cell Biol.* **136**(6), 1239–1247.
6. Ruffer C., Gerke V. (2004) The C-terminal cytoplasmic tail of claudins 1 and 5 but not its PDZ-binding motif is required for apical localization at epithelial and endothelial tight junctions. *Eur J Cell Biol.* **83**(4), 135–144.
7. Hamazaki Y., Itoh M., Sasaki H., Furuse M., Tsukita S. (2002) Multi-PDZ domain protein 1 (MUPP1) is concentrated at tight junctions through its possible interaction with claudin-1 and junctional adhesion molecule. *J Biol. Chem.*, **277**(1), 455–461.
8. Matter K., Balda M.S. (2003) Holey barrier: claudins and the regulation of brain endothelial permeability. *J Cell Biol.* **161**(3), 459–460.
9. Van Itallie C.M., Anderson J.M. (2004) The role of claudins in determining paracellular charge selectivity. *Proc Am Thorac Soc*, **1**(1), 38–41.
10. Belanger, M., Asashima, T., Ohtsuki, S., Yamaguchi, H., Ito S., Terasaki, T. (2007) Hyperammonemia induces transport of taurine and creatine and suppresses claudin-12 gene expression in brain capillary endothelial cells in vitro. *Neurochem. Int.*, **50**(1), 95–101.
11. Ohtsuki S., Sato S., Yamaguchi H., Kamoi M., Asashima T., Terasaki T. (2007) Exogenous expression of claudin-5 induces barrier properties in cultured rat brain capillary endothelial cells. *J Cell Physiol.* **210**(1), 81–86.
12. Nitta T., Hata M., Gotoh S., Seo Y., Sasaki H., Hashimoto N., Furuse M., Tsukita S. (2003) Size-selective loosening of the blood–brain barrier in claudin-5-deficient mice. *J Cell Biol.* **161**(3), 653–660.
13. Stamatovic S.M., Keep R.F., Kunkel S.L., Andjelkovic A.V. (2003) Potential role of MCP-1 in endothelial cell tight junction “opening”: signaling via Rho and Rho kinase. *J Cell Sci.* **116**, 4615–4628.
14. Stamatovic S.M., Dimitrijevic O.B., Keep R.F., Andjelkovic A.V. (2006) Protein kinase Calpha-RhoA cross-talk in CCL2-induced alterations in brain endothelial permeability. *J Biol Chem.* **281**(13), 8379–8388.
15. Stamatovic SM, Keep RF, Wang MM, Jankovic I, Andjelkovic AV. (2009) Caveolae-mediated internalization of occludin and claudin-5 during CCL2-induced tight junction remodeling in brain endothelial cells. J Biol Chem. **284**(28):19053–66.
16. Soma T., Chiba H., Kato-Mori Y., Wada T., Yamashita T., Kojima T., Sawada N. (2004) Thr(207) of claudin-5 is involved in size-selective loosening of the endothelial barrier by cyclic AMP. *Exp Cell Res.* **300**(1), 202–212.
17. Yamamoto, M., Ramirez, S.H., Sato, S., Kiyota, T., Cerny, R.L., Kaibuchi, K., Persidsky, Y., Ikezu, T. (2008). Phosphorylation of claudin-5 and occludin by Rho kinase in brain endothelial cells. Am. J. Pathol. **172**, 521–533.
18. Matsuda, M., Kubo, A., Furuse, M., Tsukita, S. (2004). A peculiar internalization of claudins, tight junction-specific adhesion molecules, during the intercellular movement of epithelial cells. J. Cell Sci. 117:1247–1257.
19. Utech, M., Ivanov, A.I., Samarin, S.N., Bruewer, M., Turner, J.R, Mrsny, R.J., Parkos, C.A., Nusrat, A. (2005). Mechanism of IFN-{gamma}-induced endocytosis of tight junction proteins: myosin II-dependent vacuolarization of the apical plasma membrane. Mol. Biol. Cell. **16**, 5040–5052.
20. Martineau, M., Galli, T., Baux, G., Mothet, J.P. (2008). Confocal imaging and tracking of the exocytotic routes for D-serine-mediated gliotransmission. Glia, **56**(12), 1271–84.
21. Zinchuk, V., Zinchuk, O., Okada, T. (2007). Quantitative colocalization analysis of multicolor confocal immunofluorescence

microscopy images: pushing pixels to explore biological phenomena. Acta Histochem Cytochem. **40**(4), 101–111.

22. Turvy DN, Blum JS. (1998): Detection of biotinylated cell surface receptors and MHC molecules in a capture ELISA: a rapid assay to measure endocytosis. J Immunol Methods. **212**(1):9–18.
23. Abedinpour, P., Jergil, B. (2003). Isolation of a caveolae-enriched fraction from rat lung by affinity partitioning and sucrose gradient centrifugation. Anal. Biochem. **313**, 1–8.
24. Beaumelle BD, Hopkins CR. High-yield isolation of functionally competent endosomes from mouse lymphocytes. Biochem J. 1989 Nov 15;**264**(1):137–49.

Chapter 23

Quantitative In Situ Analysis of Claudin Expression at the Blood–Retinal Barrier

Heping Xu and Janet Liversidge

Abstract

It is apparent that claudins are involved in signalling to and from cellular tight junctions (TJs) and control cell behaviour such as proliferation, differentiation, and migration. Methods to identify and measure specific claudins in TJs would, therefore, be useful to monitor TJ structure and functional integrity under physiological and pathological conditions. The molecular pathways involved in claudin signalling are not understood and are likely to become a focus for intensive research as better understanding of tight junction structure and function may provide opportunities for better drug delivery and absorption. In this chapter, we describe our method for quantitative analysis of specific claudins in TJ during the breakdown of the blood–retinal barrier in a mouse model of inflammatory uveitis, experimental autoimmune uveoretinitis (EAU).

Key words: Blood–retinal barrier, Blood–brain barrier, Tight junctions, Quantitative confocal microscopy, Retinal pigment epithelium

1. Introduction

Claudins are integral components of the intercellular structural contacts that characterise the tight junctions (TJs) between endothelial cells of blood vessels and epithelial cells that make up the blood–brain (BBB) and the blood–retinal barriers (BRB) (1–3). Functional properties of the BRB and BBB restrict the passage of cells and macromolecules into neural tissue, and disruption of these barriers is associated with neural degeneration and inflammation. Claudins form a family of at least 24 proteins that exhibit different expression patterns in different tissues. They are essential components of TJ structure and function as the principal barrier in epithelial cells (4). Tight junctions are found at the most apical part of the lateral surface of epithelial cell sheets and form a

Kursad Turksen (ed.), *Claudins: Methods and Protocols*, Methods in Molecular Biology, vol. 762, DOI 10.1007/978-1-61779-185-7_23, © Springer Science+Business Media, LLC 2011

continuous paracellular seal between the apical and baso-lateral parts of the intercellular junctions. Manipulations of claudin levels in vitro have established that claudins are responsible for regulating paracellular permeability, and overexpression of various claudins has demonstrated that claudin protein levels and combinations within the TJ control paracellular ion transport (5).

Various studies have indicated that claudins are arranged as hexamers within the TJ, but can under certain conditions be localised to the cytoplasm (4). Endocytic recycling of claudin proteins has been reported (6) and several claudins are known to be phosphorylated. Claudin-3 is phosphorylated by protein kinase A, claudin-4 by protein kinase C, and claudin-5 by cAMP. In their study of BRB endothelial TJ, Kobota et al. demonstrated that claudin-5 but not claudin-1 was phosphorylated demonstrating selectivity of the response. Thus, both post-transcriptional and translational regulation of claudin expression can occur. Various growth factors and cytokines have also been shown to modulate claudin expression in inflammation, and we have shown an active role for leukocytes in TJ disruption and BRB breakdown in retinal autoimmune inflammation (7, 8). Alterations in the expression of claudins have been associated with various disease entities such as multiple sclerosis, diabetes, and cancer, indicating that tissue-specific claudins could be used as therapeutic targets either to repair TJ defects or to transiently downregulate TJ barrier function to allow drug penetration (9–11).

It is apparent that claudins are involved in signalling to and from the TJs, controlling cell behaviour such as proliferation, differentiation, and migration. Methods to identify and measure claudins in TJ would, therefore, be useful to monitor TJ structure and functional integrity under physiological and pathological conditions. We have found confocal microscopy techniques to be useful as they allow analysis of protein expression within three-dimensional (3D) organ cultures or tissue flat mounts. This approach enables whole vessels within a vascular network or sheets of the epithelium and their substratum to be visualised, and the position and co-localisation of specific molecular targets to be detected. The confocal approach uses spatial filtering techniques to eliminate out-of-focus light or glare in specimens whose thickness exceeds the immediate plane of focus. Plus, the ability to collect serial optical sections from thick specimens (Z-stack) enables reconstruction of 3D images of structures identified by up to two or more fluorescently labelled antibodies and allows visualisation of complex molecular structures within tissues. Image data collected by the microscope photomultiplier (or other detector) are digitised and can be processed into increments that correspond to variations in light intensity emitted by the specifically bound fluorophore. With appropriate controls, quantitative measurements of spatial fluorescence intensity can be made.

These data can be displayed numerically in a tabular or graphical form or visualised by applying pseudo-colour scaling to indicate signal strength.

In this chapter, we describe our method for quantitative analysis of specific claudins in TJs during the breakdown of the BRB in a mouse model of inflammatory uveitis, experimental autoimmune uveoretinitis (EAU) (7, 8).

2. Materials

2.1. Sample Collection

1. CO_2 inhaler (see Note 1).
2. 2-ml syringes and 23-gauge needles (see Note 2).
3. Ophthalmic forceps and scissors.
4. 1× Phosphate-buffered saline (PBS): 8 g NaCl, 0.2 g KCl, 1.44 g $Na_2HPO_4{\cdot}7H_2O$, and 0.24 g KH_2PO_4. Add H_2O to 1 L. Adjust pH to 7.2, sterilize by autoclaving, and store at room temperature.
5. Dissecting microscope (Boom Microscopes).

2.2. Sample Preparation

1. Fixative: 100% ethanol. 100% acetone (see Note 3).
2. Blocking buffer: 5% (w/v) bovine serum albumin (BSA): 5 g BSA (Sigma–Aldrich) in 100 ml Tris-buffered saline (TBS).
3. 10× TBS: To prepare 1 L of 10× TBS: 24.2 g Tris base and 80 g NaCl; adjust pH to 7.6 with HCl (use at 1×) (see Note 4).
4. Washing buffer (TBS/T buffer): 1× TBS and 0.1% Tween-20 (Sigma–Aldrich).
5. Permeabilisation buffer: 0.3% Triton X-100 (Sigma–Aldrich) in TBS (see Note 5).
6. Eppendorf tubes: 1 ml or 0.50 ml.
7. Primary antibody: Rabbit anti-claudin-1/3 (cat. no. 71–7800; Zymed Laboratories Inc, CA). R-PE-conjugated anti-mouse F4/80 (BD Bioscience).
8. Secondary antibody: fluorescein isothiocynate (FITC) goat anti-rabbit IgG (H+L) conjugate (Zymed Laboratories Inc, CA).
9. Mounting medium: Vectashield (Victor Laboratories).

2.3. Confocal Microscope

LSM 510 META confocal microscope (Zeiss, Gottingen, Germany) (see Note 6).

2.4. Image Analysis Software

Image examiner (Zeiss).

3. Methods

3.1. Retinal Flat-Mount Preparation

1. Sacrifice the animals with CO_2 inhalation followed by cardiac puncture exsanguination (see Notes 1 and 2).
2. The eyes are carefully enucleated by cutting conjuctiva, extraocular muscles, and optic nerve (see Note 7).
3. Place the eyes in ice-cold PBS in a Petri dish. The cornea, crystalline lens, and vitreous are removed with a scissors using a dissecting microscope.
4. Transfer the eyecups into another Petri dish with fresh ice-cold PBS.
5. Make five vertical cuts (0.5–1 mm, from the edge of the eyecup towards the optic disc).
6. Peal off the retina from the RPE layer.
7. Fix the retinal tissues with 100% ethanol for 30 min at 4°C (see Note 3).
8. Transfer the tissues into pre-cold (−20°C) 100% acetone and incubate for 5 min at room temperature (see Note 3).

3.2. RPE/Choroidal Flat-Mount Preparation

1. After removing the retinal tissue, fix the remaining cups (RPE/choroidal/scleral tissue) with 100% ethanol for 30 min at 4°C (see Note 3).
2. The tissues are then incubated for 5 min at room temperature in 100% acetone that has been stored at −20°C (see Note 3).
3. Wash with PBS for 15 min × 3 at room temperature.
4. Carefully remove the extraocular tissues including the conjunctiva and extraocular muscles from the outer layer of the eyecup using a dissecting microscope.
5. Peal off the RPE/choroidal tissues from the sclera using a blade (12).

3.3. Flat-Mounts Staining

1. Place retinal or RPE/choroidal tissues into Eppendorfs.
2. Wash with TBS buffer (see Note 4) at room temperature for 15 min × 2.
3. Block and permeabilise the samples with 5% (w/v) BSA and 0.3% (v/v) Triton in TBS at room temperature (see Note 8).
4. Incubate with claudin-1 or claudin-3 primary antibody at 1:100 dilution with TBS containing 1% BSA and 0.1% Triton X-100 at 4°C overnight (see Note 9).
5. Wash the samples with TBS for 15 min × 3 at room temperature (see Note 10).
6. Incubate the samples with FITC-conjugated goat anti-rabbit immunoglobulin (Ig) (1:100) and R-PE-conjugated anti-mouse

F4/80 (1:100) for 2 h at room temperature (see Note 11). Both antibodies are diluted in 1% BSA+0.1% Triton X-100 in TBS.

7. Wash the samples with TBS for 15 min × 3 at room temperature (see Note 10).
8. Mount the tissues on glass slides with Vectashield anti-fade medium. Retinal tissues are mounted with the vitreous side facing up and photoreceptors facing down. RPE/choroidal tissues are mounted with the RPE side facing up and the choroidal side facing down.
9. Control tissues are incubated with rabbit Ig or in the absence of primary antibodies.

3.4. Confocal Microscopy of Retinal Flat-Mounts

1. LSM 510 META confocal microscope (Zeiss, Gottingen, Germany) is used to observe flat-mount samples.
2. FITC signals are detected at 488-nm excitation wavelength and with a band-pass filter of BP 505–533 nm, whereas R-PE signals are detected at 543-nm excitation wavelength and with a hand-pass filter of BP 560–615 nm.
3. Z-scanning is used to acquire stacks confocal images from different layers of tissue. For z-scanning, pinhole should be set at 1 for both 488-nm and 543-nm channels.
4. Detect gain and offset threshold are normally controlled using the range indicator function of the colour palette to avoid oversaturation of fluorescent intensity.
5. Fluorescent intensity is ranged from 0 to 250 in the range indicator. To be able to compare fluorescent intensity among samples, the samples should be processed in the same way, if possible at the same time.
6. All settings for pinhole, detect gain, and offset threshold should be kept the same for all samples.

3.5. Image Analysis

1. Reconstruction of z-stack images
   Z-stack images can be analysed in different ways using the LSM Examiner software (Carl Zeiss). It can be displayed as an image gallery, in which images from different layers are separately displayed (Fig. 1). It can also be displayed as orthogonal sections, in which cut-through (*z*-axis) sections at the *x*-axis and *y*-axis can be viewed in different parts of the image (Fig. 2). In addition, z-stack images can be reconstructed to a 2D or 3D image for further analysis.
2. Quantitative analysis of claudin distribution
   Claudin distribution at the BRB can be quantified using 2D images reconstructed from z-stack images. The fluorescent profile is the simplest way to measure fluorescent intensity in different parts of the image. Figure 3 is a reconstructed z-stack image showing claudin-1 expression at the capillaries of a

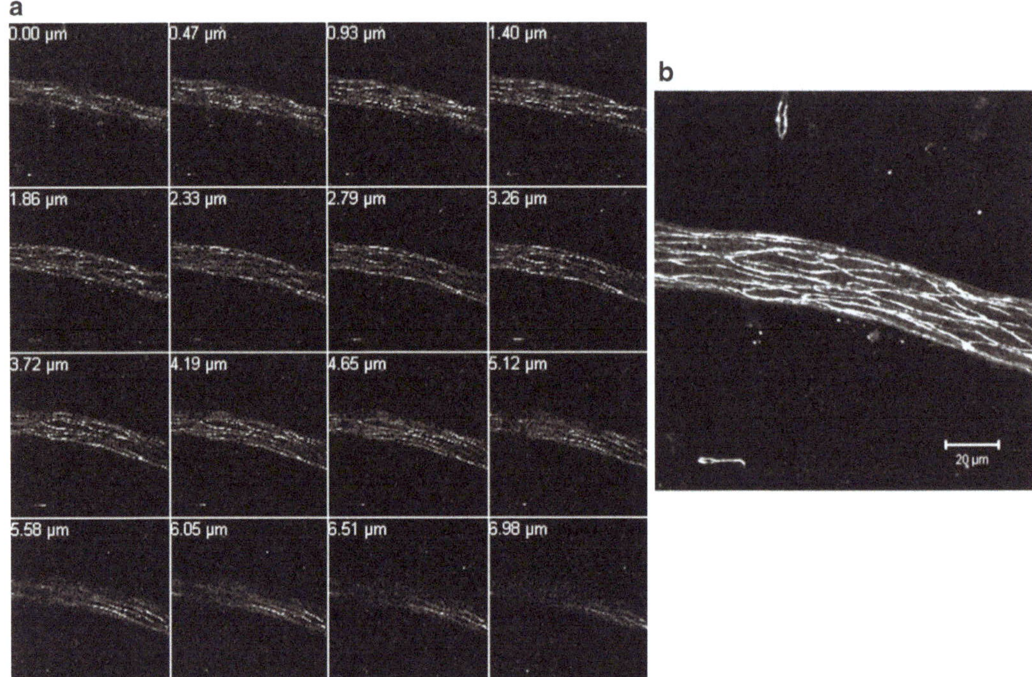

Fig. 1. Z-stack confocal images of claudin-1/3 expression at the mouse BRB. (**a**) Different layers of z-stack images. The numbers in the images show the scanning depth. (**b**) Constructed 2D image using the z-stack images shown in (**a**).

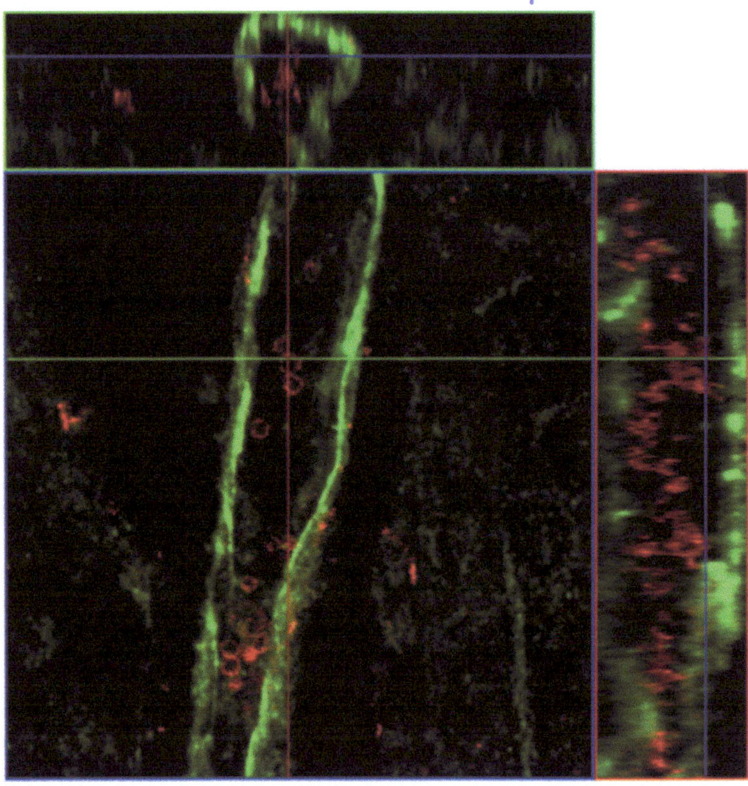

Fig. 2. Orthogonal view of z-stack images. Dual staining of retinal tissue from a mouse with retinal inflammation with claudin-1/3 (*green*) and F4/80 (*red*, a macrophage marker). Z-stack images (a total of 60 layers) is viewed through orthogonal sections. Image shows F4/80-positive cells accumulating in a retinal vessel.

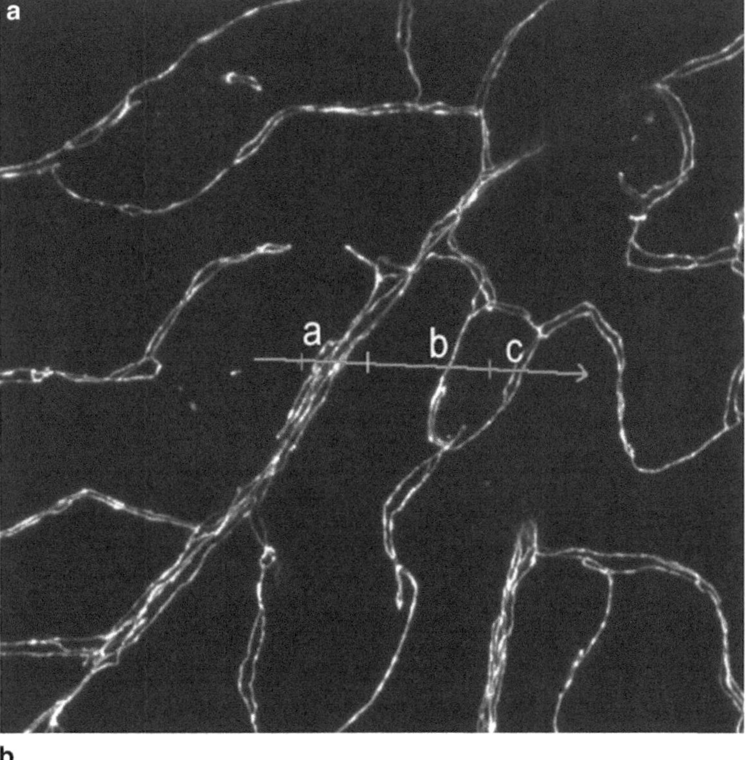

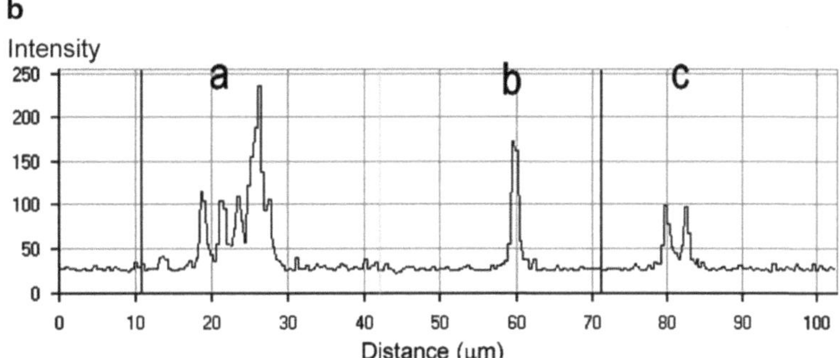

Fig. 3. Line profiles of fluorescent intensity of claudin 1/3 in a 2D reconstructed image (**a**). The fluorescent intensity profiles of the *red line* in (**a**) is shown in (**b**). The peaks in (a–c) (**b**) represent the fluorescent intensity of vessels a–c in (**a**).

mouse retina. The fluorescent intensity profile of claudin-1 along the red line is shown at the bottom of the image. Vessel-a has four fluorescent peaks, with three peaked at around 100 and one peaked at about 240. Vessel c has two peaks, both at the level of 100.

Fluorescent intensity can also be visualised using the 3D profile function of the LSM Examiner function. Figure 4a shows the same image as that in Fig. 3; however, this time we use the 3D profile command to display the distribution and the intensity of claudin-1 (Fig. 4b). In Fig. 4b, *x*- and *y*-axis shows the location

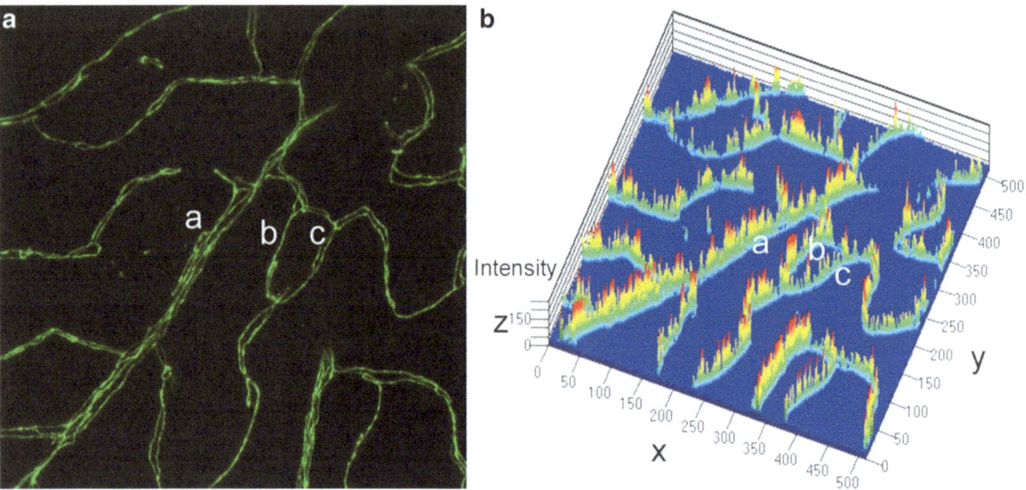

Fig. 4. 3D profiles of fluorescent intensity of claudin at the mouse BRB. (**a**) A 2D image reconstructed from z-stack images showing claudin-1/3 expression in retinal capillaries. The fluorescent intensity of caludin-1/3 is shown in (**b**) in a 3D view. *x*- and *y*-axis shows the geographical location of claudin-1/3 and the *z*-axis shows the relative fluorescent intensity of claudin-1/3. The fluorescent intensity of a–c in (**b**) correlates with vessels a–c in (**a**).

of claudin-1 signals and *z*-axis indicates the fluorescent intensity of claudin-1. Applying pseudo-colour to the grey scale image shows the highest intensity as red and the lowest as blue. As in Fig. 3, vessel-a has the highest claudin-1 expression and vessel-c has the lowest (Fig. 4b).

The 3D profile gives a more straightforward quantitative impression on the claudin-1/3 expression. Figure 5 shows the distribution of claudin-1/3 in the RPE cells of a normal mouse. Line profile indicates low levels of intracellular claudin-1/3 expression and high levels of intercellular junction expression (Fig. 5b). The 3D profile image is more informative (Fig. 5c). It not only shows the intracellular and intercellular junction distribution of claudin-1/3, but more importantly, it also clearly shows the uneven distribution of claudin-1/3 in RPE cell junctions under normal physiological conditions (Fig. 5c).

The disruption of claudin-1 expression at the BRB under pathological conditions can be directly visualised and in situ quantified in 3D profile fluorescent intensity images. Figure 6a, b shows claudin-1 expression in the RPE cells of mice with retinal inflammation EAU). Claudin-1 expression is lost in many areas of RPE junctions (Fig. 6a, b). Even in the area in which claudin-1 is still expressed, the level of expression is significantly lower than that of RPE from control noninflamed mouse eye (Fig. 5). At the inner BRB, claudin-1 expression is also significantly reduced, in particular in venule segments, where the inflammation is located (Fig. 6c, d).

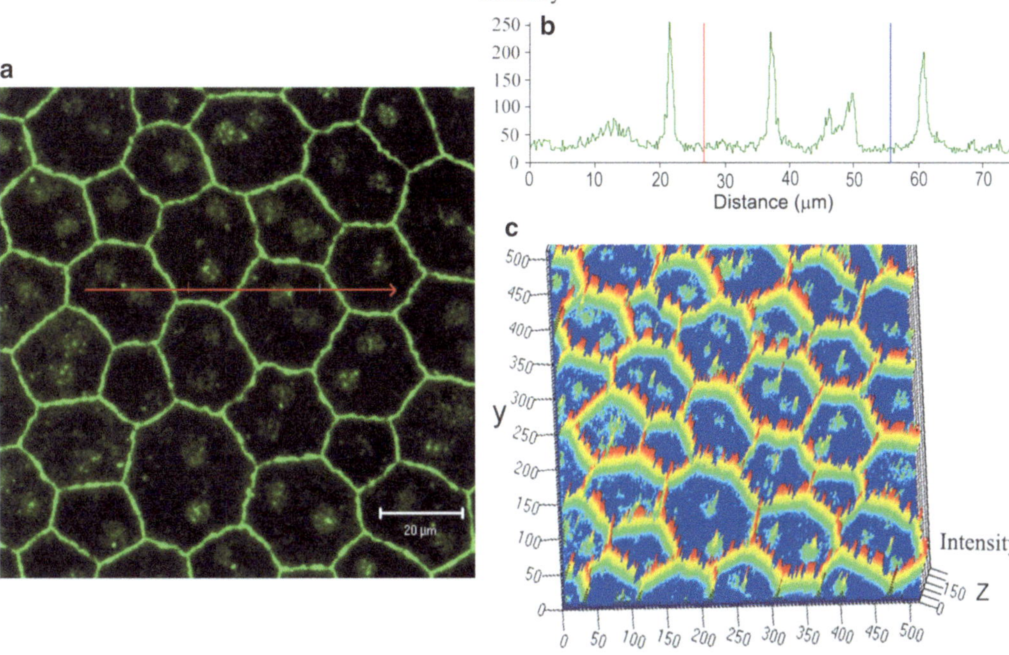

Fig. 5. Claudin-1/3 expression at the RPE cells (outer BRB). (**a**) A 2D image reconstructed from z-stack images (a total of 12 images, 5-μm thickness). (**b**) Line profile of fluorescent intensity of claudin-1/3 along the red line in (**a**). (**c**) 3D fluorescent intensity profile of claudin-1/3 shown in (**a**).

4. Notes

1. It is recommended that mice be killed by CO inhalation and cardiac puncture exsanguination. Cervical dislocation is not recommended for killing mice in this circumstance, as it may cause increased intravascular pressure and damage the integrity of BRB.
2. Normally, 0.5–0.8 ml blood can be obtained from an adult mouse (>8 weeks). If too much blood is left during cardiac puncture exsanguination, it may cause severe bleeding when removing the eye. Sometimes the bleeding might cause a problem for sample collection (e.g., damage eyes).
3. Acetone must be cooled at –20°C before using. We also tried 2% paraformaldehyde in TBS for 1–2 h at room temperature for mouse eye fixation. It normally works as well as ethanol and acetone fixation. However, prolonged fixation (>2 h) with paraformaldehyde reduces the sensitivity of claudin detection in the BRB.
4. 10× TBS buffer can be stored at room temperature for a couple of months. 1× TBS buffer needs to be prepared freshly.

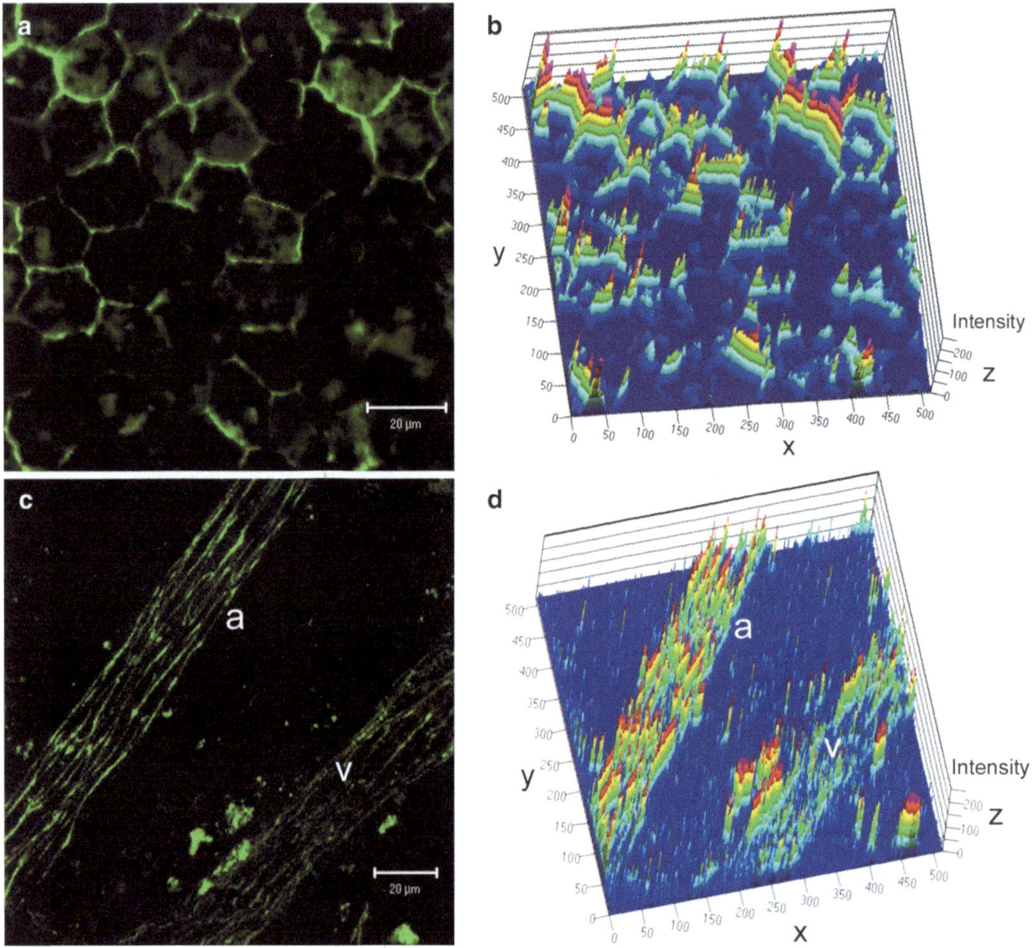

Fig. 6. Claudin-1/3 expression at the BRB in pathological condition of retinal inflammation. Retinal and RPE tissues were collected from a mouse with retinal inflammation (experimental autoimmune uveoretinitis) and stained for claudin-1/3 expression. (**a**, **c**), 2D images of RPE (**a**) and retina (**c**) reconstructed from z-stack images. (**b**, **d**) 3D fluorescent intensity profile of claudin-1/3 showing substantial claudin-1/3 defect in RPE cells (**b**) and a retinal venule (**d**). *a* artery, *v* venule.

5. 0.5–1% Triton X-100 can also be used to permeabilise retinal tissue. However, users will need to optimise the incubation conditions (e.g., the length and temperature of incubation) for their own experiments.
6. Any confocal scanning laser microscope with z-scanning function should be suitable for this work.
7. It is recommended to leave some extraocular tissues on the eyeball during this stage. The extraocular tissue will help to grab and position the eyeball during later retinal/RPE tissue dissection.

8. For blocking and permeabilisation, at least 2 h is required for retinal tissues and 0.5–1 h should be enough for RPE/choroidal samples.
9. For RPE/choroidal tissue, the incubation time can be reduced to 4 h.
10. For retinal tissue, the wash time can be extended to 1 h.
11. Only one study (Fig. 2) used F4/80 and claudin-3 dual staining in this protocol.

Acknowledgements

The authors thank Dr Mei Chen and Miss Rosemary Forsyth for technical assistance. This work was funded by the Wellcome Trust (068109).

References

1. Wolburg, H., and A. Lippoldt.(2002). Tight junctions of the blood-brain barrier: development, composition and regulation. *Vascul. Pharmacol.* 38: 323–337.
2. Morcos, Y., M. J. Hosie, H. C. Bauer, and T. Chan-Ling. (2001). Immunolocalization of occludin and claudin-1 to tight junctions in intact CNS vessels of mammalian retina. *J. Neurocytol.* 30: 107–123.
3. Williams, C. D., and L. J. Rizzolo. (1997). Remodeling of junctional complexes during the development of the outer blood-retinal barrier. *Anat. Rec.* 249: 380–388.
4. Kubota, K., M. Furuse, H. Sasaki, N. Sonoda, K. Fujita, A. Nagafuchi, and S. Tsukita. (1999). Ca(2+)-independent cell-adhesion activity of claudins, a family of integral membrane proteins localized at tight junctions. *Curr. Biol.* 9: 1035–1038.
5. Van Itallie, C. M., and J. M. Anderson. (2006). Claudins and epithelial paracellular transport. *Annu. Rev. Physiol* 68: 403–429.
6. Matsuda, M., A. Kubo, M. Furuse, and S. Tsukita. (2004). A peculiar internalization of claudins, tight junction-specific adhesion molecules, during the intercellular movement of epithelial cells. *J. Cell Sci.* 117: 1247–1257.
7. Xu, H., J. V. Forrester, J. Liversidge, and I. J. Crane. (2003). Leukocyte trafficking in experimental autoimmune uveitis: breakdown of blood-retinal barrier and upregulation of cellular adhesion molecules. *Invest Ophthalmol. Vis. Sci.* 44: 226–234.
8. Xu, H., R. Dawson, I. J. Crane, and J. Liversidge. (2005). Leukocyte diapedesis in vivo induces transient loss of tight junction protein at the blood-retina barrier. *Invest Ophthalmol. Vis. Sci.* 46: 2487–2494.
9. Minagar, A., A. Long, T. Ma, T. H. Jackson, R. E. Kelley, D. V. Ostanin, M. Sasaki, A. C. Warren, A. Jawahar, B. Cappell, and J. S. Alexander. (2003). Interferon (IFN)-beta 1a and IFN-beta 1b block IFN-gamma-induced disintegration of endothelial junction integrity and barrier. *Endothelium* 10: 299–307.
10. Choi, Y. H., W. Y. Choi, S. H. Hong, S. O. Kim, G. Y. Kim, W. H. Lee, and Y. H. Yoo. (2009). Anti-invasive activity of sanguinarine through modulation of tight junctions and matrix metalloproteinase activities in MDA-MB-231 human breast carcinoma cells. *Chem. Biol. Interact.* 179: 185–191.
11. Campbell, M., A. T. Nguyen, A. S. Kiang, L. C. Tam, O. L. Gobbo, C. Kerskens, D. S. Ni, M. M. Humphries, G. J. Farrar, P. F. Kenna, and P. Humphries. (2009). An experimental platform for systemic drug delivery to the retina. *Proc. Natl. Acad. Sci. U. S. A* 106: 17817–17822.
12. McMenamin, P. G. (2000). Optimal methods for preparation and immunostaining of iris, ciliary body, and choroidal wholemounts. *Invest Ophthalmol. Vis. Sci.* 41: 3043–3048.

Chapter 24

MMP-Mediated Disruption of Claudin-5 in the Blood–Brain Barrier of Rat Brain After Cerebral Ischemia

Yi Yang and Gary A. Rosenberg

Abstract

The blood–brain barrier (BBB) has become a major focus of attention in cerebral pathophysiology and disease progression in the central nervous system. Endothelial tight junctions, the basal lamina, and perivascular astrocytes are jointly referred to as BBB or neurovascular unit. Around the cerebral endothelial cells is the basal lamina composed primarily of laminin, fibronectin, and heparan sulfate. The basal lamina provides a structural barrier to extravasation of cellular blood elements and anchors endothelial cells to astrocytes. Barriers limiting transport into and out of the brain are found at the tight junction proteins and at the basal lamina. The relative contribution of these two sites has not been studied, but it is likely that both are disrupted to some extent in various injury scenarios. We have shown that activation of matrix metalloproteinases (MMPs) opens the BBB by degrading tight junction proteins (claudin-5 and occludin) and increases BBB permeability after stroke, and that an MMP inhibitor prevents degradation of tight junction proteins and attenuates BBB disruption.

Key words: Claudin-5, Tight junction proteins, Blood–brain barrier, Endothelial cells, Matrix metalloproteinase and stroke

1. Introduction

The blood–brain barrier (BBB) is the regulated interface between the peripheral circulation and the central nervous system. Tight junctions between endothelial cells of the BBB restrict paracellular diffusion of water-soluble substances from blood to brain (1, 2). Tight junction proteins join endothelial cells together, forming an interface between blood and brain (3–5). Transmembrane tight junction proteins consist of three integral proteins: claudins, occludin, and junctional adhesion molecules (6, 7). Zona occludens and cingulin are considered to be cytoplasmic tight junctional accessory proteins, which connect tight junctions to the actin cytoskeleton (8, 9). The extracellular

Kursad Turksen (ed.), *Claudins: Methods and Protocols*, Methods in Molecular Biology, vol. 762, DOI 10.1007/978-1-61779-185-7_24, © Springer Science+Business Media, LLC 2011

loops of occludin, claudins, and adhesion molecules originating from neighboring cells form the paracellular barrier of the tight junction, which selectively excludes most blood-borne substances from entering the brain.

Tight junctions are dynamic structures. Proteins of tight junction are subject to changes in expression, subcellular location, posttranslational modification, and protein–protein interactions under both physiological and pathophysiological conditions (2). Occludin, claudin-5, and zona occludens-1 (ZO-1), which are the main structural barrier proteins, are considered sensitive indicators of normal and disturbed functional states of the BBB (10–15). The claudins have been demonstrated to be one of the essential integral proteins in tight junction strands, and the relative composition of the claudin species in tight junctions directly determines the barrier function. Adrenomedullin, a vasodilator, improves BBB function in part through the induction of claudin-5 expression (16). Claudin-5 knockout mice have a selective permeability defect in the BBB that allows molecules of less than 800 kDa to cross the BBB (17, 18). In endothelial cell cultures and in *in vivo* BBB models, tight junction proteins are disrupted by ischemia/hypoxia (19–22). We have shown that MMPs open the BBB after stroke by degrading tight junction proteins. We have further shown that an MMP inhibitor prevents degradation of the tight junction proteins. Our findings support previous studies suggesting that tight junction proteins, including claudin-5, undergo structural changes after an ischemic injury (23) and directly implicate MMPs in stroke-induced degradation of tight junction proteins associated with BBB permeability.

Disruption of BBB tight junctions by diseases or drugs can lead to impaired BBB function and thus compromise the central nervous system. Therefore, understanding how BBB tight junctions are affected by various factors holds significant promise for the prevention and treatment of neurological diseases. Claudins are 20–24-kDa proteins; at least 24 have been identified in mammals (2). Claudins expressed in endothelial cells of neural tissues include claudin-1, -3, -5, and -12, and they are thought to be candidate molecules responsible for endothelial barrier function (24). Claudin-5 is primarily expressed in endothelial cells of BBB. There is clear evidence that the claudins constitute the backbone of tight junction strands (25). Commercial availability of antibodies to claudin-5 can be used with straightforward assay protocols to detect disruption of claudins by disease or drugs with either immunofluorescence or Western blotting. Confocal immunofluorescence detection allows confirmation by morphology of disruption of claudin-5, the location of claudin-5 in tight junctions of BBB, and the changes in expression and distribution of claudin-5 in different types of brain cells.

2. Materials

2.1. Double-Staining Immunofluorescence

1. 10× Phosphate-buffered saline (PBS, see Note 1): 80 g of NaCl (137 mM), 2.0 g of KCl (27 mM), 14.4 g of Na_2HPO_4 (14 mM), and 2.4 g of KH_2PO_4 (43 mM). Bring total volume to 1 L with deionized water (see Note 2). Adjust pH to 7.3. Sterilize by autoclaving.
2. 2% Paraformaldehyde containing 0.1 M sodium periodate and 0.075 M lysine in 100 mM phosphate buffer, pH 7.3 (2% PLP).
3. Pre-cooled acetone at 4°C.
4. Blocking buffer: 1× PBS containing 0.1% Tween 20, 1% BSA, and 5% normal serum. Blocking serum ideally should be derived from the same species in which the secondary antibody is raised.
5. Antibody dilution buffer: 1× PBS, 0.1% Tween 20, and 1% BSA.
6. Primary antibodies:
   (a) Mouse monoclonal anti-claudin-5 (22–24 kDa) (Invitrogen/Zymed, cat#: 35–2500; see Note 3).
   (b) Rabbit anti-glial fibrillary acidic protein (GFAP) antibody (Sigma–Aldrich Chemicals, St. Louis, MO) for recognizing astrocytes.
7. Secondary antibodies:
   (a) Anti-mouse IgG labeled with Cy3 (red fluorescence).
   (b) Anti-rabbit IgG labeled with FITC (green fluorescence).
8. Mounting medium: ProLong anti-fade (Molecular Probes, Eugene, OR).
9. Solvent tanks or Tek jars.
10. Pre-cleaned microscope slides (Superfrost/plus, Fisher Scientific, Pittsburgh, PA).
11. Slide racks to fit into the solvent tanks.
12. Humidifying chamber or airtight plastic container.
13. Microscope cover glass (22 × 60 mm), slightly narrower than microscope slides (Fisher Scientific, Pittsburgh, PA).
14. Pipettes and other general laboratory equipment.
15. Horizontal shaker.
16. Fluorescence and confocal microscopes.

2.2. Sodium Dodecyl Sulfate-Polyacrylamide Gel

1. 1.5 M Tris–HCl, pH 8.8 (Bio-Rad, Hercules, CA).
2. 10% Sodium dodecyl sulfate (SDS). Store at room temperature.

3. 0.5 M Tris–HCl, pH 6.8 (Bio-Rad, Hercules, CA).
4. 30% Acrylamide/bisacrylamide (29:1, 3.3% C, Bio-Rad). Acrylamide is a potent cumulative neurotoxin: wear gloves at all times.
5. *N,N,N,N′*-tetramethylethylenediamine (TEMED, Bio-Rad).
6. Ammonium persulfate: prepare 10% (100 mg/ml) solution in water immediately before use.
7. Water-saturated isobutanol. Combine equal parts of isobutanol and distilled water. Mix well and allow the phases to separate. Use the top layer. Store at room temperature.
8. Kaleidoscope pre-stained standards (Bio-Rad, Hercules, CA).
9. Electrophoresis and electrotransfer system for Western blot.
10. Orbital shaker and rocking platform. Refrigerated microcentrifuge.

2.3. Western Blotting for Claudin-5

1. Radio-immunoprecipitation assay (RIPA) buffer: 150 mM sodium chloride, 1.0% NP-40 or Triton X-100, 0.5% sodium deoxycholate, 0.1% SDS, and 50 mM Tris–HCl, pH 8.0. RIPA buffer is useful for preparing whole-cell extracts containing total nuclear, cytoplasmic, and membrane proteins.
2. Halt protease inhibitor single-use cocktail and halt phosphatase inhibitor single-use cocktail (Thermo Scientific, Rockford, IL).
3. Micro BCA protein assay kit (Thermo Scentific).
4. Laemmli 2× loading (sample) buffer: 4% SDS, 10% 2-mercaptoethanol, 20% glycerol, 0.004% bromophenol blue, and 0.125 M Tris–HCl, pH 6.8. It can also be prepared at 4× and 6× strength to minimize dilution of the samples. The 2× buffer is to be mixed in a 1:1 ratio with the sample.
5. 10× Running buffer (also called electrophoresis buffer) (1 L): 30.3 g Trizma base (0.25 M), 144 g glycine (1.92 M), and 10 g SDS (1%, add last). Dissolve the chemicals in deionized water and bring the total volume to 1 L. Do not adjust pH. Working solution: Dilute 100 ml 10× running buffer in 900 ml deionized water to make a 1× running buffer.
6. 10× Transfer buffer (1 L): 30.3 g Trizma base (0.25 M) and 144 g glycine (1.92 M), pH should be 8.3, do not adjust. To make 1 L of 1× transfer buffer: 200 ml methanol, 100 ml 10× transfer buffer, and 700 ml deionized water (see Note 4).
7. Immobilon transfer membrane: Polyvinylidene fluoride (PVDF) membrane (pore size: 0.2 μm, Millipore). 3MM chromatography paper (Whatman).
8. PBS with Tween 20 (PBS-T): 1× PBS and 0.1% Tween 20. Dilute 100 ml 10× PBS with 900 ml deionized water.

9. Blocking buffer: 5% (w/v) nonfat dry milk in PBS-T. Filter through filter paper (Qualitative Circles, Whatman).
10. Antibody dilution buffer: 5% (w/v) nonfat dry milk in PBS-T.
11. Primary antibody: Mouse anti-claudin-5 (Invitrogen, cat#: 35–2500).
12. Secondary antibody: Anti-mouse IgG conjugated to horseradish peroxidase (Cell Signaling, cat. #: 7076).
13. SuperSignal West Pico Chemiluminescent Substrate system (Pierce) and Bio-Max ML film (Kodak, Rochester, NY).
14. Stripping buffer: 0.2 M glycine, pH 2.5, 0.05% Tween 20, and 0.1% SDS. Sterile filter the solution and keep at 4°C.

3. Methods

3.1. Preparation of Brain Sections for Double Staining of Immunofluorescence

1. Spontaneously hypertensive rats (SHR) that had undergone a middle cerebral artery occlusion (MCAO) with reperfusion for 3 and 24 h are deeply anesthetized and intracardially perfused with 400 ml of 2% PLP. Brains are removed rapidly, postfixed for 24 h in 2% PLP, and cryoprotected in 30% sucrose in PBS until brains sink to the bottom of the container. Freeze PLP-fixed brains in blocks in liquid nitrogen according to standard procedures. The samples may be stored at –80°C before sectioning.
2. Before cutting sections, allow the temperature of the block to equilibrate to the temperature of the cryostat (usually –20°C). Cut 10-μm-thick sections of the brain and place the sections on a Fisher Superfrost slide. Store the slides at –80°C in a sealed slide box until ready for staining. Thaw the slides at room temperature prior to fixing and staining.
3. Fix the slides in cold acetone for 10 min and keep refrigerated at 4°C. Wash in three changes of PBS-T. Label the slides with a solvent-resistant pen and demarcate the sections if required.
4. To blocking nonspecific binding of IgG, incubate the slides in blocking buffer in a humidified slide chamber for 1 h at room temperature.
5. Remove the blocking buffer without letting the sections dry and by washing. Apply a mixture of the claudin-5 (3 μg/ml) and GFAP (1:500) primary antibodies in antibody dilution buffer. Incubate the slides with primary antibodies in a humidified slide chamber for 48 h (or two nights) at 4°C (see Note 5).
6. Rinse the slides in PBS-T for 6 min three times.

7. From this step, the slides should be put under aluminum foil and the room lights dimmed. The secondary antibodies (goat anti-mouse IgG labeled with Cy3 for claudin-5; goat anti-rabbit IgG labeled with FITC for GFAP) are diluted at 1:500 in antibody dilution buffer.
8. Incubate the slides for 2 h (see Note 5) at room temperature with secondary antibodies in a humidified slide chamber.
9. Rinse the slides in PBS-T three times, 6 min each time, and mount in ProLong anti-fade (Molecular Probes, Eugene, OR). Seal the slides around the edges of the cover glasses with nail polish or Permount (Fisher Scientific) 1 h after mounting.
10. Photograph the slides with a fluorescence microscope (Olympus BX51, Olympus Optical Co. Ltd, Japan) and a confocal microscope (Zeiss LSM 510, Carl Zeiss MicroImaging, Thornwood, NY). Sections incubated in the absence of the primary antibody, with rabbit IgGs will not exhibit immunofluorescence. Examples of the signals for claudin-5 and astrocytes (GFAP) are shown in Fig. 1.

3.2. Preparation of Samples for Assay of Claudin-5 by Western Blotting

1. SHR that had undergone an MCAO with reperfusion for 3 and 24 h are deeply anesthetized and intracardially perfused briefly with 0.9% saline to flush blood out of the brain. Remove brains rapidly and dissect the tissue of interest with clean tools, on ice preferably, as quickly as possible to prevent degradation by proteases (see Note 6).
2. Place the tissue in round-bottom microcentrifuge tubes and immerse in liquid nitrogen to "snap freeze." Store the samples at −80°C for later use or keep on ice for immediate homogenization.
3. Add RIPA buffer containing protease and phosphatase inhibitors to the tissue (1:4 tissue to RIPA buffer) and homogenize with an electric homogenizer (see Note 7). Incubate the lysates on ice for 10 min.
4. Centrifuge for 20 min at 12,000 × g at 4°C in a refrigerated microcentrifuge. Gently remove the tubes from the centrifuge and place on ice. Collect the supernatant and place in a fresh tube on ice; discard the insoluble pellet.
5. Determination of protein concentration using Micro BCA protein assay kit. Bovine serum albumin (BSA) is a frequently used protein standard. Once you have determined the concentration of each sample, you can freeze them at −20°C or −80°C for later use, or prepare for loading onto a SDS-PAGE gel.
6. Preparation of samples for loading onto gels. Add Laemmli loading buffer 1:1 with samples (50 μg total protein will be

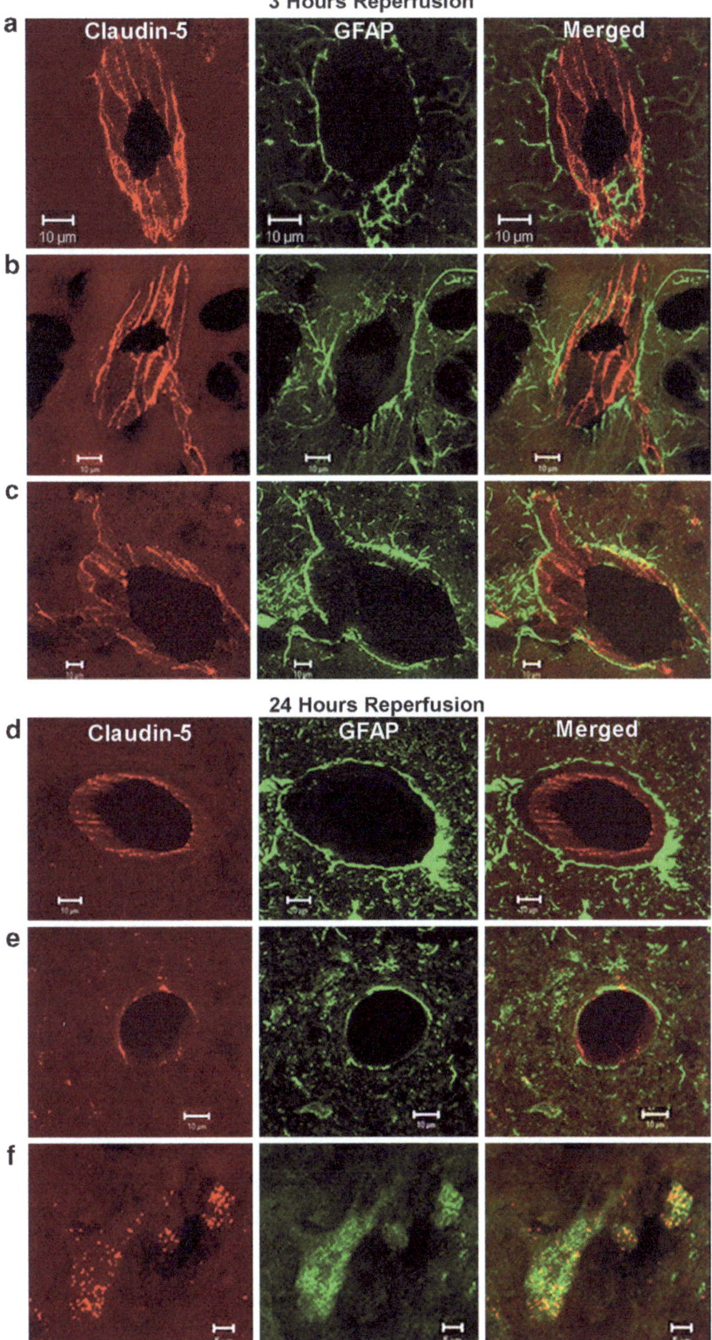

Fig. 1. Confocal micrographs showing claudin-5 immunoreactivity in the sham-operated animals and ischemic and nonischemic hemispheres after 3 h of reperfusion (**a**–**c**). The sham-operated animals and the nonischemic side show that the claudin-5 (Cy-3, *red*) in blood vessels is separated from the astrocytes (GFAP-FITC, *green*) surrounding them (**a** and **b**). The merged images show that the claudin-5 and astrocytes are separate. In the ischemic hemisphere (**c**), there is fragmentation and degeneration of the claudin-5 immunoreactivity. Co-localization of claudin-5 and GFAP was seen in the ischemic hemisphere. Confocal micrographs showing claudin-5 immunoreactivity in the ischemic and nonischemic hemispheres after 24 h of reperfusion (**d**–**f**). Immunoreactivity of claudin-5 with GFAP in brain vessels in nonischemic hemisphere (**d**). Ischemia caused a loss of claudin-5 from the blood vessels at 24 h (**e**). There was co-localization of the claudin-5 in the GFAP-positive astrocytes (**f**). These photos were taken from penumbral areas in the caudate and lateral cortex. Scale bars indicate 10 μm in (**a**–**e**). Scale bars show 5 μm in (**f**) (reproduced from ref. 23).

loaded for the claudin-5 assay). Do *not* add reducing agents (mercaptoethanol or DTT) and do *not* boil (nonreducing & denaturing conditions, see Note 8). The samples are ready for separation by SDS-PAGE.

3.3. SDS-PAGE

When separated on a polyacrylamide gel, the procedure is abbreviated as SDS-PAGE. The technique is a standard means for separating proteins according to their apparent molecular weight.

1. Preparation of PAGE gel. These instructions assume the use of a Hoefer miniVE vertical electrophoresis system. It is critical that the glass plates for the gels are scrubbed clean with distilled water followed by 95% ethanol and then air-dried before use.
2. Prepare a 1.0-mm-thick, 15% separating gel by mixing 2.5 ml of 1.5 M Tris–HCl, pH 8.8, with 5 ml acrylamide/bis solution, 100 μl of 10% SDS, 2.348 ml of water, 50 μl 10% ammonium persulfate solution (APS), and 5 μl TEMED (see Note 9). Pour the gel to about 1 cm from the bottom of the comb and carefully layer 1 ml of water-saturated isobutanol on top of the gel. Allow 30 min for the gel to polymerize (see Note 10).
3. When the gel has polymerized, pour off the isobutanol and rinse with deionized water.
4. Prepare the stacking gel by mixing 1.25 ml of 0.5 M Tris–HCl, pH 6.8, with 0.833 ml acrylamide/bis solution, 50 μl of 10% SDS, 2.837 ml water, 25 μl APS, and 5 μl TEMED. Pour the stacking gel and insert the comb. The stacking gel should polymerize within 30 min.
5. Prepare the running buffer by diluting 50 ml of the 10× running buffer with 450 ml of water in a graduated cylinder and mix.
6. When the stacking gel has polymerized, place in a gel rig and immerse in running buffer. Carefully remove the comb and flush the wells thoroughly with running buffer (see Note 11).
7. Load the samples into the wells. Include one well for pre-stained molecular weight markers. A range of molecular weight markers that will enable the determination of the protein size is also used to monitor the progress of an electrophoretic run. We use 8 μl Bio-Rad Kaleidoscope pre-stained standards (cat. #161-0324).
8. Place the safety lid on the tank and plug the leads into an approved power supply. Run the gel with constant voltage at 100 V for 75–90 min. Allow the dye front (blue, 7.6 kDa) to run to the bottom edge of the gel. The molecular weight of claudin-5 is about 22–24 kDa (see Note 12).

3.4. Western Blotting for Claudin-5

1. Preparation of membrane: Cut a piece of PVDF membrane (pore size, 0.2 µm, slightly larger than gel). Wet for 5 min in methanol on a rocker at room temperature. Remove methanol and equilibrate the membrane in 1× blotting buffer for at least 10 min.
2. When the SDS-PAGE gel has finished running, cut a corner of the gel to identify the proper orientation of the gel. Equilibrate the gel in transfer buffer for 30 min prior to blotting to facilitate the removal of electrophoresis buffer salts and detergents from the gel.
3. Assemble "sandwich" for the Hoefer blot module. Saturate the sponges and filter papers (slightly larger than gel) in 1× blotting buffer. Assemble the transfer apparatus in the following order: Black cathode (−) side of the Blot Module – Sponge – 3MM filter paper – gel – PVDF membrane – 3MM filter paper – sponge – Red Anode (+) side of the Blot Module. Ensure that no bubbles are trapped in the resulting sandwich.

   Make sure that the PVDF membrane is between the gel and the anode. It is vitally important to ensure that this orientation or the proteins will be lost from the gel into the buffer rather than transferred to the PVDF.
4. Place the cassette into the transfer tank. Set the buffer tank on top of a magnetic stirrer with a stir bar. Add the pre-cooled transfer buffer into the module such that it completely covers the sandwich. Fill the tank with distilled water, preferably chilled to 4°C, to dissipate heat from the module during the transfer.
5. Place the safety lid on the tank and plug the leads into an approved power supply. Transfer for 1 h at 25 V.
6. Once the transfer is complete, disassemble the apparatus and cut the PVDF membrane to the same size as the gel by tracing around the gel with a razor blade. The excess gel and PVDF can then be discarded. The colored molecular weight markers should be clearly visible on the membrane. *Note*: Perform all operations with the membrane quickly and do not allow it to become dry.
7. Immediately place the membrane in blocking buffer (5% (w/v) nonfat dry milk in PBS-T) at room temperature for 1 h on an orbital shaker.
8. Discard the blocking buffer.
9. Make a 1:500 dilution of the primary claudin-5 antibody in PBS-T containing 5% nonfat milk and incubate the washed membrane overnight at 4°C on a rocker in a Kapak bag.
10. Wash the membrane three times for 5 min each in a large volume of PBS-T on an orbital shaker.

11. Incubate the membrane with an anti-mouse secondary antibody at 1:5,000 dilution for 75 min at room temperature on a rocker.
12. Discard the secondary antibody and wash the membrane three times for 10 min each with a large volume of PBS-T on an orbital shaker at room temperature.
13. During the final wash, prepare 5 ml working solution of SuperSignal West Pico Chemiluminescent Substrate system. Once the final wash is removed from the blot, transfer the membrane into the working solution and incubate for 5 min.
14. Drain the detect reagent, blot the membrane with Kimwipes, and then place it between the leaves of a plastic paper protector that has been cut to the size of an X-ray film cassette.
15. In a dark room, place the membrane with the protein side up in an X-ray film cassette. Place the film on top of the membrane and expose the film for a suitable exposure time, typically, initially 10–20 s, and then re-expose for the optimal time as needed. An example of the results produced is shown in Fig. 2.

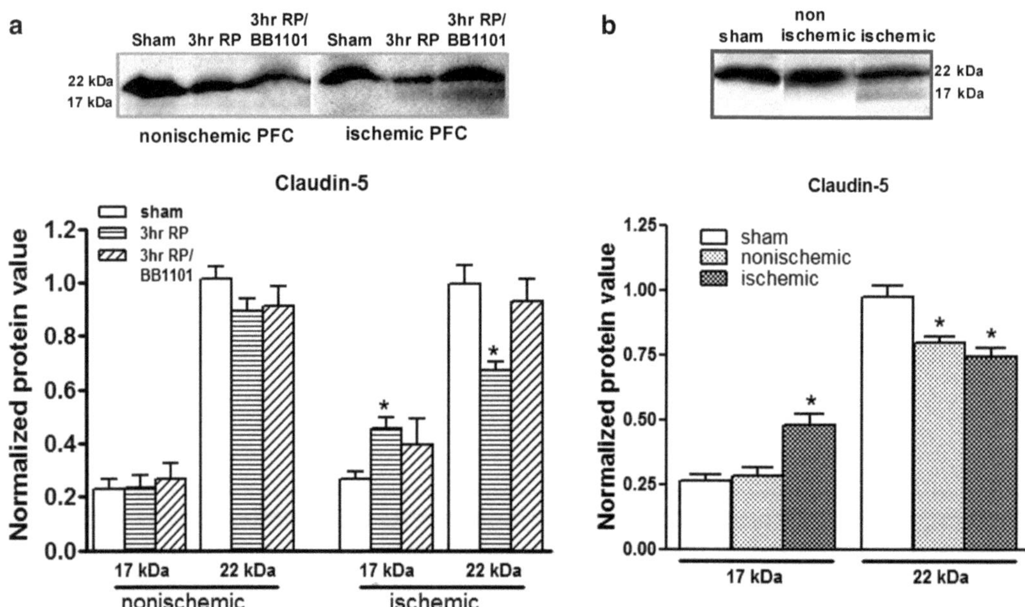

Fig. 2. (**a**) Western blots for claudin-5. The Western blots were performed with whole-cell extracts from rat brains with a 90-min MCAO followed by 3 hours reperfusion (3hr RP). The Western blot showed significant reductions ($p<0.01$, $n=6$) of claudin-5 (22 kDa) at 3 h in the ischemic piriform cortex (PFC). In addition, a new 17-kDa band for claudin-5 was found in the ischemic hemisphere at 3 h post-reperfusion. The 17-kDa protein showed a significant increase ($p<0.01$). Treatment with BB-1101, a broad-spectrum MMP inhibitor (3hr RP/BB1101), significantly inhibited degradation of claudin-5 (22 kDa) ($p<0.02$, $n=6$). (**b**) Western blots showing degradation of claudin-5 after a 90-min MCAO followed by 24-h reperfusion. A decrease in the higher molecular weight form (22 kDa) of claudin-5 was observed in both the ischemic and nonischemic hemispheres compared to sham-operated animals ($p<0.05$, $p<0.01$, respectively, $n=6$). A significant increase in the 17-kDa claudin-5 was seen in the ischemic hemisphere ($p<0.002$) (reproduced from ref. 23).

3.5. Stripping and Re-probing Blots for Other Tight Junction Proteins or Loading Control (See Notes 13 and 14)

1. Rinse the membrane with 1× PBS-T.
2. Cut the membrane to a slightly bigger size than the blot and place in a Kapak bag. Add 10 ml of stripping buffer and seal the bag.
3. Immerse in an 80°C water bath for 20 min. Rinse the membrane in 1× PBS-T. Save the membrane in PBS-T and store at 4°C until re-probing. Before re-probing, block the membrane for 1 h in blocking buffer.

4. Notes

1. There are numerous recipes for PBS. We recommend the recipe listed here because we found it to produce minor background for immunohistochemistry.
2. All solutions should be prepared in deionized water that has a resistivity of 18.2 MΩ-cm and a total organic content of less than five parts per billion.
3. We have found this claudin-5 antibody to be excellent for both Western blotting and immunofluorescence. Similar antibodies are available from other commercial sources.
4. For transfer of proteins larger than 80 kDa, it is recommended that SDS is included at a final concentration of 0.1%. If your protein of interest is small, consider removing SDS from the transfer buffer and keep the methanol concentration at 20%.
5. Claudin-5 and other claudins are 20–24-kDa integral membrane proteins with four transmembrane domains. We have found that 48-h incubation of primary antibody and 2-h incubation of secondary antibody generated very clear immunostaining of claudin-5.
6. As soon as lysis occurs, proteolysis, dephosphorylation, and denaturation begin. These events can be slowed down tremendously by keeping the samples on ice at all times and adding protease and phosphatase inhibitors to the lysis buffer immediately before use.
7. Volumes of lysis buffer must be determined in relation to the amount of tissue present (protein extract should not be too diluted to avoid loss of protein and large volumes of samples to be loaded onto gels). Ideally, protease and phosphatase inhibitors should be added to RIPA buffer immediately before adding to tissue samples. The buffer (with inhibitors) should be ice cold prior to homogenization.
8. We have found this antibody against claudin-5 to produce a specific band of 22 kDa when samples are prepared under nonreducing and non-denaturing conditions.

9. TEMED is best stored at room temperature in a desiccator. Buy small bottles as it may decline in quality (gels will take longer to polymerize) after opening.
10. Acrylamide is a potent cumulative neurotoxin: wear gloves at all times.
11. Gels can be purchased precast, or produced in the laboratory (recipes can be found in laboratory handbooks). Either way, choose carefully the percentage of your gel as this will determine the rate of migration and degree of separation between proteins. The smaller the size of the protein of interest, the higher the percentage of acrylamide that is required to achieve adequate separation and focusing of protein bands (see http://www.abcam.com/technical for details).
12. Proceed to the transfer step immediately since proteins that have been electrophoretically separated may slowly diffuse within gels after electrophoresis.
13. Loading controls are required when a comparison is to be made between the levels of a protein in different samples. The loading control bands can be used to quantify the protein amounts in each lane and check for even transfer from the gel to the membrane across the whole gel. For whole-cell/cytoplasmic samples, beta-actin, GAPDH, and tubulin are markers commonly used as loading controls. In our case, we used Brilliant Blue R (Sigma–Aldrich) staining on the same PVDF membranes to normalize protein loading and transfer, since many control proteins commonly used in conventional relative protein level analysis, such as GAPDH and beta-actin, have been reported to be affected by ischemia (26). The results are reported as normalized band intensity for quantifying relative protein levels (Fig. 2) (23). As the Brilliant Blue R stain is not reversible, stain membranes with Brilliant Blue R only when there are no further plans to do Western blots.
14. This protocol can be adapted for other claudins and other tight junction proteins, such as occludin (65 kDa) and ZO-1 (220 kDa), when a gradient polyacrylamide gel (4–20%) is used.

References

1. Vracko, R., *Basal lamina scaffold-anatomy and significance for maintenance of orderly tissue structure.* Am J Pathol, 1974. 77(2): p. 314–46.
2. Hawkins, B.T. and T.P. Davis, *The blood-brain barrier/neurovascular unit in health and disease.* Pharmacol Rev, 2005. **57**(2): p. 173–85.
3. Ballabh, P., A. Braun, and M. Nedergaard, *The blood-brain barrier: an overview: structure, regulation, and clinical implications.* Neurobiol Dis, 2004. **16**(1): p. 1–13.
4. Brightman, M.W. and T.S. Reese, *Junctions between intimately apposed cell membranes in the vertebrate brain.* J Cell Biol, 1969. **40**(3): p. 648–77.
5. Liebner, S., et al., *Correlation of tight junction morphology with the expression of tight junction proteins in blood-brain barrier endothelial cells.* Eur J Cell Biol, 2000. **79**(10): p. 707–17.
6. Papadopoulos, M.C., et al., *Occludin expression in microvessels of neoplastic and non-neoplastic*

human brain. Neuropathol Appl Neurobiol, 2001. **27**(5): p. 384–95.

7. Furuse, M., et al., *Occludin: a novel integral membrane protein localizing at tight junctions*. J Cell Biol, 1993. **123**(6 Pt 2): p. 1777–88.
8. Haskins, J., et al., *ZO-3, a novel member of the MAGUK protein family found at the tight junction, interacts with ZO-1 and occludin*. J Cell Biol, 1998. **141**(1): p. 199–208.
9. Itoh, M., et al., *Involvement of ZO-1 in cadherin-based cell adhesion through its direct binding to alpha catenin and actin filaments*. J Cell Biol, 1997. **138**(1): p. 181–92.
10. Hirase, T., et al., *Regulation of tight junction permeability and occludin phosphorylation by Rhoa-p160ROCK-dependent and -independent mechanisms*. J Biol Chem, 2001. **276**(13): p. 10423–31.
11. Hirase, T., et al., *Occludin as a possible determinant of tight junction permeability in endothelial cells*. J Cell Sci, 1997. **110 (Pt 14)**: p. 1603–13.
12. Kanda, T., Y. Numata, and H. Mizusawa, *Chronic inflammatory demyelinating polyneuropathy: decreased claudin-5 and relocated ZO-1*. J Neurol Neurosurg Psychiatry, 2004. **75**(5): p. 765–9.
13. Wen, H., et al., *Selective decrease in paracellular conductance of tight junctions: role of the first extracellular domain of claudin-5*. Mol Cell Biol, 2004. **24**(19): p. 8408–17.
14. Dobrogowska, D.H. and A.W. Vorbrodt, *Immunogold localization of tight junctional proteins in normal and osmotically-affected rat blood-brain barrier*. J Mol Histol, 2004. **35**(5): p. 529–39.
15. Matter, K. and M.S. Balda, *Holey barrier: claudins and the regulation of brain endothelial permeability*. J Cell Biol, 2003. **161**(3): p. 459–60.
16. Honda, M., et al., *Adrenomedullin improves the blood-brain barrier function through the expression of claudin-5*. Cell Mol Neurobiol, 2006. **26**(2): p. 109–18.
17. Furuse, M., et al., *Claudin-1 and -2: novel integral membrane proteins localizing at tight junctions with no sequence similarity to occludin*. J Cell Biol, 1998. **141**(7): p. 1539–50.
18. Nitta, T., et al., *Size-selective loosening of the blood-brain barrier in claudin-5-deficient mice*. J Cell Biol, 2003. **161**(3): p. 653–60.
19. Huber, J.D., R.D. Egleton, and T.P. Davis, *Molecular physiology and pathophysiology of tight junctions in the blood-brain barrier*. Trends Neurosci, 2001. **24**(12): p. 719–25.
20. Mark, K.S. and T.P. Davis, *Cerebral microvascular changes in permeability and tight junctions induced by hypoxia-reoxygenation*. Am J Physiol Heart Circ Physiol, 2002. **282**(4): p. H1485-94.
21. Witt, K.A., et al., *Effects of hypoxia-reoxygenation on rat blood-brain barrier permeability and tight junctional protein expression*. Am J Physiol Heart Circ Physiol, 2003. **285**(6): p. H2820-31.
22. Yamagata, K., et al., *Hypoxia-induced changes in tight junction permeability of brain capillary endothelial cells are associated with IL-1beta and nitric oxide*. Neurobiol Dis, 2004. **17**(3): p. 491–9.
23. Yang, Y., et al., *Matrix metalloproteinase-mediated disruption of tight junction proteins in cerebral vessels is reversed by synthetic matrix metalloproteinase inhibitor in focal ischemia in rat*. J Cereb Blood Flow Metab, 2007. **27**(4): p. 697–709.
24. Wolburg, H., et al., *Localization of claudin-3 in tight junctions of the blood-brain barrier is selectively lost during experimental autoimmune encephalomyelitis and human glioblastoma multiforme*. Acta Neuropathol (Berl), 2003. **105**(6): p. 586–92.
25. Forster, C., *Tight junctions and the modulation of barrier function in disease*. Histochem Cell Biol, 2008. **130**(1): p. 55–70.
26. Nishida, Y., et al., *Screening for control genes in mouse hippocampus after transient forebrain ischemia using high-density oligonucleotide array*. J Pharmacol Sci, 2006. **101**(1): p. 52–7.

Chapter 25

Claudin-5 Expression in In Vitro Models of the Blood–Brain Barrier

Itzik Cooper, Katayun Cohen-Kashi-Malina, and Vivian I. Teichberg

Abstract

Claudins are transmembrane proteins that form the backbone of the tight junctions (TJs) at the blood–brain barrier (BBB). TJs are cellular structures that physically obstruct the inter-endothelial space and restrict the paracellular diffusion of blood-borne substances from the peripheral circulation into the CNS. TJs are also dynamic structures that rapidly respond to external signals that produce changes in BBB permeability.

We focus here on the biochemical and immunohistochemical properties of claudin-5 as expressed in three in vitro models of the BBB, and show that the contact co-culture of endothelial cells with glial cells significantly increases claudin-5 expression.

Key words: BBB, In vitro model, Claudin-5, Tight junctions, Immunocytochemistry, Western blot, Endothelial cells, Glia

1. Introduction

The blood–brain barrier (BBB) is a selective barrier formed by the endothelial cells (EC) that line cerebral capillaries, together with perivascular elements such as the closely associated astrocytic end-feet processes, perivascular neurons, and pericytes (1). In the last two decades, extensive efforts were made to design suitable and reproducible in vitro models of the BBB which mimic as closely as possible the real barrier, and enable the study of the physiology, pharmacology, and pathology of the BBB (2). Cerebral EC are unique in that they form, in addition to cell–cell adherens junctions, complex tight junctions (TJs) via the interactions of several transmembrane proteins that effectively seal the paracellular pathway. The main proteins contributing to the formation of tight junctions are occludin, claudins, and the junctional adhesion molecules (JAM) (1).

Kursad Turksen (ed.), *Claudins: Methods and Protocols*, Methods in Molecular Biology, vol. 762,
DOI 10.1007/978-1-61779-185-7_25,

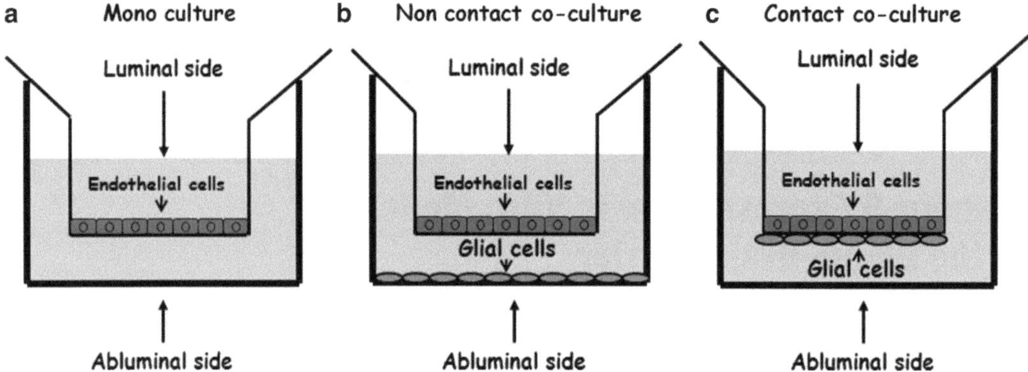

Fig. 1. In vitro models of BBB. Schematic drawings of the three in vitro BBB culture systems used. A monolayer of EC forming tight junctions is grown on a collagen-coated microporous membrane filter culture insert. The monolayer organizes in an asymmetric fashion while the upper side of the inserts corresponds to the blood side (luminal) and the lower side to the brain side (abluminal).

The family of claudin proteins comprises more than 20 members. Some claudins have limited distribution. For instance, claudin-5 is restricted to EC (3) and it is known that claudin-5, -3, and -12 are localized at the BBB (4, 5), while the presence of claudin-1 is still controversial (6).

As claudin-5 plays a crucial role in the functionality of the BBB, monitoring its expression pattern and levels could be of great benefit for determining the quality and integrity of an in vitro BBB model.

In this chapter, we will describe two biochemical methods, Western blotting and immunocytochemistry, for the examination of the properties of claudin-5 expression in three types of BBB in vitro models: a monoculture of pig brain capillary endothelial cells, a noncontact co-culture, and a contact co-culture with rat brain glial cells (Fig. 1) (7).

2. Materials

2.1. Cell Culture

1. Plating medium: 10% new born calf serum (NBCS), 2 mM l-glutamine, 100 units/ml penicillin, 0.1 mg/ml streptomycin, and 0.1 mg/ml gentamicin (all from Biological Industries, Israel) in Earl's medium 199 (Sigma).
2. Assay medium: 2 mM l-glutamine, 100 units/ml penicillin, 0.1 mg/ml streptomycin, and 550 nm hydrocortisone in DME/Ham's F12 medium.
3. Astrocytes medium: 10% fetal calf serum (FCS, Biological Industries, Israel), 0.25 mg/ml gentamicin, 1 mM pyruvate (Sigma), and 2 mM l-glutamine in Dulbecco's modified Earl's medium (DMEM, Gibco).

2.2. Immunocytochemistry for the Detection of Claudin-5 in In Vitro BBB Models

1. Transwell inserts cat# 3401 (Costar) coated with rat tail collagen: 100 μl/insert of 27 μg/ml sterile stock (stored at 4°C, stable for at least 2 years).
2. Phosphate-buffered saline (PBS, Sigma).
3. 4% Cold paraformaldehyde (PFA, Bio-Lab, Israel).
4. Blocking solution: 20% normal horse serum (NHS)/0.1% Triton X-100/PBS.
5. Primary antibody: monoclonal anti-claudin-5 (Zymed) diluted 1:200 in 3% NHS/0.1% Triton X-100/PBS.
6. Secondary antibody: Alexa Fluor 546-conjugated goat anti-mouse (Molecular Probes) diluted 1:200 in 3% NHS/0.1% Triton X-100/PBS.
7. Nuclear stain: 1 μg/ml Hoechst 33342 reagent (Sigma), diluted from 1 mg/ml stock stored at 4°C.
8. Mounting medium: Aqua Poly/Mount (cat# 18606, Polysciences, Inc.).

2.3. SDS-PAGE

2.3.1. SDS-Polyacrylamide Gel Electrophoresis

1. Separating buffer 4×: 1.5 M Tris–HCl, pH 8.8. For 100 ml: 18.2 g Tris–HCl is dissolved in DDW and pH is then adjusted to 8.8 with HCl 32%. Store at 4°C.
2. Stacking buffer 4×: 0.5 M Tris–HCl, pH 6.8. For 100 ml: 6.1 g Tris is dissolved in DDW and pH is then adjusted to 6.8 with HCl 32%. Store at 4°C.
3. 30% (w/v) acrylamide-bis solution (37.5:1) (this is a neurotoxin when unpolymerized, so care should be taken when handling) and *N*,*N*,*N*,*N*′-tetramethylethylenediamine (TEMED, Bio-Rad).
4. Ammonium persulfate (APS): to be prepared as 10% in DDW and freezed in single-use aliquots at –20°C.
5. Running buffer: Prepare solution by mixing 3 g Tris with 14.4 g glycine and 1 g SDS. Add 1 L DDW in a measuring cylinder and mix with a magnetic stirrer.

2.3.2. SDS-Polyacrylamide Gel Electrophoresis

1. Transfer buffer: Prepare solution by mixing 1.9 g Tris with 9 g glycine. Add 1 L DDW in a measuring cylinder and mix with a magnetic stirrer.
2. Nitrocellulose membrane and 3-mm chromatography paper from Whatman.
3. Tris buffer saline with 0.1% Tween 20 (TTBS): Prepare ×10 stock with 1.37 M NaCl, 27 mM KCl, and 250 mM Tris–HCl, pH 7.4. Add 0.1% Tween 20 to the ×1 solution before use. The ×1 solution can be stored for several weeks at 4°C.
4. Blocking buffer: 5% skimmed milk/TTBS (see Note 1).
5. Primary antibody: mouse anti-claudin-5 (Zymed) diluted 1:10,000 in 1% skimmed milk/TTBS (see Note 2).

6. Secondary antibody: IRDye 800CW (LI-COR Biosciences)-conjugated goat anti-mouse IgG diluted 1:10,000 in 1% skimmed milk/TTBS (see Note 2).

3. Methods

The formation of TJ proteins is seen as one of the key factors for the evaluation of the quality of an in vitro BBB model. As claudin-5 is one of the major TJ proteins at the BBB, the determination of its expression pattern and levels may be used for this purpose. In the assay described herein, we use Western blotting and immunocytochemistry to assess the expression levels and pattern of claudin-5 in brain endothelial cells (Fig. 2).

We apply these methods while examining porcine EC in three types of in vitro BBB models (see Fig. 1). The described assays can be applied for similar in vitro BBB models using porous membranes as a platform for seeding the EC (see Note 3).

3.1. Immunocytochemistry for the Detection of Claudin-5 in the In Vitro BBB Models

1. Porcine brain endothelial cells (PBEC) are grown to confluency till they reach a typical high transendothelial electrical resistance (TEER; i.e., above 200 Ω cm^2); 2–3 days in plating medium and 1–2 days in assay medium (see Note 3).
2. In a fume hood, the inserts are transferred to a 12-well plate filled with cold PFA (1 ml/well), and then the upper chamber

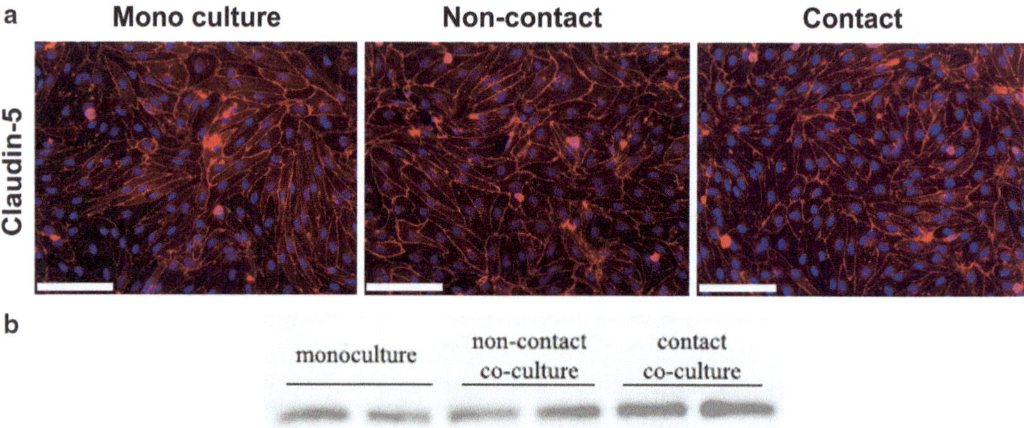

Fig. 2. Claudin-5 expression pattern and level in the three in vitro BBB models. (**a**) PBEC were fixed and stained with specific monoclonal antibody directed against claudin-5. Nuclei were counterstained with the Hoechst reagent. (**b**) Western blot analysis showing the expression level of claudin-5 in the three in vitro models in duplicate, where each band represents a separate Transwell filter. The intensity of the bands is increased in the co-culture systems showing the highest intensity in the contact co-culture model. Glial cells did not display any immunoreactivity to the anti-claudin-5 antibody (*data not shown*). Reproduced from (7) with permission from Elsevier.

medium is rapidly removed and replaced with 300 μl/well of cold PFA for 10 min.

3. The PFA is discarded into a hazardous waste container and the inserts are washed three times for 5 min with PBS.
4. The insert membranes are cut using a scalpel and transferred (cells facing up) into a 24-well plate filled with 200 μl/well of blocking solution for 1 h at room temperature.
5. The blocking solution is removed and replaced with the primary anti-claudin-5 antibody solution (200 μl/well) for overnight (ON) incubation at 4°C with gentle agitation.
6. The primary antibody is removed and the samples are washed three times, for 5 min each, with PBS (0.5 ml/well).
7. The secondary antibody is added (200 μl/well) for 1 h at room temperature with gentle agitation. All steps from here onward are carried out with dimmed lights/under aluminum foil.
8. The secondary antibody is discarded and the samples are washed twice for 5 min each with PBS (0.5 ml/well). Then the Hoechst reagent is added (200 μl/well) for 4 min.
9. The Hoechst reagent is removed and the samples are washed three times for 5 min each with PBS (0.5 ml/well).
10. The membranes are inverted (cells facing down) into a drop of mounting medium on a microscope slide. On top of each membrane, a 13-mm cover slide is placed.
11. The samples can be viewed immediately or stored at 4°C for several months. The slides are viewed under fluorescence microscope with the appropriated excitations for claudin-5 (546 nm) and Hoechst reagent for nuclei detection (346 nm).

3.2. Preparation of Protein Samples for the Determination of Claudin-5 Expression Levels by Western Blotting (See Note 4)

1. PBEC are grown, as described above, on rat tail collagen-coated inserts in the three types of in vitro models till they reach a typical high TEER (i.e., above 200 Ω cm^2 (see Note 3)).
2. The inserts are rapidly washed twice with PBS by aspiration of the medium both in the upper and lower sides of the insert.
3. The inserts are transferred to a new 12-well plate and the PBS in the upper side of the insert is aspirated and 60 μl of lysis buffer is added. The lysed cells are immediately transferred to Eppendorf tubes and put on ice.
4. Protein concentrations in the lysates are determined using the bicinchoninic acid (BCA, Pierce) method and equal amounts of proteins from the samples are transferred to new tubes where sample buffer ×2 is added to the samples in equal volumes 1:1. The samples are boiled for 5 min and kept on ice till the separation by SDS-PAGE (see Note 5).

3.3. SDS-PAGE

1. These instructions are for mini-gels with a Bio-Rad system.
2. Prepare a 12.5% gel by mixing the following in a 15-ml tube: 4.17 ml acrylamide-bis solution, 2.5 ml separating gel buffer ×4, 3.29 ml DDW, 0.1 ml 10% SDS solution, 30 μl 10% APS solution, and 15 μl TEMED. Add the TEMED immediately before pouring the gel between the two glasses (make sure the glasses are clean). Pour the gel, leaving space for a stacking gel and overlay with DDW. The gel should polymerize in about 20 min.
3. Prepare the stacking gel by mixing 0.66 μl acrylamide-bis solution with 1 ml stacking gel buffer ×4, 2.32 ml DDW, 40 μl 10% SDS solution, 12 μl 10% APS solution, and 6 μl TEMED. Pour the DDW and add the stacking gel. Insert the comb immediately. The gel should polymerize in about 20 min.
4. Once the stacking gel has set, carefully remove the comb and add the running buffer to the upper and lower chambers of the gel unit.
5. Load the protein samples from the three in vitro BBB models including the astrocytes controls (see Note 4). Include one well for pre-stained molecular weight markers.
6. Complete the assembly of the gel unit and connect to a power supply. Run the gel for about 1.5 h at 125 V. Follow the dye fronts till they reach the end of the gel.

3.4. Western Blotting for Claudin-5

1. These instructions are for a Bio-Rad's Trans-Blot system.
2. Pre-wet the sponges, filter papers, and nitrocellulose membranes in the transfer buffer for at least 20 min.
3. The transfer cassette is assembled as follows: Sponge – filter paper – gel – membrane – filter paper – sponge. Make sure that no bubbles are left by rolling a pipette on the upper filter paper before closing the cassette.
4. The transfer unit is put on ice and the transfer is accomplished at 0.2 A (about 80 V) for 1.5 h.
5. Once the transfer is complete, take out the cassette, discard the gel and filter papers, and remove the nitrocellulose membrane into a suitable container filled with excess of Ponceau S (Sigma) solution for 2 min with agitation (see Note 6). The nitrocellulose is then rinsed with tap water and the total proteins should be visible (see Note 7). For economic reasons and convenience, the nitrocellulose can be cut using the visible red-stained proteins as markers.
6. The membrane is washed twice with TTBS until no Ponceau S remains.
7. The nitrocellulose is then incubated in 3–5 ml blocking solution for 1 h at room temperature (or overnight at 4°C) on a rocking platform.

8. The blocking buffer is discarded and the primary antibody solution is added to cover the nitrocellulose fully and incubated overnight at 4°C on a rocking platform.
9. The primary antibody is then removed (see Note 2) and the membrane is washed four times for 5 min each with excess TTBS.
10. The secondary antibody solution is added and the membrane is incubated under aluminum foil for 1 h at room temperature on a rocking platform.
11. The secondary antibody is then removed (see Note 2) and the membrane is washed four times for 5 min each with excess TTBS.
12. The membrane is scanned with the Odyssey infra-red scanner (LI-COR, Biosciences) at the 800-nm channel, with resolution and quality set to 169 and medium, respectively, and the bands are quantified using the densitometry software of the Odyssey (see Note 8).

4. Notes

1. Blocking buffer (BB) from LI-COR (cat# 927–4000) is recommended for use with infra-red (IR) systems such as the Odyssey. This BB can be diluted 1:1 with PBS, stored at 4°C, and reused at least five times.
2. The primary and secondary antibody solutions can also be prepared in the BB mentioned in Note 1. For economic reasons, both the primary and the secondary (IR) antibodies can be saved for subsequent experiments by addition of 0.01% final concentration of sodium azide (conveniently done by dilution from a 5% stock solution; exercise caution since azide is highly toxic) and stored at 4°C.
3. The protocols described here could be applied for endothelial cells from different sources/species that are seeded on porous membranes. However, we use porcine-origin endothelial cells that are co-cultured with rat glial cells. A detailed protocol for the preparation of these cells could be found in references (7, 8).
4. When using a contact co-culture BBB in vitro model, one should use a control sample made of inserts devoid of endothelial cells with only glial cells at the bottom (i.e., the glia cells are seeded at the lower side of the chamber (see Fig. 1) and one should take the samples from the upper side where glia end-feet processes might reach). Also, a pure glia sample should be checked to verify that no immunoreactivity exists with the anti-claudin-5 antibody.

5. We avoid using a cytoskeleton protein (like actin, for example) as a normalization standard, since the expression of such proteins can also be changed in different BBB conditions as described here. Loading equal amounts of proteins is preferred in this case.
6. Ponceau S solution is stored at room temperature and can be reused as long as the solution is clean.
7. Staining the membrane with the Ponceau S solution is done for three reasons:
   (a) To make sure that the proteins indeed are transferred to the nitrocellulose membrane.
   (b) For economic reasons and convenience: after this step, the lanes are visible and it is possible to cut the membrane to fit adequate containers. This can save you antibody solution.
   (c) This staining can be used for a rough estimation regarding the quality of the loading of the samples.
8. Other software such as the Image J can be used to quantify the resultant bands.

Acknowledgments

This work was supported by grants from the Weizmann-Negri Fund, the Nella and Leon Benoziyo Center for Neurological Diseases, and the Israel Science Foundation. VIT was the incumbent of the Louis and Florence Katz-Cohen Chair of Neuropharmacology.

References

1. Cecchelli, R., Berezowski, V., Lundquist, S., Culot, M., Renftel, M., Dehouck, M. P., and Fenart, L. (2007) Modelling of the blood-brain barrier in drug discovery and development. *Nat Rev Drug Discov 6*, 650–61.
2. Deli, M. A., Abraham, C. S., Kataoka, Y., and Niwa, M. (2005) Permeability studies on in vitro blood-brain barrier models: physiology, pathology, and pharmacology. *Cell Mol Neurobiol 25*, 59–127.
3. Bazzoni, G., and Dejana, E. (2004) Endothelial cell-to-cell junctions: molecular organization and role in vascular homeostasis. *Physiol Rev 84*, 869–901.
4. Nitta, T., Hata, M., Gotoh, S., Seo, Y., Sasaki, H., Hashimoto, N., Furuse, M., and Tsukita, S. (2003) Size-selective loosening of the blood-brain barrier in claudin-5-deficient mice. *J Cell Biol 161*, 653–60.
5. Zlokovic, B. V. (2008) The blood-brain barrier in health and chronic neurodegenerative disorders. *Neuron 57*, 178–201.
6. Lee, S. W., Kim, W. J., Choi, Y. K., Song, H. S., Son, M. J., Gelman, I. H., Kim, Y. J., and Kim, K. W. (2003) SSeCKS regulates angiogenesis and tight junction formation in blood-brain barrier. *Nat Med 9*, 900–6.
7. Cohen-Kashi Malina, K., Cooper, I., and Teichberg, V. I. (2009) Closing the gap between the in-vivo and in-vitro blood-brain barrier tightness. *Brain Res 1284*, 12–21.
8. Franke, H., Galla, H. J., and Beuckmann, C. T. (1999) An improved low-permeability in vitro-model of the blood-brain barrier: transport studies on retinoids, sucrose, haloperidol, caffeine and mannitol. *Brain Res 818*, 65–71.

Chapter 26

HIV-1-Induced Alterations of Claudin-5 Expression at the Blood–Brain Barrier Level

Ibolya E. András and Michal Toborek

Abstract

HIV-1 crosses the blood–brain barrier (BBB) early in the course of systemic infection and resides in brain macrophages and microglia. The integrity of the brain endothelium is regulated by intercellular tight junctions, which also play a critical role in HIV-1-entry into the brain. Disruption of tight junctions, including changes in claudin-5 expression, is common in HIV-1-infected patients. Recent evidence indicates that both exposure to HIV-1 and HIV-1 specific proteins, such as Tat protein, can contribute to alterations of expression and distribution of claudin-5 in brain endothelial cells and brain microvessels.

Key words: Claudin-5, Brain microvascular endothelial cell, Blood–brain barrier, HIV-1

1. Introduction

HIV-1 crosses the blood–brain barrier (BBB) early in the course of systemic infection and resides in brain macrophages and microglia (1–5). However, the mechanisms of these effects are not fully understood and no productive HIV-1 infection of brain endothelial cells was observed in HIV-1-infected patients. It has been hypothesized that HIV-1 trafficking into the brain may be mediated through a "Trojan horse" mechanism, in which HIV-infected CD4+ T-lymphocyte and/or circulating monocytes enter the CNS through the BBB via interaction with inflammatory mediators (1, 2). In an experimental in vitro setting, this mechanism is mimicked by an exposure of brain endothelial cells to HIV-1-infected monocytes.

HIV-1 can induce a variety of inflammatory mediators in brain endothelial cells and other cell components of the CNS. Inflammatory responses were observed in the brain tissue of patients with HIV-1-related encephalitis (6). HIV-1-induced

Kursad Turksen (ed.), *Claudins: Methods and Protocols*, Methods in Molecular Biology, vol. 762,
DOI 10.1007/978-1-61779-185-7_26, © Springer Science+Business Media, LLC 2011

upregulation of chemokines, CCL2 in particular, may stimulate paracellular transendothelial migration of both normal and HIV-infected leukocytes via disrupted tight junctions between adjacent brain endothelial cells (7, 8).

Well-developed tight junctions are the most prominent feature of the brain endothelium and are responsible for the integrity of the BBB. They limit paracellular flux and restrict permeability across the brain endothelium (9–11). Transmembrane tight junction proteins, such as claudin-5, are responsible for sealing together brain endothelial cells. Disruption of tight junctions was described in HIV-1-infected patients (12–14). These changes were associated with the accumulation of HIV-1-infected macrophages in the brain, fibrinogen leakage, and apoptosis (12, 14).

HIV-1 does not infect rodent cells. Therefore, HIV-1-specific proteins, instead of live virus, are frequently used in experimental research. Several vascular effects of HIV-1 observed in the CNS can be reproduced by treatment with HIV-1 protein Tat. Tat, produced by HIV-1-infected cells, is found circulating in the blood of HIV-1-infected patients (15). Tat can be taken up by host cells, enter the nucleus, and transactivate genes linked to the HIV-1 long terminal repeat (LTR) (16). In addition, Tat is a strong pro-inflammatory agent that can recruit and induce transendothelial migration of monocytes (17). Tat also affects endothelial cell actin microfilament dynamics, causing actin cytoskeletal rearrangements such as membrane ruffling, peripheral retraction, and subsequent cytoskeletal disassembly (18).

The mechanisms of Tat interaction with the cell surface are not fully understood. It was suggested that Tat can mimic extracellular matrix proteins and bind to αvß3 and α5ß1 integrins through the arginine–glycine–aspartate domain (19). In addition, the VEGF receptor-2 (VEGFR-2) has been proposed to serve as a high-affinity receptor for Tat in endothelial cells (20). This notion was supported by our data that indicate that blocking VEGFR-2 can attenuate Tat-induced disruption of tight junctions (21). VEGFR-2 and a variety of other cell surface receptors that belong to the large family of G-protein-coupled receptors (GPCRs) are localized to caveolae in endothelial cells. Activation of these receptors can stimulate the heterotrimeric G-proteins, which then induce the Ras signaling (22, 23). Indeed, Tat binding to cell membranes resulted in activation of the Ras/MAPK signaling pathway and contributed to cell progress through the G1 phase in response to mitogen in primary and immortalized human umbilical endothelial cells (19). Our research identified that Tat-induced activation of Ras results in stimulation of the downstream kinases, leading to alterations of tight junction proteins, including claudin-5 (24, 25).

2. Materials

2.1. HIV-1 and HIV-1-Specific Proteins

2.1.1. HIV-1 Stock

1. PYK-JRCSF plasmid DNA (clone of HIV-$1_{JR\text{-}CSF}$, a primary isolate from cerebral spinal fluid obtained from an AIDS patient) available from the NIH AIDS Research & Reference Reagent Program (http://www.aidsreagent.org).
2. HB101 competent cells, 5 × 200 μL (Promega, Madison, WI).
3. LB broth base (Invitrogen, Carlsbad, CA).
4. LB agar (Invitrogen).
5. Dulbecco's modified Eagle medium (DMEM) (Sigma, St Louis, MO).
6. RPMI 1640 medium supplemented with 10% FBS, 100 U/mL penicillin, and 100 μg/mL streptomycin.
7. HEK 293 T/17 cells (American Type Culture Collection, Manassas, VA) cultured in DMEM with 10% FBS, 100 U/mL penicillin, and 100 μg/mL streptomycin.
8. Calcium phosphate transfection kit (Invitrogen).
9. Sac I restriction enzyme (Promega).
10. p24 ELISA kit (ZeptoMetrix, Buffalo, NY).
11. Agarose (Fisher Scientific).
12. Tris/borate/EDTA (TBE) buffer.
13. Plasmid isolation kit (Mini) (Qiagen, Valencia, CA).
14. Plasmid isolation kit (Maxi) (Qiagen).

2.1.2. Infection of U937 Cells

1. U937 cells (American Type Culture Collection, Manassas, VA) cultured in RPMI 1640 (Invitrogen, Carlsbad, CA) with 10% FBS (Hyclone), 100 U/mL penicillin, and 100 μg/mL streptomycin.
2. Phosphate-buffered saline (PBS) solution prepared from 10× PBS (80.6 mM Na_2HPO4, 19.4 mM KH_2PO4, 27 mM KCl, and 1.37 M NaCl, pH 7.4; USB Corp., Cleveland, OH).
3. RIPA cell lysis buffer (TBS, 1% Nonidet P-40, 0.5% sodium deoxycholate, 0.1% SDS, 0.004% sodium azide, PMSF in DMSO, protease inhibitor cocktail, and sodium orthovanadate).
4. 1% Triton X-100 in lysis buffer to inactivate HIV-1.
5. p24 ELISA kit (ZeptoMetrix, Buffalo, NY).
6. T-25 flasks (Corning Costar, Corning, NY).

7. 6-well Transwell plates (pore size 0.4 μm, Corning Costar, Corning, NY).
8. Cell counting tubes with cap (TPP-US, St. Louis, MO).
9. Polyethylene cell lifters (Corning Inc., Corning, NY).

2.1.3. HIV-1-Specific Proteins

Several HIV-1-specific proteins are commercially available, e.g., from Diatheva (Fano, Italy). We produce and isolate HIV-1 Tat protein as described by Ma and Nath (26). Lyophilized Tat is then stored in siliconized Eppendorf tubes at −20 C. Stock solution of Tat (100 mM) is prepared by dissolving lyophilized Tat in sterile conditions in cell culture medium without serum and then stored at −80°C.

2.2. Analysis of Claudin-5 Expression by Western Blotting

2.2.1. SDS–Polyacrylamide Gel Electrophoresis

1. Ready 4–15% Tris–HCl mini gels with 50-μL 10-well comb (Bio-Rad, Hercules, CA).
2. Running buffer (10×): 250 mM Tris, 1.92 M glycine, and 1.0% (w/v) SDS, pH 8.6 (USB Corp.).
3. Pre-stained molecular weight marker: kaleidoscope marker (Bio-Rad).

2.2.2. Western Blotting

1. Transfer buffer (10×): 250 mM Tris, and 1.92 M glycine, pH 8.3 (USB Corp.).
2. Nitrocellulose membrane and blotting filter paper (Bio-Rad).
3. TBS-T solution: dilute 20× TBS (500 mM Tris, 60 mM KCl, and 2.8 M NaCl, pH 7.4 [USB Corp.]) to 1× TBS and add 0.05% Tween-20.
4. Blocking buffer: 5% milk in TBS-T. Dissolve skim milk (Difco BD, Sparks, MD) in TBS-T. Store for 1 week maximum at 4°C.
5. Primary antibody: mouse anti-claudin-5 antibody (Invitrogen/Zymed, cat No. 18–7364, Carlsbad, CA).
6. Secondary antibody: anti-mouse IgG conjugated to horseradish peroxidase (Santa Cruz, Santa Cruz, CA).
7. Anti-actin antibody conjugated to horseradish peroxidase (Sigma, St Louis, MO).
8. Enhanced chemiluminescence (ECL) reagent (GE Healthcare, Buckinghamshire, UK) and blue basic autoradiographic film (ISC Bioexpress, Belgium).

2.3. Analysis of Claudin-5 Expression by Immunofluorescence Microscopy

2.3.1. Cell Cultures

1. Collagen I-coated chambered glass slides (BD Biocoat, Bedford, MA). Permanox plastic 4-well chambered slides (BD, Bedford, MA) can also be used; however, fluorescence signal is stronger and autofluorescence of the plastic slides is avoided when cells are grown on glass slides. Both types of dishes are additionally coated with collagen IV and fibronectin as described in Subheading 3.2.1.

2. Microscope coverslips (1½ 22 mm²) (Corning).
3. Absolute ethanol (Sigma) stored at 4°C.
4. Blocking buffer: 3% (w/v) bovine serum albumin (BSA, Sigma) in PBS, sterile filtered and stored at 4°C.
5. Primary antibody: mouse anti-claudin-5 antibody (Invitrogen/Zymed, cat No. 18-7364).
6. Secondary antibody: anti-mouse IgG conjugated to FITC or Texas Red (Santa Cruz Biotechnology) or anti-mouse IgG conjugated to AlexaFluor 488 or AlexaFluor 568 (Invitrogen).
7. Mounting medium: Anti-fade mounting medium with DAPI for visualizing the nuclei (Invitrogen).

2.3.2. Microvessels

1. ProbeOn™ Plus microscope slides (charged and pre-cleaned, Fisher Scientific).
2. Liquid blocker super Pap pen (Daido Sangyo, Japan).
3. Paraformaldehyde: 4% (w/v) solution in PBS freshly prepared for each experiment from a 10% paraformaldehyde solution (Fisher Scientific).
4. Permeabilization solution: 0.1% (v/v) Triton X-100 in PBS, stored at 4°C.
5. Blocking buffer: 1% (w/v) BSA (Sigma) in PBS, sterile filtered and stored at 4°C.
6. Microscope coverslips (22×40–1) (Fisher Scientific, cat No.12-548-5C).
7. PBS, mounting medium, primary antibody, and secondary antibody are the same as that used for cell cultures.

3. Methods

3.1. Exposure to HIV-1 and HIV-1-Specific Proteins

Endothelial cells lack a CD4 receptor (27) and are not productively infected with HIV-1. Therefore, the experiments are performed using co-cultures of human brain endothelial cells (e.g., hCMEC/D3 cells) with HIV-1-infected leukocytic cells (e.g., primary human monocytes, Jurkat cells, or U937 cells). We first describe the preparation of HIV-1 stock and infection of U937 cells, followed by co-culture of hCMEC/D3 cells with HIV-1-infected U937 cells. Both hCMEC/D3 and mouse brain endothelial cells are susceptible to Tat treatment (see Note 1), which will also be described.

3.1.1. Generation of HIV-1 Stock

1. Prepare LB agar plates (20 mL agar with 100 μg/mL ampicillin per one P100 plate) and dry them at 37°C overnight with the lids off. Prepare also LB broth containing 100 μg/mL ampicillin.

2. Transformation of HB101 cells:
   (a) Cool down a polypropylene tube on ice,
   (b) Remove HB101 cells from a freezer and place them on ice for 5 min,
   (c) Transfer 100 μL cell suspension to the polypropylene tube, add 1–4 μg plasmid DNA, and leave on ice for 10 min,
   (d) Heat shock the cells for 50 s at 42°C in a water bath without shaking,
   (e) Put the tube on ice for 2 min,
   (f) Add 900 μL cold LB broth and place the tube at 37°C for 1 h on a shaker.
3. Spread the transformed HB101 cells on the LB agar plates and incubate at 37°C overnight.
4. Next morning, choose at least five clones from the bacterial colonies (mark the colony locations on the bottom of the plate or on the lid) and place bacteria using sterile pipette tips from these colonies in 5 mL LB broth with ampicillin. Shake overnight at 37°C, seal the plates, and store at 4°C.
5. Extract plasmid DNA from the chosen expanded clones using the Plasmid Isolation Mini kit. Digest the plasmid DNA with Sac I restriction enzyme according to the instruction provided by the manufacturer, and perform electrophoresis (0.7% agarose gel in 0.5× TBE buffer; 100 V for 45 min) to identify the proper bacterial clones.
6. With a sterile loop, transfer bacteria from the selected clones into 250 mL LB broth with ampicillin and incubate on a shaker at 37°C (150 rpm) for 24–48 h.
7. Extract plasmid DNA with the Plasmid Isolation Maxi kit.
8. Transfect HEK 293T/17 cells at ~70% confluence with the calcium phosphate transfection kit following the protocol provided by the manufacturer.
9. Following transfection, incubate the cells in RPMI 1640 medium with 10% FBS and penicillin plus streptomycin for 24 h in a cell culture incubator. Collect cell culture supernatant, pass it through 0.45-μm filters (Millipore, Bedford, MA), and freeze in aliquots at −80°C.
10. HIV-1 p24 levels in the supernatant are determined by ELISA (ZeptoMetrix) as the indicator of HIV-1 infection (see Note 2).

3.1.2. Infection of U937 Cells

1. Suspend 5×10^6 U937 cells in 10 mL RPMI 1640 complete medium and place in a T-25 flask. Add viral isolate from the HIV-1 stock (we routinely use 80 ng p24/mL). Control

cultures of U937 cells are prepared by adding RPMI 1640 complete medium instead of viral isolate.

2. The cells are incubated overnight at 37°C and 5% CO_2, followed by centrifugation and resuspension in 10 mL of complete RPMI 1640 medium.
3. The next day, count the cell number of HIV-1-infected and uninfected U937 cells. Use the cell counting tubes with the cap due to HIV-1 biohazard.
4. Measure p24 levels by ELISA in a 0.5-mL aliquot as the indicator of HIV-1 infection.

3.1.3. Exposure to HIV-1

hCMEC/D3 cells are exposed to HIV-1-infected U937 cells in 6-well Transwell systems (see Note 3). Uninfected control or HIV-1-infected U937 cells are cultured on the filter inserts in 1.5 mL of complete media, and hCMEC/D3 cells are cultured in the bottom chamber. The ratio of hCMEC/D3 cells to U937 cells is 1:1. Following HIV-1 exposure, the filter inserts are removed and the media are discarded. This procedure can only be performed in an HIV-1-designated biosafety cabinet and all waste materials are discarded into a bleach solution. The cells are then washed twice with PBS, each time removing PBS into the bleach solution. Following the final wash, hCMEC/D3 cells are lysed by adding 100 μL/well of RIPA cell lysis buffer (with freshly added PMSF, sodium orthovanadate, and protease inhibitor cocktail) and 1% Triton X-100 to inactivate HIV-1. Cells are scraped off the dish; the lysates are pipetted up and down ten times on ice, and then left for 30 min on ice. After use, the scrapers are dipped into the bleach solution and discarded in the biohazard cardboard box.

Because HIV-1 is inactivated in the cell lysates, the sample tubes (disinfected from the outside with 70% ethanol) can be safely taken out from the HIV-1 biosafety cabinet and then centrifuged at 15,000 × *g* for 15 min at 4°C. The supernatants can be stored at −80°C until analysis.

3.1.4. Exposure to HIV-1-Specific Proteins in the Example of HIV-1 Tat

Prior to Tat treatment, cells are washed twice with sterile PBS solution, followed by replacing normal cell culture medium with a serum-free media (see Note 3). Cells are typically exposed to Tat at the final concentration of 1–100 nM; Tat is added directly into the media. Gentle shaking of the dishes helps to distribute Tat evenly in the media. Repeated freezing-thawing of Tat solution results in loss of biological activity of Tat and should be avoided. Negative controls include exposure to heat-inactivated Tat (100°C for 5 min, followed by centrifugation of denaturated protein) and immunoabsorbtion by Tat anti-sera conjugated to protein-A beads.

3.2. Analysis of HIV-1-Induced Changes in Claudin-5 Expression by Western Blotting

We routinely analyze claudin-5 expression in rat and mouse brain endothelial cells (RBMEC and MBMEC, respectively) and in the human cell line (hCMEC/D3 cells). Handling of rodent and human cells is different and will be described separately.

3.2.1. Cell Culture and Lysis

1. Primary cultures of MBMEC (the same approaches are used for RBMEC). Cells are cultured on dishes coated with collagen type IV and fibronectin. Coating solution constitutes of 800 μL sterile water, 100 μL fibronectin, and 100 μL collagen type IV (both from Sigma, St Louis, MO). The whole surface of cell culture dishes is covered with the coating solution and removed immediately before seeding the microvessel fragments (28). MBMEC are cultured in DMEM (Sigma, St Louis, MO) with 20% plasma-derived bovine serum (PDBS, Animal Technologies Inc., Tyler, TX), 40 mg/mL endothelial cell growth supplement (BD Biosciences Pharmingen, San Jose, CA), 100 mg/mL heparin, 2 mmol/L glutamine, 5 mg/mL insulin, 5 mg/mL human transferrin, 5 ng/mL Na-selenite (Insulin-transferrin-sodium selenite media supplement, Sigma, St Louis, MO), and 25 mg/mL gentamicin (see Note 4).
2. hCMEC/D3 cultures. hCMEC/D3 cells are cultured on dishes coated with collagen type I (BD Biosciences). Collagen I is diluted 1:50 in sterile distilled water, sterile filtered with a 0.45-μm syringe filter, and stored at 4°C. The whole surface area of cell culture dishes is covered with collagen I solution, quickly removed, incubated with the complete medium for at least 1 h, and discarded immediately prior to seeding the cells. hCMEC/D3 cells are cultured in EBM-2 medium (Lonza, Walkersville, MD) supplemented with the bullet kit (1 kit/2 L medium) supplied with the EBM-2 medium (29). The bullet kit contains vascular endothelial growth factor, insulin-like growth factor-1, epidermal growth factor, basic fibroblast growth factor, hydrocortisone, ascorbate, gentamicin, and 2.5% fetal bovine serum (FBS). hCMEC/D3 cells are passaged with trypsin-EDTA (0.25%; Invitrogen, Carlsbad, CA), typically twice a week from one dish to four dishes of equal size or from one P100 dish to four 6-well plates. In order to pass the cells, a confluent culture in a P100 dish is rinsed twice with PBS, followed by adding trypsin-EDTA solution to cover the whole area (we use 1 mL trypsin–EDTA per one P100 dish and remove the excess with a Pasteur pipette). The dish is then placed in a cell culture incubator at 37°C for 5 min. When the cells are detached, 12 mL of complete EBM-2 medium is added; the cells are washed off the dish by repeated pipetting and transferred in equal aliquots into new dishes.

3.2.2. Sample Preparation for Claudin-5 Western Blotting

One well of a 6-well plate or one 35-mm dish of confluent MBMEC or hCMEC/D3 cells provides sufficient sample size to determine claudin-5 expression by Western blot. After HIV-1 or Tat treatment and measuring protein concentrations, add 6× gel loading buffer to each sample, followed by boiling for 5 min. Before boiling the samples, make a hole in the tube cap with a syringe needle. Place the tubes from ice directly into the boiling water. After boiling, cool the samples on ice. Following centrifugation at 15,000 × *g* for 3 min at room temperature, the samples can be used for SDS–polyacrylamide gel electrophoresis (SDS–PAGE).

3.2.3. SDS–PAGE

1. The analyses are performed using the ready 4–15% Tris–HCl mini gels and the Mini-Protean electrophoresis system (both from Bio-Rad).
2. Prepare the running buffer by diluting 100 mL of the 10× running buffer with 900 mL of water.
3. Carefully remove the comb from the mini gel. Place the gel in the holder and in the tank.
4. Add running buffer to the upper and lower chambers of the gel unit and wash the wells with the running buffer from the tank using a 10-mL syringe with a 22-gauge needle.
5. Load 20 μL of sample into each gel well. Include one well for a pre-stained molecular weight marker. Do not place sample in the last well; load it with the loading buffer instead.
6. Complete the assembly of the gel unit and connect it to a power supply. Perform electrophoresis for 1 h at 100 V (see Note 5).

3.2.4. Western Blotting

1. The samples separated by SDS–PAGE are then transferred onto pure nitrocellulose membranes using the Mini TransBlot system (Bio-Rad). A tray of transfer buffer has to be large enough to lay out a transfer cassette with a sponge and a blotting paper pad on the top of the black side of the cassette. The second sponge, the nitrocellulose membrane, and the blotting paper are submerged in the transfer buffer in separate smaller trays. The nitrocellulose membrane is cut to the size of the blotting paper.
2. The gel unit is disconnected from the power supply and disassembled. The gel is laid on the top of the blotting paper; the nitrocellulose membrane is placed on the gel, followed by another sheet of the blotting paper and the second sponge. Avoid the formation of any air bubbles between these layers. A plastic pipette (or a similar device) is used as a rolling pin to press out any air between the layers. The amount of transfer buffer in the tray should be sufficient to cover the blot entirely. The cassette is carefully closed after the assembly.

3. The cassette is placed into the transfer tank with transfer buffer in such a way that its black part is facing the black side of the transfer unit. It is important to ensure this orientation; otherwise, proteins will not be transferred onto the nitrocellulose membrane. Then, an icepack is placed in the transfer system. The entire transfer tank is placed in a bigger tray filled with ice to ensure low temperature during transfer. The lid is put on the transfer tank and the protein transfer is performed at constant 250 mA for 1 h 45 min.
4. Once the transfer is complete, the cassette is taken out of the tank and carefully disassembled (black side down). The nitrocellulose membrane is carefully removed and placed in a smaller tray with TBS-T buffer and washed for 5 min with gentle shaking. The colored molecular weight markers should be visible as sharp and strong bands on the membrane. The gel can be discarded.
5. The membrane is then incubated in 10 mL blocking buffer (5% milk in TBS-T) for 1 h at room temperature on a shaker.
6. The blocking buffer is discarded and the membrane is incubated with the primary anti-claudin-5 antibody diluted in 5% milk in TBS-T (antibody dilution 1:1,000) overnight at 4°C on a shaker.
7. The primary antibody is removed and the membrane is washed three times for 10 min with ~50 mL TBS-T.
8. The membrane is incubated with the secondary antibody (diluted 1:2,000 in 5% milk in TBS-T) for 1 h at room temperature on a rocking platform.
9. The secondary antibody is discarded and the membrane is washed three times for 10 min with TBS-T. Note that the next steps are performed in a dark room where the film can be quickly placed over the membrane.
10. The washed membrane is placed on a sheet of plastic foil; 1 mL of reagent A and 1 mL of reagent B of the ECL reagent are mixed together and poured over the blot to cover it completely for 1 min.
11. The blot is drained from the ECL reagents and placed upside down on a sheet of plastic foil (avoid the formation of air bubbles and wrinkles), wrapped in the foil, and placed in an X-ray film cassette.
12. One of the film corners is cut for easier orientation, and the film is placed promptly over the membrane (without moving it over the membrane to avoid shifted ECL signals) for the duration of the exposure time (typically a few minutes). An example of the results produced is shown in Fig. 1.

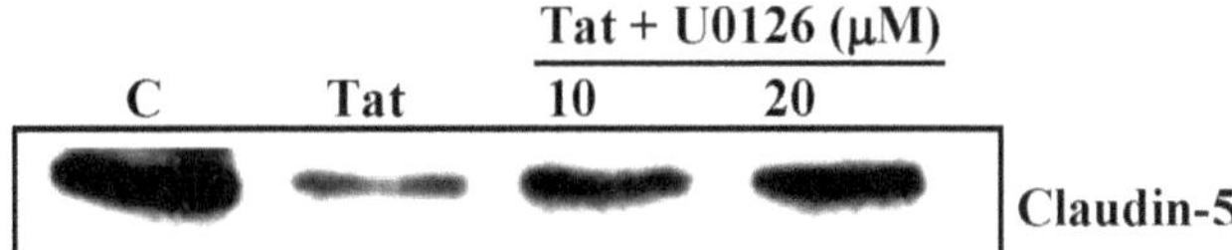

Fig. 1. Effects of Tat treatment on claudin-5 expression in brain microvascular endothelial cells. Confluent cultures were treated with vehicle (control) or 100 nM Tat for 24 h, and claudin-5 protein levels were detected by Western blot. In addition, selected cultures were pretreated for 15 min with U0126 (10 or 20 μM; an inhibitor of mitogen-activated protein kinase kinase1/2 [MEK1/2]), followed by a co-exposure to 100 nM Tat for 24 h (24).

3.3. Analysis of HIV-1-Induced Changes in Claudin-5 Expression by Immunofluorescence Microscopy

3.3.1. Immunofluorescence Microscopy for Cell Cultures

1. Use only confluent cultures to detect positive claudin-5 immunoreactivity at the cell–cell borders. Following exposure to HIV-1 or Tat protein, experimental media are removed and the cells are rinsed twice with PBS (gently add PBS on the side walls of the chambers to avoid washing off the cells) (see Note 6).
2. Add cold absolute ethanol into each chamber (1 mL/chamber for a 4-well chambered slide) to cover the cells entirely and place the slides for 30 min at 4°C for fixation and permeabilization. No additional permeabilization step is necessary. Fixation with absolute ethanol also inactivates HIV-1; however, before the chambers can be safely removed from the biosafety cabinet, the outside walls have to be decontaminated with 70% ethanol.
3. The chambers and the gaskets are removed using the plastic tool provided together with the chambered slides. It is important to avoid touching the cell layer. The slides are drained by placing a corner on a paper tissue, then the slides are placed in 100-mm cell culture dishes containing 10 mL PBS (1 slide per dish, cells facing up). The slides are washed three times with PBS for 5 min with gentle shaking on a platform. Place the slides with the cells facing up on a paper tissue while removing PBS solution when washing the cells. Use the same technique for all washing procedures throughout the procedure (see Note 7).
4. Add blocking solution (3% BSA in PBS) for 30 min at room temperature (1 mL/slide to cover it completely).
5. After removing the blocking solution, the cells are incubated with anti-claudin-5 antibody diluted in blocking buffer (100 μL/chamber) overnight at 4°C in a humidified chamber to prevent cells from drying out. Use 1:100 dilution if the secondary antibody is FITC conjugated, or 1:500 if the secondary antibody is AlexaFluor 488 or AlexaFluor 568 conjugated. To prepare a humidified chamber, place the slides on a platform in a plastic box with a cover and add distilled water to the bottom of the chamber. After closing the lid, the chamber is sealed with parafilm and placed overnight in a refrigerator.

6. The next day, the antibody solution is removed and the slides are washed three times (5 min each) with PBS with very gentle shaking on a platform. Then, add anti-mouse FITC (1:100) or AlexaFluor-conjugated secondary antibody (1:1,000) diluted in blocking buffer (typically 100 μL per chamber) for 2 h at room temperature. This step can be done in the same chamber as that used for an overnight incubation with primary antibody. However, the chamber should be covered with aluminum foil to protect the slides from exposure to light.
7. The secondary antibody solution is removed and the slides are washed three times with PBS for 5 min each, with gentle shaking. Washing is done in a box covered with aluminum foil for light protection.
8. Following the final wash (avoid completely drying the cells), the samples are ready to be mounted. Place two drops of mounting medium containing DAPI on each slide and carefully place two coverslips per slide. Avoid formation of air bubbles. Excess mounting medium will make the coverslip move over the cells. Do not move or press the coverslip. The slides can be viewed immediately or stored in the dark at 4°C. Cells can be examined under an epifluorescence or confocal microscope. A variety of specialized software can be used to merge the fluorescence images for claudin-5 and DAPI. Examples of claudin-5 immunoreactivity are shown in Fig. 2.

3.3.2. Immunofluorescence Microscopy for Microvessels

1. Freshly isolated brain microvessels are spread onto glass microscope slides by pipetting 5–15 μL of the microvessel suspension on one side of a slide (mark the slides in advance with a pencil) and smearing the microvessels with a second slide by moving it continuously toward the other side of a slide to achieve a fine layer. The area of spread microvessels is outlined with a liquid blocker pen.
2. The samples are fixed by placing the slides on a preheated platform (95°C) for 10 min (see Note 8), followed by incubation with freshly prepared 4% paraformaldehyde in PBS (1 mL/slide) for 10 min at room temperature. Remove paraformaldehyde by placing a corner of the slide on paper tissues and washing the slides five times with PBS (1 mL/slide each time).
3. The samples are permeabilized with 0.1% Triton X in PBS for 30 min at room temperature (1 mL/slide). Discard the solution as above and wash the slides five times with PBS. After the final wash, the samples are blocked with 1% BSA solution in PBS for 30 min at room temperature.
4. The blocking solution is removed and the slides are incubated with anti-claudin-5 antibody (1:500 dilution in 1% BSA in PBS; typically, 500 μL per slide covering the microvessel smear) overnight at 4°C. A humidified plastic chamber

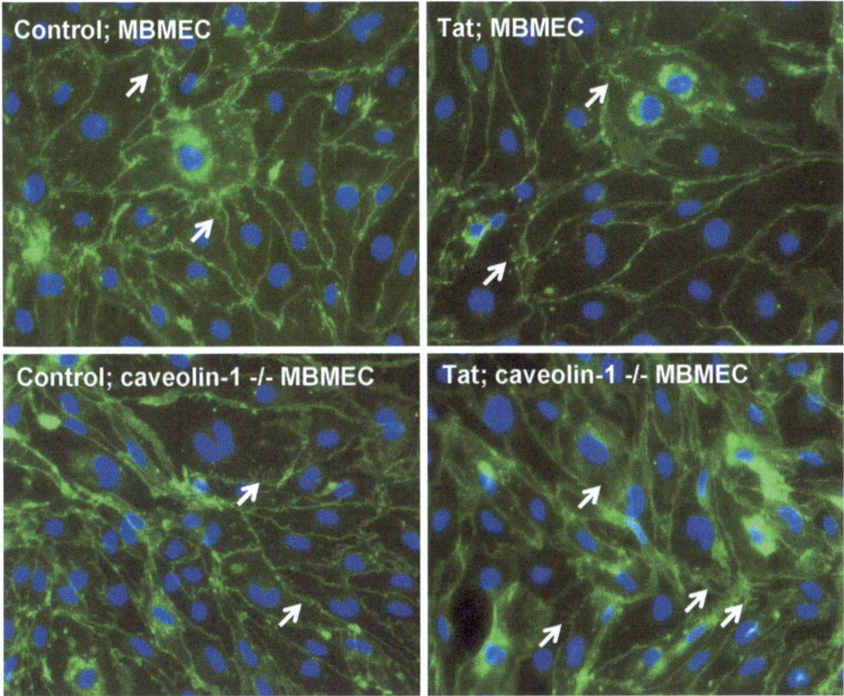

Fig. 2. Effects of Tat treatment on claudin-5 expression in mouse brain microvascular endothelial cells (MBMEC) as determined by immunofluorescence microscopy. Confluent primary cultures of MBMEC were derived from normal or from caveolin-1-deficient (caveolin-1–/–) mice and treated with 100 nM Tat for 24 h. In control cultures (*upper left panel*), claudin-5 immunoreactivity (*green staining*) was restricted to junctional areas and some punctuate staining in the cytoplasm. In addition, short branches and/or ramifications from the cell border toward the cytoplasm were observed (*arrows*). Exposure of normal MBMEC to Tat (*upper right panel*) resulted in a slightly weaker junctional and cytoplasmic pattern of claudin-5 immunoreactivity. The short branching pattern of claudin-5-positive staining (*arrows*) was similar to that of the control cells. MBMEC isolated from caveolin-1–/– mice exhibited markedly enhanced ramification (*lower left panel, arrows*) of claudin-5 immunoreactivity as compared to control cells. This effect was further potentiated by Tat treatment (*lower right panel, arrows*; 100 nM for 24 h) resulting in a dense brush-like image of the claudin-5 immunoreactivity. DAPI (*blue staining*) was used to visualize the nuclei.

(described in Subheading 3.3.1) should be used to avoid drying the slides (see Note 9).

5. The slides are washed five times with PBS and incubated with anti-mouse AlexaFluor 488- or AlexaFluor 568-conjugated secondary antibody (1:1,000) diluted in the blocking solution (500 μL per slide) for 2 h at room temperature. Cover the box with aluminum foil to protect the slides from light.

6. The slides are washed five times with PBS. Following the final wash (avoid completely drying the slides), the samples are ready to be mounted. Place a drop of mounting medium with DAPI in the center of each slide and carefully place a coverslip over it (avoid formation of air bubbles). Do not move or press the coverslip. Analyze the samples as described in Subheading 3.3.1. Examples of claudin-5 immunoreactivity in brain microvessels are shown in Fig. 3.

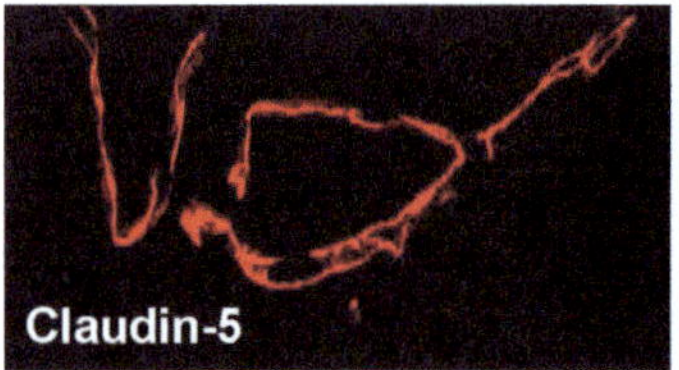

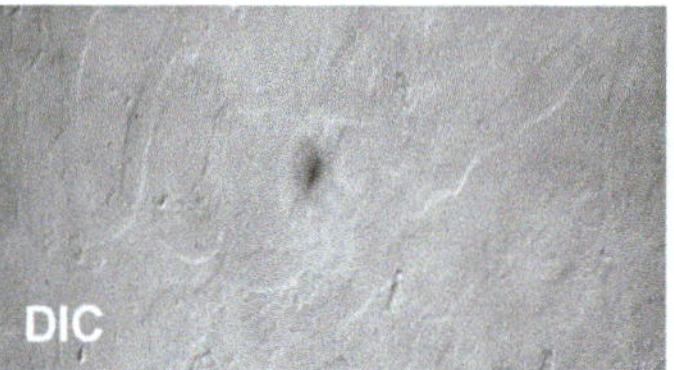

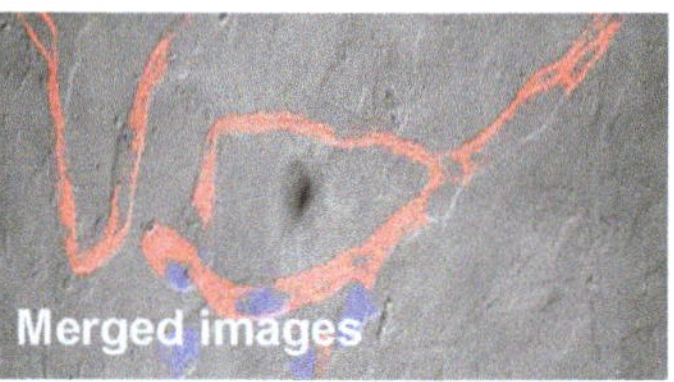

Fig. 3. Claudin-5 expression in microvessels isolated from mouse brain. Claudin-5 immunoreactivity (*red staining*) exhibits a linear, sharply defined pattern along the endothelial cell–cell borders and some weaker immunoreactive patterns in the cytoplasm (*left panel*). DAPI (*blue staining*) was used to visualize the nuclei. Differential interference contrast (DIC) image was taken to visualize microvessels (*middle panel*). *Right panel* illustrates merged claudin-5 immunoreactivity and DIC images.

4. Notes

1. Exposure to Tat is performed in media without serum or other supplements to avoid binding of Tat to serum proteins. In addition, working with Tat requires the use of low-adhesion pipette tips (Corning, Pittsburgh, PA) in order to prevent adherence of Tat to the plastic.
2. It is necessary to dilute the samples ~5,000× because of high p24 levels resulting from this protocol.
3. To synchronize cellular responses, hCMEC/D3 cells can be serum starved overnight prior to HIV-1 or Tat exposure. However, the primary cells are more sensitive and do not withstand such conditions. When removing culture medium and PBS from HIV-1-exposed cells, pipette the entire solution into bleach.
4. All experiments are performed only on primary cultures, because primary endothelial cells lose their phenotypic properties even after one passage. Nevertheless, MBMEC at passage one express high immunoreactivity for tight junction proteins, including claudin-5, as determined by immunofluorescence microscopy.
5. If claudin-5 bands are not sharply and uniformly visualized using the described electrophoresis conditions, shorten the running time to approximately 45 min.
6. Throughout the procedure, when removing solutions and/or washing the cells with PBS, do not dry out the samples.
7. When using plastic chambered slides, place them upside down during the washing procedure. The plastic slides float in PBS; therefore, the cells have to face down in order to be washed.

8. Leave the frosted part of the slide outside the heating platform, so it will not get hot and you can remove the slide quickly after fixation.
9. Place the slides in the humidified chamber first and then add the antibody solution.

Acknowledgments

This work was supported by MH63022, MH072567, and NS39254.

References

1. Wu DT, Woodman SE, Weiss JM, McManus CM, D'Aversa TG, Hesselgesser J, Major EO, Nath A, Berman JW (2000) Mechanisms of leukocyte trafficking into the CNS. J Neurovirol 6(Suppl 1):S82–85
2. Gonzalez-Scarano F, Martin-Garcia J (2005) The neuropathogenesis of AIDS. Nat Rev Immunol 5:69–81
3. Toborek M, Lee YW, Flora G, Pu H, Andras IE, Wylegala E, Hennig B, Nath A (2005) Mechanisms of the blood-brain barrier disruption in HIV-1 infection. Cell Mol Neurobiol 25:181–199
4. Banks WA, Ercal N, Price TO (2006) The blood-brain barrier in neuroAIDS. Curr HIV Res 4:259–266
5. Persidsky Y, Poluektova L (2006) Immune privilege and HIV-1 persistence in the CNS. Immunol Rev 213:180–194
6. Yadav A, Collman RG (2009) CNS inflammation and macrophage/microglial biology associated with HIV-1 infection. J Neuroimmune Pharmacol 4:430–447
7. Eugenin EA, Osiecki K, Lopez L, Goldstein H, Calderon TM, Berman JW (2006) CCL2/monocyte chemoattractant protein-1 mediates enhanced transmigration of human immunodeficiency virus (HIV)-infected leukocytes across the blood-brain barrier: a potential mechanism of HIV-CNS invasion and NeuroAIDS. J Neurosci 26:1098–1106
8. Dhillon NK, Williams R, Callen S, Zien C, Narayan O, Buch S (2008) Roles of MCP-1 in development of HIV-dementia. Front Biosci 13:3913–3918
9. Abbott NJ, Rönnbäck L, Hansson E (2006) Astrocyte-endothelial interactions at the blood-brain barrier. Nat Rev Neurosci 7:41–53
10. Hawkins BT, Davis TP (2005) The blood-brain barrier/neurovascular unit in health and disease. Pharmacol Rev 57:173–185
11. Zlokovic BV (2008) The blood-brain barrier in health and chronic neurodegenerative disorders. Neuron 57:178–201
12. Dallasta LM, Pisarov LA, Esplen JE, Werley JV, Moses AV, Nelson JA, Achim CL (1999) Blood-brain barrier tight junction disruption in human immunodeficiency virus-1 encephalitis. Am J Pathol 155:1915–1927
13. Boven LA, Middel J, Verhoef J, De Groot CJ, Nottet HS (2000) Monocyte infiltration is highly associated with loss of the tight junction protein zonula occludens in HIV-1-associated dementia. Neuropathol Appl Neurobiol 26:356–360
14. Persidsky Y, Heilman D, Haorah J, Zelivyanskaya M, Persidsky R, Weber GA, Shimokawa H, Kaibuchi K, Ikezu T (2006) Rho-mediated regulation of tight junctions during monocyte migration across the blood-brain barrier in HIV-1 encephalitis (HIVE). Blood 107:4770–4780
15. Xiao H, Neuveut C, Tiffany HL, Benkirane M, Rich EA, Murphy PM, Jeang KT (2000) Selective CXCR4 antagonism by Tat: implications for in vivo expansion of coreceptor use by HIV-1. Proc Natl Acad Sci USA 97:11466–11471
16. Nath A, Geiger J (1998) Neurobiological aspects of human immunodeficiency virus infection: neurotoxic mechanisms. Prog Neurobiol 54:19–33
17. Toborek M, Lee YW, Pu H, Malecki A, Flora G, Garrido R, Hennig B, Bauer HC, Nath A (2003) HIV-Tat protein induces oxidative and inflammatory pathways in brain endothelium. J Neurochem 84:169–179

18. Wu RF, Gu Y, Xu YC, Mitola S, Bussolino F, Terada LS (2004) Human immunodeficiency virus type 1 Tat regulates endothelial cell actin cytoskeletal dynamics through PAK1 activation and oxidant production. J Virol 78: 779–789
19. Toschi E, Bacigalupo I, Strippoli R, Chiozzini C, Cereseto A, Falchi M, Nappi F, Sgadari C, Barillari G, Mainiero F, Ensoli B (2006) HIV-1 Tat regulates endothelial cell cycle progression via activation of the Ras/ERK MAPK signaling pathway. Mol Biol Cell 17:1985–1994
20. Albini A, Soldi R, Giunciuglio D, Giraudo E, Benelli R, Primo L, Noonan D, Salio M, Camussi G, Rockl W, Bussolino F (1996) The angiogenesis induced by HIV-1 tat protein is mediated by the Flk-1/KDR receptor on vascular endothelial cells. Nat Med 2: 1371–1375.
21. Andras IE, Pu H, Tian J, Deli MA, Nath A, Hennig B, Toborek M (2005) Signaling mechanisms of HIV-1 Tat-induced alterations of claudin-5 expression in brain endothelial cells. J Cereb Blood Flow Metab 25:1159–1170
22. Wennerberg K, Rossman KL, Der CJ (2005) The Ras superfamily at a glance. J Cell Sci 118:843–846
23. Hancock JF (2003) Ras proteins: different signals from different locations. Nat Rev Mol Cell Biol 4:373–384
24. Andras IE, Pu H, Tian J, Deli MA, Nath A, Hennig B, Toborek M (2005) Signaling mechanisms of HIV-1 Tatinduced alterations of claudin-5 expression in brain endothelial cells. J Cereb Blood Flow Metab 25:1159–1170
25. Zhong Y, Smart EJ, Weksler B, Couraud PO, Hennig B, Toborek M (2008) Caveolin-1 regulates human immunodeficiency virus-1 Tat-induced alterations of tight junction protein expression via modulation of the Ras signaling. J Neurosci 28:7788–96
26. Ma M, Nath A (1997) Molecular determinants for cellular uptake of Tat protein of human immunodeficiency virus type 1 in brain cells. J Virol 71:2495–2499
27. Kanmogne GD, Schall K, Leibhart J, Knipe B, Gendelman HE, Persidsky Y (2007) HIV-1 gp120 compromises blood-brain barrier integrity and enhances monocyte migration across blood-brain barrier: implication for viral neuropathogenesis. J Cereb Blood Flow Metab 27:123–134
28. Deli MA, Szabo CA, Dung NTK, Joo F (1997) Immunohistochemical and electron microscopy detections on primary cultures of rat cerebral endothelial cells. In: Drug transport across the blood–brain barrier: in vivo and in vitro techniques (de Boer AG, Sutanto W, eds), Amsterdam: Harwood Academic Publishers, 23–8
29. Weksler BB, Subileau EA, Perriere N, Charneau P, Holloway K, Leveque M, Tricoire-Leignel H, Nicotra A, Bourdoulous S, Turowski P, Male DK, Roux F, Greenwood J, Romero IA, Couraud PO (2005) Blood-brain barrier-specific properties of a human adult brain endothelial cell line. FASEB J 19:1872–1874

Chapter 27

Enhanced Immunohistochemical Resolution of Claudin Proteins in Glycolmethacrylate-Embedded Tissue Biopsies

Jane E. Collins, Adam Kirk, Sara K. Campbell, Juan Mason, and Susan J. Wilson

Abstract

There are a number of disadvantages with conventional tissue immunohistochemistry for accurate localisation of claudin proteins. Traditionally, tissue cryopreservation or formaldehyde fixation with wax embedding is utilised prior to sectioning and antibody localisation. Wax embedding gives better morphological preservation than frozen tissue, but the required use of chemical cross-linking fixatives renders many antigens inaccessible to antibody binding or results in subsequent disruption of antibody localisation patterns due to the use of harsh antigen retrieval methods. Use of frozen or wax-embedded tissue also requires the cutting of relatively thick >6-μm sections, making the interrogation of serial sections very limited.

The use of glycolmethacrylate (GMA) tissue embedding with fixation in acetone is compatible with epitope preservation for many antibody reagents that are often destroyed by chemical cross-linking fixatives. GMA is a water-miscible embedding resin that maintains tissue hydration during processing, thus reducing tissue shrinkage, while embedding and cutting in the polymerised resin physically supports the tissue, thus improving morphology. This method also facilitates the cutting of 2-μm sequential sections for analysis of multiple antigens and maximises the information available from small tissue biopsies from human clinical sources.

Key words: Claudin, Immunohistochemistry, Glycolmethacrylate, Serial sections, Biopsies

1. Introduction

Claudins are tight junction proteins that comprise a family of at least 24 members whose differential expression and properties determine the paracellular permeability of epithelia (1–4). When claudins are overexpressed in fibroblasts, they polymerise in the membrane and reconstitute paired intramembranous networks of strands and complementary grooves, as seen by freeze-fracture electron microscopy (5), appearing similar to structures seen in

Kursad Turksen (ed.), *Claudins: Methods and Protocols*, Methods in Molecular Biology, vol. 762,
DOI 10.1007/978-1-61779-185-7_27, © Springer Science+Business Media, LLC 2011

mature tight junctions (6). The demonstration of heterogeneous claudins in tight junctions (7) and claudin copolymers in transfected cells (8) suggests that variations in the tightness of individual paired strands are determined by the combinations and ratios of the claudin types (9, 10). In addition, claudins have a dynamic relationship with tight junctions, where claudin concentrations may be regulated by physiological, homeostatic, and inflammatory stimuli (11–14).

Recognising the heterogeneity of claudin content in tight junctions and variations in their cellular distribution, it is essential to be able to localise claudin expression and distribution accurately in tissues. Conventional histology relies on the use of tissue cryopreservation or formaldehyde fixation with wax embedding, prior to sectioning and antibody localisation. There are a number of disadvantages with these approaches in relation to claudin staining. Use of frozen or wax-embedded tissue requires cutting of relatively thick >6-μm sections, making the interrogation of serial sections very limited. Retention of epithelial layers in sections is problematical in tissues where the epithelium is superficial, such as airway and intestine. This is particularly relevant with frozen tissue, which often shows loss of epithelium in sections due to processing and poor preservation of tissue morphology. Wax embedding gives better morphological preservation than frozen tissue, but requires the use of chemical cross-linking fixatives, such as formaldehyde, and renders many antigens inaccessible to antibody binding. Subsequent antigen retrieval methods destroy the delicate patterns of claudin expression, making mapping and interpretation of the original distribution unreliable.

In the light of the above considerations, we have used the method of glycolmethacrylate (GMA) tissue embedding (15) to map claudin protein distribution in addition to other junctional adhesion molecules (16). This method provides a number of advantages over conventional histological strategies. It utilises fixation in acetone with protease inhibitors, making it compatible with epitope preservation for many antibody reagents that are often destroyed by chemical cross-linking fixatives. GMA is a water-miscible embedding resin that maintains tissue hydration during processing, thus reducing tissue shrinkage, while embedding and cutting in the polymerised resin physically supports the tissue, thus improving morphology. Figure 1a–c shows comparisons of three antigen retrieval methods used on formalin-fixed wax-embedded material with the same claudin-2 antibody, demonstrating the loss of resolution in claudin antigen distribution with diffusion of peroxide diaminobenzidene reaction product. Figure 1d shows the same antibody used at a fourfold lower concentration on GMA-embedded sectioned tissue.

The resin also facilitates the cutting of thin (2 μm) sections, which when viewed sequentially may impart information from

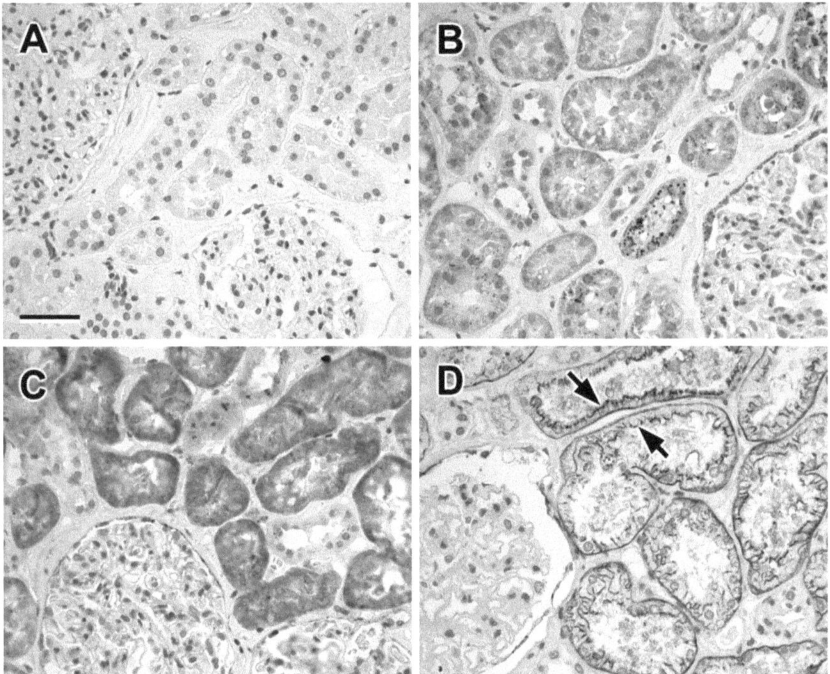

Fig. 1. Comparisons of immunohistochemical antigenicity to same mouse anti-claudin-2 antibody in wax or glycolmethacrylate-embedded tissue shown in photomicrographs of human kidney sections. Dewaxed formalin-fixed tissue sections (**a–c**) were incubated in 0.1% trypsin for 9 min at 37°C (**a**), heat treated in a pressure cooker for 20 min with 0.01 M citrate buffer pH 6.0 (**b**) or 1 mM EDTA buffer pH 8.0 (**c**) prior to staining as per standard immunohistochemical protocols with avidin–biotin block similar to those outlined for glycolmethacrylate. Superior claudin-2 antigen preservation and resolution was seen at tight junctions with glycolmethacrylate embedding and staining protocols as outlined above (*arrows* in **d**). Scale bar 100 μm.

many of the same cells in a tissue preparation, in addition to allowing the cutting of multiple sections from small biopsies. The latter is particularly relevant from clinical human biopsies from sites such as gut, lung, and kidney where tissue availability is very limited.

Figure 1a–c shows comparisons of three antigen retrieval methods used on formalin-fixed wax-embedded kidney tissue with the same claudin-2 antibody, demonstrating lack of staining in (a) and loss of resolution in claudin antigen distribution with diffusion of peroxide/diaminobenzidene reaction product in (b) and (c). Figure 1d shows claudin-2 staining in similar proximal tubules in GMA-processed kidney tissue; however, the staining is discrete at tight junction and basolateral locations, giving a more accurate picture of antigen distribution. Figure 2 shows the application of sequential GMA sections for comparison of claudin staining patterns with occludin (f) as a marker of intact tight junctions, and the distal tubule and collecting duct marker calbindin (e). In distal and collecting duct, claudin-4 and -8 staining (b and c) was mostly located at discrete junctional sites, with

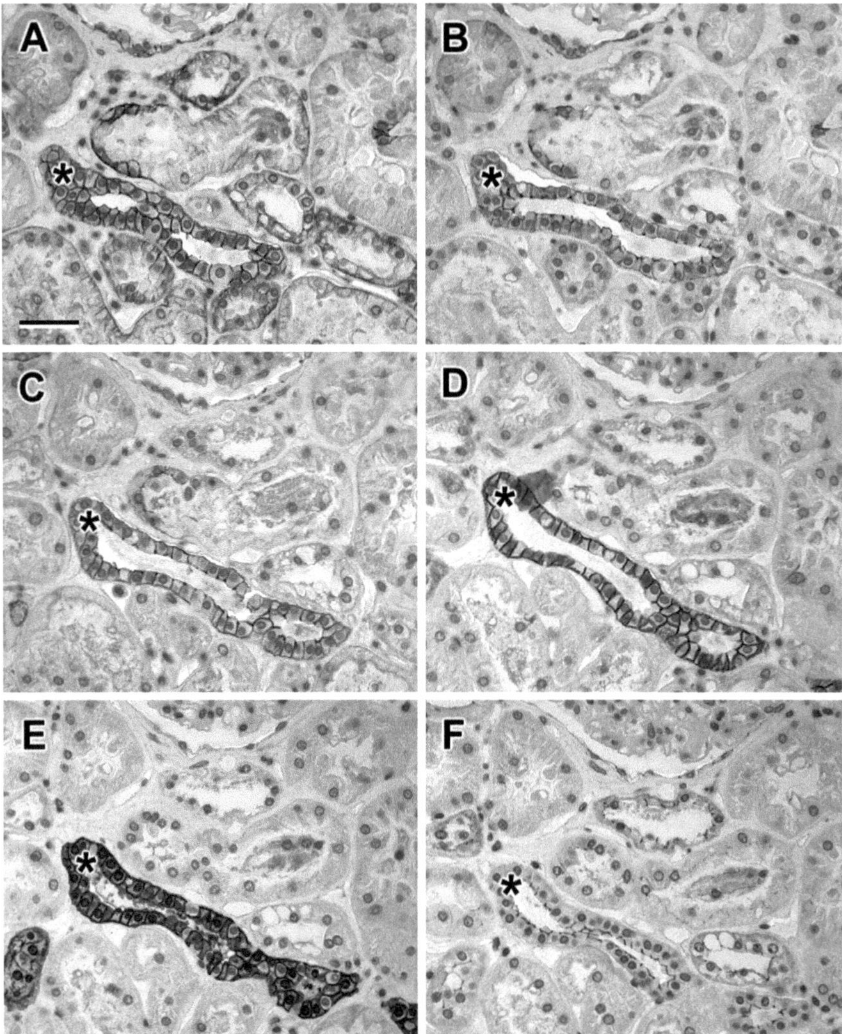

Fig. 2. Sequence of serial 2-μm glycolmethacrylate-embedded tissue sections showing staining of claudin-3 (**a**), -4 (**b**), -8 (**c**), -7 (**d**), calbindin (**e**), and occludin (**f**). Claudin-4 and -8 (**b** and **c**) were observed to be mostly in discrete tight junction staining patterns, whereas claudin-3 and -7 (**a** and **d**) showed strong basolateral cell membrane staining in addition to tight junction staining. In corresponding tubules, occludin (**f**) was stained in a discrete pattern consistent with tight junction localisation (*asterisks mark corresponding tubules*). Calbindin staining (**e**) shows the localisation of distal tubule and collecting duct in the sections. These data illustrate the use of GMA serial sections to compare antigen localisation. Scale bar 100 μm.

claudin-3 and -7 observed junctionally and basolaterally (a and d). The basolateral staining pattern is unlikely to be an artefact of fixation or staining since discrete junctional occludin with no basolateral staining is seen in the sequential section in (f).

In conclusion, our experience with staining of claudins using acetone fixation and GMA embedding has been very informative and it is the preferred method in our laboratory for these proteins.

2. Materials

2.1. Tissue Fixation and GMA Processing

Analar-grade reagents are used throughout, unless otherwise stated.

1. Dry acetone: place molecular sieve 4A (Fisher Scientific, Loughborough, Leicestershire, UK) in the bottom of 1-L storage bottle.
2. Protease inhibitors: iodoacetamide (Sigma, Poole, Dorset, UK), phenylmethyl-sulphonyl fluoride (PMSF) (Sigma).
3. Acetone fixative: 20 mM iodacetamide (370 mg/100 mL) and 2 mM PMSF 35 mg/100 mL) in dry acetone. Gloves and face mask should be worn when handling dry protease inhibitors. Fixative can be made in advance and stored, and aliquoted (5 mL) in glass vials at −20°C for up to 3 months.
4. Methyl benzoate, laboratory reagent grade (Fisher Scientific).
5. JB4 kit, includes monomer solution A, solution B, and benzoyl peroxide (Poly-sciences [Warrington, PA], cat.no. 0226). Monomer solution A (0226A) is available separately.
6. Glass vials with snap-on lids, ~10 mL (e.g., G060, Taab, Reading, UK).
7. Polythene bottles: must be airtight and resin proof (recommended B053, Taab).
8. Taab 8-mm flat-bottomed polythene-embedding capsules (cat. No C094, Taab).
9. Rack for above capsules (cat. No. C054, Taab).
10. N-DEX blue nitrile gloves, disposable (Fisher Scientific).
11. Airtight plastic storage boxes suitable for −20°C.
12. Silica gel (Fisher Scientific).

2.2. GMA Cutting

1. Resin microtome with a binocular microscope head and capable of sectioning at 2 μm.
2. Glass knives, 25-degree, made from 6-mm glass strips (Leica, Wetzlar, Germany) using a glass knife maker (e.g. Leica).
3. Poly-L-lysine (PLL)-coated microscope slides: dilute PLL (Sigma) 1:10 with distilled water and place in a trough. Place the glass microscope slides into a slide rack and immerse in PLL for 5 min; allow to dry overnight, replace into boxes, and store until required.
4. Ammonia.
5. Toludine blue stain: 1 g toluidine blue (Sigma) and 1 g borax (Sigma) in 100 mL distilled water.
6. N-DEX blue nitrile gloves, disposable (Fisher Scientific).

2.3. GMA Immunohistochemistry

1. Endogenous peroxidase inhibitor: 10 mL 0.1% aqueous sodium (Na) azide plus 100 μL 30% hydrogen peroxide. Prepare fresh before use. 0.1% aqueous sodium azide can be prepared as a stock solution and stored at room temperature (RT).
2. Avidin–biotin blocking kits (Vector Laboratories, Burlingame, CA, USA).
3. Culture medium-blocking solution: 20% new born calf serum, 1% bovine serum albumin in Dulbecco's modified essential medium (basic grade). This can be prepared in advance, stored, and aliquoted (5 mL amounts) at -20°C.
4. Tris-buffered saline (TBS), pH 7.6: 80 g NaCl, 6.05 g Tris, and 38 mL 1 M hydrochloric acid. Dissolve and mix in 1 L distilled water and adjust to give a final pH of 7.6: This is a 10× stock and should then be made up to 10 L. Store at RT for up to 1 week.
5. Diaminobenzidene (DAB) substrate kit (Biogenex Laboratories Inc, San Ramon, CA, USA). Prepared according to the manufacturer's instructions with the addition of 100 μl of 15% sodium azide to assist in blocking of endogeneous peroxidase.
6. Mayer's hematoxylin.
7. Permanent aqueous mounting medium (AbD Serotec/MorphoSys UK Ltd, Oxford, UK).
8. Primary Abs: This is discussed in Subheading 3.3.1.
9. Secondary Abs: Biotinylated rabbit anti-mouse Fab_2 fragments for monoclonal antibodies and biotinylated swine anti-rabbit for Fab_2 fragments for polyclonal antibodies (cat. Nos. E0413 and E0431, Dako, Glostrup, Denmark).
10. Vectastain elite ABC peroxidase kit (Vector Labs).

3. Methods

3.1. GMA Processing

Caution: steps 4–7 of this procedure must be performed in a fume extraction hood, and N-DEX blue nitrile gloves should be worn when handling resin components, because there is a risk of developing a contact hypersensitivity.

1. Fresh tissue samples should be cut into pieces no larger than 2 mm^3 and placed immediately into ice-cold acetone fixative (see Note 1).
2. Fix overnight at -20°C.
3. Replace fixative with acetone at RT, for 15 min.
4. Methyl benzoate at RT, 15 min.

5. Infiltrate with processing solution: 5% methyl benzoate in GMA solution A at 4°C, 3 × 2 h (see Note 2).
6. Prepare embedding solution immediately before use: dissolve 70 mg benzoyl peroxide in 10 mL GMA solution A and then add 250 μL GMA solution B (see Note 3).
7. Embed specimens in freshly prepared embedding solution in Taab flat-bottomed capsules, placing tissue sample in the bottom of the capsule, filling to the brim with resin and closing the lid to exclude air. A pencil-written label should also be placed in the capsule (see Note 4).
8. Polymerize for 48 h at 4°C.
9. Store in airtight boxes at –20°C (see Note 5).

3.2. GMA Cutting

3.2.1. Initial Assessment

1. Remove blocks from capsules (see Note 6).
2. Trim away excess resin to form a trapezium shape around the tissue (see Note 7).
3. Cut 2-μm sections, float out on water, and pick up on PLL slide.
4. Dry on hot plate, 10 min.
5. Stain with toluidine blue, 2 min.
6. Wash sections to remove excess stain.
7. Blot dry and mount in Pertex (Leica Biosystems, Peterborough, UK).
8. Examine under light microscope to check sample quality.

3.2.2. Sectioning for IHC

1. Cut 2-μm sections and float out on ammonia water (1 mL ammonia in 500 mL distilled water) for 1–1.5 min (see Note 8).
2. Pick sections up onto labelled PLL slides.
3. Dry for at least 1 h at RT.
4. Commence IHC staining the same day or wrap the slides back to back in aluminium foil and store at –20°C; use within 2 weeks.

3.3. GMA Immunohistochemistry

3.3.1. Primary Abs

There are a number of companies producing Abs directed against the claudins, with Invitrogen having the largest catalogue of reliable antibodies. Table 1 shows those that we have found to give good staining in GMA. Biotinylated lectins (Vector Labs) also work well in GMA for identification of cell subpopulations on serial sections.

3.3.2. Titration of Abs

Before Abs can be used on test sections, the optimum working dilution needs to be established by titration. Initially, five double dilutions should be tried: If the concentration of the Ab is 1 mg/mL,

Table 1
Claudin antibody reagents used in immunohistochemical staining of GMA-embedded tissue biopsies

Antigen	Supplier	Catalogue number	Antibody host	Dilution[a]
Claudin 2	Invitrogen, Life Technologies	32-5600	Mouse	1:800
Claudin 3	Invitrogen, Life Technologies	34-1700	Rabbit	1:400
Claudin 4	Invitrogen, Life Technologies	32-9400	Mouse	1:300
Claudin 7	Invitrogen, Life Technologies	34-9100	Rabbit	1:100
Claudin 8	Invitrogen, Life Technologies	40-2600	Rabbit	1:350
Claudin 10	Abcam, Cambridge, UK	Ab24792	Rabbit	1:150
Claudin 11	Invitrogen, Life Technologies	36-4500	Rabbit	1:50
Claudin 14	Invitrogen, Life Technologies	36-4200	Rabbit	1:25
Claudin 16	Invitrogen, Life Technologies	34-5400	Rabbit	1:50

[a]This is the current working dilution being used in the author's laboratory and should be used as a guide only. All new batches of antibodies should be titrated (see Subheading 3.3.2)

this usually has a working dilution of approximately 1:100; therefore, when titrating an Ab of this concentration, the following dilution series would be used: 1:50, 1:100, 1:200, 1:400, and 1:800. Commercial Abs are usually very reliable with little batch-to-batch variation, and therefore, once a working dilution has been established, new batches only need checking by using the current working dilution and a higher or lower dilution.

When initially establishing an IHC system, the second-stage Abs will also require titration and for this purpose, checkerboard titrations are required in which the concentration of two Abs are varied against each other. Directly labelled biotinylated lectins can be detected using procedures identical to those described here for biotinylated secondary antibodies. Titration of concentrations of lectins required is also strongly recommended.

3.3.3. Controls

The use of appropriate controls is essential in IHC. Positive controls should be included to ensure that the technique is working, and negative controls to ensure that there is no nonspecific staining. When staining for claudins in epithelia, useful positive controls include monoclonal cytokeratin antibodies such as CAM 5.2 and rabbit polyclonal anti-EGF receptor antibody. Two types of negative controls are routinely used: omission of the primary Ab and isotype-matched IgGs (Sigma). The latter should be used at the same concentration as the strongest test Ab in place of the primary Ab.

3.3.4. IHC Procedure

If sections have been stored at −20°C, they should be removed from the freezer, unwrapped, and laid out to allow the condensation to evaporate. Prior to commencement of staining, the sections should be circled with a diamond marker pen or a PAP pen (Dako) so that they can be visualised during the IHC procedure. The sections should not be allowed to dry out during the procedure and the staining tray should be covered during the blocking, Ab, and substrate stages.

1. Inhibit endogenous peroxidase with peroxidase inhibitor solution, for 30 min.
2. Wash with TBS, 3 × 5 min.
3. Drain slides and apply avidin D blocking solution, for 30 min, if appropriate (see Note 9).
4. Wash with TBS, 3 × 5 min.
5. Drain the slides and apply biotin blocking solution, for 30 min.
6. Wash with TBS, 3 × 5 min.
7. Drain the slides and apply culture medium-blocking solution, for 30 min (see Note 10).
8. Drain the slides and apply primary Abs diluted in TBS at appropriate dilutions, cover with coverslips, and incubate monoclonal antibodies at RT and polyclonal antibodies at 4°C, overnight (at least 16 h) (see Notes 11 and 12).
9. Wash off coverslips with TBS.
10. Wash with TBS, 3 × 5 min.
11. Drain the slides and apply biotinylated second-stage Ab, diluted in TBS, at appropriate dilution, for 2 h.
12. Wash with TBS, 3 × 5 min.
13. Drain the slides and apply avidin–biotin peroxidase complexes, diluted in TBS, at appropriate dilution, for 2 h (see Note 13).
14. Wash with TBS, 3 × 5 min.
15. Drain the slides and apply diaminobenzidene substrate solution, for 5 min at RT.

16. Rinse with TBS, place in staining racks, and wash in running tap water, for 5 min.
17. Counterstain sections with Mayer's hematoxylin (~2 min, depending on age of solution) and blue in running tap water (see Note 14).
18. Drain the slides and apply permanent aqueous mount (AbD Serotec) (one drop/section) and bake at 80°C for approximately 15 min until set.
19. Allow the slides to cool and mount in Pertex.

4. Notes

1. When collecting samples for GMA processing, the acetone fixative must be kept cold on ice at all times and the size of the samples must not be >2 mm^3.
2. Ten millilitres of processing solution should be allowed per pot to be processed.
3. One millilitre embedding solution should be allowed per sample to be embedded. When solution B is added to the embedding solution, the colour of the solution will change to a pale straw colour; if agitated too much and for too long after this, the solution will turn brown and should then be discarded as it will not polymerize properly. If a large number of samples are to be embedded (>10), batches of embedding solution up to the addition of solution B should be prepared and kept on ice. Solution B should be added to a batch and this used immediately to embed samples, which should then be transferred to 4°C before the next batch is used.
4. The label identifying the biopsy must be written in pencil; pen-written labels will run when in contact with the resin. The label is then coiled, writing outermost, and placed inside the Taab capsule just above halfway. When the resin polymerizes, the label then becomes an integral part of the block.
5. Once the blocks are polymerized, there is usually a small amount of unpolymerised resin remaining in the very top of the capsule: This should be cleaned out with tissue (wearing in N-DEX blue nitrile gloves) before storing the blocks. The blocks, still in their capsules, are placed in small containers (e.g. empty coverslip boxes) which are then placed in larger airtight plastic storage boxes with silica gel at the bottom. This is then stored at −20°C.
6. *Caution*: N-DEX blue nitrile gloves should be worn when removing the blocks from the capsules because any unpolymerised resin may cause contact hypersensitivity.

7. When trimming the blocks, the trapezium should be close to the tissue on the sides and top but ~1–2-mm resin should be left at the bottom to allow for picking up the sections with the forceps. The edges of the trimmed block face should be at an angle of about 45°; if the edges are at too steep an angle, the block will be unstable and could lead to cutting difficulties.
8. When cutting sections for claudin staining, sequential sections may be used to map relative expression in tissues where several claudins are known to be co-expressed. Two sections should be cut for each marker. Place one section on each glass slide for several serial sections, then cut a second series and place it below the first on each slide. The number of serial sections that can be compared depends on the tissue and how close the comparisons need to be for reliable interpretation. Judicious use of repeated stains within the sequence and the use of two series on the slides as described can greatly enhance the data obtained.
9. Some tissues contain binding sites for avidin or biotin because they have high levels of endogenous biotin, biotin-binding proteins, or lectins, which increase nonspecific binding to the biotinylated antibody, avidin, or avidin system constituents. Such tissues, notably kidney, require these sites to be preblocked with avidin D, followed by biotin, which blocks the remaining biotin-binding sites on the avidin.
10. For the polyclonal anti-claudin antibodies, 5% serum (same species as the second-stage host antibody) should be added to the culture medium-blocking solution to further reduce nonspecific staining.
11. 150 μL diluted Ab should be allowed per slide.
12. Coverslips are placed over the sections with the primary Abs to prevent evaporation and to enable small quantities of Ab to be used.
13. The avidin–biotin peroxidase components need to be mixed at least 30 min in advance of use to allow time for the complexes to form.
14. In areas where the water supply is not alkaline, Scotts tap water substitute (potassium bicarbonate 2 g and magnesium sulphate 20 g, made up to one litre with distilled water) will be required to blue the hematoxylin.

References

1. Krause G, Winkler L, Mueller SL, Haseloff RF, Piontek J, Blasig IE (2008) Structure and function of claudins. Biochim Biophys Acta **1778**:631–645
2. Angelow S, Yu AS (2007) Claudins and paracellular transport: an update. Curr Opin Nephrol Hypertens **16**:459–464

3. Anderson JM, Van Itallie CM, Fanning AS (2004) Setting up a selective barrier at the apical junction complex. Curr Opin Cell Biol **16**:140–145
4. Turksen K, Troy TC (2004) Barriers built on claudins. J Cell Sci **117**:2435–2447
5. Furuse M, Sasaki H, Fujimoto K, Tsukita S (1998) A single gene product, claudin-1 or -2, reconstitutes tight junction strands and recruits occludin in fibroblasts. J Cell Biol **143**:391–401
6. Staehelin LA (1974) Structure and function of intercellular junctions. Int Rev Cytol **39**:191–283.:191–283
7. Kirk A, Campbell S, Bass P, Mason J, Collins J (2010) Differential expression of claudin tight junction proteins in the human cortical nephron. Nephrol Dial Transplant. doi:10.1093/ndt/gfq006
8. Tsukita S, Furuse M (2000) Pores in the wall: claudins constitute tight junction strands containing aqueous pores. J Cell Biol **149**:13–16
9. Furuse M, Furuse K, Sasaki H, Tsukita S (2001) Conversion of zonulae occludentes from tight to leaky strand type by introducing claudin-2 into Madin-Darby canine kidney I cells. J Cell Biol **153**:263–272
10. Angelow S, Schneeberger EE, Yu AS (2007) Claudin-8 expression in renal epithelial cells augments the paracellular barrier by replacing endogenous claudin-2. J Membr Biol **215**:147–159
11. Prasad S, Mingrino R, Kaukinen K, Hayes KL, Powell RM, MacDonald TT, Collins JE (2005) Inflammatory processes have differential effects on claudins 2, 3 and 4 in colonic epithelial cells. Lab Invest **85**: 1139–1162
12. Capaldo CT, Nusrat A (2009) Cytokine regulation of tight junctions. Biochim Biophys Acta **1788**:864–871
13. Gonzalez-Mariscal L, Namorado MC, Martin D, Sierra G, Reyes JL (2006) The tight junction proteins claudin-7 and -8 display a different subcellular localization at Henle's loops and collecting ducts of rabbit kidney. Nephrol Dial Transplant **21**:2391–2398
14. Tipsmark CK, Baltzegar DA, Ozden O, Grubb BJ, Borski RJ (2008) Salinity regulates claudin mRNA and protein expression in the teleost gill. Am J Physiol Regul Integr Comp Physiol 294:R1004-R1014
15. Britten KM, Howarth PH, Roche WR (1993) Immunohistochemistry on resin sections: a comparison of resin embedding techniques for small mucosal biopsies. Biotech Histochem **68**:271–280
16. Wilson SJ, Holgate ST. Immunohistochemical Analysis of Adhesion Molecules in Airway Biopsies. Asthma, mechanisms and protocols editors K Fan Chung and Ian Adcock. **17**: 227–240. 2000. Humana Press, Totowa, New Jersey.

Chapter 28

Claudin-16/Paracellin-1, Cloning, Expression, and Its Role in Tight Junction Functions in Cancer and Endothelial Cells

Tracey A. Martin and Wen G. Jiang

Abstract

Metastatic spread of cancer is the major reason for patient mortality in human breast cancer. We examined expression of claudin-16 in human breast cells and tissues to identify a possible link between expression and aggressiveness in cells and between claudin-16 levels and patient prognosis. Forced expression of claudin-16 in breast cancer cells resulted in a less aggressive phenotype and reduced in vivo tumour volume. Claudin-16 expression was reduced in human breast cancer, particularly in patients with aggressive tumours and high mortality (Martin et al. J Cell Biochem 105:41–52,2008). This suggests that claudin-16 plays a role beyond that of an initial metastasis repressor in this cancer type.

Key words: Paracellin-1, Claudin-16, Tight junctions, Breast cancer, Hepatocyte Growth Factor, Metastasis

1. Introduction

Recent studies have shown that several tight junction (TJ) components are directly or indirectly involved in breast cancer progression and metastasis (1–6). The mutual adhesiveness of cancer cells is significantly weaker than that of normal cells and so reduced cell–cell interaction results in a loss of normal tissue architecture. We have previously demonstrated that the expression of TJ molecules in breast cancer led to this current study examining the effect of claudin-16 overexpression in human breast cancer cells and the expression and distribution of claudin-16 in human breast cancer tissues (7, 8). The loss or reduction of TJ molecules in cancer is becoming a frequently found occurrence. That claudins are key in the maintenance of TJ function begs investigation into the changes that can occur during the metastatic process. Overexpression of claudin-16 in human breast cancer cells allows us to evaluate the effect that claudin can exert

Kursad Turksen (ed.), *Claudins: Methods and Protocols*, Methods in Molecular Biology, vol. 762,
DOI 10.1007/978-1-61779-185-7_28, © Springer Science+Business Media, LLC 2011

using a variety of in vitro assays on cell behaviour, together with in vivo tumour growth. Quantitative PCR technology and immunohistochemistry allow us to quantify and observe the expression and distribution of claudin-16 in human breast tissues.

2. Materials

2.1. Cell Culture, Protein, and RNA Extraction

1. Dulbecco's modified Eagle's medium (DMEM) (Sigma–Aldrich Ltd, Poole, Dorset, UK) supplemented with 10% fetal bovine serum (FBS), 100 units/ml penicillin, and 100 μg/ml streptomycin (Sigma–Aldrich Ltd, Poole, Dorset, UK) (see Notes 1 and 2).
2. Human breast cancer cell line MDA-MB-231 (American Type Culture collection (ATCC), Maryland, USA) and the human endothelial cell line, HECV (ICLC Genova, Italy) (see Note 3).
3. Solution of trypsin (0.25%) and ethylenediaminetetraacetic acid (EDTA) (1 mM) from Gibco/BRL (Pasley, N. Ireland, UK).
4. Human recombinant hepatocyte growth factor (HGF) was kindly provided by Dr T Nakamura, Osaka, Japan, and is dissolved at 1 mg/mL in DMEM and stored in single-use aliquots at −80°C. Working solutions are prepared by dilution in 100 μg/mL bovine serum albumin (BSA).
5. Disposable cell scrapers, 25-cm^2 and 75-cm^2 tissue culture flasks or six-well plates (Greiner, Gloucester, England, UK).
6. Inhibitor buffer (3×). Dissolve 2.76 g sodium nitrate, 5.58 g EDTA, 630 mg sodium fluroide, 10 g $Na_2H_2PO_4$, and 300,000 Units of aprotonin made up to 1 L with distilled water. Store at 4°C. All chemicals from Sigma (Sigma–Aldrich, Dorset, UK).
7. Cell lysis buffer (2×). Dissolve 0.48 g Tris–HCl, 0.87 g NaCl, 2 ml of Triton x-100, 0.5 g sodium deoxycholate, 0.02 g sodium azide, and 0.27 g sodium orthovanadate in 100 ml of distilled water. Store at 4°C. A working solution is prepared by diluting 5 ml of the stock solution in 3.33 ml of inhibitor buffer, 100 μl of PMSF (Sigma–Aldrich, Dorset, UK), 40 μl of 10 mM CaCl, and 1.5 ml of 10% Triton X-100 made up to 10 ml with distilled water (see Note 4).
8. Sample buffer (2×): Mix 50 ml 0.5 M Tris–HCl (pH = 6.8), 5 ml 2-mercaptoethanol, 20 ml glycerol, 20 ml 10% SDS, 2 ml 1% bromophenol blue, and 3 ml distilled water. Store in aliquots at −20°C (see Note 5).
9. 1% Bromophenol blue – Dissolve 0.1 g bromophenol blue in 10 ml of 70% ethanol.

10. Fluorescein solution – Prepare fresh by dissolving 3 mg fluorescein (Sigma–Aldrich, Dorset, UK) in 10 ml of acetone (see Note 6).
11. Sodium phosphate buffer – Dissolve 0.89 g $Na_2HPO_4{\cdot}2H_2O$ and 0.78 g $NaH_2PO_4{\cdot}2H_2O$ in 100 ml of distilled water. Stir vigorously and adjust pH to 8.0 with NaOH. Store at room temperature.
12. BSA (Sigma–Aldrich, Dorset, UK) (see Note 7).

2.2. SDS-Polyacrylamide Gel Electrophoresis

1. Separating buffer: 1.5 M Tris–HCl, pH 8.8. Store at room temperature.
2. Stacking buffer: 0.5 M Tris–HCl, pH 6.8. Store at room temperature.
3. 30% Acrylamide/0.8% bisacrylamide solution.
4. *N,N,N,N'*-tetramethylethylenediamine (TEMED, Bio-Rad, Hercules, CA).
5. Ammonium persulphate (APS): prepare 10% solution in distilled water and store at 4°C (see Note 8).
6. Running buffer (10×): 125 mM Tris–HCl, 960 mM glycine, and 0.5% (w/v) SDS. Store at room temperature.
7. High molecular weight markers (Bio-Rad, Hercules, CA).
8. 8% Resolving gels (10 ml): 4.6 ml of water, 2.7 ml 30% acrylimide/0.8% bisacrylamide, 2.5 ml of separating buffer, 0.1 ml 10% SDS, and 0.1 ml 10% APS. Add 6 μl TEMED (Sigma–Aldrich, Dorset, UK).
9. 5% Stacking gels (2 ml): 1.4 ml water, 0.33 ml 30% acrylimide/0.8% bisacrylamide, 0.25 stacking buffer, 0.02 ml 10% SDS, and 0.02 ml 10% APS. Add 2 μl TEMED (see Note 9).
10. Microcentrifuge (Biofuge 13, Heraeus, Sepatech, UK).
11. Microtitre plate (Greiner Laboratories, Germany).
12. Fluorescent plate reader (Denly WelFluor 7620, Sussex, UK).
13. Protein sample buffer; 50 mM Tris–HCl, 10% glycerol, 2% SDS, 0.1% bromophenol blue, and 5% β-mercaptoethanol (Sigma-Aldrich, Dorset, UK) (see Note 10).
14. Micro Protean II Gel system (Bio-Rad Laboratories, Richmond, CA, USA).

2.3. Western Blotting for Claudin-16

1. Transfer buffer: 25 mM Tris–HCl (do not adjust pH), 190 mM glycine, and 20% (v/v) methanol. Store at room temperature (see Note 11).
2. Nitrocellulose membrane (from Millipore, Bedford, MA) and 3MM chromatography paper (from Whatman, Maidstone, UK) (see Note 12).

3. Tris–HCl-buffered saline with Tween (TBS-T): Prepare 10× stock with 1.37 M NaCl, 27 mM KCl, 250 mM Tris–HCl, pH 7.4, and 1% Tween-20. Dilute 100 mL with 900 mL water for use.
4. TBS buffer (10×). Dissolve 121 g Tris base, 400.3 g NaCl in 5 L of distilled water. Store at room temperature. A working solution is prepared by diluting 500 ml of the stock solution in 4,500 ml of distilled water.
5. Ponceau S: Mix 20 ml of Ponceau S concentrate (Sigma) in 180 ml of distilled water. Store at room temperature.
6. Blocking buffer: 10% (w/v) nonfat dry milk in TBS-T. Make fresh as required.
7. Wash buffer: 3% (w/v) nonfat dry milk in TBS-T. Make fresh as required.
8. Primary antibody dilution buffer: TBS-T supplemented with 2% (w/v) fraction V BSA. Antibodies supplied from the manufacturer are diluted in 0.1% BSA in BSS buffer and stored in 100 μl aliquots at −20°C.
9. Anti-claudin-16 and actin antibodies (PharMingen International and Santa Cruz Biotechnologies Inc., respectively) (see Note 13).
10. Secondary antibody: Anti-mouse IgG and anti-rabbit IgG conjugated to horseradish peroxidase (Sigma–Aldrich Ltd (Poole, UK)).
11. SuperSignal West Dura Extended Duration Substrate Chemiluminescence system (Perbio Science UK Ltd., Cramlington, UK) (see Note 14).
12. Amido black: Dissolve 0.1 g in napthalene black, 10 ml acetic acid, and 25 ml ethanol and make up to 100 ml with distilled water. Store at room temperature.
13. Amido black destain: Mix 100 ml acetic acid and 250 ml ethanol, and make up to 1 L with distilled water.
14. Biometra FastBlot apparatus (Whatman-Biometra, Kent, UK).
15. Power supply (Bio-Rad model 1000/500, Bio-Rad Laboratories, Richmond, CA, USA).
16. 50 ml polypropylene tubes (Falcon, Greiner, Germany).
17. Universal container (Steralin, UK).
18. CCD UVI Prochemi System (UVItec Ltd, Cambridge, UK).

2.4. RNA Isolation

1. RNA isolation solution (Advanced Biotechnologies Ltd, Surrey, UK).
2. Diethyl polycarbonate (DEPC) solution – Add 0.1 ml DEPC to 100 ml of the solution to be treated, shake vigorously, and incubate for 12 h at 37°C. Autoclave for 15 min to remove any trace of DEPC.

2.5. RT-PCR and PCR

1. Loading dye is prepared as a 10× concentrate. Dissolve 0.125 g bromophenol blue and 20 g sucrose in 50 ml distilled water. Dispense to microcentrifuge tubes (Fisher-Scientific, UK) and freeze until needed.
2. RT-PCR kit (AB Gene Reverse Transcription System, Surrey, UK).
3. Primer set Paracell F11 5′-atgacctccaggaccccact-3′ and Paracell R11 5′-cacccttgtgtctacagcat-3′ (synthesised by Invitrogen, Pasley, N. Ireland).
4. Conventional PCR mix (ABgene, Epsom, Surrey, UK) was used to screen a number of human breast, prostate, colorectal, and bladder cancer cell lines. Normal human kidney cDNA was used as a positive control.
5. Ethidium bromide (EtBr) staining solution: Dissolve 0.1 g EtBr in 10 ml distilled water. Mix well and wrap in aluminium foil (see Note 15).
6. PCR thermocycler (GeneAmp 9700, Perkin Elmer).
7. Biofuge 13 microcentrifuge (Heraeus Sepatech, UK).
8. Isopropanol (Sigma–Aldrich, Dorset, UK).
9. DEPC (Sigma–Aldrich, Dorset, UK).
10. DNA/RNA spectrophotometer (WPA UV1101, Biotech, Wolf, UK).
11. Horizontal gel electrophoresis apparatus (Gibco BRL, Life Technologies, UK).
12. 1,000 base-pair DNA ladder (Pharmacia biotech, USA).
13. Power pack (model 250 EX Gibco BRL, Life Technologies, UK).
14. UV transilluminator (UVP, Cambridge, UK).
15. Thermal printer (Mitsubishi, Japan).
16. Qiaquick gel extraction kit and Qiagen mini plasmid extraction kit (Qiagen, UK).
17. Luria–Bertani agar (LB agar) (Sigma–Aldrich, Dorset, UK).
18. Electroporation cuvettes (Eurpgented, Southampton, UK).
19. Equibio Easyjet Plus electroporation equipment (Flowgen, UK).

2.6. Agarose Gel Electrophoresis

1. TBE electrophoresis buffer – (5×). Dissolve 540 g Tris–HCl, 275 g boric acid, and 46.5 g of disodium EDTA (or 20 ml of 0.5 M EDTA; pH 8.0) in a final volume of 10 L distilled water. Store at room temperature. Discard any buffers that develop a precipitate. A working solution is prepared by diluting 200 ml of the stock solution in 800 ml of distilled water.

2.7. Cloning and Expression

1. Normal human breast tissue was screened for endogenous expression of paracellin-1/claudin-16. An invasive cell line, MDA-MB-231 (MDAWT) which was claudin-16 negative, was chosen for introduction of the claudin-16 gene. The gene, after amplification from normal breast tissue cDNA, was T-A cloned into an NT GFP-TOPO (Invitrogen) plasmid before electroporation into the breast cancer cells.
2. G418 (Invitrogen, Paisley, N. Ireland, UK).
3. G418 selection media – add 200 μl of G418 (200 mg/ml) to a bottle of complete cell culture medium. Use 75 μl of G418 for maintenance medium. Protect from light by wrapping in aluminium foil (see Note 16).

2.8. QPCR

1. Amplofluor™ Uniprimer™ system (Intergen Company, Oxford, UK) and Thermo-Start® Q-master mix (ABgene, Epsom, Surrey, UK).
2. Specific primer pairs for claudin-16 were designed by the authors using a Beacon Designer software (Biosoft, Palo Alto, California, USA) and manufactured by Invitrogen (Invitrogen Life Technologies, Paisley, Scotland, UK), each amplifying a region that spans at least one intron (primer details given in supplement 1), generating an approximately 100-base pair product from both the control plasmid and the cDNA. PCR primers: ParazF AGCCACGTTACTAATAGCAG and ParazR ACTGAACC-TGACCGTACAATTGTGCAAAACCAAAGTAG.
3. Thermostart® (AB gene, Epsom, Surrey, UK).
4. Icycler IQ System (Bio-Rad, UK).

2.9. Immunohistochemistry

1. Fixative – 50:50 solution of alcohol:acetone.
2. Buffer – TBS automation wash buffer 20× (Menapath, A. Menarini Diagnostics, Florence, Italy).
3. Claudin-16 antibody (Santa Cruz Biotechologies Inc., Santa Cruz, CA, USA).
4. Biotinylated secondary antibody – at 1:100 dilution (Universal secondary, Vectastain Elite ABC, Vector Laboratories Inc., Burlingham, CA, USA) in horse serum/buffer solution.
5. Avidin/biotin complex – as provided by the manufacturer (Vectastain, Vecta Labs, Peterborough, UK).
6. Diaminobenzadine used as a chromogen to visualise the antibody/antigen complex.
7. Counterstain – Gill's haematoxylin (Vectastain, Vecta Labs, Peterborough, UK).

2.10. Immunofluorescence

1. FITC-conjugated anti-mouse and anti-rabbit IgG were from Santa-Cruz Biotechnologies (Santa-Cruz, CA, USA).
2. BSS buffer: Dissolve 794.6 g of NaCl, 22.35 g KCl, 21.45 g KH_2PO_4, and 113.6 g Na_2HPO_4 (Sigma–Aldrich, Dorset, UK) in 10 L of distilled water. Store at room temperature.
3. Triton X-100 (Sigma–Aldrich, Dorset, UK).
4. FluorSave (Calbiochem-Novabiochem Ltd, Nottingham, UK).
5. Olympus BX51 microscope with Hammamatsu Orca-ER digital camera (Olympus, UK).

2.11. Transepithelial Resistance and Paracellular Permeability

1. Carbonate filter inserts with pore size of 0.4 μm (for 24-well plates) were from Greiner Bio-One Ltd (Stonehouse, Glos., UK).
2. FITC-conjugated dextran (40 kDa) was obtained from Molecular Probe Inc. (Eugene, OR).
3. EVOM Voltohmmeter (EVOL, World Precision Instruments, Aston, Herts, UK).
4. STX-2 chopstick electrodes (WPI, Sarasota, Florida, USA).

2.12. Cytodex, Invasion, Adhesion, and Growth Assays

1. Cytodex-2 beads: 114–198-μm dextran micro-carrier beads (GE healthcare, Sigma–Aldrich, Poole, UK). Swell 1 g/50 ml BSS and autoclave at 120°C for 15 min (see Note 17).
2. Matrigel, a reconstituted basement membrane, was purchased from Collaborative Research Products (Bedford, MA) or Becton-Dickinson (Bristol, UK) (see Note 18).
3. Carbonate filter inserts with pore size of 0.8 μm (for 24-well plates) were from Greiner Bio-One Ltd (Stonehouse, Glos., UK).
4. 4% Paraformaldehyde solution: Dissolve 4 g of paraformaldehyde in 90 ml of distilled water. Heat in water bath at 75°C for 2–3 h, stirring occasionally. Once dissolved, allow the solution to cool before adding 10 ml of concentrated PBS (dissolve 1 PBS tablet (Sigma) in 10 ml distilled water). Store at 4°C.
5. 1% Crystal violet solution: Dissolve crystal violet powder (Sigma–Aldrich, Poole, UK) in distilled water. Filter to remove large particles.

2.13. ECIS

1. Electric cell–substrate impedance sensing system – ECIS 1600R (Applied Biophysics, Troy, New York, USA).
2. Gold electrodes – were purchased from Applied BioPhysics (Troy, NY).
3. L-Cysteine (10 mM) – 400 ml/well for 40 min (Applied Biophysics, Troy, New York, USA).

2.14. Tumour Model

1. Matrigel, a reconstituted basement membrane, was purchased from Collaborative Research Products (Bedford, MA) or Becton-Dickinson (Bristol, UK).
2. Athymic nude mice (CD-1, Charles River Laboratories, Kent, UK).

3. Methods

Six assays were used to determine what, if any, effect overexpression of Claudin-16 would have on the human breast cancer cell line MDA-MB-231.

3.1. Sample Preparation

1. Collect breast tissue samples (124 tumour and 33 matched background) and immediately freeze in liquid nitrogen before processing – a portion of each sample for quantitative PCR analysis, a portion for immunohistochemical analysis, and a portion for routine histological examination.
2. Maintain all cells in a humidified incubator at 37°C with 5% CO_2. All cells are grown in either 25-cm^2 or 75-cm^2 tissue culture flasks or six-well plates depending on the application.
3. For cell stimulation studies, change confluent cells to serum-free media for 3 h and grow either in the presence or in the absence of HGF at 25 ng/ml (used as a stimulator of motility and invasive behaviour).
4. Prepare cell extracts from 90% confluent cells harvested from six-well plates by aspiration of the medium, adding 100 μl of HCMF buffer plus 0.5% SDS, 0.5% Triton X-100, 2 mM $CaCl_2$, 100 μg/ml phenylmethylsulphonyl fluoride, 1 mg/ml leupeptin, 1 mg/ml aprotinin, and 10 mM sodium orthovanadate, and pipette into a labelled microcentrifuge tube.
5. Incubate the samples at 4°C for 40 min with continuous rotation in order to ensure appropriate lysis. Centrifuge the tubes at 13,000 × *g* for 10 min to remove cellular debris, and collect the protein. Transfer the supernatant to a clean microcentrifuge tube and determine the protein concentration.
6. From each lysate, transfer 40 μl of protein into separate wells of a 96-well microtitre plate. Serial dilute a BSA stock (100 mg/ml) in cell lysis buffer (40 μl/well) to give a working concentration range between 50 and 0.79 mg/ml. Add 80 μl of sodium phosphate buffer *to* each well. Then add 100 μl of fluorescamine solution to each well. The fluorescamine (4-phenylspiro [furan-2(3H),1′-phthalan]-3, 3′-dione) reacts with protein amino groups to yield a highly fluorescent

product at an alkaline pH. Place the microtitre plate into a fluorescence plate reader and measure fluorescence at an excitation wavelength of 480 nm and an emission wavelength of 590 nm.

7. Construct a standard protein curve from the BSA and determine the unknown protein concentrations of the cell lysate samples from this plot. Adjust protein concentrations to a working range of 1–2 mg/ml by diluting in cell lysis buffer and add a portion of this protein sample to be analysed 1:1 (v/v) with a sample buffer.
8. Denature crude cell lysate samples by boiling at 100°C for 5 min in a heating block with a hole poked in the cap using a 26-gauge syringe needle.
9. Allow to cool for a further 5 min at room temperature, before flash spinning to bring down condensation.
10. Load the samples on a SDS-PAGE gel or store at –20°C until required.

3.2. Sodium Dodecyl Sulfate-Polyacrylamide Gel Electrophoresis

1. The Mini Protean II gel system apparatus in accordance with the manufacturer's instructions was used for sodium dodecyl sulfate polyacrylamide gel electrophoresis (SDS-PAGE).
2. Prepare 10 ml of 8% resolving gel and pour into the prepared gel cassette, leaving space for the 5% stacking gel, and overlay with water. The gel should polymerise in about 30 min. Once set, pour off the water, prepare the stacking gel, quickly add to the top of the resolving gel, and insert the comb. The stacking gel should polymerise within 30 min. Once the stacking gel has set, carefully remove the comb and use a 3-ml syringe fitted with a 22-gauge needle to wash the wells with running buffer (see Note 19).
3. Fill the electrophoresis chamber with 300 ml of running buffer.
4. Load equal amounts of protein from each cell sample (controls and those treated with HGF) into the wells (12 μl per well) of the SDS-PAGE gel using a syringe with a flat-tipped needle. Also load 10 μl of pre-stained high molecular weight standard.
5. Add the running buffer to the upper and lower chambers of the gel unit and the unit assembly completed. Current was applied at a constant current of 30 mA at 180 V (15 mA per gel). Electrophoresis was stopped when the dye front was about 1 cm from the bottom of the gel. Usual running time was about 45–60 min. Complete the assembly of the gel unit and connect to a power supply (see Note 20).

3.3. Western Blotting

1. ElectroBlotting – protein transfer from gel to membrane – A piece of nitrocellulose membrane was cut to the dimensions of the gel (6 × 9 cm), the bottom right corner was marked, and the membrane immersed in transfer buffer for about 10–20 min. Similarly, four sheets of filter paper were cut to the same dimensions (6 × 9 cm) and soaked in transfer buffer for 10–20 min. Upon completion of electrophoresis, the stacking gel was removed to ensure that the gel would be oriented correctly in the transfer apparatus. The gel was soaked in transfer buffer (Tris–HCl, glycine, and methanol) for about 5 min (see Note 21).
2. After gel electrophoresis, proteins resolved by SDS-PAGE were transferred to nitrocellulose membrane using Biometra Fastblot apparatus.
3. Place two pieces of pre-soaked filter paper on the bottom electrode graphite anode base and place the pre-soaked membrane on top of the filter papers.
4. Carefully remove the gel from the tray of buffer and transfer on top of the nitrocellulose membrane, taking care not to trap air bubbles between the gel and membrane.
5. Cover the gel with two more sheets of wet filter paper, creating a sandwich of paper – nitrocellulose – gel – paper and roll a glass rod across the surface to remove any air bubbles and insure good contact between the gel and nitrocellulose.
6. Position the cover containing the cathode plate firmly on top of the transfer "sandwich" and connect the high voltage leads to the power supply. Apply a constant current of 500 mA, at 5 V, and 9.5 W for 30–40 min.
7. After transfer, remove the membrane and mark the lower right corner of membrane to indicate orientation of gel.
8. Immerse the membranes in Ponceau S stain for approximately 60 s at room temperature.
9. Rinse the membranes with distilled water until bands are visible and mark the position of the molecular weight markers on the membranes using a pencil.
10. After Ponceau S staining, transfer the membrane from the staining tray into clean 50-ml polypropylene tubes, ensuring that the membrane surface that was in contact with the gel is facing upward. Add 15 ml of blocking buffer (10% horse serum in TBS) and incubate the blotted membrane with agitation for 60 min at room temperature or overnight at 4°C in a tray, in order to block nonspecific protein binding of the antibody.
11. Pour off the blocking solution, add 3 ml of primary diluted antibodies in wash buffer (3% milk/TBS/Tween 20), and incubate the blot on a spiramix for 1 h at room temperature.

Remove unbound primary antibodies by washing the membrane with 15 ml of wash buffer for 30 min with agitation, changing the wash buffer every 10 min.

12. Prepare 3 ml of diluted horseradish peroxidase-conjugated secondary antibody and add to the blots.
13. Incubate on a spiramix at room temperature for 1 h with continuous agitation. Unbound secondary antibodies were removed by washing the membrane with 15 ml of wash buffer for 30 min with agitation, changing the buffer every 10 min. This was followed by two 15-min washes in 20 ml of 0.2% Tween solution (in TBS). Finally, the membranes were washed twice using 20 ml per wash of TBS buffer to remove residual detergent and transferred to weighing boats containing TBS solution using forceps until ready for chemiluminescent detection.
14. Mix 4 ml of SuperSignal West Dura Extended Duration Substrate chemiluminescent system reagent A and 4 ml of reagent B in a universal container and pour into a weighing boat.
15. Remove excess TBS buffer by draining the membrane over a piece of folded tissue paper. Immerse the membranes, protein side down, in the chemiluminescent solution and agitate for about 1 min. Use tweezers to remove the membrane from the tray, and remove excess solution onto tissue paper as before and transfer the membrane onto CCD UVIprochemi system.
16. After developing, immerse the membranes in Amido black stain for approximately 15 s, destain in a tray containing Amido black destain solution, and agitate until bands are clearly observed.
17. Transfer the membrane to a second tray containing fresh destain, rinse thoroughly with water, and allow to dry on some tissue for a permanent record of protein positions.

3.4. RNA Isolation

1. Extract RNA from cells of approximately 90% confluency using a disposable cell scraper and add 1 ml of RNA isolation reagent to the cells lysed by repetitive pipetting.
2. Transfer the resultant homogenates to 1.5-ml microcentrifuge tubes and incubate on ice for 5 min to permit complete dissociation of nucleoprotein complexes.
3. Add 0.2 ml of chloroform per millilitre of RNA isolation reagent and invert the sample tubes vigorously for 15 s.
4. After 5-min incubation on ice, centrifuge the samples at 13,000 rpm for 15 min at 4°C.
5. Following centrifugation, transfer the aqueous phase to a fresh tube and precipitate the RNA from the aqueous phase

by mixing with 0.5 ml of isopropanol. Keep the samples on ice for 10 min and then centrifuge at 13,000 rpm for 10 min at 4°C (see Note 22).

6. Remove the supernatant and add 1 ml of 75% ethanol to the RNA pellet.
7. Mix the RNA pellet by vortexing and precipitate by centrifugation at 7,500 rpm for 5 min at 4°C.
8. Dry at 55°C for 5 min and dissolve in 50 μl DEPC-treated water by vortexing.
9. Measure RNA concentration by absorbance at A260 nm wavelength and purity by ratio of A260 nm/A280 nm. Store RNA at −80°C until required.

3.5. Reverse Transcription PCR

1. Reverse Transcription PCR (RT-PCR) was performed using a reverse transcription kit according to the manufacturer's instructions.
2. Pipette no more than 250 ng of total RNA into a thin-walled PCR tube with 1 μl of anchored oligo-dT (100 μg/ml) and adjust the volume with sterile water to 13 μl.
3. Complete the reaction mixture by adding 4 μl of first-strand reaction buffer (5× concentrate), 2 μl of 5 mM dNTP mix, 1 μl of 5 U of RNAse inhibitor, and 1 μl of 25 U of M-MLV.
4. Incubate the reaction mix at 42°C for 50 min to commence cDNA synthesis; followed by incubation at 75°C for 10 min (for inactivation of the RTase).
5. Dilute the cDNA 1:2 in sterile water ready to be used as a template for amplification via PCR or stored at −20°C until required.

3.6. Polymerase Chain Reaction

1. All polymerase chain reactions (PCRs) were performed to a total volume of 20 μl, with 1 μl of cDNA template, 1 μl of each primer (at a concentration of 10 picomoles), 10 μl of MasterMix (2× concentrate from ABgene, Surrey, UK), and 7 μl of sterile water. PCRs were carried out using a PCR thermal cycler and cycling parameters as follows: enzyme activation 95°C for 5 min, 1 cycle, followed by 36 cycles of denaturing: 95°C for 15 s; annealing: 55°C for 15 s; and extension: 72°C for 2 min. Amplified products were visualised on agarose gels.

3.7. Agarose Gel Electrophoresis

1. DNA PCR products were separated using horizontal gel electrophoresis apparatus according to the manufacturer's instructions.
2. PCR products were separated on 2% agarose gels and run using TBE (Tris–HCl, borate, and EDTA) buffer.

3. Add 5 μl of a 1,000 base-pair ladder, prepared according to the manufacturer's instructions in a loading buffer to a well on the gel (see Note 23).
4. Add 5 μl of PCR products into separate wells of the gel.
5. Run the gel at a constant voltage of 110 V.
6. Stain using 5 μl of ethidium bromide stock solution (10 mg/ml stock concentration) for 5 min on a rocker and destain in distilled water for 20 min, if required.
7. PCR DNA products were visualized using an UV transilluminator and photographed using a thermal printer.

3.8. Agarose Gel DNA Extraction

1. The extraction and purification of DNA from agarose gels were performed using the QIAquick gel extraction kit in accordance with the manufacturer's instructions.
2. Remove the PCR products from the gel using a scalpel under UV light.
3. Weigh the gel slice and add 3 volumes of buffer QG (provided in the kit) according to the weight of gel (100 mg = 100 μl) in a 1.5-ml microcentrifuge tube.
4. Incubate the tube in a heat block set at 50°C for 10 min and vortex occasionally to dissolve the gel.
5. Once the gel slice had completely dissolved, transfer the sample to a QIAquick column to bind the DNA and microcentrifuge for 1 min at 13,000 rpm.
6. After centrifugation, wash the DNA twice with 750 μl of Qiagen buffer PE and elute into a clean 1.5-ml microcentrifuge tube using 30 μl of Qiagen buffer EB. Store DNA at −20°C until required.

3.9. Cloning and Expression of the Claudin-16 Gene

1. Cloning was performed using TOPO TA Cloning® kit in accordance with the manufacturer's instructions.
2. Mix 2 μl of PCR product extracted from agarose gel (see Subheading 3.8 gel extraction protocol) with 0.5 μl of high salt solution (provided in kit) and 0.5 μl of TOPO GFP fusion plasmid (that carries a GFP reporter gene and neomycin-resistance gene), in sterile 750-μl microcentrifuge tubes.
3. Incubate for 5 min at room temperature before placing the reaction on ice.
4. Mix with 24 μl of One Shot® TOP10 chemically competent *Escherichia coli* and incubate on ice for 30 min (see Note 24).
5. Heat-shock by incubating the tubes for 30 s in a 42°C water bath and place back on ice.
6. Add 250 μl of room temperature SOC medium to the tubes before transferring the mixture to a universal container.

7. Incubate at 37°C for 1 h on an orbital shaker (220 rpm).
8. Spread 50 μl of each transformation onto pre-warmed ampicillin-containing LB agar plates and incubate overnight at 37°C.
9. Screen bacterial colonies for a plasmid incorporating the PCR product in the correct orientation using RT-PCR and primers for the PCR product (see Subheading 3.6). Cells positive for claudin-16 are designated MDA^{CL-16}.
10. Transfer separately the positive colonies from a LB agar plate and inoculate into 10 ml of LB medium (containing antibiotic) in a sterile 50-ml polypropylene Falcon tube.
11. Incubate overnight at 37°C with a shaking speed of 220 rpm, and pellet the bacteria by centrifuging at 3,000 × *g* for 10 min at 4°C.
12. Plasmids were purified in accordance with protocol from the Qiagen mini purification kit.
13. Resuspend the pellet in 0.3 ml of buffer P1, add 0.3 ml of buffer P2, and mix gently.
14. Incubate the mixture at room temperature for 5 min, add 0.3 ml of chilled buffer P3, mix, and incubate on ice for 15 min.
15. Centrifuge in a microcentrifuge at 13,000 rpm for 10 min at 4°C, remove the supernatant, and transfer to a pre-equilibrated Qiagen-tip 20.
16. Wash the Qiagen-tip with 2× 1 ml of buffer QC and elute DNA with 0.8 ml of buffer QF. Precipitate DNA by adding 0.56 ml of room temperature isopropanol to the DNA fraction and centrifuge at 10,000 rpm for 30 min at 4°C.
17. Wash the DNA pellet twice with 1 ml of room temperature 70% ethanol and centrifuge at 10,000 rpm for 10 min.
18. Air-dry and dissolve the pellet in 50 μl sterilised water. Determine DNA concentrations by UV spectrophotometry.
19. Harvest mammalian cells from 80 to 90% confluent cultures by trypsinisation to release adherent cells (Subheading 2.1). Centrifuge the cells at ~1100 × *g* for 5 min, resuspend the cell pellet in 1 ml of complete medium, and store on ice.
20. Transfer 400 μl aliquots of the cell suspension into electroporation cuvettes and place on ice. Set the Equibio easyjet plus electroporation equipment parameters at 310 V and a capacitance value of 1,050 μF (see Note 25).
21. Add 10–30 μg of plasmid DNA in a volume of up to 20 μl to each cuvette containing cells, and mix by gently pipetting the solution up and down. Keep the curvette at room temperature for 2 min, before transferring to the electroporator.

22. After discharge, transfer the electroporated cells to a 25-cm^2 culture flask containing 4 ml of pre-warmed complete medium using a micropipetter equipped with a sterile tip and incubate in a humidified incubator at 37°C at an atmosphere of 5% CO_2.
23. The GFP expression vector shows only transient expression of the fluorescent protein and the cells were examined within 24–96 h after electroporation using a microscope under 450–490-nm illumination.
24. To isolate stable transfectants, cells are grown in complete medium containing 100 μg/ml of G418 antibiotic. Change the selective medium every 2–4 days for 2–3 weeks to remove the debris of dead cells and allow colonies of resistant cells to grow until sufficient numbers of cells are available for assays.

3.10. Quantitative PCR

1. Samples are made up to a total volume of 7 μl as per the manufacturer's instructions and using Thermostart PCR mastermix and specific primer pairs for claudin-16.
2. An Icycler IQ system, which incorporates a gradient thermocycler and a 96-channel optical unit, is used to quantify RNA transcript levels.
3. Plasmid standards and breast cancer cDNA are simultaneously assayed in duplicated reactions using a standard hot-start Q-PCR master mix.
4. Q-PCR conditions are as follows: Enzyme activation 95°C for 12 min, 1 cycle, followed by 60 cycles of denaturing: 95°C for 15 s; annealing: 55°C for 40 s; and extension: 72°C for 25 s.
5. Using purified plasmids as internal standards, the level of cDNA (copies/50 ng RNA) in the breast cancer samples was calculated. Q-PCR for β-actin was also performed on the same samples, to correct for any residual differences in the initial level of RNA in the specimens (in addition to spectrophotometry).
6. Results must then be normalised using cytokeratin-19 levels in the same tissues (see Fig. 1).
7. The products of Q-PCR are visualized on agarose gels to verify the reactions.

3.11. Immunohistochemistry

1. Use a cryostat to cut 6-μm sections of frozen tissue, place on Super Frost Plus slides, air-dry, and fix in a 50:50 solution of alcohol:acetone.
2. The sections are then air-dried again and stored at −20°C until used.
3. Immediately before commencing immunostaining, the sections are washed in buffer for 5 min and treated with horse serum for 20 min as a blocking agent to nonspecific binding.

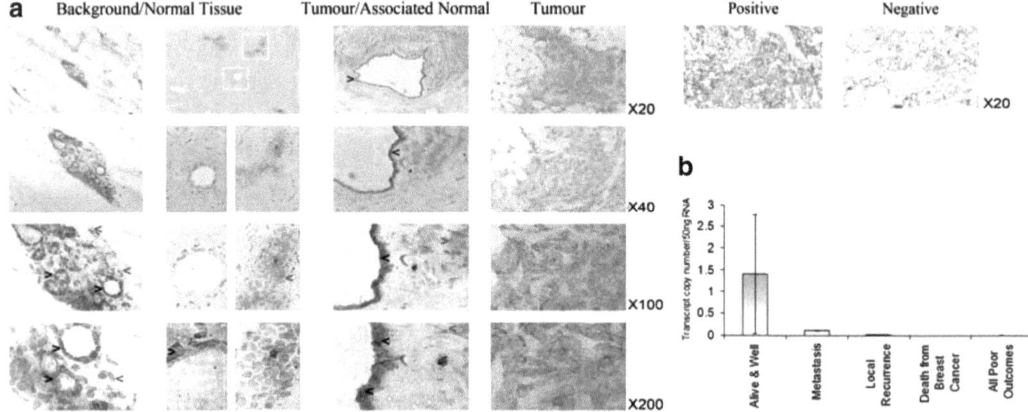

Fig. 1. Comparison of human breast cancer and background tissues using Q-PCR and immunohistochemistry. (**a**) Demonstrates the difference in expression of claudin-16 in background and tumour breast tissues ($n = 10$) at consecutively increasing magnification. Regions of claudin-16 expression located at the TJ area are indicated by *black arrows* (showing epithelial cells lining vessels and residual breast epithelia). Stromal cells are indicated by *grey arrows*, and show little/no staining. The positive and negative controls are also shown. (**b**) Shows the results of Q-PCR analysis for mRNA levels of the total patient cohort with claudin-16 and correlation with patient outcome. Levels are expressed as transcript copy number/50 ng RNA.

4. Sections are stained using claudin-16 antibody. Primary antibodies are used at 1:100 dilution for 60 min and washed in buffer.
5. The secondary biotinylated antibody at 1:100 dilution is added (in horse serum/buffer solution) for 30 min, followed by numerous washings.
6. The avidin/biotin complex is added for 30 min, again followed with washes.
7. Diaminobenzadine is used as a chromogen to visualise the antibody/antigen complex. Sections were counterstained in Mayer's haematoxylin for 1 min, dehydrated, cleared, and mounted in DPX. Following this, the sections were analyzed for staining intensity as previously described (9), see examples in Fig. 1.

3.12. Immunofluorescent Staining of Human Breast Cancer Cells

1. For immunofluorescence staining, cells were grown in 16-well chamber slides (see Note 26).
2. Trypsinise cells from culture flacks, pellet by centrifugation at 1600 rpm, and adjust the cell numbers to give 300–400,000 cells/ml in complete medium.
3. Add 100 μl of cell suspension to duplicate wells (30,000 cells/well) in the presence or absence of HGF/SF in an additional 100 μl of complete medium and incubate in a 37°C/5% incubator for a set period of time (0–24 h).

4. After incubation, aspirate the culture medium, rinse the wells with balanced salt solution (BSS) buffer using a disposable pastette, and fix the cells in 200 μl of 100% ethanol for 20 min at −20°C.
5. After fixation, wash the cells twice using BSS buffer and permeabilise by the addition of 200 μl of 0.1% Triton X-100 detergent in PBS for 5 min at room temperature.
6. Rinse twice with BSS buffer, add 200 μl of blocking buffer (10% horse serum in TBS) to each well, and incubate the chamber slide for 60 min at room temperature on a bench rocker.
7. Wash the wells once with wash buffer (3% horse serum in TBS buffer containing 0.1% Tween 20) and prepare 100 μl of primary antibodies (1:200) in wash buffer and add to the appropriate wells (see Note 27).
8. Incubate the chamber slide on the rocker for a further 60 min at room temperature.
9. Wash the wells twice with TBS buffer (with 0.1% Tween 20) and incubate the cells in 100 μl of secondary antibodies (fluorescein isothiocyanate (FITC), diluted in the same manner as the primary antibodies) for 50 min.
10. Wrap the chamber slide in foil to prevent light reaching the conjugates (see Note 28).
11. Finally, rinse the wells twice with wash buffer, once in BSS buffer, detach the wells, and carefully remove the gasket.
12. Mount fluorescently stained cells with FluorSave reagent, add coverslip, and visualise using an Olympus BX51 microscope with a Hamamatsu Orca ER digital camera at ×100 using oil immersion lens (Fig. 2 shows examples of immunofluorescence carried out on cells and human breast tissues).

3.13. Transepithelial Resistance

1. Transepithelial Resistance (TER) is measured with an EVOM voltohmmeter, equipped with a pair of STX-2 chopstick electrodes (8).
2. MDA-MB cells are seeded into the 0.4-μm pore size insert (upper chamber) and allowed to reach full confluence in a volume of 500 μl. 1 ml of medium is added to the outside of the well (lower chamber).
3. Replace the medium before further experiments. Inserts without cells, inserts with cells in medium, and inserts with cells with HGF (25 ng/ml) are tested every 30 min for a minimum period of 2 h.
4. Electrodes are placed at the upper and lower chambers and resistance measured with the voltohmmeter. At least three readings are obtained for each time point (Fig. 3).

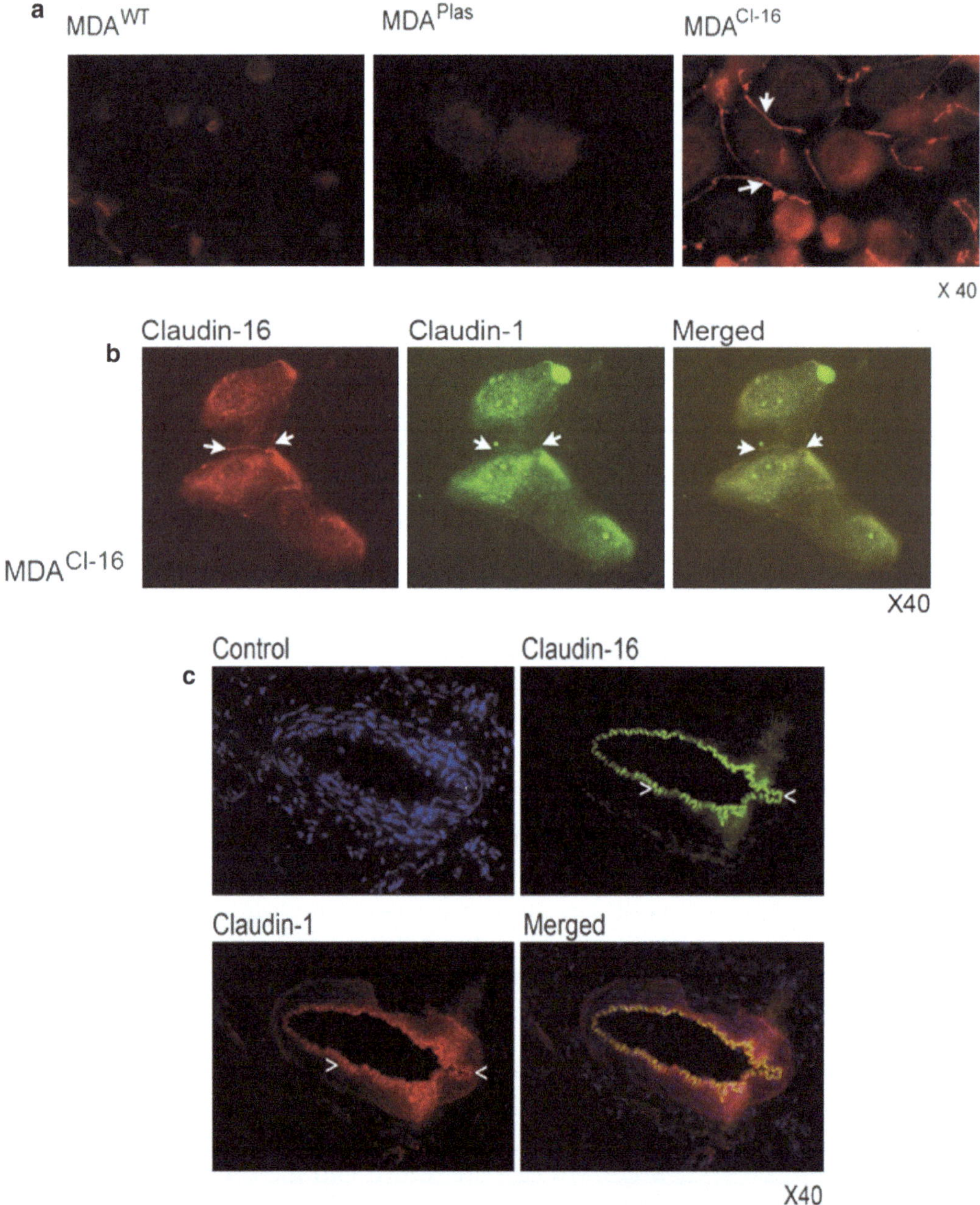

Fig. 2. Examples of the use of immunofuorescence imaging for claudin-16 distribution. (**a**) Immunofluorescence imaging of claudin-16 location at the TJ of transfected MDA-MB-231 human breast cancer cells. There is a lack of expression for claudin-16 in both wild-type and plasmid control cells, with a concurrent change in cell morphology. (**b**) Co-localisation of claudin-16 and claudin-1 in transfected MDA-MB-231 cells, with *white arrows* indicating localisation of both proteins to the junctional region. (**c**) Claudin-16 and claudin-1 (*white arrows*) dual staining to show co-localization at the junctional area of epithelial cells lining vessels in human breast tissue.

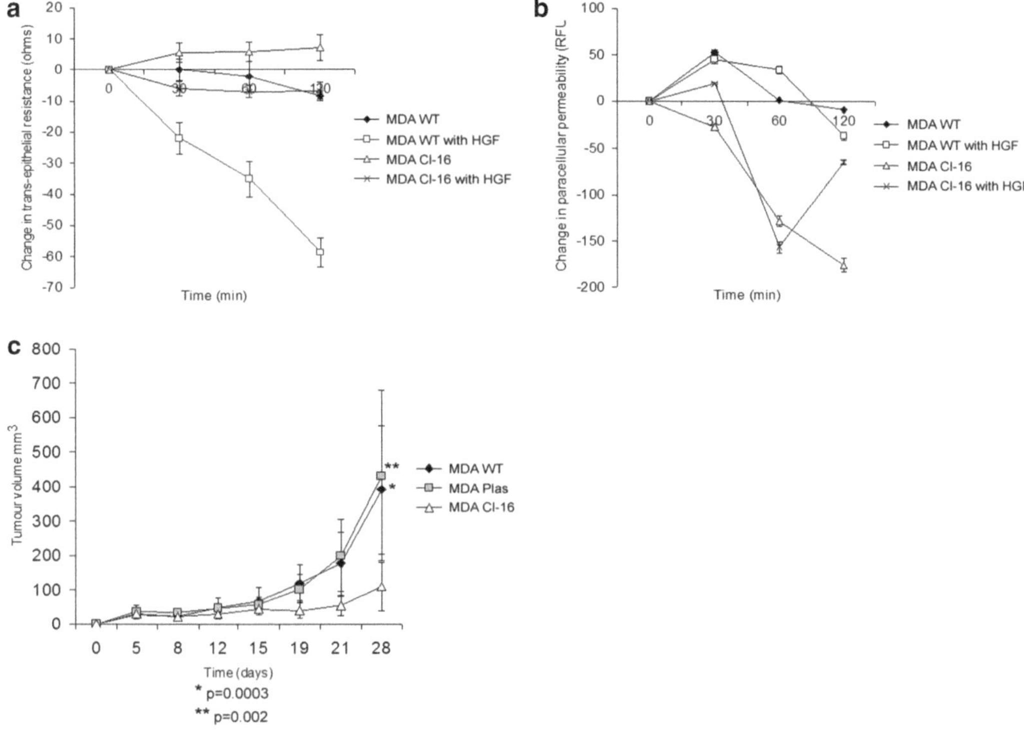

Fig. 3. Examples of three assays demonstrating that claudin-16-transfected MDA-MB-231 cells have a less aggressive phenotype than wild-type cells. (**a**) Graph depicting the change in transepithelial resistance of these cells and to their response to HGF (25 ng/ml). (**b**) Graph showing the change in paracellular permeability of these cells. (**c**) In vivo tumour growth over a 28-day period (mm^3).

3.14. Transepithelial Cell Permeability

1. Paracellular permeability (PCP) is determined using fluorescently labeled dextran FITC-dextran 40, molecular weight 40 kDa.
2. Human breast cancer cells are prepared and treated as in the TER study, but with the addition of dextran-40 to the upper chamber (10 μl).
3. At each time point, remove 50 μl medium from the lower chamber and retain.
4. The relative fluorescence from these collections is then read on a multichannel fluorescence reader (Fig. 3).

3.15. Cytodex-2 Bead Motility Assay

1. Trypsinize cells and resuspend to a count of 10,000 cells/ml.
2. Cells are pre-coated onto cytodex-2 carrier beads for 2 h in complete medium.
3. After the medium is aspirated and the cells washed (×2 in complete medium), aliquot them into wells of a 96-well plate in triplicate (300 μl/well).
4. Add HGF (25 ng/ml) to the cells and incubate overnight.

5. Wash the beads off in medium.
6. The cells that migrated onto the floor of each well are fixed (4% formaldehyde) and stained with crystal violet. The cells are then counted microscopically (×40, ten fields of view).

3.16. Invasion Assay

1. Invasiveness of MDA-MB-231 breast cancer cell line was assessed using the following in vitro assay. Place Transwell chambers equipped with 6.5-mm-diameter polycarbonate filter (pore size 8 μm) into a 24-well plate using tweezers. 100 μl of pre-cooled Matrigel (250 μg/ml solution), sufficient to coat the base of the insert, was added to each pre-chilled cell culture inserts and the 24-well plate placed in a 37°C incubator overnight to allow the Matrigel to set. After incubation, the Matrigel was rehydrated by the addition of 100 μl of DMEM/F12 media and incubation for 1 h in a 37°C/5% incubator.
2. After membrane rehydration, 15,000 cells were aliquoted into each insert with/without HGF. 100 μl of media containing HGF/SF was added to half of the inserts with DMEM/F12 to the remaining inserts. After 96 h co-culture noninvasive cells were removed with cotton swabs.
3. Return the plate to the 37°C/5% incubator for approximately 96 h. After incubation, the noninvasive cells were removed from the inside of the insert using a cotton swab.
4. The invasive cells on the underside of the insert were fixed in 4% formaldehyde, for 10 min at room temperature and stained with crystal violet, followed by microscopic counting (20 fields/insert).
5. The invading cells were identified using a light microscope and the numbers of invading cells per field determined using a 20× lens objective.

3.17. Cell Matrix Adhesion Assay

The cell–matrix attachment assay is carried out as previously reported (10).

1. Briefly, Matrigel (10 μg/well) is added to a 96-well plate, which is incubated for 24 h to allow binding of matrix protein to the surface of the well.
2. Wash the plates in medium and block with 5% BSA.
3. Add the cells at 10,000/well and incubate for 30 min, followed by aspiration and washing.
4. The number of attached cells is determined by direct counting under microscope (seven counts per experimental setting).

3.18. Growth Assay

Standard growth assays using crystal violet were carried out over 5 days to determine whether the insertion of the claudin-16 gene changed the growth rate of the MDA-MB-231 breast cancer cells.

1. Seed 10,000 cells/well into a 96-well plate at a volume of 300 ml, with/without 25 ng/ml HGF.
2. Prepare one 96-well plate for each day's growth.
3. Remove a plate for each time point, remove the medium, and fix in 4% formaldehyde.
4. Cells were stained with crystal violet and counted at ×40 using a standard light microscope.

3.19. ECIS Testing of Claudin-16

Electric cell–substrate impedance sensing measurements – The electric cell–substrate impedance sensing system was used to measure transepithelial impedance in the transfected and control cells to ascertain differences in cell migration and in the presence or absence of HGF (25 ng/ml) as a motogen.

1. Gold electrodes (Fig. 4).
2. Electrodes were treated with L-cysteine (10 mM, 400 ml/well for 40 min) in order to prepare the electrodes for the cells. This treatment removes any impurities from the electrode that might else interfere with adherence of the cells. Wash the well with 400 μl medium.
3. Cells were seeded at 2 × 105 cells/well in 400 μl medium and allowed to recover for 24 h at 37°C. The cell layer was checked under ×40 objective in order to observe confluency.
4. The array was placed in the array holder and electrode checked. This ascertains if the electrode is connected correctly and if there is any damage to the array.
5. The ECIS is then run for 24 h using a continuous electrical current. Attached cells on the electrode act as insulating particles; therefore, the current will flow through cell–substrate spaces beneath the cell that can eventually lead to the paracellular space. Changes in cell morphology manifest as changes in impedance as the paracellular and/or cell–substrate spaces changes, and the contribution of each of these is accounted for separately in the ECIS system.
6. Apply a 1-VAC signal at 4,000 Hz to the samples through a 1-mV resistor.
7. Monolayers cultured on ECIS plates can be wounded by submitting an elevated voltage pulse of 40-kHz frequency, 3.5-V amplitude, and 30-s duration, which leads to death and detachment of cells present on the small active electrode, resulting in a wound.
8. The cells are left for 24 h to undergo healing by cells surrounding the small active electrode that have not been submitted to the elevated voltage pulse.
9. This is assessed by continuous resistance measurements.

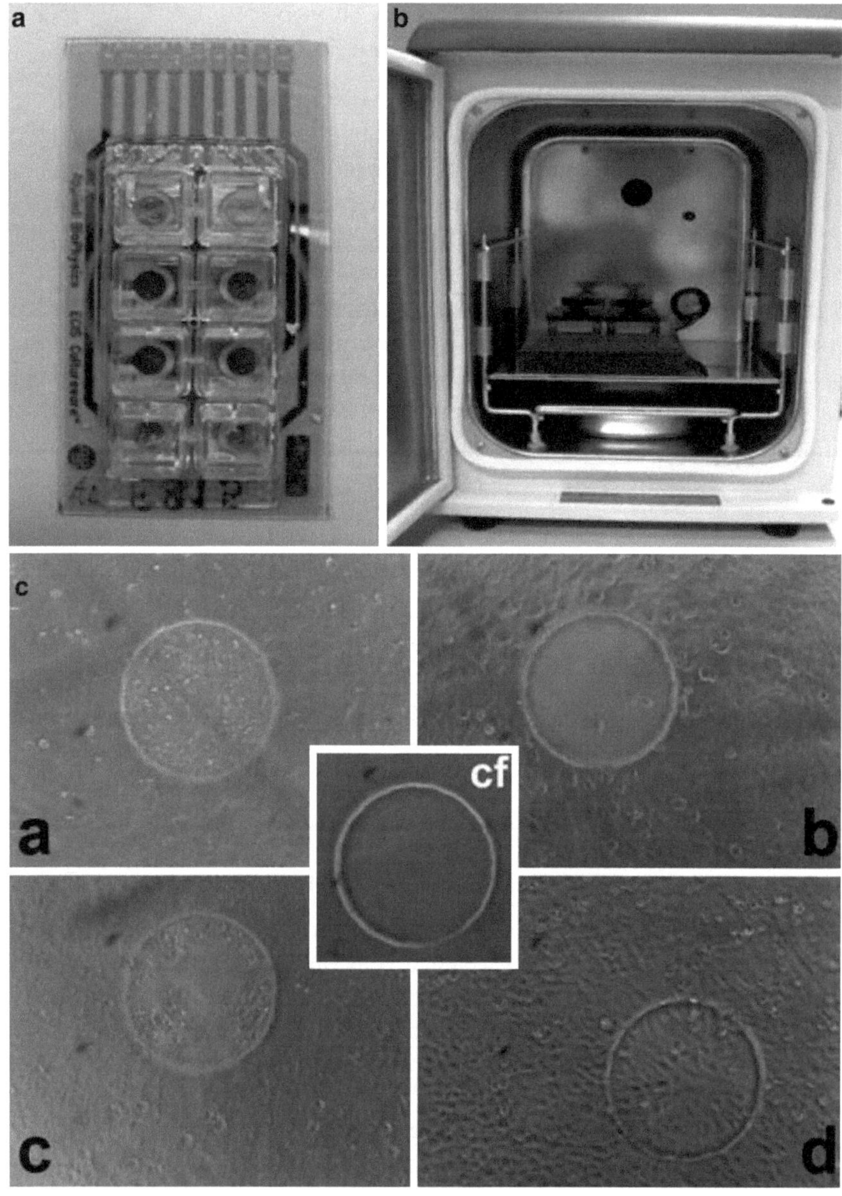

Fig. 4. ECIS array and array holder in situ (**a** and **b**). Cells on the ECIS array showing the electrodes in the centre (**c**). (**c**) a – before wounding the cell layer; b –immediately after wounding (6 V for 30 s); c – 3 h after wounding showing cells that migrated into the wounded space; d – 4 h after wounding showing the completely healed cell monolayer; *cf* – cell-free electrode.

3.20. In Vivo Growth Assay Analysing Claudin-16-Transfected Breast Cancer Cells

1. Mix 2×10^6 cells in 100 μl in a 0.5 mg/ml Matrigel suspension.
2. Inject subcutaneously into the left and right flanks of 4–6-week-old athymic nude mice and allow the cells to develop.
3. The mice ($n=5$) are maintained in filter-top units and weighed, and the tumour size measured weekly using Vernier calipers under sterile conditions.

4. Humane end points are (a) mice that suffer 25% weight loss and (b) development of tumours exceeding 1 cm^3 (subject to schedule 1 method according to the United Kingdom Home Office and the United Kingdom Coordinating Committee on Cancer Research (UKCCCR) guidelines).
5. At the conclusion of the experiment, the tumours ($n = 10$) are measured, the mice humanely killed under schedule 1, and the tumours removed and weighed.
6. Tumour volume is determined using the following formula: tumour volume (mm^3) = $0.523 \times \text{width}^2 \times \text{length}$.

3.21. Statistical Analysis

Statistical analysis was performed by MINITAB version 13.32 (Minitab Inc. State College, PA, USA) using a two-sample student *t*-test and the nonparametric Mann–Whitney confidence interval and test or Kruskal–Wallis, where appropriate. In addition, Microsoft Excel Professional was used to calculate correlation coefficients.

4. Notes

1. All solutions were prepared using deionised, double-distilled water, unless stated otherwise.
2. The calf serum, penicillin, and streptomycin can be made as per the manufacturers' instructions and aliquoted for storage at −20°C until required.
3. HECV cells adhere strongly to the culture vessel floor. Trypsinization is aided by incubation of the cells in trypsin/EDTA at 37°C for 10 min.
4. PMSF is highly toxic and should be weighed and prepared in a fume hood. It must be dissolved in a solvent (isopropanol or DMSO) as it is not water soluble.
5. Mercaptoethanol should be opened in a fume hood to minimize the unpleasant odor.
6. Wrap the stock solution in foil to prevent exposure to light.
7. Aliquot the stock solution and store at −20°C.
8. APS solution should be prepared immediately before use as its reactivity decreases over time.
9. Store the acrylamide and bisacrylamide solution at 4°C and TEMED at room temperature.
10. The sample buffer once prepared can be stored at −20°C until use.
11. The transfer buffer can be reused up to five times.

12. Handle the membrane carefully, use gloves and try not to touch the surface. Mark on the face on the membrane that will face the gel with pencil only.
13. The claudin-16 antibody used proved excellent for Western blotting, immunohistochemistry, and immunofluorescence.
14. CCD camera systems require an enhanced chemiluminescent system in order to pick up the signal for bound proteins. Other products are available.
15. Ethidium bromide is extremely harmful and should be handled with great care.
16. G418 is toxic; wear gloves when handling, weigh and prepare in a fume hood.
17. The beads once prepared and sterilised can be stored for up to a year at 4°C. Ensure to mix thoroughly before use.
18. Defrost Matrigel on ice at it quickly sets at room temperature. Chill all tips and cultureware before use.
19. Always prepare freshly prepared reagents to minimize the risk of gels not setting correctly. Gels can be prepared in the cassettes and stored overnight at 4°C in running buffer, if necessary. However, ensure that the gels are at room temperature before running samples and replace the storage running buffer.
20. Make sure the power supply is turned off before dismantling the apparatus.
21. Carefully remove the gel and place on the membrane, making sure the orientation of the gel is correct. Mark the membrane if necessary to avoid confusion later.
22. When transferring the aqueous phase, do not aspirate the interphase layer, which contains cell debris, etc. This will interfere in downstream reactions.
23. The DNA ladder can be aliquoted and stored in sample buffer until required, at –20°C.
24. Do not pipette the mixture up and down to mix; instead, gently tap the solution to avoid damaging the bacteria.
25. Electroporation parameters are dependent on the robustness of the cell line used. MDA-MB-231 and HECV cells are relatively robust and can withstand up to 450 v. Cell lines should be tested at varying voltages (from 140 to 450) and cultured overnight to determine the voltage required that does not kill all the cells.
26. For tight junctions to form fully, the cell layer must be fully confluent before commencing experimentation.
27. Tween 20 is highly viscous; use 1-ml syringe to measure small quantities.

28. Pick up and handle the chamber slides only using the glass base; the chambers are easy to release from the silicon glue. Before mounting the coverslips, remove as much of the silicon as possible. Use an adequate amount of FluorSave as the slides dry out easily. Also ensure there are no air bubbles. The slides should be covered in foil and left at 4°C to set.

References

1. Martin, T. A., Harrison, G. M., Watkins, G., and Jiang, W. G. (2008) Claudin-16 reduces the aggressive behavior of human breast cancer cells, *J Cell Biochem* ***105***, 41–52.
2. Martin, T. A., Das, T., Mansel, R. E., and Jiang, W. G. (2006) Synergistic regulation of endothelial tight junctions by antioxidant (Se) and polyunsaturated lipid (GLA) via Claudin-5 modulation, *J Cell Biochem* ***98***, 1308–1319.
3. Martin, T. A., Das, T., Mansel, R. E., and Jiang, W. G. (2007) Enhanced tight junction function in human breast cancer cells by antioxidant, selenium and polyunsaturated lipid, *J Cell Biochem* ***101***, 155–166.
4. Grone, J., Weber, B., Staub, E., Heinze, M., Klaman, I., Pilarsky, C., Hermann, K., Castanos-Velez, E., Ropcke, S., Mann, B., Rosenthal, A., and Buhr, H. J. (2007) Differential expression of genes encoding tight junction proteins in colorectal cancer: frequent dysregulation of claudin-1, -8 and -12, *Int J Colorectal Dis* *22*, 651–659.
5. Hewitt, K. J., Agarwal, R., and Morin, P. J. (2006) The claudin gene family: expression in normal and neoplastic tissues, *BMC Cancer* ***6***, 186.
6. Morin, P. J. (2005) Claudin proteins in human cancer: promising new targets for diagnosis and therapy, *Cancer Res* ***65***, 9603–9606.
7. Martin, T. A., Watkins, G., Mansel, R. E., and Jiang, W. G. (2004) Loss of tight junction plaque molecules in breast cancer tissues is associated with a poor prognosis in patients with breast cancer, *Eur J Cancer* ***40***, 2717–2725.
8. Martin, T. A., Watkins, G., Mansel, R. E., and Jiang, W. G. (2004) Hepatocyte growth factor disrupts tight junctions in human breast cancer cells, *Cell Biol Int* ***28***, 361–371.
9. Martin, T. A., Goyal, A., Watkins, G., and Jiang, W. G. (2005) Expression of the transcription factors snail, slug, and twist and their clinical significance in human breast cancer, *Ann Surg Oncol* ***12***, 488–496.
10. Hiscox, S., and Jiang, W. G. (1999) Ezrin regulates cell-cell and cell-matrix adhesion, a possible role with E-cadherin/beta-catenin, *J Cell Sci* ***112 Pt 18***, 3081–3090.

Chapter 29

Dynamics of Claudins Expression in Colitis and Colitis-Associated Cancer in Rat

Yoshiaki Arimura, Kanna Nagaishi, and Masayo Hosokawa

Abstract

Claudins comprise a multigene family of 24 species and have been shown to constitute the backbone of tight junction strands in simple epithelial cells and to be directly involved in their barrier functions. Apical-most tight junction protein complexes (TJs) are implicated in inflammatory bowel disease (IBD) pathophysiology. Except for claudin-8, TJs explored in this study (including ZO-1, claudin-1, -2, -3, -7, -12, and -15) were found to be expressed in rat colonic tissues. ZO-1 and claudin-7 were ubiquitously expressed in all study groups. As depicted in Fig. 1b, expressions of claudin-2, -12, and -15 significantly diminished after combined treatment with dextran sulfate sodium (DSS) and busulfan (BU) (lane 5), compared with either agent alone (lanes 2 and 4). Despite the lack of significance, there appeared to be a subtle dose-dependent decrease with DSS treatment (lanes 2 and 3). In contrast to these results obtained from DSS colitis, expression of claudin-1 was significantly downregulated, while expression of claudin-15 was upregulated in colitis-associated cancer tissues in the azoxymethane (AOM)/DSS model (Fig. 2b). It is very intriguing that claudins' expression dynamics were mutually exclusive between colitis and colitis-associated cancer in rats. However, the biological significance of disease-specific claudin expression profiles will remain elusive until the specific expression and function of each claudin in a tissue- and cell-type relationship are comprehensively clarified. Currently, the physiologic consequences of the diversity of TJ barrier function resulting from multiple combinations of claudins are only beginning to be recognized. Full unraveling of these complexities could inspire a new paradigm of inflammation and cancer, and eventually translate to clinical practice on IBD.

Key words: Inflammatory bowel disease, DSS colitis, Azoxymethane, Colitis-associated cancer, Confocal immunofluorescence, Busulfan

1. Introduction

Intestinal epithelial cells, via highly regulated cellular turnover and tight junction protein complexes (TJs), form a physical and dynamic barrier against bacterial activation of the mucosal immune system. TJs are the most apical components of intercellular junctional complexes and establish cell polarity and act as major

Kursad Turksen (ed.), *Claudins: Methods and Protocols*, Methods in Molecular Biology, vol. 762, DOI 10.1007/978-1-61779-185-7_29, © Springer Science+Business Media, LLC 2011

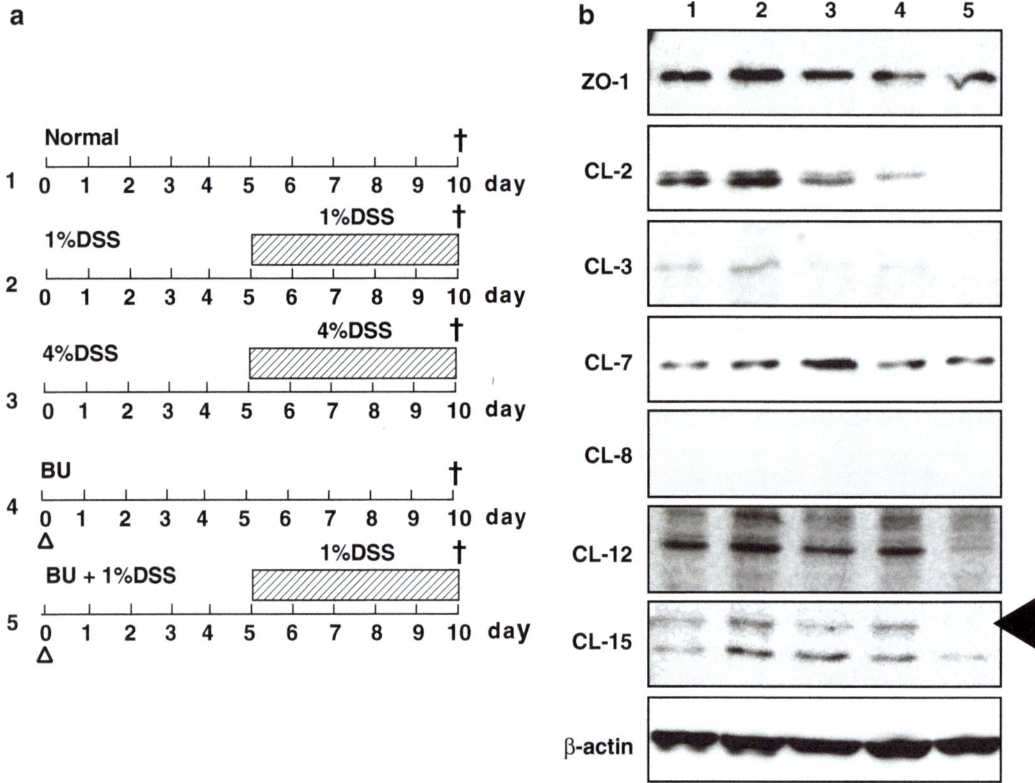

Fig. 1. DSS colitis with/without BU-induced bone marrow hypoplasia and claudins expression. (**a**) Design of inducing DSS colitis with/without acute bone marrow hypoplasia. *Open arrowheads* represent intraperitoneal BU administration (20 mg/kg) at day 0. The rats were killed on day 10, depicted by † in the panel. (**b**) Western blot analysis on claudin expression. *The lane numbers* correspond to the digit indicating the study design in (**a**). Claudin-2, -12, and -15 are abbreviated as CL-2, CL-12, and CL-15. A *closed triangle* indicates the CL-15 blots.

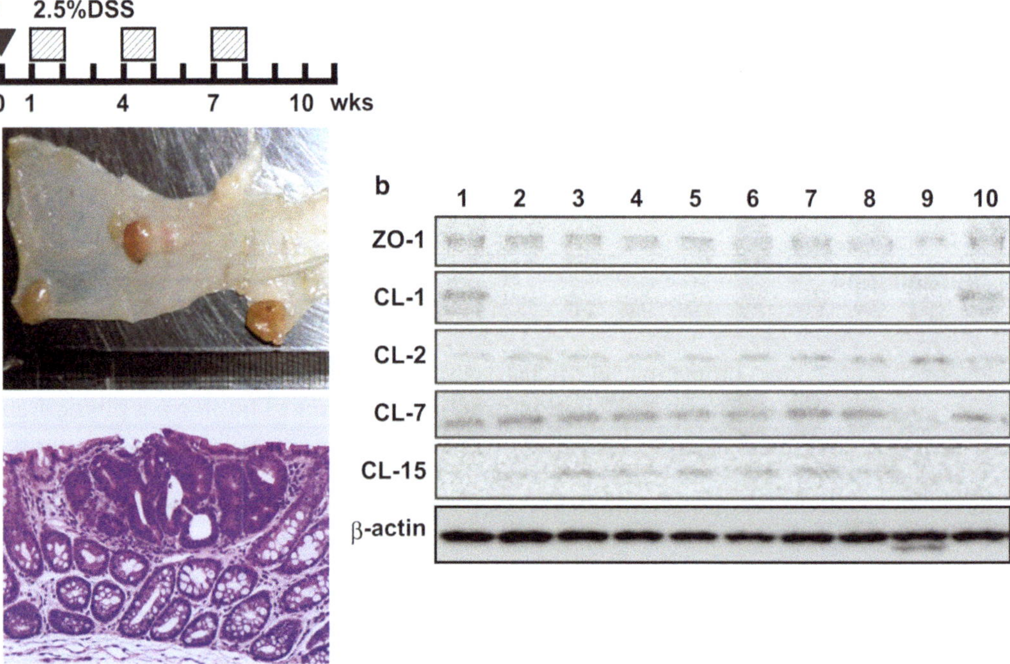

determinants of epithelial barrier function; evidence is mounting that impaired function of TJs is critical in inflammatory bowel disease (IBD) pathophysiology (1) and in experimental colitis (2), such as dextran sulfate sodium (DSS) colitis (3). DSS, a long-chain polymer of glucose, is believed to be directly toxic to gut epithelial cells and impairs the epithelial barrier integrity with subsequent Th1 skewed immune-mediated inflammation. Among proteins constituting TJs, claudins, by adjusting their TJ strands, are major determinants of the paracellular permeability of the barrier, including size and charge selectivities (4).

Colorectal cancer is a life-threatening disease that can develop spontaneously or it may complicate IBD. Rodent models provide a highly reliable experimental model to study inflammation-dependent tumor growth. It is based on the mutagenic agent azoxymethane (AOM) combined with DSS colitis, which causes rapid growth of multiple colon tumors per rat within 10 weeks (5).

Because the dynamics of claudins expression in DSS colitis and AOM/DSS models in rats remain poorly understood (6), we here explore TJs expression by Western blotting and confocal immunofluorescence.

2. Materials

2.1. DSS Colitis and Colitis-Associated Cancer [Azoxymethan (AOM)/DSS] Model in Rats

1. Sex- and age-matched 6- to 8-week-old rats of body weight >150 g, all of the same strain, namely Lewis (Charles River Laboratories). All animal experiments must conform to local and national animal care regulations.
2. DSS (MW 40 kDa, Sigma, St. Louis, MO) is dissolved in deionized distilled water (see Note 1) to obtain a 1.0–4.0% (wt/vol) solution. DSS solution can be stored at 4°C for up to 1 week (see Note 2).
3. Busulfan (1,4-butanediol dimethanesulfonate, BU, Sigma) was reconstituted as follows: 500 mg of BU was dissolved in 2.0 ml dimethylsulfoxide (DMSO, Sigma); 100 μl of this solution was then mixed with 900 μl DMSO and 4 ml warm (37°C) phosphate-buffered saline (PBS) (pH 7.4) to create a 5 mg/ml solution. Rats were weighed and 20 mg/kg of BU was injected intraperitoneally.

Fig. 2. TJs in colitis-associated cancer. (**a**) Design of AOM/DSS model. A *closed triangle* indicates intraperitoneal injection of AOM working solution (15 mg per kg body weight) and *three rectangles* represent three inductive cycles of colitis by giving drinking water containing 2.5% DSS ad libitum for a week. Three tumors developed in the rectum during 10 weeks, and the representative HE stain of colitis-associated cancer is shown. (**b**) Western blot analysis on the TJs expression. Lane 1 normal, lanes 2–9 tumors developed in AOM/DSS models, and lane 10 normal tissues in AOM/DSS models.

4. AOM (Sigma) may cause cancer and heritable genetic damage in humans. It is toxic, flammable, and can cause burns. When handling AOM, suitable protective clothing, gloves, and face protection are mandatory. All handling, especially the injections, should be performed by skilled experimenters. To prepare AOM stock solution, dissolve AOM in water at 10 mg/ml. Aliquot and store at −20°C. Avoid repeated freeze and thaw cycles. AOM is more stable in water or saline than in buffered PBS, and is more stable in glassware than in plastic tubes. Working solutions are prepared by thawing aliquots directly before use and diluting the stock solution 1:10 in sterile isotonic saline (i.e. to obtain a working concentration of 1 mg/ml). Rats were weighed and 15 mg/kg of AOM was injected intraperitoneally.

2.2. Preparation of Samples for Immunoblotting

1. Radio-immuno precipitation assay (RIPA) buffer for cell lysis: 20 mM Tris–HCl, pH 7.4, 150 mM sodium chloride, 1 mM EDTA, pH 8.0, 0.1% (w/v) sodium dodecyl sulfate (SDS), 0.1% sodium deoxycholate, and 1% Triton X-100. One tablet complete Mini™, protease inhibitor cocktail tablets (Roche Diagnostics, Mannheim, Germany), was added immediately before use to 7 ml of cold RIPA. The 10% sodium deoxycholate stock solution (5 g into 50 ml) must be protected from light and should be stored in aliquots at −20°C (Table 1).
2. Perform a Bradford assay using Bio-Rad protein assay kit (Bio-Rad, Hercules, CA) (7). All kit components have a 1-year shelf life at 4°C. Standards are provided in a 0.9% NaCl and 0.05% NaN_3 solution. 1× Dye reagent: 1 L of dye solution containing methanol and phosphoric acid. One bottle of dye reagent is sufficient for 200 assays using the standard 5 ml procedure or 4,000 assays using the microplate procedure. BSA standard, 2 mg/ml: provided in 2-ml tubes. This diluted reagent may be used for about 2 weeks when kept at room temperature. Bovine gamma-globulin standard, 2 mg/ml: provided in 2-ml tubes. Bovine serum albumin (BSA) standard set: set of 7 concentrations of BSA (2, 1.5, 1, 0.75, 0.5, 0.25, 0.125 mg/ml) in 2-ml tubes. Bovine gamma-globulin standard set: set of 7 concentrations of gamma-globulin (2, 1.5, 1, 0.75, 0.5, 0.25, 0.125 mg/ml) in 2-ml tubes.
3. The linear range of microassay protocol for BSA is 125–1,000 μg/ml, whereas with gamma-globulin, the linear range is 125–1,500 μg/ml. Protein solutions are normally assayed in duplicate.

2.3. SDS-Polyacrylamide Gel Electrophoresis

1. 1.5 M Tris–HCl (pH 8.8), 0.5 M Tris–HCl (pH 6.8), and 10% SDS. Store at room temperature.
2. 30% Acrylamide/bis solution (37.5:1 with 2.6% C) (this is a neurotoxin when unpolymerized and so care should be taken

Table 1
Composition of SDS-PAGE and Western blotting buffers

RIPA buffer	100 ml	Working concentration
1 M Tris–Cl (pH7.4)	2 ml	20 mM
5 M NaCl	3 ml	150 mM
0.5 M EDTA (pH 8.0)	0.2 ml	1 mM
SDS	100 mg	0.1%
DOC[a]	100 mg	0.1%
Triton X-100	1 ml	1%
5× Laemmli buffer	10 ml	Working concentration
0.5 M Tris–Cl (pH 6.8)	2.5 ml	125 mM
10% SDS	4 ml	4%
Glycerol	2 ml	20%
2-Mercaptoethanol	1 ml	10%
BPB	4 mg	0.04%
10× Running buffer	1,000 ml	Working concentration
Tris-base	30 g	250 mM
Glycine	144 g	1,920 mM
SDS	10 g	1%
10× Transfer buffer	1,000 ml	Working concentration
Tris-base	30 g	250 mM
Glycine	144 g	1,920 mM
Transfer buffer	1,000 ml	Working concentration
10× Transfer buffer	100 ml	10%
Methanol	200 ml	20%

Diluted to 100 ml for RIPA, 10 ml for Laemmli, and 1,000 ml for both running and transfer buffers with water
[a]DOC: sodium deoxycholate

to avoid exposure) and *N*,*N*,*N*,*N*′-tetramethylethylenediamine (TEMED, Bio-Rad).

3. Ammonium persulfate: prepare 10% solution in water and freeze immediately for one time use as 200 μl aliquots at −20°C.
4. Water-saturated isobutanol. Shake equal volumes of water and isobutanol in a glass bottle and allow them to separate. Use the top layer and store at room temperature.
5. Laemmli sample buffer (5×): 125 mM Tris–HCl, 4% (w/v) SDS, 20% glycerol, 10% 2-mercaptoethanol, and 0.04% bromophenol blue (BPB) (8). Note that BPB is added prior to use.
6. Running buffer (10×): 250 mM Tris-base, 1,920 mM glycine, and 1% (w/v) SDS. Store at room temperature (Table 1).
7. 12% Separating gel: Prepare four sheets of gel with a 0.75-mm-thick, 12% gel (for claudins) by mixing 5.0 ml of 1.5 M

Table 2
Composition of separating and stacking gel for four pieces

Separating gel	15%	12%	10%	7.5%	Stacking gel
Water	4.7 ml	6.7 ml	8.0 ml	9.7 ml	6.1 ml
1.5 M Tris–Cl (pH 8.8)	5.0 ml	5.0 ml	5.0 ml	5.0 ml	2.5 ml (0.5 M Tris–Cl, pH 6.8)
10% SDS	200 μl	200 μl	200 μl	200 μl	100 μl
30% Acrylamide	10 ml	8 ml	6.7 ml	5.0 ml	1.33 ml
10% APS	200 μl	200 μl	200 μl	200 μl	100 μl
TEMED	20 μl	20 μl	20 μl	20 μl	20 μl

Total volume of solution is approximately 20 ml. TEMED is added just prior to use

Tris–HCl (pH 8.8), 200 μl of 10% SDS, 8 ml 30% acrylamide/bis solution, 6.7 ml water, 200 μl ammonium persulfate solution, and 20 μl TEMED (Table 2).

8. Stacking gel: Prepare four sheets of the stacking gel by mixing 2.5 ml of Tris–HCl (pH 6.8), 1.33 ml 30% acrylamide/*bis* solution, 6.1 ml water, 100 μl ammonium persulfate solution, and 20 μl TEMED (Table 2).
9. Ponceau S staining solution: 0.1% (w/v) Ponceau S and 5.0% (w/v) acetic acid.
10. Coomassie staining solution: 40% water, 10% acetic acid, and 50% methanol for protein fixation, and the same mixture of water/acetic acid/methanol with 0.25% (w/v) Coomassie brilliant blue R-250 (CBB).
11. Pre-stained molecular weight markers: Kaleidoscope markers (Bio-Rad).

2.4. Western Blotting for Tight Junctions

1. Transfer buffer (10×): 250 mM Tris-base (do not adjust pH) and 1,920 mM glycine (Table 1).
2. Transfer buffer (1×): transfer buffer (10×) plus 20% (v/v) methanol. Store in the transfer apparatus at room temperature (with cooling during use, see Note 3). (Table 1).
3. Polyvinylidene difluoride (PVDF) membrane from Millipore, Bedford, MA, and 3MM Chr chromatography paper from Whatman (Maidstone, UK).
4. Tris-buffered saline with Tween-20 (TBST): prepare 10× stock with 1.37 M NaCl, 27 mM KCl, 250 mM Tris–Cl (pH 7.4), and 1% Tween-20. Dilute 100 ml with 900 ml water for use.
5. Blocking buffer: 5% (w/v) nonfat dried milk in TBST.
6. Primary antibody dilution buffer: TBST.

7. Primary antibodies against claudin-1, -2, -3, -7 (kindly gifted by professor Chiba), -8 (Invitrogen, CA), -12 (IBL, Takasaki, Japan), -15, and ZO-1 (Invitrogen).
8. Secondary antibody: anti-mouse or anti-rabbit IgG conjugated to horseradish peroxidase (HRP) (GE Healthcare, Buckinghamshire, UK).
9. Enhanced chemiluminescent (ECL) reagents from GE Healthcare (9) and ImageQuant™ LAS 3000 mini Control Software (FUJI FILM, Tokyo, Japan) (see Note 4).

2.5. Confocal Immunofluorescence for Tight Junctions

1. Microscope coverslips and slides ($24 \times 40 \times 0.15$ mm) from Matsunami Glass, Osaka, Japan.
2. PBS (10×): prepare 0.1 M stock with 1.37 M NaCl, 27 mM KCl, 100 mM Na_2HPO_4, and 18 mM KH_2PO_4 (adjust to pH 7.4 with HCl if necessary) and autoclave before storage at room temperature. Prepare working solution by diluting one part with nine parts of water.
3. Paraformaldehyde (Fisher Scientific, Loughborough, UK): prepare a fresh 4% (w/v) solution in PBS for each experiment. The solution may need to be carefully heated (use a stirring hot plate in a fume hood) to dissolve, and then cooled to room temperature for use.
4. 15% Sucrose in PBS: 500 ml sterile PBS and 75 g sucrose. Mix above and filter sterilize with a disposable Nalgene filtration unit type S (0.45 μm) (NALGENE Labware, Rochester, NY). Store at 4°C.
5. Antibody dilution buffer: 3% (w/v) BSA in PBS.
6. Secondary antibody: anti-mouse or anti-rabbit IgG conjugated to Texas Red (Molecular Probes, Eugene, OR).
7. Nuclear stain and mounting medium: VECTASHIELD® mounting medium with 4,6-diamidino-2-phenylindole (DAPI) (Vector Laboratories, Burlingame, CA).

3. Methods

3.1. DSS Colitis Model

1. Marrow hypoplasia was induced by a single intraperitoneal injection of BU (10), and DSS colitis (11) was induced by giving drinking water containing 1% or 4% DSS ad libitum for the indicated (5–7) days (Fig. 1a).
2. Animals were observed daily for fluid intake, weight changes, and for major symptoms such as loose stools, mucous diarrhea, and hematochezia. At the end of each cycle, the rats were killed by inhaling a lethal dose of diethylether.

3.2. AOM/DSS Model

1. Inject intraperitoneally AOM working solution for the experimental group or sterile isotonic saline for the control group with a sterile syringe in a laminar airflow hood.
2. Prepare 2.5% (w/v) DSS solution (see Subheadings 2.1); at this concentration, DSS gives good results in the Lewis rat and other strains (see Note 5).
3. Fill the water bottles of the mouse cages with DSS-containing water for 3 days and provide 10–12 ml DSS solution per 100 g rat body weight per day. Mount the bottle lids properly and make sure the tips are not obstructed.
4. Empty the remaining DSS solution from the water bottles of the cage at day 4 and refill with fresh DSS solution for another 2 days.
5. Empty the remaining DSS solution from the bottles at day 6 and refill with DSS solution for another 2 days.
6. Empty the remaining DSS solution from the bottles at day 8 and refill with water.
7. Repeat steps 2–6 at days 22–29.
8. Evaluate for precursor lesions at day 40 as an option. This can be done without killing the animal.
9. Repeat steps 2–6 at days 53–60.
10. Evaluate for tumor development at day 80 (Fig. 2a).

3.3. Preparation of Samples for Assay of Tight Junctions by Western Blotting

1. Dissect the colon tissue with clean instruments, preferably on ice, and as quickly as possible to prevent degradation by proteases.
2. Place the tissue in Eppendorf tubes and immerse in liquid nitrogen to "snap freeze." Store samples at –80°C for later use or keep on ice for immediate homogenization. For a 5-mm sized piece of tissue, add 150 μl RIPA buffer rapidly to the tube, homogenize with a Teflon homogenizer, and maintain constant agitation for 0.5 h at 4°C (e.g., place on an orbital shaker in the fridge). Volumes of lysis buffer must be determined in relation to the amount of tissue present (protein extract should not be over-diluted to avoid loss of protein and large sample volumes to be loaded onto gels. The minimum concentration is 0.1 mg/ml and the optimal concentration is 1–5 mg/ml).
3. Centrifuge for 10 min at 14,000 rpm 15,560 g at 4°C in a microcentrifuge. Gently remove the tubes from the centrifuge, place on ice, aspirate the supernatant, and place in a fresh tube kept on ice; discard the pellet. The buffer (with inhibitors) should be ice-cold prior to homogenization.
4. Determine protein concentration by a Bradford assay (see Subheading 2.2, item 2). Pipet 10 μl of each standard and

sample solution into separate microtiter plate wells. Add 200 μl of diluted dye reagent to each well. Mix the sample and reagent thoroughly using a microplate mixer. Alternatively, use a multichannel pipet to dispense the reagent. Depress the plunger repeatedly to mix the sample and reagent in the well. Replace with clean tips and add reagent to the next set of wells. Incubate at room temperature for at least 5 min. Absorbance will increase over time; the samples should be incubated at room temperature for no more than 1 h. Measure absorbance at 595 nm.

5. Prepare loading sample by diluting one part sample with four parts of 5× Laemmli sample buffer (see Subheading 2.3, item 5 and Table 1). The tubes are closed and then boiled for a further 5 min. After cooling to room temperature, they are ready for separation by SDS-PAGE (see Note 6).

3.4. SDS-PAGE

1. These instructions assume the use of a Mini-Protean Tetra Cell (Bio-Rad). They are easily adapted to other formats. It is critical that the glass plates for the gels are scrubbed clean with a rinsable detergent after use (e.g., Alconox, Alconox, New York, NY) and rinsed extensively with water. They can be kept clean until use in a plastic rack in 30% nitric acid (use caution when removing). Rinse the plates with water followed by 95% ethanol to remove the acid, and air-dry.
2. Prepare four sheets of gel. Pour the gel, leaving space for a stacking gel, and overlay with water-saturated isobutanol. The gel should polymerize in about 30 min. For ZO-1, a 7.5% gel was suitable (Table 2).
3. Pour off the isobutanol and rinse the top of the gel twice with water.
4. Prepare four sheets of the stacking gel. Use about 0.5 ml of this to rinse the top of the gel quickly and then pour the stack and insert the comb (15-well, 0.75 mm). The stacking gel should polymerize within 30 min (Table 2).
5. Prepare the running buffer by diluting 100 ml of the 10× running buffer with 900 ml of water (Table 1).
6. Once the stacking gel has set, carefully remove the comb and gently wash the wells with running buffer.
7. Add the running buffer to the upper and lower chambers of the gel unit and load <10 μl of each sample (40 μg protein) in a well. Include one well for pre-stained molecular weight markers.
8. Complete the assembly of the gel unit and connect to a power supply. The minigel can be run at 150 V for approximately 1 h. Once the BPB dye fronts reach the bottom of the separating gel, disconnect the power supply.

9. Visualization of proteins in gels: visualization of protein at this stage is useful to determine if proteins have migrated uniformly and evenly. Use Ponceau S stain if you plan to transfer the separated proteins to a membrane, as the Coomassie stain is not reversible. Use the Coomassie stain on gel only post-transfer to check the efficiency of the transfer, or if you have no plans to transfer and just want to observe the results of the SDS-PAGE separation, Ponceau S staining (12) is a simple method for visualizing proteins on PVDF membranes. Although relatively insensitive, Ponceau staining permits a quick visual inspection of the blot to verify transfer and to mark the position of the molecular weight standards. The stain readily washes away and does not interfere with subsequent immunoblotting.
10. Ponceau S stain (reversible protein detection): Stain solution can be reused up to ten times. Store at room temperature. Briefly rinse freshly electrophoresed gels in water (30 s maximum), immerse the gel in Ponceau S staining solution, and stain for a minute. During rinsing the gel with water, proteins should be visualized and then the gel rapidly immersed in an aqueous solution of 0.1 M NaOH. Protein bands will start to disappear after 10–30 s. Rinse the gel with running water for 2–3 min. Move the gel to a dish of transfer buffer before proceeding with transfer according to the transfer apparatus manufacturer's instructions.
11. Coomassie stain (total protein detection): as soon as the power is turned off, the separated protein bands will begin to diffuse (they are freely soluble in aqueous solution). To prevent this diffusion, treat the gel with a fixative solution which causes almost all proteins to precipitate (become insoluble). To visualize the fixed proteins, place the gel in the CBB staining solution. Incubate from 1 h to overnight at room temperature on a shaker. Transfer the gel (save the dye mixture; it can be reused many times) to a mixture of 67.5% water, 7.5% acetic acid, and 25% methanol; place on a shaker; and replace with fresh rinse mixture until the excess dye has been removed. The stain will not bind to the acrylamide and will wash out (leaving a clear gel). However, it remains strongly bound to the proteins in the gel, and these take on a deep blue color.

3.5. Western Blotting for Tight Junctions

1. The samples that have been separated by SDS-PAGE are transferred to PVDF membranes electrophoretically using tank transfer systems (13).
2. Preparing PVDF membranes: cut the membrane to the same size as the gel, plus 1–2 mm on each edge. Immerse for 1–2 s in 100% methanol in a polypropylene plastic tray. Note that PVDF membranes are hydrophobic and will not wet simply

from being placed in water or transfer buffer. Decant the methanol and equilibrate for 5 min with transfer buffer. Do not let them dry out at any time. If this occurs, wet the membrane once again with methanol and transfer buffer, as described above.

3. These directions assume the use of a Mini-Protean Tetra Modules system (Bio-Rad). Fill the Bio-Ice cooling unit with water and store it in your laboratory freezer at −20°C until ready to use. After use, return the cooling unit to the freezer for storage. Prepare the transfer buffer (see Subheading 2.4 and Table 1 for buffer formulation). Using buffer chilled to 4°C will improve heat dissipation. Cut the membrane and the filter paper to the dimensions of the gel. Always wear gloves when handling membranes to prevent contamination. Equilibrate the gel and soak the membrane, filter paper, and fiber pads in transfer buffer (15 min to 1 h depending on gel thickness).
4. The gel unit is disconnected from the power supply and disassembled. The stacking gel is removed and discarded and one corner cut from the separating gel to identify its orientation. The separating gel is then laid on top of the PVDF membrane.
5. Preparing the gel sandwich. Place the cassette, with the gray side down, on a clean surface. Build the sandwich by adding, in the following order, one pre-wetted fiber pad on the gray side of the cassette, a sheet of filter paper on the fiber pad, the equilibrated gel on the filter paper, and the pre-wetted membrane on the gel. Complete the sandwich by placing a piece of filter paper on the membrane. Add the last fiber pad. Next, remove all air bubbles that may have formed during the above procedure. This is vital to obtain good results. Use a glass tube to gently roll air bubbles away.
6. Close the cassette firmly, being careful not to move the gel and filter paper sandwich. Close and lock the cassette with the white latch.
7. The cassette is placed into the transfer tank such that the PVDF membrane is between the gel and the anode. It is essential to ensure that this orientation or the proteins will be lost from the gel into the buffer rather than transferred to the PVDF (12). Furthermore, it is also important to identify the front and back sides of the PVDF membrane.
8. Add the frozen Bio-Ice cooling unit. Place in the tank and completely fill the tank with buffer to maintain a temperature between 10 and 15°C.
9. The lid is put on the tank and the power supply activated. Transfers can be accomplished at either 30 V overnight or 100 V for 1 h.

10. Once the transfer is complete, the cassette is taken out of the tank and carefully disassembled, with the top sponge and sheets of 3 MM paper removed. The gel is left in place on top of the membrane and together they are laid on a glass plate so that the shape of the gel (including the cut corner for orientation) can be cut into the membrane using a razor blade. The colored molecular weight markers should be clearly visible on the membrane (see Note 7).
11. The membrane is then incubated in 50 ml blocking buffer for 1 h at room temperature on a rocking platform (see Note 8).
12. The blocking buffer is discarded and the membrane quickly rinsed three times with TBST for 5 min each rinse prior to adding an indicated dilution of the anti-claudin antibody in TBST for overnight at 4°C on a rocking platform. If some suggestions are written about diluents on the antibody data sheet, follow the instructions (Table 3).
13. The primary antibody is then removed (see Note 9) and the membrane washed with TBST for 15 min and subsequently twice more for 5 min each.
14. The HRP-linked secondary antibody is freshly prepared for each experiment as a 1:2,000–3,000-fold dilution in blocking buffer and added to the membrane for 1 h at room temperature on a rocking platform.
15. The secondary antibody is discarded and the membrane washed with TBST for 15 min and subsequently twice more for 5 min each.
16. During the final wash, 4 ml aliquots of each portion of the ECL reagent are warmed separately to room temperature,

Table 3
Features of antibodies against TJs used in this study

Antibody	Clonality	Immunized animal	Dilution for WB	Dilution for IF	Manufacturer	Code no.
Claudin-1	MAb	Mouse	500	100	Invitrogen	37-4900
Claudin-2	PAb	Rabbit	500	200	Invitrogen	51-6100
Claudin-3	PAb	Rabbit	500	200	Gifted	NA
Claudin-7	PAb	Rabbit	500	200	Gifted	NA
Claudin-12	PAb	Rabbit	200	20	IBL	18801
Claudin-15	PAb	Rabbit	1,000	20	Invitrogen	38-9200
ZO-1	PAb	Rabbit	250	100	Invitrogen	40-2200

WB Western blotting, *IF* immunofluorescence

and once the final wash is removed from the blot, the ECL reagents are mixed together and then immediately added to the blot, which is then rotated by hand for 1 min to ensure even coverage. Again, it is vitally important to ensure the orientation of the blot, namely, the front side of the membrane should be sufficiently immersed in the ECL reagents (see the above Subheading 3.6).

17. The blot is removed from the ECL reagents, blotted with Kimwipes and wrapped in plastic wrap without any wrinkles. The desired positions of ladders in the colored markers in the analysis are marked with a lumino marker on the wrapped blot and then the blot is placed on the sample tray in the ImageQuant™ LAS 3000 mini.
18. Scan the blot and analyze by ImageQuant™ Control software for a suitable exposure time, typically a few minutes. An example of the results produced is shown in Fig. 1b and Fig. 2b.

3.6. Confocal Immunofluorescence for Tight Junctions

3.6.1. Preparation of Samples for Immunofluorescence

1. Perfusion fixation through the heart: set up perfusion pump (L/S® Variable speed modular drives, Cole-Parmer, IL); attach perfusion set and perfusion needle. First, run about 100 ml of 0.01 M PBS through the tubing to remove any residue. Then place the open end of the perfusion tube in a beaker filled with cold 4% paraformaldehyde (in ice box). The 200–300 ml of solution will usually be sufficient for one rat. Open valve and adjust to a slow steady drip (25 ml/min) and then close the valve (see Note 10).
2. Set up surgery site with scissors, forceps, and clamps. Give an appropriate amount of anesthetic to rat. Once the rat is anesthetized, place it back down on the operating table. Use tape to restrain the legs so that the rat is securely fixed.
3. Use pinch-response method to determine the depth of anesthesia. Rat must be unresponsive before proceeding with the following steps.
4. Make incision with scalpel through abdomen to the length of the diaphragm. Use sharp scissors to cut through the connective tissue at the bottom of diaphragm to allow access to rib cage.
5. Use large scissors, blunt side down, to cut through the ribs just left of the rib cage midline.
6. Make one center or two end horizontal cuts through the rib cage, and open the thoracic cavity. Clamp open to expose the heart and provide drainage for blood and fluids.
7. Steady the heart with forceps (it should still be beating) and insert the needle directly into the apex of the left ventricle and direct it straight up about 5 mm. Be careful not to advance the needle too far, as it may pierce an interior wall and compromise circulation of solutions. Secure the needle position by

clamping it in place near the point of entry. Release the valve to allow a slow, steady flow of around 25 ml/min of PBS.

8. Open the atrium with sharp scissors and make sure the solution is flowing freely. If not or if fluid is coming from the animal's nostrils or mouth, reposition the needle.
9. When blood has been cleared from body, change to 4% paraformaldehyde solution (200–300 ml). Take care not to introduce air bubbles while transferring from one solution to the other.

   It is best to wear protective eye goggles during the whole perfusion process, as the connecting tubes occasionally become undone and may spray fixating solution into your eyes. Perfusion is almost complete when spontaneous movement (formalin dance) and lightened color of the liver are observed.
10. Stop the perfusion and excise the tissues of interest. Place them in vials containing the same fixation fluid and fix at 4°C overnight.
11. Immerse the tissue in sterile 5% sucrose/PBS for 1 h at 4°C, transfer the tissue to 10% sucrose/PBS for another 1 h at 4°C, and finally immerse the tissue in 15% sucrose/PBS, leave overnight at 4°C.
12. Embed the tissue in OCT (Sakura Finetek Japan, Tokyo) in plastic embedding molds. The tissue should be oriented in the block appropriately for sectioning (cross-section, longitudinal, etc.). Freeze tissue block in liquid nitrogen. Place the bottom third of the block into the liquid nitrogen, allow to freeze until all but the center of the OCT is frozen, and complete freezing on dry ice. Store tissue blocks at −80°C in a sealed container or move to a cryostat.

3.6.2. Confocal Laser Scanning Microscopy

1. 3 μm-thick OCT-embedded circumferential sections are prepared for immunofluorescent determination of TJs.
2. The samples are blocked by incubation in antibody dilution buffer for 1 h at room temperature.
3. The blocking solution is removed and replaced with the primary antibody (1:20–100) in antibody dilution buffer for 1 h at room temperature (Table 3).
4. The primary antibody is removed and the sample washed three times for 5 min each with PBS. The sample is then put under aluminium foil and the room lights dimmed for the subsequent steps.
5. The secondary antibody is prepared at 1:250 in antibody dilution buffer and added to the samples for 30 min at room temperature.
6. The secondary antibody is discarded. The samples are washed five times for 10 min each with PBS and then aspirated dry from one corner.

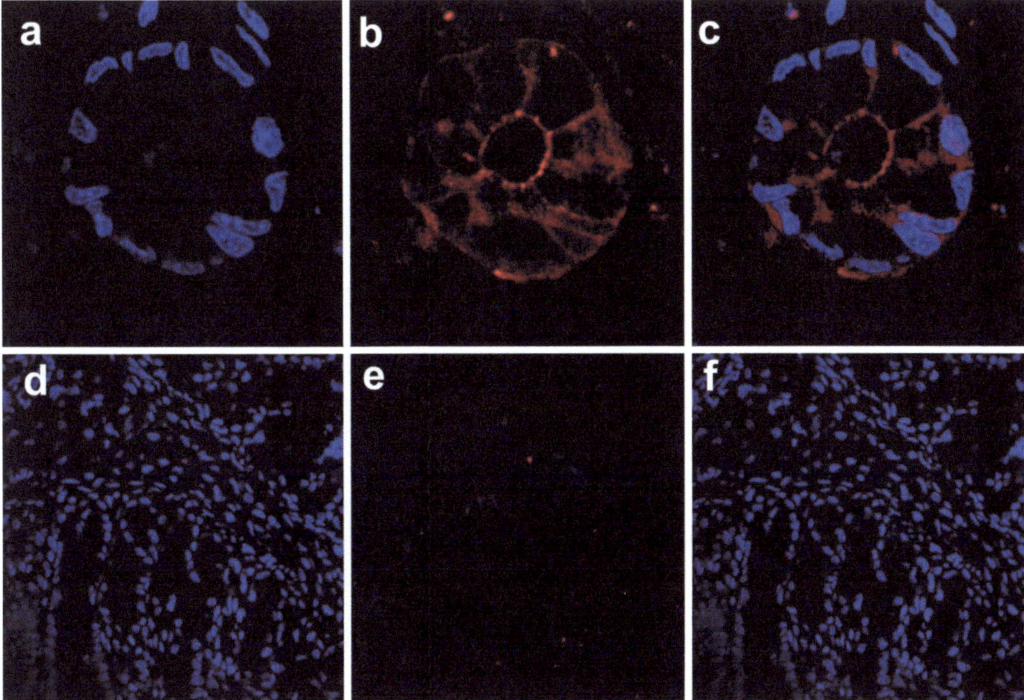

Fig. 3. Confocal immunofluorescence of claudin-2. (**a–c**) Normal rat, (**d–f**) rat with BU + 1% DSS colitis, (**a** and **d**) DAPI, (**b** and **e**) CL-2 by Texas Red immunofluorescence, (**c** and **f**) merged image.

7. The samples are then ready to be mounted. If they are on a coverslip, then the coverslip is carefully inverted into a drop of mounting medium including DAPI to stain the DNA and identify the nuclei on a microscope slide. Nail varnish is used to seal the sample. The sample can be viewed immediately once the varnish is dry, or can be stored in the dark at 4°C.
8. The slides are viewed under confocal microscopy (LSM 510 META ConfoCor 3, Carl Zeiss, Germany). Excitation at 543 nm induces the Texas Red fluorescence (red emission) for claudins, while excitation at 364 nm induces DAPI fluorescence (blue emission). Software can be used to overlay the phase contrast and fluorescence images. Examples of the signals for claudin-2 are shown in Fig. 3.

4. Notes

1. Unless stated otherwise, all solutions should be prepared in deionized distilled water. Deionized distilled water is referred to as "water" in this text.
2. Colitis induced by 40-kDa DSS is more severe in the distal colon than that induced by 5 kDa, which induces colitis

predominantly in the cecum and proximal colon. A 500-kDa DSS dose does not increase the severity of colitis in the rodents.

3. Transfer buffer can be used for up to five transfers within 1 week so long as the voltage is maintained constant for each successive run (the current will increase each time). Adequate cooling by use of a refrigerated bath to keep the buffer at or less than room temperature is essential to prevent heat-induced damage to the apparatus and compromise of the experiment (14).
4. Quantification of data may be desired and the chemiluminescent signal can be captured digitally with an instrument such as ImageQuant™ LAS-3000 mini.
5. Some strains may be more susceptible to DSS-induced colitis, leading to strain-to-strain differences in colitis and tumor development. Consequently, depending on the strain, it might be necessary to decrease or increase the DSS concentration between 1% and 3% (w/v). Alternatively, the period of DSS administration could be adjusted (e.g., 4–5 days instead of 7 days).
6. Some transmembrane proteins that span the entire biological membrane may aggregate and precipitate in boiling sample buffers. They require to be denatured at lower temperature and longer time, for example, for 20 min at room temperature or 4°C overnight. Although claudins are 20–27-kDa tetraspan proteins with a short cytoplasmic N-terminus, two extracellular loops, and a C-terminal cytoplasmic domain, boiling claudin proteins in the sample buffer improves subsequent SDS-PAGE.
7. Membranes can be dried and stored in re-sealable plastic bags at 4°C for 1 year or longer. Before use, dried PVDF membranes must be wetted in 100% methanol and rinsed in water to remove the methanol.
8. A variety of agents are currently used to block binding sites on the membrane. Of these, nonfat dried milk in TBST usually achieves the best result. However, nonfat dried milk should be substituted for 5% BSA to analyze phosphoprotein because otherwise endogenous casein in nonfat dried milk will be detected instead.
9. The primary antibody can be saved for subsequent experiments by addition of 0.02% final concentration sodium azide (conveniently done by dilution from a 10% stock solution; exercise caution since azide is highly toxic) and storage at 4°C. These primary antibodies have been used for up to 20 blots over several months, with the only adjustment required being increasing the length of exposure at the ECL detection step (14).

10. In general, an adult rat will require around 30–60 min of perfusion time, but this may vary depending on the size of the animal and technique.

Acknowledgments

The author would like to thank Professor H. Chiba, Department of Basic Pathology, Fukushima Medical University, for providing claudins antibodies and Ms. K Fujii, Research Assistance, First Department of Internal Medicine, for technical assistance. We would like to thank Dr Peter M. Olley, Emeritus Professor of Sapporo Medical University, for linguistic assistance. This study was supported by a Health and Labour Sciences Research Grant-in-aid for research on intractable disease from the Ministry of Health, Labor and Welfare of Japan (to KI).

References

1. Zeissig, S., Burgel, N., Gunzel, D., Richter, J., Mankertz, J., Wahnschaffe, U., et al. (2007) Changes in expression and distribution of claudin 2, 5 and 8 lead to discontinuous tight junctions and barrier dysfunction in active Crohn's disease. *Gut* **56,** 61–72.
2. Zwiers, A., Fuss, I. J., Leijen S., Mulder, C. J., Kraal, G., Bouma, G. (2008) Increased Expression of the Tight Junction Molecule Claudin-18 A1 in Both Experimental Colitis and Ulcerative Colitis. *Inflamm Bowel Dis.* **14,** 1652–1659
3. Kong, J., Zhang, Z,. Musch, M. W., Ning, G, Sun, J., Hart, J., Bissonnette, M, Li, Y. C. (2008) Novel role of the vitamin D receptor in maintaining the integrity of the intestinal mucosal barrier. *Am J Physiol Gastrointest Liver Physiol.* **294,** 208–216.
4. Tsukita, S., Furuse, M., Itoh, M. (2001) Multifunctional strands in tight junctions. *Nat Rev Mol Cell Biol.* **2,** 285–293.
5. Neufert, C., Becker, C. and Neurath, M. F. (2007) An inducible mouse model of colon carcinogenesis for the analysis of sporadic and inflammation-driven tumor progression. *Nat Prot* **2,** 1998–2004.
6. Yabana, T., Arimura, Y., Tanaka, H., Goto, A., Hosokawa, M., Nagaishi, K., et al. (2009) Enhancing epithelial engraftment of rat mesenchymal stem cells restores epithelial barrier integrity. *J Pathol.* **218,** 350–359.
7. Bradford, M. M. (1976) A rapid and sensitive method for the quantitation of microgram quantities of protein utilizing the principle of protein-dye binding. *Anal Biochem.* **72,** 248–254.
8. Laemmli, U. K. (1970) Cleavage of structural proteins during the assembly of the head of bacteriophage T4. *Nature* **227,** 680–685.
9. Whitehead, T. P., Kricka, L. J., Carter, T. J., Thorpe, G. H. (1979) Analytical luminescence: its potential in the clinical laboratory. *Clin Chem.* **25,** 1531–46.
10. Pugsley, C. A. J., Forbes, I. J., Morley, A. A. (1978) Immunologic abnormalities in an animal model of chronic hypoplastic marrow failure induced by busulfan. *Blood* **51,** 601–610.
11. Okayasu, I., Hatakeyama, S., Yamada, M. (1990) A novel method in the induction of reliable experimental acute and chronic ulcerative colitis in mice. *Gastroenterology* **98,** 694–702.
12. Salinovich, O., and Montelaro, R. C. (1986) Reversible staining and peptide mapping of proteins transferred to nitrocellulose after separation by sodium dodecyl sulfate-polyacrylamide gel electrophoresis. *Anal Biochem.* **156,** 341–347.
13. Bolt, M. W. and Mahoney, P.A. (1997) High-efficiency blotting of proteins of diverse sizes following sodium dodecyl sulfate polyacrylamide, gel electrophoresis. *Anal Biochem.* **247,** 185–192.
14. Mattingly, R. R. (2003) Mitogen-activated protein kinase signaling in drug-resistant neuroblastoma cells. *Methods Mol Biol.* **218,** 71–83.

Chapter 30

Anti-claudin-4-Conjugated Highly Luminescent Nanoparticles as Biological Labels for Pancreatic Cancer Sensing

Ken-Tye Yong

Abstract

Anti-claudin-4, whose corresponding antigen receptors are known to be overexpressed in both primary and metastatic human pancreatic cancer, is utilized for targeted delivery and imaging of pancreatic cancer. In this protocol, we describe the use of quantum dots (QDs) as sensitive optical contrast agent for imaging pancreatic cancer in vitro and in vivo by using anti-claudin-4 as targeting ligands. The claudin-4-mediated targeting is demonstrated in using both in vitro confocal microscopy and in vivo tumor imaging system. This targeted QD platform will be further modified for the purpose of developing as an early detection imaging tool for pancreatic cancer.

Key words: Claudin-4, Quantum dots, Bioconjugation, Cancer, Bioimaging, Targeted delivery

1. Introduction

Quantum dots (QDs) are semiconductor crystals in the size range of 3–10 nm with size-tunable optical and electronic properties (1, 2). These QDs can be made in either polar or nonpolar media followed by surface functionalization for targeted bioimaging and biosensing (3, 4). The use of QDs as highly luminescence labels for numerous biological and biomedical applications has become an area of intense research focus over the last few years (5). Such materials have significant advantages over traditional organic dyes and fluorescent dye proteins (6). QDs with different emission colors can be simultaneously excited with a single light source, with minimal spectral overlap, providing significant advantages for multiplexed imaging of molecular targets (7). QDs have been

Kursad Turksen (ed.), *Claudins: Methods and Protocols*, Methods in Molecular Biology, vol. 762,
DOI 10.1007/978-1-61779-185-7_30,

shown to remain brightly emissive in biological environment even after long periods of excitation, whereas organic dyes or fluorescence proteins are photobleached quickly (8). Moreover, the emission peak of the QDs can be systematically tuned from visible to near-infrared wavelength by systemically tailoring the size and composition of the nanocrystals, whereas new architecture of organic dyes or fluorescence proteins must be synthesized to shift their emission wavelength to the desirable detectable wavelengths for an in vitro or in vivo system imaging (9). More importantly, this tunability allows the synthesis of near-infrared-emitting QDs for imaging deep-sitting tumors such as pancreatic cancer (10). Thus, due to the size-tunable luminescence emission from the visible to infrared wavelength, broad absorption spectra, and high levels of brightness and photostabilty of QDs, these nanocrystals have recently attracted much attention as new-generation probes for diagnostic imaging and sensing of human cancers (11).

To date, there are several factors for engineering functionalized QDs for cancer research application, including: (1) creating biocompatible surface coating without loss of quantum efficiency; (2) development of efficient, reliable, and reproducible techniques for labeling of cellular targets interest; (3) preparing non-aggregated bioconjugated QDs with stable brightness; and above all, (4) the use of bioconjugated QDs with no minimal cellular and tissue cytotoxicity (12). Recently, our group has developed a versatile method for preparing lysine-coated CdSe/CdS/ZnS QDs by a simple solution-phase synthesis method (13). The main advantages offered by CdSe/CdS/ZnS nanocrystals lies in the robustness of the core–shell–shell structure versus the core–shell structure in CdSe/ZnS, as well as the better passivation of the CdSe core nanocrystals by the double shell layers. These QDs can be easily dispersed in aqueous systems using a simple ligand exchange reaction with mercaptoundecanoic acid (MUA) followed by lysine cross-link with the mercapto ligands (14). The cross-linked lysine groups on the QD surface allow conjugation with biomolecules for specific targeted delivery to pancreatic cancer in vitro and in vivo (15).

Pancreatic cancer is the fourth most common cause of cancer-related mortality in the USA and a disease of near-uniform lethality (16). The vast majority of patients present with locally advanced or distant metastatic disease, rendering their malignancy surgically inoperable (17). The best option currently available for ameliorating pancreatic cancer survival is to diagnose the neoplasm at an early, and therefore, potentially curable stage. It is crucial that new generation of robust image contrast agents be developed, which can be specifically targeted to pancreatic cancer in vivo, leading to improved diagnosis at an early stage (18). Nanoparticles, linked with cancer-specific targeting ligands, are ideally suited for this purpose. Targeting of these imaging modalities to pancreatic cancer at an early stage of pancreatic cancer development can be achieved with the help of an antibody directed against overexpressed surface receptors on the cancer cells/tissues, such as the antigen claudin-4 (19, 20).

Claudin-4 is classified as an integral constituent of tight junctions. Recent reports have shown that they are highly expressed in pancreatic cancer cell lines such as Panc-1, MiaPaCa, and Colo-357 (21). More recently, effects of claudin-4 on invasion and growth in in vitro and in vivo pancreatic cancer have been intensively investigated for the last few years. Tight junctions are biological components of intercellular junctional complexes. Their main functions are establishing cell polarity and serving as major determinants of paracellular permeability. The family of claudins, which form integral constituents of tight junctions, consists of at least transmembrane proteins and represents a major factor in establishing the intercellular barrier. Many members of the claudin family show a distinct organ-specific distribution pattern within the human body. Claudin-4 consists of 209 amino acids and contains four putative transmembrane segments. In previous studies, many groups have identified claudin-4 as overexpressed in pancreatic cancer in various expression profiling approaches using representational difference analysis and DNA array technology.

On the basis of our previous reports and findings, the aim of the present protocol is to use anti-claudin-4-conjugated CdSe/CdS/ZnS QDs as effective reliable probes to label pancreatic cancer in vitro and in vivo and differentiate them from the healthy ones. This technique will serve as a powerful platform to evaluate and determine the metastatic behavior of pancreatic cancer.

2. Materials

2.1. Fabrication of CdSe/CdS/ZnS QDs Using Hot Colloidal Synthesis Approach

1. Cadmium oxide, 99% (Sigma).
2. 8.6 g selenium powder is dissolved in 100 mL trioctylphosphine and stored in the dark at room temperature. This mixture is referred as TOP-Se.
3. 8.6 g sulfur powder is dissolved in 100 mL trioctylphosphine and stored in the dark at room temperature. This mixture is referred as TOP-S.
4. Oleic acid (sigma).
5. Trioctylphosphine oxide (Sigma).
6. Tetradecylphosphonic acid (Strem Chemicals).
7. Hexane and toluene are used as dispersion or quenching solvents (Sigma).
8. Ethanol is used as the polar solvent to precipitate QDs and remove excess surfactants in the reaction mixture (Sigma).

9. 100-mL three-neck flask is employed for the QD synthesis at high temperature (VWR).
10. Stir plate (VWR).
11. Stir bar (VWR).
12. 10-mL glass syringe is used to inject reaction precursors into the reaction flask (VWR).

2.2. Preparation of Lysine-Coated QDs

1. 80 mg of CdSe/CdS/ZnS dispersed in 2 mL of chloroform at room temperature.
2. MUA.
3. 15 mmol DL-lysine is dissolved in 25 mL HPLC water and stored in a single vial at room temperature. The final concentration is 14 mM.
4. 30 mmol *N*,*N*′-dicyclohexylcarbodiimide (DCC) (Sigma) is dissolved in dimethyl sulfoxide (DMSO; Sigma) at room temperature.
5. Ammonium hydroxide solution, 30% (Sigma).
6. 0.45-μm pore size of syringe filters (VWR) is used to remove unreacted or byproduct agglomerates.

2.3. Conjugation of QDs with Anti-claudin-4

1. Refrigerated Anti-claudin-4 antibody solution at −7°C is removed from freezer and defrost for 30 minutes (Invitrogen).
2. (1-Ethyl-3-[3-dimethylaminopropyl]carbodiimide hydrochloride) (EDC) is dissolved in 10 mL of HPLC water at room temperature. The final concentration of this stock solution is 1 mM.
3. Lysine-coated CdSe/CdS/ZnS QD stock solution.
4. 0.45-μm pore size of syringe filters (VWR) is used to remove unreacted or byproduct agglomerates.

2.4. Cell Culture

1. Human pancreatic cancer cell lines Panc-1, MiaPaCa-2, and CoLo-357 are obtained from American Type Tissue Collection (ATTC).
2. Dulbecco's modified Eagle's medium (DMEM) (Gibco) supplemented with 10% fetal bovine serum.

2.5. Confocal Microscopy Imaging of Pancreatic Cancer Cells Labeled with Functionalized QDs

1. Confocal microscopy images are obtained using a Leica TCS SP2 AOBS spectral confocal microscope (Leica Microsystems Semiconductor GmbH, Wetzler, Germany) with laser excitation at 442 nm. All images are taken in exact same conditions of laser power, aperture, gain, offset, scanning speed, and scanning area.

2.6. In Vivo Tumor Imaging with Functionalized QDs

1. The mice are imaged using the Maestro in vivo imaging system (CRI, Inc., Woburn, MA; excitation filter = 445–490 nm, emission filter = 515-nm long pass).
2. The Maestro optical system consists of an optical head that includes a liquid-crystal tunable filter (with a bandwidth of 20 nm and a scanning wavelength range of 500–950 nm) with a custom-designed, spectrally optimized lens system that relays the image to a scientific-grade megapixel CCD. The CCD captures the images at each wavelength.
3. The captured images (spectral cube, containing a spectrum at every pixel) can be loaded into the vendor software and analyzed. Spectra from the autofluorescence (from the skins, tissues, and food, coded green) and QD-associated luminescence signals (coded red) can be unmixed using the vendor software, as shown in Fig. 1.
4. In this study, the scanning wavelength range between 500 and 900 nm was used as recommended by the CRI, Inc., instrument manual.

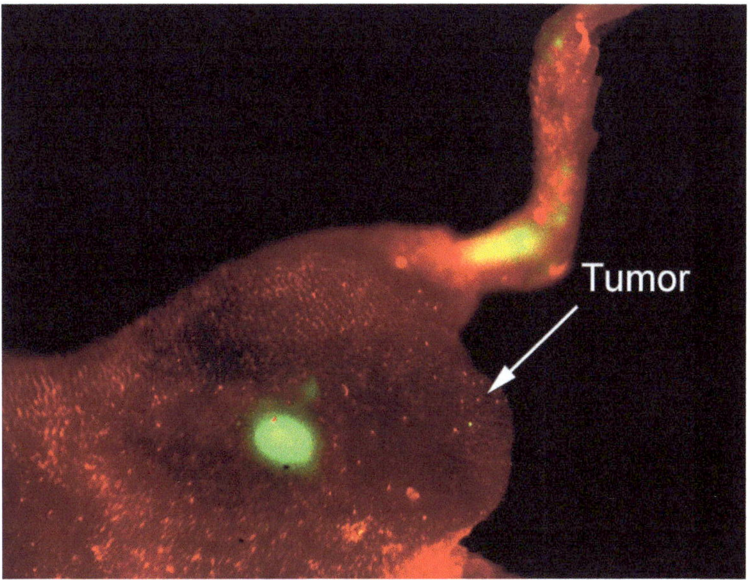

Fig. 1. In vivo luminescence image of pancreatic tumor-bearing mouse intravenously injected with anti-claudin-4 antibody-conjugated QDs. The bioconjugated QDs were accumulated in the tumor due to the receptor-mediated process between claudin-4 receptors and anti-claudin-4 antibody. *Green* represents emission from the QDs and *red* represents auto-fluorescence from the mouse.

3. Methods

3.1. Preparation of CdSe QDs

1. 1 mmol cadmium oxide, 3 mmol TDPA, and 5 g TOPO are loaded into a 100-mL three-necked flask. Next, the reaction mixture is slowly heated under an argon atmosphere to 290–300°C. After 10–15 min of heating, a clear homogeneous solution is obtained.
2. The reaction mixture is maintained at 300°C for another 5 min, then 1 mL of 1 M TOP-Se is rapidly injected. After the injection, the colorless reaction mixture turns to red color, indicating the formation of CdSe nanocrystals. The reaction was stopped after 2–3 min by removing the heating mantle.
3. The reaction mixture is allowed to cool to room temperature and 5–10 mL of toluene is added into the reaction flask to prevent solidification of the mixture.
4. Various emission colors of QDs ranging from green (~550 nm) to deep red (~680 nm) can be obtained by simply withdrawing the aliquot from the reaction mixture at different reaction time. In general, as reaction time proceeds, a red shift in the QD emission is expected.
5. The QDs are separated from the surfactants solution by addition of ethanol and centrifugation. The reddish QD precipitate could be readily re-dispersed in various organic solvents (hexane, toluene, and chloroform).

3.2. Coating CdS/ZnS Double-Layer Shells on CdSe QDs

1. CdSe QD solution is prepared by dissolving 0.2 g of CdSe QDs in 5 mL of chloroform. Separately, 2 mmol of cadmium oxide, 4 mmol of zinc acetate, 5 g of TOPO, and 10 mL of oleic acid are loaded into 100-mL three-neck reaction flask.
2. The reaction mixture is heated to 170°C for 30 min under an argon flow and then the CdSe QD solution is injected slowly under stirring into the hot reaction mixture. The reaction mixture is held at 170°C, with a needle outlet that allowed the toluene to evaporate. After 10 min of heating, the needle is removed and then the reaction temperature is raised to 200–210°C. Upon reaching the desired temperature, 2 mL of TOP-Sulfur was added drop wise into the reaction mixture.
3. The reaction mixture was then held at 210°C for 10–15 min. Next, an aliquot is removed via syringe and is injected into a large volume of toluene at room temperature, thereby quenching any further growth of the QDs.
4. The CdSe/CdS/ZnS QDs are separated from the toluene solution by addition of ethanol and centrifugation. The supernatant is discarded and the red precipitate is re-dispersed into the chloroform for further use.

3.3. Instrumentation for Characterization of Functionalized QDs

1. UV-visible absorbance: The absorption spectra are collected using a Shimadzu model 3101PC UV-Vis-NIR scanning spectrophotometer over a wavelength range from 300 to 800 nm. The samples are measured against chloroform as reference. All samples are dispersed in chloroform and loaded into a quartz cell for measurements.
2. Photoluminescence (PL) spectroscopy: The emission spectra are collected using a Fluorolog-3 spectrofluorometer (Jobin Yvon; fluorescence spectra). All the samples are dispersed in chloroform and loaded into a quartz cell for measurements. Fluorescence quantum yields (QYs) of the QD chloroform solutions are determined by comparing the integrated emission from the nanocrystals to Rhodamin 6 G dye solutions of matched absorbance. The samples are diluted so that they are optically thin.
3. Transmission electron microscopy (TEM): TEM images were obtained using a JEOL model JEM-100CX microscope at an acceleration voltage of 80 kV. The specimens are prepared by drop-coating the sample dispersion onto an amorphous carbon-coated 300-mesh copper grid, which is placed on a filter paper to absorb excess solvent.
4. X-ray diffraction (XRD): X-ray powder diffraction patterns are recorded using a Siemens D500 diffractometer, with Cu Kα radiation. A concentrated QD dispersion is drop-cast onto a quartz plate for measurement.
5. Dynamic light scattering (DLS): Hydrodynamic diameter distribution of QD sample is recorded using a Brookhaven Instruments 90Plus particle size analyzer.
6. Zeta potential measurement: Zeta potential of the QD sample is measured using a Brookhaven Instruments 90Plus analyzer.

3.4. Preparation of Lysine Cross-linked MUA CdSe/CdS/ZnS QDs

1. 3 mmol MUA is dissolved in 10 mL of chloroform under vigorous stirring. After stirring for 10–15 min, 2 mL of concentrated (~40 mg/mL) CdSe/CdS/ZnS QRs solution was added to this mixture. Approximately 1 min later, to this vigorously stirred solution, 2 mL of ammonium hydroxide was added. This solution was stirred overnight at room temperature.
2. After overnight stirring, the QDs are separated from the surfactant solution by addition of ethanol and centrifugation.
3. The QD precipitate is re-dispersed in 15 mL DMSO for further lysine cross-linking process.
4. The lysine cross-linked MUA QRs were obtained by mixing the DMSO QD solution with both DCC (30 mmol) and lysine (15 mmol) under vigorously stirring for 2 h.

5. After 2 h of stirring, the lysine-coated QDs were precipitated from the solution by addition of ethanol and centrifugation.
6. The red-brownish precipitate was re-dispersed in 10 mL HPLC water and the solution was further filtered using a syringe filter with a nominal pore diameter of 0.45 μm. The lysine-coated QDs have relatively good colloidal stability and no precipitation was observed after several months. This solution was kept in the refrigerator at 4°C for further use.

3.5. Conjugation of Cross-linked MUA CdSe/CdS/ZnS QDs with Anti-claudin-4 Antibody

1. Lysine-coated QD stock solution is diluted to an optical density of 0.27–0.3 using a Shimadzu model 3101PC UV-Vis-NIR scanning spectrophotometer over a wavelength range from 300 to 800 nm. Based on the optical density, the concentration can be estimated using Beer's law.
2. 150 μL lysine-coated QD stock solution is mixed with 2 mL of 2.5 mM EDC solution and gently stirred for 2–3 min.
3. Next, 7 μL of anti-claudin-4 antibody solution is added to this mixture and left undisturbed at room temperature for 2 h to allow the antibody to covalently bond to the lysine-coated QD surface.
4. The bioconjugated QD sample is further purified by centrifugation. The precipitate is re-dispersed in water for cell-labeling purpose.

3.6. Cell Labeling with Anti-claudin-4-Conjugated QD Formulation

1. For pancreatic cancer cell imaging with bioconjugated QDs, the human pancreatic cancer cell line Panc-1 or MiaPaCa is cultured in Dulbecco's minimum essential media with 10% fetal bovine serum (FBS), 1% penicillin, and 1% amphotericin B.
2. The day before treatment, the cells were seeded in 35-mm culture dishes at a confluency of 70–80%.
3. On the treatment day, the cells in serum-supplemented media are treated with the anti-claudin-4-conjugated QDs at a final concentration of ~3 μg/mL for 2 h at 37°C. After 2 h of incubation, the cells were washed thrice with PBS and directly imaged using a confocal microscope.
4. Unconjugated QDs are incubated with human pancreatic cancer cell lines and served as additional control. After 2 h of incubation, the cells were washed thrice with PBS and directly imaged using a confocal microscope (see Fig. 2).

3.7. Preparation of Xenograph Tumor Model

1. Panc-1 cells are maintained in DMEM high glucose medium (Sigma–Aldrich, St. Louis, MO) containing 10% FBS (Sigma–Aldrich), 1 mM l-glutamine (Sigma–Aldrich), 100 μg/mL kanamycin (Sigma–Aldrich), and 0.5 μg/mL amphotericin B

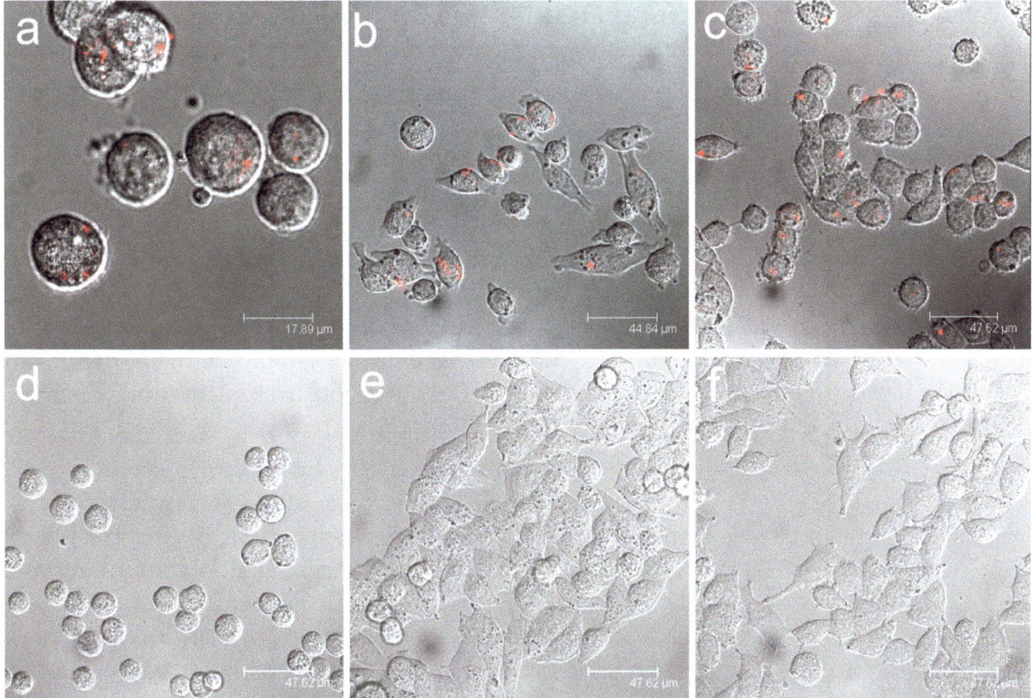

Fig. 2. Confocal microscopic images of (**a**) Colo-357 cells treated with anti-claudin-4-conjugated QDs; (**b**) MiaPaCa cells treated with anti-claudin-4-conjugated QDs; (**c**) Panc-1 cells treated with anti-claudin-4-conjugated QDs; (**d**) Colo-357 cells treated with unconjugated QDs; (**e**) MiaPaCa cells treated with unconjugated QDs; and (**f**) Panc-1 cells treated with unconjugated QDs. In all cases, red represents emission from QDs.

(Sigma–Aldrich) in a 75-cm^2 nunclon delta-treated flask (Nunc A/S, Rockslide, Denmark).

2. For xenograph preparation, an 85–90% confluent cell flask is rinsed three times with sterile DPBS (Sigma–Aldrich). 5 mL of a trypsin-EDTA (Sigma–Aldrich) solution is added to the flask and incubated at 37°C and 5% CO_2 until the cells are released. Once the cells are released, 5 mL of medium is added to the trypsin, and the entire flask content is removed and placed in a sterile 50-mL centrifuge tube (Becton Dickenson Labware, Franklin Lakes, NJ). The cells are spun down at 2,000 rpm for 5 min and the supernatant is removed. The cell pellet is resuspended in 5 mL of fresh media and a cell count using trypan blue (Sigma–Aldrich) is performed. Based on the concentration of cells needed per injection, 1:1 ratio of cell suspension and Matrigel (BD Biosciences, Bedford, MA) is made.
3. 5–6-week-old female athymic nude mice (Hsd:Athymic Nude-Foxn1nu) are obtained from Harlan Laboratories, Inc., and allowed an acclimation period of 1 week. The nude mice are

housed in sterile M.I.C.E caging (Animal Care Systems, Centennial CO.) that contains sterile bedding, food, and water.

4. Human pancreatic tumor models are created in the athymic mice by injecting Panc-1 (ATCC, CRL-1469) cells at a concentration of $2–3 \times 10^6$ cells in a 100 μL suspension of Matrigel (BD Biosciences) and medium mixture (1:1) maintained at 4°C. This mixture is then injected subcutaneously in one scapular region of the mice using a 1 mL Monoject tuberculin syringe with a 25 g × 5/8 in. detachable needle (Tyco Healthcare Group, LP, Mansfield, MA). Tumor growth is monitored every 24–48 h until a tumor size of approximately 5 mm^2 is obtained 10–14 days posttransplantation of cells.
5. Once the tumors reach the appropriate size, the mice are injected with functionalized QDs by tail vein injection at a volume of 150–200 μL per injection. After injection, the mice are anesthetized with Aerrane:isoflurane, USP (Baxter Healthcare Corporation, Deerfield, IL) at an induction concentration of 5% isoflurane/1 L O_2. Anesthesia maintenance concentrations are 2–3% isoflurane/1 L O_2. Once proper plane of anesthesia is reached, the mice are imaged using the Maestro in vivo optical imaging system (CRI, Inc., Woburn, MA.) at specific time points to monitor progress of the QDs.

4. Notes

1. Unless stated otherwise, HPLC water is used for preparation of aqueous solution samples. This standard is referred to as "water" in this text.
2. During the synthesis of lysine-coated QDs, the process generates an unpleasant smell in the laboratory. We, therefore, routinely prepare the QDs in fume hood to minimize the smell (see section 3.4).
3. DCC is a toxic and harmful compound. When using the chemical, the user should perform all the experiments in the fume hood (see section 3.4).
4. Lysine cross-linked MUA CdSe/CdS/ZnS QD aqueous stock solution sample is best stored in refrigerator at 4°C (see section 3.4).
5. QD–anti-claudin-4 bioconjugates can be used for cell-labeling studies within 2 days as long as the sample is kept at room temperature (see section 3.5).
6. This protocol can be adapted for other cancer cell line-labeling and sensing studies. For cancer cells that are overexpressed with claudin-4, these cells can be labeled with functionalized QDs (see section 3.6).

7. For economy, only 7 μL of diluted antibody per QD conjugation sample needs to be used in this step.
8. EDC solution needs to be used within 20 min of preparation because the reactivity of EDC is time dependent.
9. During the conjugation of antibody to lysine-coated QDs, it is important to remove excess EDC from the bioconjugated QD sample, or else this will result in unwanted cell death.
10. To produce high-quality QDs, the reaction flask needs to be cleaned thoroughly using aqua regia. After cleaning the reaction flask with aqua regia, the flasks need to be repeatedly washed with ethanol and water for the next round of QD synthesis.
11. To purify the antibody–QD bioconjugates, 2,000–3,000 rpm centrifuge speed is used to precipitate bioconjugates. The supernatant consists of free antibody, and excess EDC solution is discarded (see section 3.5).
12. During the tail vein injection, air bubbles are undesirable in the bioconjugated QD solution sample for in vivo tumor targeting and imaging applications (see section 3.7).
13. For in vivo tumor imaging studies, bioconjugated QDs with emission peak above 650 nm are preferable in this targeted delivery because auto-fluorescence signals can be significantly reduced during the imaging process. This will enable one to differentiate the QDs from the background signals clearly (see section 3.7).

Acknowledgment

This work was supported by the AACR-Pancreatic Cancer Action Network Fellowship for Pancreatic Cancer Research.

References

1. Akerman, M. E., Chan, W. C. W., Laakkonen, P., Bhatia, S. N., and Ruoslahti, E. (2002) Nanocrystal targeting in vivo. *Proceedings of the National Academy of Sciences* **99**, 12617–12621.
2. Michalet, X., Pinaud, F. F., Bentolila, L. A., Tsay, J. M., Doose, S., Li, J. J., Sundaresan, G., Wu, A. M., Gambhir, S. S., and Weiss, S. (2005) Quantum Dots for Live Cells, in Vivo Imaging, and Diagnostics. *Science* **307**, 538–544.
3. Alivisatos, P. (2004) The use of nanocrystals in biological detection. *Nat Biotech* **22**, 47–52.
4. Yong, K. T., Sahoo, Y., Choudhury, K. R., Swihart, M. T., Minter, J. R., and Prasad, P. N. (2006) Shape Control of PbSe Nanocrystals Using Noble Metal Seed Particles. *Nano Lett.* **6**, 709–714.
5. Yong, K.-T., Qian, J., Roy, I., Lee, H. H., Bergey, E. J., Tramposch, K. M., He, S., Swihart, M. T., Maitra, A., and Prasad, P. N. (2007) Quantum Rod Bioconjugates as Targeted Probes for Confocal and Two-Photon Fluorescence Imaging of Cancer Cells. *Nano Letters* 7, 761–765.
6. Bruchez, M., Jr., Moronne, M., Gin, P., Weiss, S., and Alivisatos, A. P. (1998) Semiconductor Nanocrystals as Fluorescent Biological Labels. *Science* **281**, 2013–2016.
7. Yong, K.-T., Roy, I., Swihart, M. T., and Prasad, P. N. (2009) Multifunctional nano-

particles as biocompatible targeted probes for human cancer diagnosis and therapy. *Journal of Materials Chemistry* **19**, 4655–4672.

8. Chan, W. C. W. and Nie, S. (1998) Quantum Dot Bioconjugates for Ultrasensitive Nonisotopic Detection. *Science* **281**, 2016–2018.
9. Ballou, B., Lagerholm, B. C., Ernst, L. A., Bruchez, M. P., and Waggoner, A. S. (2003) Noninvasive Imaging of Quantum Dots in Mice. *Bioconjugate Chemistry* **15**, 79–86.
10. Yong, K.-T., Roy, I., Ding, H., Bergey, E. J., and Prasad, P. N. (2009) Biocompatible Near-Infrared Quantum Dots as Ultrasensitive Probes for Long-Term in vivo Imaging Applications. *Small* **5**, 1997–2004.
11. Erogbogbo, F., Yong, K.-T., Roy, I., Xu, G., Prasad, P. N., and Swihart, M. T. (2008) Biocompatible Luminescent Silicon Quantum Dots for Imaging of Cancer Cells. *ACS Nano* **2**, 873–878.
12. Gao, X., Cui, Y., Levenson, R. M., Chung, L. W. K., and Nie, S. (2004) In vivo cancer targeting and imaging with semiconductor quantum dots. *Nat Biotech* **22**, 969–976.
13. Qian, J., Yong, K.-T., Roy, I., Ohulchanskyy, T. Y., Bergey, E. J., Lee, H. H., Tramposch, K. M., He, S., Maitra, A., and Prasad, P. N. (2007) Imaging Pancreatic Cancer Using Surface-Functionalized Quantum Dots. *The Journal of Physical Chemistry B* **111**, 6969–6972.
14. Xu, G., Yong, K.-T., Roy, I., Mahajan, S. D., Ding, H., Schwartz, S. A., and Prasad, P. N. (2008) Bioconjugated Quantum Rods as Targeted Probes for Efficient Transmigration Across an in Vitro Blood–Brain Barrier. *Bioconjugate Chemistry* **19**, 1179–1185.
15. Yong, K.-T., Hu, R., Roy, I., Ding, H., Vathy, L. A., Bergey, E. J., Mizuma, M., Maitra, A., and Prasad, P. N. (2009) Tumor Targeting and Imaging in Live Animals with Functionalized Semiconductor Quantum Rods. *ACS Applied Materials & Interfaces* **1**, 710–719.
16. Hruban, R. H., Maitra, A., Kern, S. E., and Goggins, M. (2007) Precursors to Pancreatic Cancer. *Gastroenterology Clinics of North America* **36**, 831–849.
17. Yong, K.-T. (2009) Mn-doped near-infrared quantum dots as multimodal targeted probes for pancreatic cancer imaging. *Nanotechnology* **20**, 015102.
18. Yong, K.-T., Ding, H., Roy, I., Law, W.-C., Bergey, E. J., Maitra, A., and Prasad, P. N. (2009) Imaging Pancreatic Cancer Using Bioconjugated InP Quantum Dots. *ACS Nano* **3**, 502–510.
19. Nichols, L. S., Ashfaq, R., and Iacobuzio-Donahue, C. A. (2004) Claudin 4 Protein Expression in Primary and Metastatic Pancreatic Cancer Support for Use as a Therapeutic Target *American Journal of Clinical Pathology* **121**, 226–230.
20. Yong, K.-T., Roy, I., Pudavar, H. E., Bergey, E. J., Tramposch, K. M., Swihart, M. T., and Prasad, P. N. (2008) Multiplex Imaging of Pancreatic Cancer Cells by Using Functionalized Quantum Rods. *Advanced Materials* **20**, 1412–1417.
21. Swierczynski, S. L., Maitra, A., Abraham, S. C., Iacobuzio-Donahue, C. A., Ashfaq, R., Cameron, J. L., Schulick, R. D., Yeo, C. J., Rahman, A., Hinkle, D. A., Hruban, R. H., and Argani, P. (2004) Analysis of novel tumor markers in pancreatic and biliary carcinomas using tissue microarrays. *Human Pathology* **35**, 357–366.

Index

Kursad Turksen (ed.), *Claudins: Methods and Protocols*, Methods in Molecular Biology, vol. 762,
DOI 10.1007/978-1-61779-185-7, © Springer Science+Business Media, LLC 2011

D

E

F

G

H

I

L

M

N

O

P

Q

V

W

X

Z

MIX
Papier aus verantwortungsvollen Quellen
Paper from responsible sources
FSC® C105338

If you have any concerns about our products,
you can contact us on
ProductSafety@springernature.com

In case Publisher is established outside the EU,
the EU authorized representative is:
Springer Nature Customer Service Center GmbH
Europaplatz 3, 69115 Heidelberg, Germany

Printed by Libri Plureos GmbH
in Hamburg, Germany